# Handbook of Electronic Components

## Volume I

# Handbook of Electronic Components Volume I

Edited by **Annie Kent**

NY RESEARCH PRESS
New York

Published by NY Research Press,
23 West, 55th Street, Suite 816,
New York, NY 10019, USA
www.nyresearchpress.com

**Handbook of Electronic Components: Volume I**
Edited by Annie Kent



International Standard Book Number: 978-1-63238-239-9 (Hardback)



Printed in China.

# Contents

# Preface

Humans have an indispensable relationship with electricity. The two are co-existent. But, mostly we humans are dependent on electricity in every sphere of our life. We may take electricity for granted but the very next second it leaves us in its awe. Electricity powers our world through a wide spectrum of machines. Physics classifies these electronic components into active components and passive components based on their source of energy. Active components are a combination of circuits that have the capability to control the electron flow; whereas the passive components cannot do the same. This book elucidates a wide variety of topics with vital importance; which revolve around these electronic components. Constant researches have shown that these components can play a remarkable role in the progress of science.

Based on these two primary categorizations; this book presents some of the most phenomenal studies and researches of modern times. Circuit implementation, operation, and simulation of multivalued non-volatile static random access memory, novel pseudo-PMOS integrated ISFET device for water quality monitoring, potential and quantum threshold voltage modeling of Gate-All-Around nanowire MOSFETs and series resistance in Gate-Recessed nanoscale SOI MOSFETs are essential devices for assessing electron components.

This book is a result of the invariable hard work of all our researchers and contributors. I wish to thank them for their contribution, co-operation and unfailing efforts towards this book. I also wish to acknowledge their hard work towards maintaining the time parameters set for this publication. Lastly, I wish to thank the publisher for considering me worthy of this task and giving me an opportunity to connect with internationally renowned experts and share their expertise.

**Editor**

# Cascadable Current-Mode First-Order and Second-Order Multifunction Filters Employing Grounded Capacitors

**Jiun-Wei Horng, Chun-Li Hou, Ching-Yao Tseng, Ryan Chang, and Dun-Yih Yang**

*Department of Electronic Engineering, Chung Yuan Christian University, Chung-Li 32023, Taiwan*

Correspondence should be addressed to Jiun-Wei Horng, jwhorng@cycu.edu.tw

Academic Editor: Daisaburo Takashima

A configuration for realizing low input and high output impedances current-mode multifunction filters using multiple output second-generation current conveyors (MOCCIIs) is presented. From the proposed circuit configuration, first-order allpass, highpass, lowpass and second-order allpass, notch, bandpass filters can be obtained. The simulation results confirm the theoretical analysis.

## 1. Introduction

Current conveyors (CCs) are receiving much attention for their potential advantages such as inherent wider signal bandwidths, simpler circuitry and larger dynamic range [1, 2]. Current-mode active filters with low input impedance and high output impedance are of great interest because they can be directly connected in cascade to implement higher order filters [3, 4]. Several current-mode first-order allpass filters using various active components have been reported. Some circuits use two second-generation current conveyors (CCIIs) to realize such a first-order allpass filter function with high output impedance [5]. However, the number of passive components they used are not canonical. Some first-order circuits use one active component [6, 7]. However, their input impedances are not low. Some first-order circuits use one active element, one capacitor, and one resistor [8–10]. However, these circuits have not the advantage of low input impedance and the capacitors they used are not grounded. In 2007, Metin et al. [11] propose a current-mode first-order allpass filter using two CCIIs, two grounded resistors and one grounded capacitor with low input and high output impedances. In 2009, Lahiri and Chowdhury [12] proposed a current-mode first-order allpass filter using one current differencing transconductance amplifier (CDTA) and one grounded capacitor with low input and high output impedances. As the function of CDTA can also be replaced by two CCII and one operational transconductance amplifier (OTA) [12], if the structure of [12] is constructed by CCIIs and OTA, it requires two CCIIs and one OTA.

In this paper, a new current-mode circuit configuration with low input and high output impedances uses two multiple output second-generation current conveyors (MOCCIIs) is presented. The first-order allpass, highpass, and lowpass filters can be obtained from the proposed circuit configuration.

Several current-mode universal biquadratic filters with single-input and multioutput were presented in the literature [13–15]. However, the input impedances of the circuits in [13, 14] are high. Moreover, at least three CCIIs are required in [14, 15]. Several current-mode universal biquadratic filters with multi-input terminals using current conveyors are presented [16, 17]. However, the input impedances of these circuits are high at most of their input terminals.

In this paper, from the proposed circuit configuration, the second-order allpass, notch, and bandpass filters can also be obtained. All proposed circuits use only grounded capacitors that are attractive for integrated circuit implementation [19].

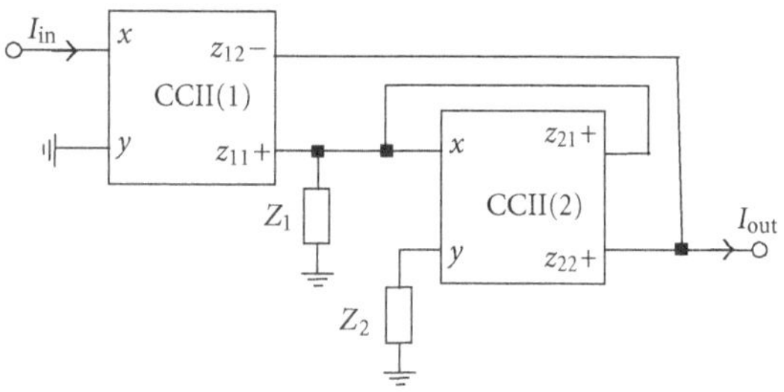


FIGURE 1: The proposed cascadable current-mode circuit configuration.

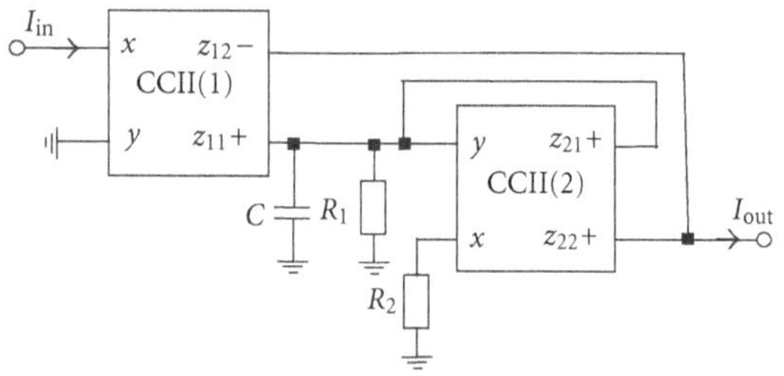


FIGURE 2: The current-mode first-order filter.

## 2. Proposed Circuits

Using standard notation, the port relations of a MOCCII can be characterized by $v_x = v_y$, $i_{zk} = \pm i_x$ and $i_y = 0$. Considering the proposed current-mode circuit in Figure 1, the current transfer function can be expressed as

$$\frac{I_{out}}{I_{in}} = \frac{Z_2 - 2Z_1}{Z_2 - Z_1}. \quad (1)$$

If the impedances of Figure 1 are taken to be $Z_1 = R_1/(1+sCR_1)$ and $Z_2 = R_2$, a first-order circuit is obtained as shown in Figure 2. The current transfer function of Figure 2 is

$$\frac{I_{out}}{I_{in}} = \frac{sCR_1R_2 + R_2 - 2R_1}{sCR_1R_2 + R_2 - R_1}. \quad (2)$$

For $R_1 = R$ and $R_2 = 1.5R$, (2) becomes

$$\frac{I_{out}}{I_{in}} = \frac{s - 1/(3CR)}{s + 1/(3CR)}. \quad (3)$$

This represents a first-order allpass function.

For $R_1 = R$ and $R_2 = 2R$, (2) becomes

$$\frac{I_{out}}{I_{in}} = \frac{s}{s + 1/(2CR)}. \quad (4)$$

This represents a first-order highpass function.

In Figure 2, if the $z_{12}-$ output terminal of CCII(1) is eliminated, the transfer function becomes

$$\frac{I_{out}}{I_{in}} = \frac{-R_1}{sCR_1R_2 + R_2 - R_1}. \quad (5)$$

For $R_1 = R$ and $R_2 = 2R$, (5) becomes

$$\frac{I_{out}}{I_{in}} = \frac{-1/(2RC)}{s + 1/(2CR)}. \quad (6)$$

This represents a first-order lowpass function.

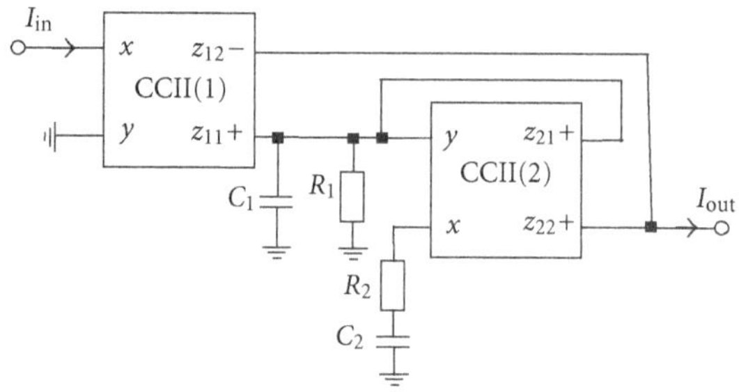


FIGURE 3: The current-mode second-order filter.

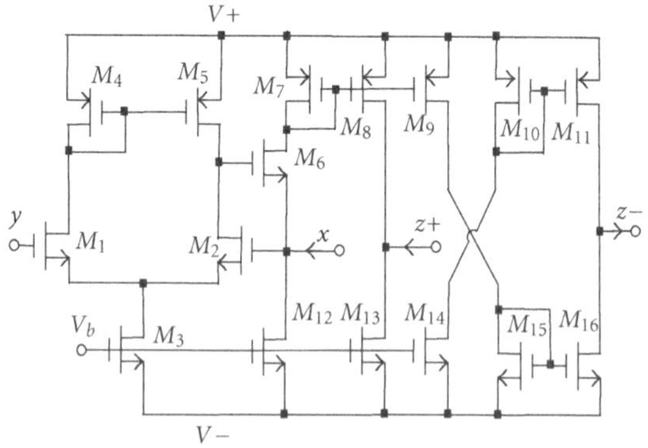


FIGURE 4: The CMOS MOCCII implementation.

If the impedances of Figure 1 are taken to be $Z_1 = R_1/(1+sC_1R_1)$ and $Z_2 = (sC_2R_2 + 1)/sC_2$, a second-order circuit can be obtained as shown in Figure 3. The current transfer function of Figure 3 is

$$\frac{I_{out}}{I_{in}} = \frac{s^2C_1C_2R_1R_2 + s(C_1R_1 + C_2R_2 - 2C_2R_1) + 1}{s^2C_1C_2R_1R_2 + s(C_1R_1 + C_2R_2 - C_2R_1) + 1}. \quad (7)$$

For $R_1 = 2R$, $R_2 = R$ and $C_1 = C_2 = C$, (7) becomes

$$\frac{I_{out}}{I_{in}} = \frac{2s^2C^2R^2 - sCR + 1}{2s^2C^2R^2 + sCR + 1}. \quad (8)$$

This represents a second-order allpass function.

For $R_1 = R_2 = R$ and $C_1 = C_2 = C$, (7) becomes

$$\frac{I_{out}}{I_{in}} = \frac{s^2C^2R^2 + 1}{s^2C^2R^2 + sCR + 1}. \quad (9)$$

This represents a second-order notch function.

In Figure 3, if the $z_{12}-$ output terminal of CCII(1) is eliminated, the transfer function becomes

$$\frac{I_{out}}{I_{in}} = \frac{-sC_2R_1}{s^2C_1C_2R_1R_2 + s(C_1R_1 + C_2R_2 - C_2R_1) + 1}. \quad (10)$$

This represents a second-order bandpass function.

Because the input terminal of Figure 1 is connected directly to the $x$-terminal of CCII(1) and the $y$-terminal of CCII(1) is grounded, the input terminal has the advantage of low input impedance. Because the $I_{out}$ output terminal is taken out directly from the $z_{12}$ and $z_{22}$ terminals, the $I_{out}$ output terminal has the advantage of high output impedance.

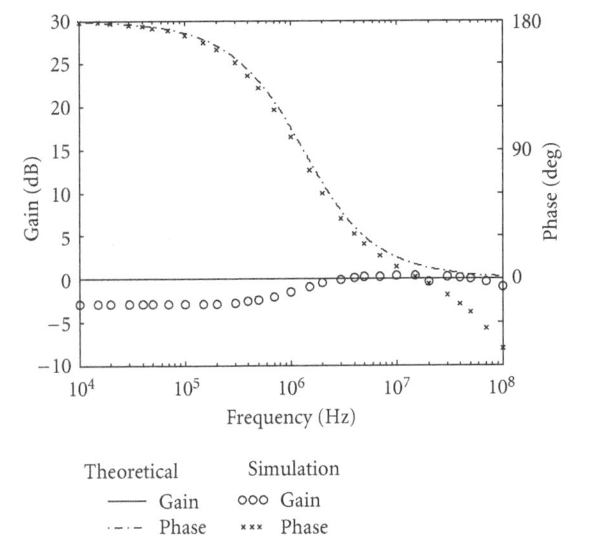


FIGURE 5: Simulation results of the proposed current-mode first-order allpass filter.

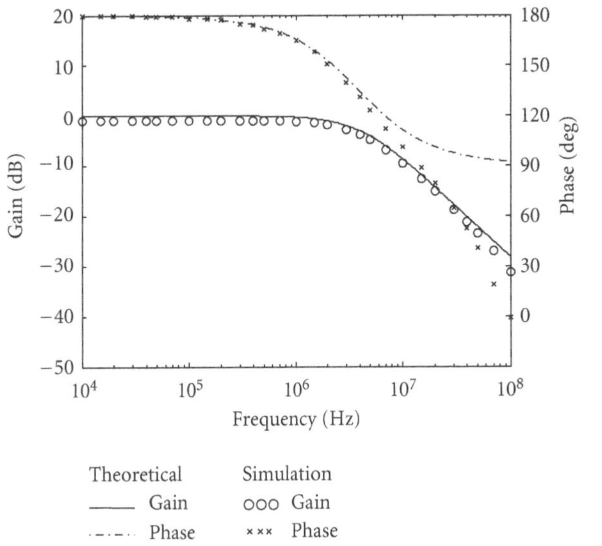


FIGURE 7: Simulation results of the proposed current-mode first-order lowpass filter.

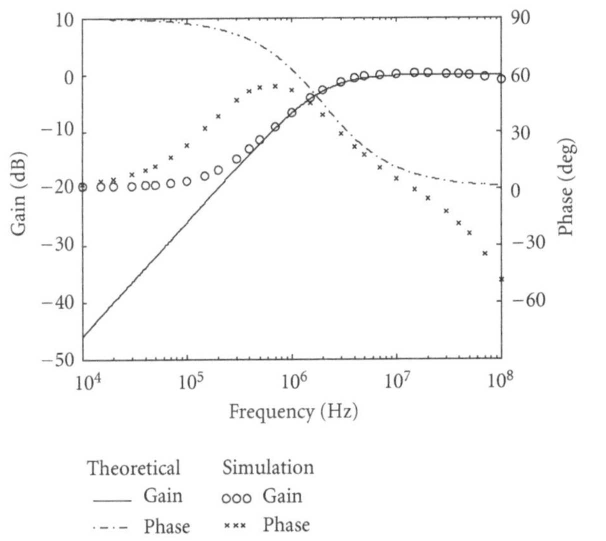


FIGURE 6: Simulation results of the proposed current-mode first-order highpass filter.

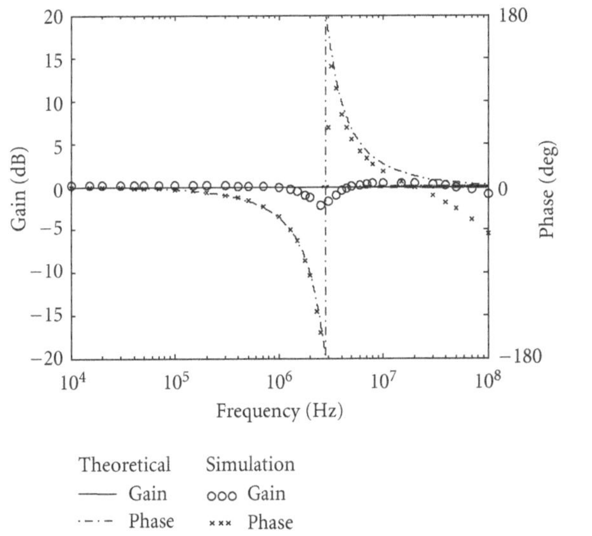


FIGURE 8: Simulation results of the proposed current-mode second-order allpass filter.

The proposed first-order and second-order filters employ only grounded capacitors. The use of grounded capacitors is attractive for integrated circuit implementation [19].

## 3. Nonideality Analysis of the MOCCIIs

Taking into consideration the MOCCII nonidealities, the port relations of MOCCII can be expressed as

$$v_x = \alpha v_y, \qquad i_z = \pm\beta_k i_x, \tag{11}$$

where $\beta_k = 1 - \varepsilon_{ki}$ and $\varepsilon_{ki}(|\varepsilon_{ki}| \ll 1)$ denotes the current tracking error, $\alpha = 1-\varepsilon_v$ and $\varepsilon_v(|\varepsilon_v| \ll 1)$ is the input voltage tracking error of a MOCCII. Reanalysis of the filter circuit in Figure 2 yields the following modified transfer functions:

$$\frac{I_{\text{out}}}{I_{\text{in}}} = \frac{sCR_1R_2\beta_{12} + R_2\beta_{12} - R_1\alpha_2(\beta_{12}\beta_{21} + \beta_{11}\beta_{22})}{sCR_1R_2 + R_2 - R_1\alpha_2\beta_{21}}. \tag{12}$$

The cutoff angular frequency is obtained by

$$\omega_c = \frac{R_2 - R_1\alpha_2\beta_{21}}{CR_1R_2}. \tag{13}$$

The active and passive sensitivities are low and obtained as

$$S_{C}^{\omega_c} = -1,$$
$$S_{R_1}^{\omega_c} = \frac{-R_2}{R_2 - R_1\alpha_2\beta_{21}},$$
$$S_{R_2}^{\omega_c} = -S_{\alpha_2,\beta_{21}}^{\omega_c}$$
$$= \frac{R_1\alpha_2\beta_{21}}{R_2 - R_1\alpha_2\beta_{21}}. \quad (14)$$

Reanalysis of the filter circuit in Figure 3 yields the following modified transfer functions:

$$\frac{I_{\text{out}}}{I_{\text{in}}} = \frac{s^2C_1C_2R_1R_2\beta_{12} + s(C_1R_1\beta_{12} + C_2R_2\beta_{12} - C_2R_1\alpha_2(\beta_{11}\beta_{22} + \beta_{12}\beta_{21})) + \beta_{12}}{s^2C_1C_2R_1R_2 + s(C_1R_1 + C_2R_2 - C_2R_1\alpha_2\beta_{21}) + 1}. \quad (15)$$

The resonance angular frequency and quality factor are obtained by

$$\omega_o = \frac{1}{\sqrt{C_1C_2R_1R_2}},$$
$$Q = \frac{\sqrt{C_1C_2R_1R_2}}{C_1R_1 + C_2R_2 - C_2R_1\alpha_2\beta_{21}}. \quad (16)$$

The active and passive sensitivities are low and obtained as

$$S_{C_1,C_2,R_1,R_2}^{\omega_o} = -\frac{1}{2},$$
$$S_{C_1}^{Q} = \frac{-C_1R_1 + C_2R_2 - C_2R_1\alpha_2\beta_{21}}{2(C_1R_1 + C_2R_2 - C_2R_1\alpha_2\beta_{21})},$$
$$S_{C_2}^{Q} = \frac{C_1R_1 - C_2R_2 + C_2R_1\alpha_2\beta_{21}}{2(C_1R_1 + C_2R_2 - C_2R_1\alpha_2\beta_{21})},$$
$$S_{R_1}^{Q} = \frac{-C_1R_1 + C_2R_2 + C_2R_1\alpha_2\beta_{21}}{2(C_1R_1 + C_2R_2 - C_2R_1\alpha_2\beta_{21})},$$
$$S_{R_2}^{Q} = \frac{C_1R_1 - C_2R_2 - C_2R_1\alpha_2\beta_{21}}{2(C_1R_1 + C_2R_2 - C_2R_1\alpha_2\beta_{21})},$$
$$S_{\alpha_2,\beta_{21}}^{Q} = \frac{C_2R_1\alpha_2\beta_{21}}{C_1R_1 + C_2R_2 - C_2R_1\alpha_2\beta_{21}}. \quad (17)$$

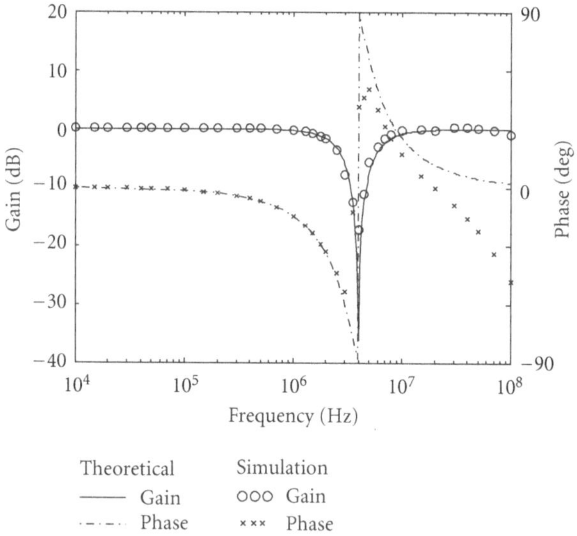


Figure 9: Simulation results of the proposed current-mode second-order notch filter.

## 4. Simulation Results

HSPICE simulations were carried out to demonstrate the feasibility of the proposed circuits using 0.18 $\mu$m, level 49 MOSFET from TSMC. The MOCCII was realized by the CMOS implementation in Figure 4 [20] with the NMOS and PMOS transistor aspect ratios $W/L = 4.5\,\text{u}/0.9\,\text{u}$ and $W/L = 9\,\text{u}/0.9\,\text{u}$, respectively.

Figure 5 represents the magnitude and phase responses of the first-order allpass filter in Figure 2, designed with $f_c = 1.326$ MHz: $C = 20$ pF, $R_1 = 2\,\text{k}\Omega$ and $R_2 = 3\,\text{k}\Omega$. The power supply was $\pm 1.25$ V. The bias voltage is $V_b = -0.6$ V. Figure 6 represents the magnitude and phase responses of the first-order highpass filter, designed with $f_c = 1.989$ MHz: $C = 20$ pF, $R_1 = 2\,\text{k}\Omega$, and $R_2 = 4\,\text{k}\Omega$. Figure 7 represents the magnitude and phase responses of the first-order lowpass filter, designed with $f_c = 3.979$ MHz: $C = 20$ pF, $R_1 = 1\,\text{k}\Omega$, and $R_2 = 2\,\text{k}\Omega$.

Figure 8 represents the magnitude and phase responses of the second-order allpass filter in Figure 3, designed with $f_o = 2.814$ MHz: $C_1 = 20$ pF, $C_2 = 20$ pF, $R_1 = 4\,\text{k}\Omega$, and $R_2 = 2\,\text{k}\Omega$. Figure 9 represents the magnitude and phase responses of the second-order notch filter, designed with $f_o = 3.979$ MHz: $C_1 = 20$ pF, $C_2 = 20$ pF, $R_1 = 2\,\text{k}\Omega$, and $R_2 = 2\,\text{k}\Omega$. Figure 10 represents the magnitude and phase responses of the second-order bandpass filter, designed with $f_o = 3.979$ MHz: $C_1 = 20$ pF, $C_2 = 20$ pF, $R_1 = 2\,\text{k}\Omega$, and $R_2 = 2\,\text{k}\Omega$.

A nonideal MOCCII model is shown in Figure 11 [18]. It is shown that the real MOCCII has parasitic resistors and capacitors from the $y$- and $z$-terminals to the ground, and also, a series resistor at the input terminal $x$. The parasitic resistances of MOCCII become effective at very low frequency. This can explain why Figures 6 and 10 have nonideal frequency responses at low frequencies. Because there are parasitic resistance and capacitance at the output terminal $z$, they introduce a pole at high frequency. This can explain why

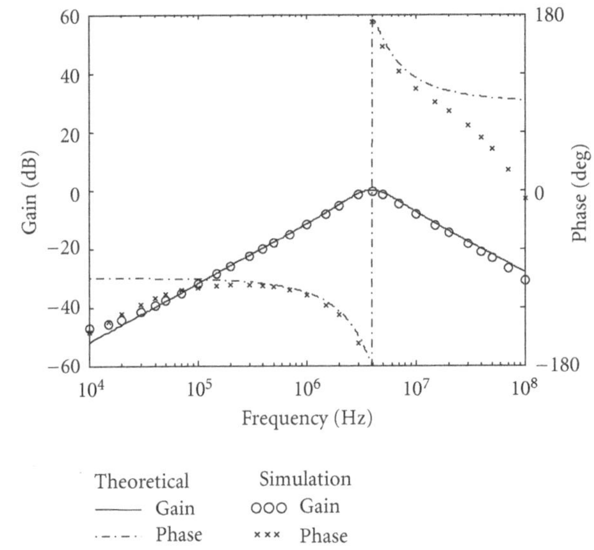


FIGURE 10: Simulation results of the proposed current-mode second-order bandpass filter.

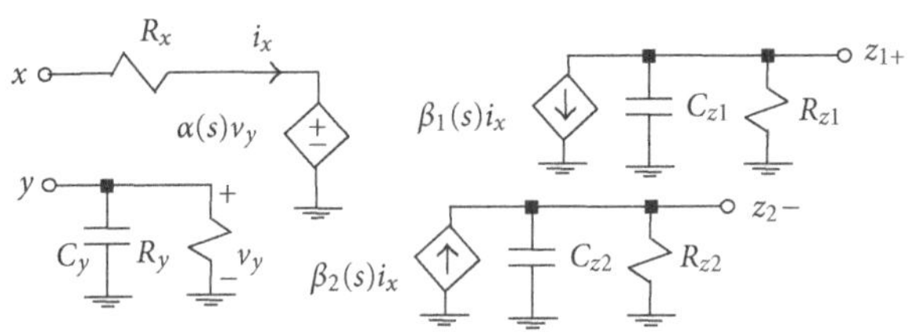

FIGURE 11: The nonideal MOCCII model [18]. $R_x = 30\,\Omega$, $R_y = 17\,\mathrm{M}\Omega$, $R_{z1} = 577\,\mathrm{k}\Omega$, $R_{z2} = 572\,\mathrm{k}\Omega$, $C_y = 1.4\,\mathrm{pF}$, $C_{z1} = 1.2\,\mathrm{pF}$, $C_{z2} = 1.2\,\mathrm{pF}$.

the simulations have nonideal frequency responses at high frequencies.

## 5. Conclusions

A new cascadable current-mode circuit configuration using two MOCCIIs is presented. The proposed circuit has the advantages of low input and high output impedances. By choosing appropriate passive components, the first-order allpass, lowpass, highpass or second-order allpass, notch, bandpass filters can be obtained. All proposed filters use only grounded capacitors.

## Acknowledgment

The authors would like to thank the reviewers for their suggestions.

## References

[1] C. Toumazou, F. J. Lidgey, and D. G. Haigh, *Analog IC Design: The Current-Mode Approach*, Peter Peregrinus, London, UK, 1990.

[2] J. W. Horng, "Voltage/current-mode universal biquadratic filter using single CCII+," *Indian Journal of Pure and Applied Physics*, vol. 48, no. 10, pp. 749–756, 2010.

[3] A. M. Soliman, "Current mode universal filter," *Electronics Letters*, vol. 31, no. 17, pp. 1420–1421, 1995.

[4] J. W. Horng, "Current-mode and transimpedance-mode universal biquadratic filter using multiple outputs cciis," *Indian Journal of Engineering and Materials Sciences*, vol. 17, no. 3, pp. 169–174, 2010.

[5] J. W. Horng, C. L. Hou, C. M. Chang, W. Y. Chung, H. L. Liu, and C. T. Lin, "High output impedance current-mode first-order allpass networks with four grounded components and two CCIIs," *International Journal of Electronics*, vol. 93, no. 9, pp. 613–621, 2006.

[6] S. Minaei and M. A. Ibrahim, "General configuration for realizing current-mode first-order all-pass filter using DVCC," *International Journal of Electronics*, vol. 92, no. 6, pp. 347–356, 2005.

[7] S. Maheshwari, "Novel cascadable current-mode first order all-pass sections," *International Journal of Electronics*, vol. 94, no. 11, pp. 995–1003, 2007.

[8] A. Toker, S. Ozoguz, O. Cicekoglu, and C. Acar, "Current-mode all-pass filters using current differencing buffered amplifier and a new high-Q bandpass filter configuration," *IEEE Transactions on Circuits and Systems II: Analog and Digital Signal Processing*, vol. 47, no. 9, pp. 949–954, 2000.

[9] S. Kilinç and U. Çam, "Current-mode first-order allpass filter employing single current operational amplifier," *Analog Integrated Circuits and Signal Processing*, vol. 41, no. 1, pp. 47–53, 2004.

[10] S. Maheshwari, "A new current-mode current-controlled all-pass section," *Journal of Circuits, Systems and Computers*, vol. 16, no. 2, pp. 181–189, 2007.

[11] B. Metin, K. Pal, and O. Cicekoglu, "All-pass filter for rich cascadability options easy IC implementation and tunability," *International Journal of Electronics*, vol. 94, no. 11, pp. 1037–1045, 2007.

[12] A. Lahiri and A. Chowdhury, "A novel first-order current-mode all-pass filter using CDTA," *Radioengineering*, vol. 18, no. 3, pp. 300–305, 2009.

[13] M. Kumngern, W. Jongchanachavawat, and K. Dejhan, "New electronically tunable current-mode universal biquad filter using translinear current conveyors," *International Journal of Electronics*, vol. 97, no. 5, pp. 511–523, 2010.

[14] A. Ü. Keskin and U. Cam, "Insensitive high-output impedance minimum configuration SITO-type current-mode biquad using dual-output current conveyors and grounded passive components," *International Journal of Electronics and Communications*, vol. 61, no. 5, pp. 341–344, 2007.

[15] O. Çiçekoğlu, N. Tarim, and H. Kuntman, "Wide dynamic range high output impedance current-mode multifunction filters with dual-output current conveyors," *Archiv fur Elektronik und Ubertragungstechnik*, vol. 56, no. 1, pp. 55–60, 2002.

[16] J. Zhao, J. G. Jiang, and J. N. Liu, "Design of tunable biquadratic filters employing CCCIIs: state variable block diagram approach," *Analog Integrated Circuits and Signal Processing*, vol. 62, no. 3, pp. 397–406, 2010.

[17] J.-W. Horng, "Current-mode universal biquadratic filter with five inputs and one output using three ICCIIs," *Indian Journal of Pure and Applied Physics*, vol. 49, no. 3, pp. 214–217, 2011.

[18] A. Fabre, O. Saaid, and H. Barthelemy, "On the frequency limitations of the circuits based on second generation current conveyors," *Analog Integrated Circuits and Signal Processing*, vol. 7, no. 2, pp. 113–129, 1995.

[19] M. Bhushan and R. W. Newcomb, "Grounding of capacitors in integrated circuits," *Electronic Letters*, vol. 3, no. 4, pp. 148–149, 1967.
[20] W. Surakampontorn, V. Riewruja, K. Kumwachara, and K. Dejhan, "Accurate CMOS-based current conveyors," *IEEE Transactions on Instrumentation and Measurement*, vol. 40, no. 4, pp. 699–702, 1991.

# 2

# The Application of Approximate Entropy Theory in Defects Detecting of IGBT Module

**Shengqi Zhou, Luowei Zhou, Suncheng Liu, Pengju Sun, Quanming Luo, and Junke Wu**

*State Key Laboratory of Power Transmission Equipment & System Security and New Technology, University of Chongqing, Chongqing 400044, China*

Correspondence should be addressed to Shengqi Zhou, shengqi.z@cqu.edu.cn

Academic Editor: Jiun-Wei Horng

Defect is one of the key factors in reducing the reliability of the insulated gate bipolar transistor (IGBT) module, so developing the diagnostic method for defects inside the IGBT module is an important measure to avoid catastrophic failure and improves the reliability of power electronic converters. For this reason, a novel diagnostic method based on the approximate entropy (ApEn) theory is presented in this paper, which can provide statistical diagnosis and allow the operator to replace defective IGBT modules timely. The proposed method is achieved by analyzing the cross ApEn of the gate voltages before and after the occurring of defects. Due to the local damage caused by aging, the intrinsic parasitic parameters of packaging materials or silicon chips inside the IGBT module such as parasitic inductances and capacitances may change over time, which will make remarkable variation in the gate voltage. That is to say the gate voltage is close coupled with the defects. Therefore, the variation is quantified and used as a precursor parameter to evaluate the health status of the IGBT module. Experimental results validate the correctness of the proposed method.

## 1. Introduction

The survey carried out in [1] shows that the reliability of power electronic converters deserves increasing interest from industry in recent years, but the satisfaction level with reliability monitoring is low at 50%. This suggests more research effort is needed. One of the major challenges is the health management of power electronic devices due to their significant impacts on the degradation of power electronic converters. At present, the IGBTs are the most used power electronic devices in many industrial applications as well as some safe-critical fields such as traction and renewable energy [2, 3], where stringent reliability is required. Though the reliability of IGBTs has been substantial developed recently, the failure rate of IGBTs dropped from 1,000 FITs in 1995 to 20 FITs in 2000, and to only a few FITs currently where 1 FIT is equal $1 \times 10^{-9}$ failures per device-hour [4], the IGBTs used in those safe-critical fields mentioned above may undergo severe electrical and thermal stress. Such as wind power generation, the IGBTs are subject to large-junction temperature swings during normal operation due to the inherent intermittent nature of the wind speed and low-converter-modulation frequencies [5]. Therefore, monitoring reliability degradation of IGBTs becomes an important issue and still remains as an open topic for research [6].

Reliability is the ability of a product or system to perform as intended for a specified time, in its life-cycle environment [7]. Traditional reliability prediction methods rely on the collection of failure data of the components, but are not applicable for IGBTs because of vast time consumption. More and more researches on detecting reliability degradation of IGBTs are reported in recent years [4, 5, 8–14]. Most of those researches are based on the external indication of aging or impending failure and physics of failure. However, since many of the physical wear-out mechanisms give little external indication [15] or it is very difficult to sense the junction temperature during operation [5], those methods are still under development [6]. Relevant researches indicate that the reliability degradation of IGBTs depends on the fatigue accumulated over time, there are some localized defects induced inside the IGBT module before catastrophic failure, such as the bond wire liftoff and the solder layer cracking [16]. Besides significantly reducing the reliability of

IGBTs, these localized defects may also affect some intrinsic parasitic elements of the IGBT module and consequently change the gate voltage. This provides a practical way to monitor those defects inside the IGBT module by identifying the variation of the gate voltage. Unlike post-fault detections [17–19], monitoring defects inside the IGBT module allows operators to find out an unhealthy IGBT module in a prefault condition and replace it timely to avoid burnout and subsequently collateral damage to the rest of the converter.

The gate circuit consists of intrinsic junction capacitances and associated parasitic inductances and resistances of packaging materials, so it is firmly coupled with those intrinsic parasitic elements inside the IGBT module and sensitive to the occurring of defects. But, sensing the gate voltage during converter operation may be affected by many uncertain factors such as noises, which would lead the conclusion to be ambiguous. To circumvent this limit, a diagnostic method based on the approximate entropy theory is proposed in this paper, which can provide statistical diagnosis [20]. Such statistical diagnosis can be well understood and accepted.

The rest of this paper is organized as follows. Firstly, the failure mechanisms of IGBTs and the influences of defects over the performance of the gate circuit are discussed, a linearization of the gate circuit over a particular time period of the turn-on process of the IGBT module is adopted to eliminate disturbance caused by the nonlinear gate to collector capacitor $C_{GC}$, then the approximate entropy theory is overviewed and the calculation of the cross ApEn is also carried out. Finally, a diagnostic method based on the variation of the cross ApEn of the gate voltages before and after the occurring of defects is proposed to find out the possible abnormality, and a confirmatory experiment is also carried out by adopting a special unencapsulated IGBT module to verify the correctness of the diagnostic method proposed in this paper.

## 2. Failure Mechanisms and Defects of IGBTs

*2.1. Failure Mechanisms of IGBTs.* Modern IGBTs are characterized by numerous advantages such as low on-state resistance, good switching speed and excellent safe-operating area. They are widely used in low-to-medium-power and high-frequency applications, and nowadays even replaced the gate turn-off (GTO) thyristors in high power range. There are a variety of configurations of IGBTs from discrete IGBTs to IGBT modules. Among those configurations, the IGBT modules are very popular in high-power applications. Though upcoming semiconductor materials, such as silicon carbide has attracted more and more attention, most commercial available high-power IGBT modules are still fabricated using silicon as the base material. Being limited by the material properties of silicon, the power density of the IGBT chip is restricted, thus many IGBT chips are connected in parallel inside the IGBT module, together with diode chips antiparallel placed serving as freewheeling diodes (FWD) at present, in order to enhance the current conducting capability as shown in Figure 1. In this case, there are two IGBT switches in the IGBT module. Each switch consists of two parallel IGBT chips and two antiparallel FWD chips.

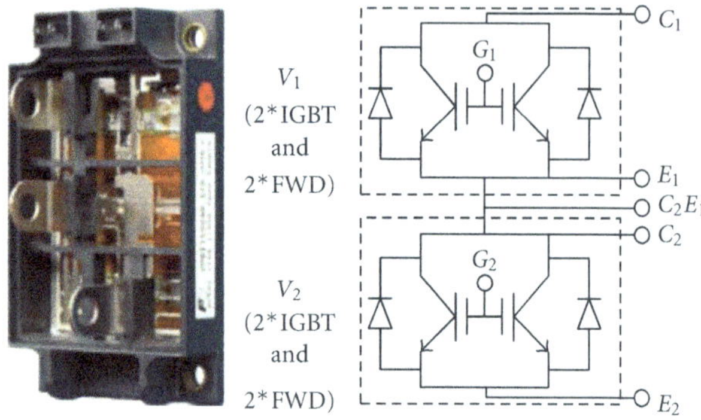

Figure 1: A typical half bridge module and its equivalent circuit.

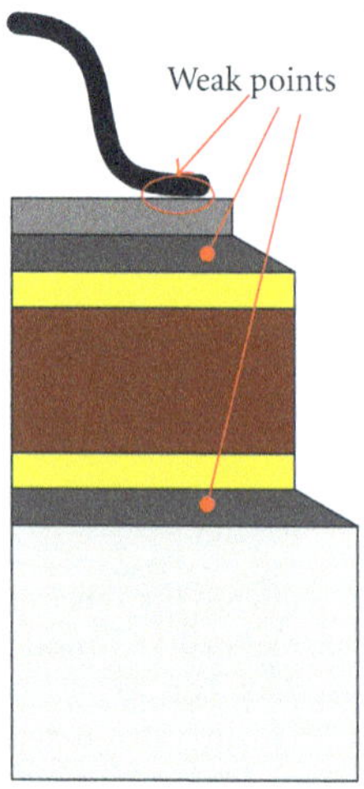


| Material | CTE (ppm/°C) |
|---|---|
| Al | 22 |
| Si | 3 |
| Solder | Compliant |
| Cu | 16.8 |
| $Al_2O_3$ or AlN | 7 or 4 |
| Cu | 16.8 |
| Solder | Compliant |
| Cu or AlSiC | 17 or 8 |

Figure 2: Cross section of an IGBT Module and related CTEs of its component materials.

The IGBT module is complex multilayered structure with different materials in intimate contact with each other, which provide mechanical stability, electrical insulation, and thermal conduction. Thermal interactions between these materials play an important role in the long-term reliability of the IGBT module [16, 21]. The schematic cross-section of the IGBT module and related coefficients of thermal expansion (CTE) of its component materials are represented in Figure 2. From top to bottom are bond wire, silicon chip, direct bonded copper ceramic substrate (DBC), and baseplate. From Figure 2, it can be seen that there are two weak points within the IGBT module due to the CTE mismatch, one is the interface between bond wire and silicon chip, the other is solder layer. Thermal stress caused by large junction temperature swings (up to 80°C) occurred in some fields has a severe effect on the reliability of IGBTs [4, 5]. Bond-wire fatigue and solder fatigue have been identified as the main failure mechanisms [9, 22].

*2.2. Defects and Their Influences.* There are two main defects observed inside the IGBT module: the liftoff of bond wire

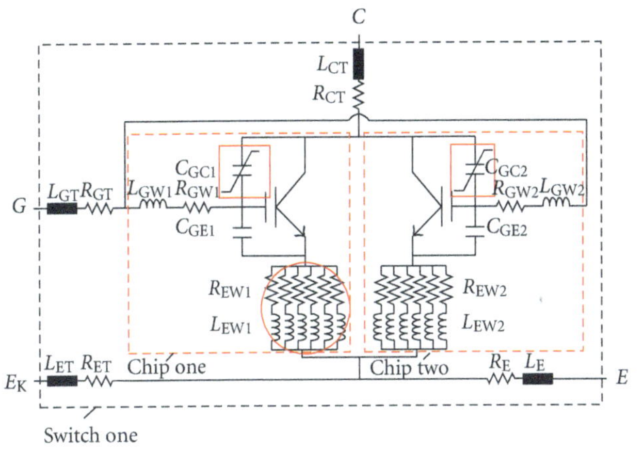


FIGURE 3: Parasitic elements of one switch inside an IGBT module.

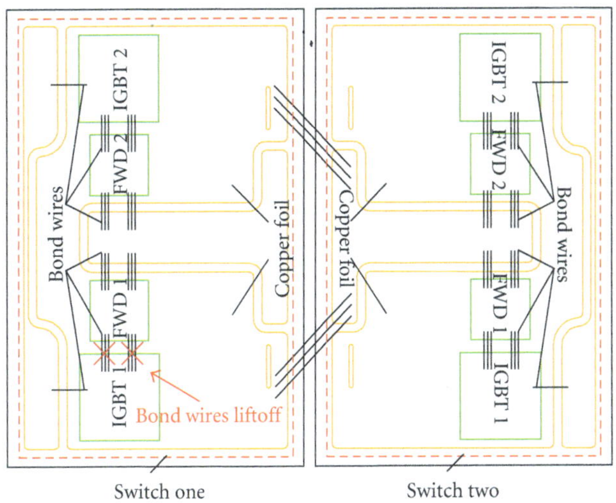


FIGURE 4: Bond wires liftoff inside an IGBT module.

and the crack of solder layer as a results of aging. These defects would affect some intrinsic parasitic elements of the IGBT module like parasitic inductance and resistance of the emitter bond wires or junction capacitances, as shown in Figure 3 [23, 24], where $L_{GT}$ and $R_{GT}$ are the parasitic inductance and resistance of the gate lead respectively, $L_{GW}$ and $R_{GW}$ are the parasitic inductance and resistance of the gate bond wire respectively, $C_{GE}$ is the gate-emitter capacitance, $C_{GC}$ is the gate-collector capacitance, $L_{EW}$ and $R_{EW}$ are the parasitic inductance and resistance of the emitter bond wires respectively, $L_{ET}$ and $R_{ET}$ are the parasitic inductance and resistance of the auxiliary emitter lead respectively, $L_E$ and $R_E$ are the parasitic inductance and resistance of the emitter lead respectively, $L_{CT}$ and $R_{CT}$ are the parasitic inductance and resistance of the collector lead respectively, and the numbers in those symbols indicate different chips in the switch. Note that the collector-emitter capacitance $C_{CE}$ is not included in Figure 3 because it has little effect on the switching characteristics of the IGBT module [25].

The liftoff of bond wire can change not only the parasitic inductance and resistance of emitter bond wires, but also the layout of the IGBT module. For example, if six bond wires connecting the emitter of chip one are liftoff as shown in Figure 4, this chip will be to failure, then the layout of the IGBT module is altered that two paralleled chips become one. Consequently, the equivalent total gate-emitter parasitic capacitance is also changed. Cracks at silicon-substrate or substrate-baseplate solder layers can reduce the effective area for heat conduction and increase the thermal resistance from junction to case, which will lead to higher junction temperature [6] and ionize more hot carriers to change the junction capacitance. In addition, the gate electrostatic discharge and the collector over-voltage can also ionize hot carriers and affect the junction capacitance, but they should be accidental factors. On the whole, defects would affect some intrinsic parasitic elements involved in gate circuit.

*2.3. Linearization.* From Figure 3, it can be seen that the gate voltage $V_{GE}$ is electrically coupled with the collector-emitter voltage $V_{CE}$ by the Miller capacitance $C_{GC}$ including $C_{GC1}$ and $C_{GC2}$. This Miller capacitance consists of the series combination of the gate-drain-overlap oxide capacitance $C_{OXD}$ and the gate-drain-overlap depletion capacitance $C_{GDJ}$, is a typical nonlinear intrinsic parasitic element due to the voltage dependence of the capacitance $C_{GDJ}$ [26].

$$C_{GC} = \begin{cases} C_{OXD} & V_{CE} < V_{GE} - V_{GE(th)} \\ \dfrac{C_{OXD} C_{GDJ}}{C_{OXD} + C_{GDJ}} & V_{CE} \geq V_{GE} - V_{GE(th)} \end{cases}, \tag{1}$$

$$C_{GDJ} = \frac{A_{GD}\varepsilon_{Si}}{\sqrt{2\varepsilon_{Si}\left(V_{CE} - V_{GE(th)}\right)/q/N_B}},$$

where $A_{GD}$ is the gate-drain overlap area, $\varepsilon_{Si}$ is the dielectric constant of silicon, $V_{GE(th)}$ is the gate threshold voltage, and $N_B$ is the base doping concentration. The significant change in $C_{GC}$ can be as large as a factor of 10 to 100 [25]. Since $V_{CE}$ differs in different applications, the direct use of the gate voltage would lead the diagnostic results to be ambiguous. So a linearization of the gate circuit over a particular time period of the turn-on process of the IGBT module is employed. As shown in Figure 5, during the beginning stage of the turn-on process, that is, the time interval from $t_0$ to $t_1$, the collector-emitter voltage $V_{CE}$ remains at the high DC voltage as long as the free-wheeling diode is conducting, $C_{GC}$ remains at a small constant value because of the high and unchanged $V_{CE}$ and almost equals to $C_{GDJ}$ which is far less than $C_{GE}$. So the beginning from $t_0$ to $t_1$, is the most relevant stage, where the nonlinear influence caused by $C_{GDJ}$, and $V_{CE}$ is eliminated and the whole intrinsic parasitic elements involved in the gate circuit can be seen as linear elements because rest-associated intrinsic parasitic inductances and resistances like $L_{EW1}$ and $R_{EW1}$ are linear elements and not affected by $V_{CE}$.

Then the gate voltage can be seen as a step response of a second-order linear system as shown in Figure 6, where $V_1$ and $V_2$ are negative and positive supply voltage of the gate circuit, respectively, $S_1$ is the control signal, $R$ represents the total external resistance, $L$ represents the total external inductance, $R_G$ represents the total intrinsic parasitic

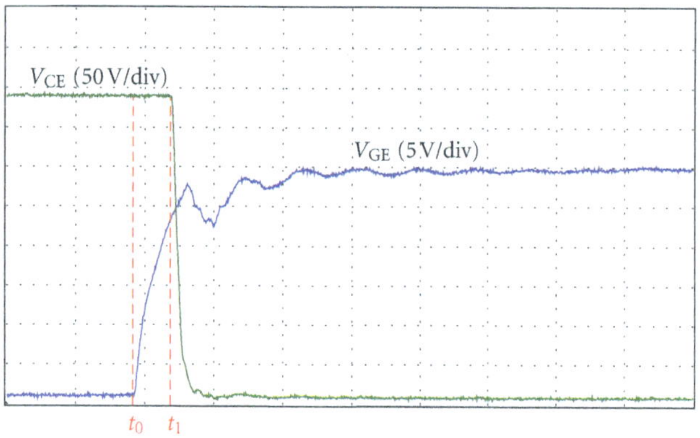


FIGURE 5: Turn-on waveforms of the IGBT module.

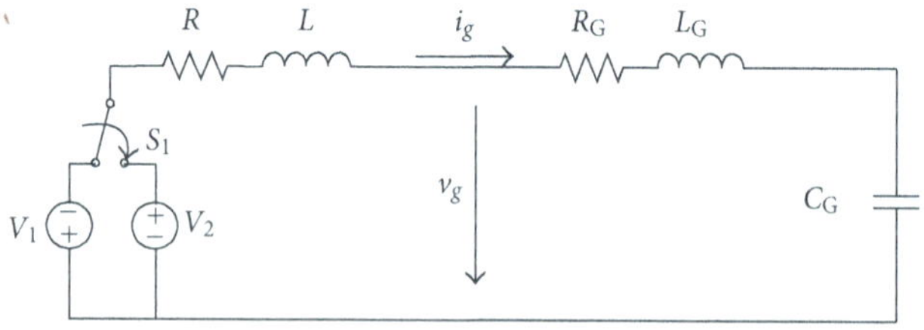


FIGURE 6: A simplified gate-drive circuit of an IGBT.

resistance involved in the gate circuit, $L_G$ represents the total intrinsic parasitic inductance involved in the gate circuit, $C_G$ represents the total intrinsic parasitic capacitance involved in the gate circuit. In most cases, the external components of the gate circuit are fixed. Therefore, the variation of the gate voltage is mainly contributed by the change of intrinsic elements. The purpose of this paper is to extract precursor information from variation of the gate voltage at the beginning stage of the turn-on process.

## 3. Approximate Entropy Theory and Defects Diagnosis

From analysis mentioned above, it is clear that the variation of the gate voltage at the beginning stage of the turn-on process is closely coupled with defects inside the IGBT module. So this variation can be used as precursor parameter to develop a diagnostic methodology for IGBT defects. However, sensing the gate voltage during converter operation may be affected by many uncertain factors such as noises, which may have impact on the diagnostic results. To overcome this limit, a diagnostic method based on the approximate entropy theory is proposed in this paper.

*3.1. Theory of ApEn.* The ApEn is developed from statistics, and is used to quantify the amount of system complexity and the unpredictability of variations over time-series data, which has the advantages of lower computational demand, small data samples, less noise effects [20]. The definition and algorithm for the ApEn of the gate voltage are introduced in the following.

(1) Given a time series of the gate voltage data $U = \{u_i, i \in N\}$ from measurements equally spaced in time. The value $N$ presents the length of the time series.

(2) Fix $m < N$, an integer, and $r$, a positive real number. The value $m$ presents the length of compared run of data, and $r$ presents a tolerance level. Typically choose $m = 2$, and $r = 0.1 \sim 0.2SD(U)$, $SD(U)$ is the standard deviation of $U$.

(3) Form a sequence of vectors $X = [u(i), u(i+1), \ldots, u(i+m-1)]$, in $R^m$, real $m$-dimensional space.

(4) Define for each $i$, $i = 1 \sim N - m + 1$,

$$C_i^m(r) = \frac{(\text{number of } x(j) \text{ such that } d[x(i), x(j)] \le r)}{(N - m + 1)}, \quad (2)$$

where, $d[x(i), x(j)] = \max_{k=1,2,\ldots,m}(|u(i+k-1) - u(j+k-1|)$.

(5) Define $\Phi^m(r) = (N - m + 1)^{-1} \sum_{i=1}^{N-m+1} \log(C_i^m(r))$.

(6) Form $X \in R^{m+1}$, and repeat (4) ~ (5), get $\Phi^{m+1}(r)$.

(7) Define approximate entropy as

$$\text{ApEn}(m, r, N) = \Phi^{m+1}(r) - \Phi^m(r). \quad (3)$$

*3.2. Cross ApEn of Gate Voltage Developed from the Theory of ApEn.* In order to distinguish the gate-voltage data sampled at different stages of the life cycle, the algorithm for the cross ApEn is also provided as follow.

(1) Given one time series of the gate-voltage data $U = \{u_i, i \in N\}$ from measurements equally spaced in time as references, and another one $V = \{v_i, i \in N\}$ of real-time samples. The value $N$ presents the length of the time series.

(2) Fix $m < N$, an integer, and $r$, a positive real number. The value $m$ presents the length of compared run of data, and $r$ presents a tolerance level. Typically choose $m = 2$, and $r = 0.1 \sim 0.2COV(U, V)$, $COV(U, V)$ is the covariance of $U$ and $V$.

(3) Form a sequence of vectors $X = [u(i), u(i+1), \ldots, u(i+m-1)]$, in $R^m$, real $m$-dimensional space.

(4) Form a sequence of vectors $Y = [v(i), v(i+1), \ldots, v(i+m-1)]$, in $R^m$, real $m$-dimensional space.

(5) Define for each $i$, $i = 1 \sim N - m + 1$,

$$C_i^m(r) = \frac{(\text{number of } y(j) \text{ such that } d[x(i), y(j)] \le r)}{(N - m + 1)}, \quad (4)$$

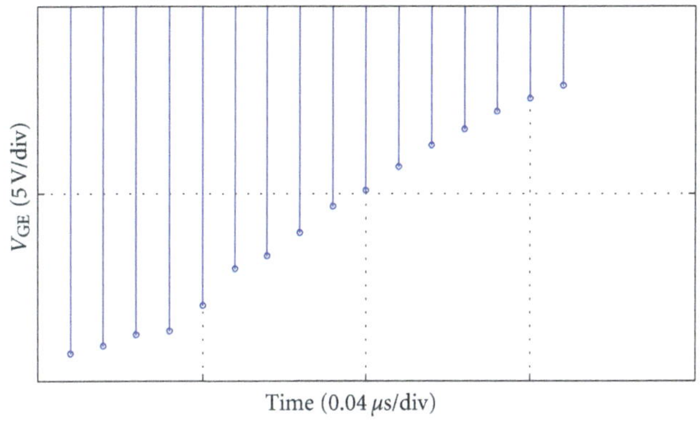


FIGURE 7: One data acquisition of the gate voltage.

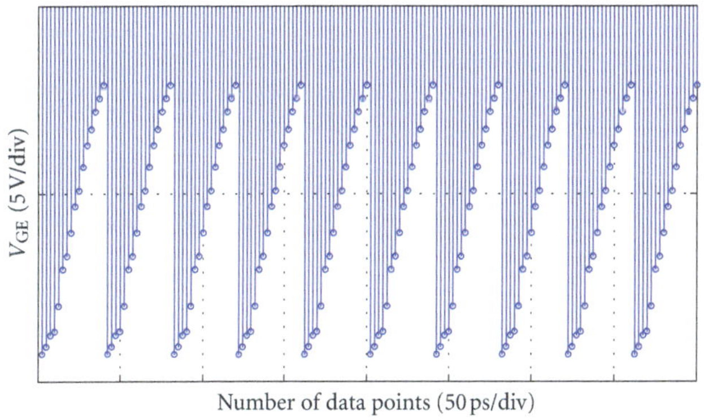


FIGURE 8: One sample of the gate voltage.

where, $d[x(i), y(j)] = \max_{k=1,2,\ldots,m}(|u(i+k-1) - v(j+k-1|)$.

(6) Define $\Phi^m(r) = (N-m+1)^{-1}\sum_{i=1}^{N-m+1}\log(C_i^m(r))$.

(7) Form $X \in R^{m+1}$, and repeat (4) ~ (5), get $\Phi^{m+1}(r)$.

(8) Define cross ApEn as

$$\text{CApEn}(m, r, N) = \Phi^{m+1}(r) - \Phi^m(r). \quad (5)$$

*3.3. Sample of Gate Voltage.* The time interval from $t_0$ to $t_1$ shown in Figure 5 is very short. Generally, this time interval is less than 1 $\mu$s. But, the highest sampling rate of an analog-to-digital converter at a reasonable price is about 125 M. This implies one data acquisition is not enough to carry out the algorithm for the ApEn and the cross ApEn as shown in Figure 7.

To circumvent this limit, a number of successive cycles of the gate voltage are measured and reassembled to one sample as shown in Figure 8, in this case 10 cycles.

## 4. Valiadation and Discussion

*4.1. Validation.* In order to examine the diagnostic method provided in this paper, a test rig is built in lab as shown in Figure 9. An especially uncovered 2MBI150U4H-170 IGBT module provided by FUJI as shown in Figure 1 is included, together with a power source (Agilent 6813B), a control

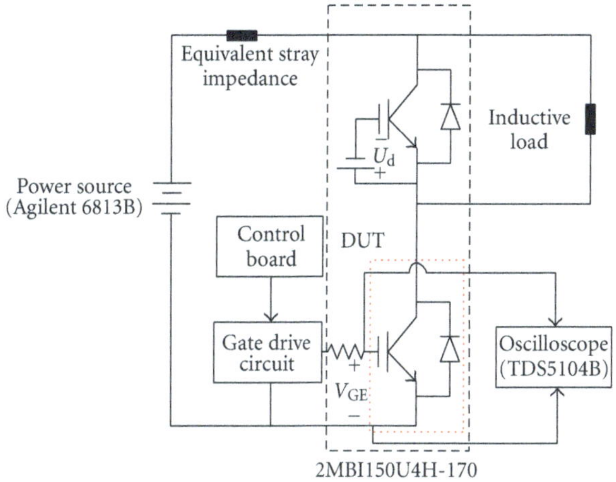


FIGURE 9: A test rig built for verification.

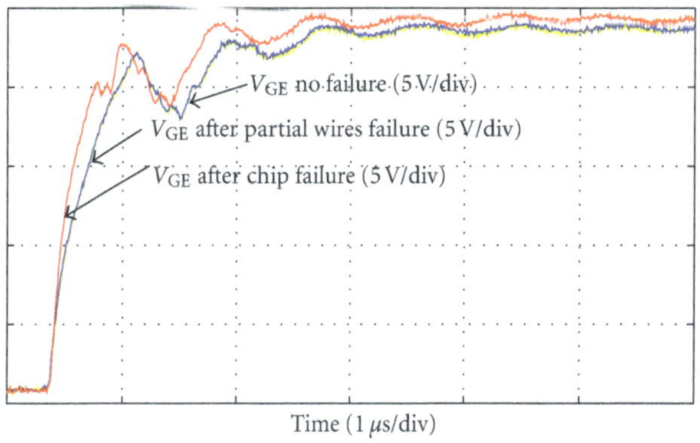


FIGURE 10: Turn-on waveforms of the gate voltage according to different states.

board, a gate driver, and a oscilloscope (TDS5104B). The inductive load has 10 Ω resistance in series with 300 $\mu$H inductance, and the switching frequency of the device under test (DUT) is 10 kHz, and the sampling rate of TDS5104B is set to 125 MS/s, at approximately the same rate of an analog-to-digital converter AD9600ABCPZ-125 from ADI. Two levels of bond-wire liftoff defect are simulated, the first one is three bond wires are liftoff inside the DUT, and the other one is chip failure, namely, six bond wires which are connected to the emitter of chip one inside the DUT are all liftoff as shown in Figure 4. The experiment results are shown in the following. Here, the gate voltage data at the health state is used as reference.

From the experiment results detailed in Figures 10 and 11, it is clear that partial bond wires failure contributes little to the variations of the gate voltage. This is because the values of parasitic inductances and resistances of bond wires are very small, so they do not play an important role in deciding the gate voltage as shown in Figure 10. But, after the chip failure, the significant changes can be found as shown in Figures 10 and 11, and the corresponding inference results by ApEn model also suggests that the cross ApEn of the gate voltage has a remarkable change from 0.189 to 0.457,

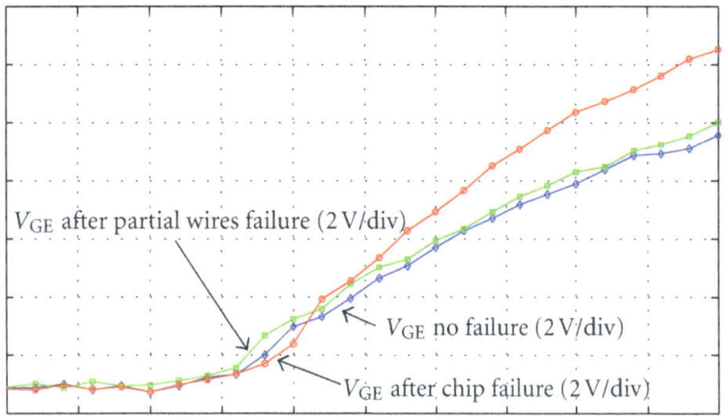


FIGURE 11: Detailed waveforms of the gate voltage according to different states.

TABLE 1: Predicted results.

| Defect level | ApEn | CApEn |
|---|---|---|
| No failure | 0.163 | 0.167 |
| Wire failure | 0.284 | 0.189 |
| Chip failure | 0.231 | 0.457 |

comparing to partial bond wire failure which just changes from 0.167 to 0.189 as shown in Table 1. Though the results show that the ApEn of the gate voltage also changes with defects, the trend is not obvious. Hence, the cross ApEn of the gate voltage will be better for defects diagnosis. Due to the fact that enormous number of chips in an IGBT module in high-power case, a single or two chips failure would not lead to an catastrophic failure to the entire module, in other words, the IGBT module is still operable, and the post-fault detection methods may do not work here. Therefore, the method proposed in this paper can be considered as prefault detection, which gives the operators a chance to replace the defective modules and avoid a sudden breakdown, and the maintenance can also be scheduled.

## 5. Conclusion

This paper proposed an ApEn-based method to monitor defects in an IGBT module using gate voltage at beginning state. This method can be seen as prefault detection, and is able to accurately detect chip defect in an IGBT module, that may cause the rest chips over current and the whole module damaged. The method may give enough time for the operators to replace the defective IGBT modules. Experimental results are provided to verify the effectiveness of the method. And it is hopeful to improve the operational reliability of power converter. The future work should consider online acquisition of the gate voltage, and so forth.

## Acknowledgments

This work was supported in part by the International S&T Cooperation Program of China (ISTCP) under Grant no. 2010DFA72250, and in part by the National Natural Science Foundation of China under Grant no. 51077137. Besides, the authors would like to thank FUJI Electric Device Technology Hong Kong Co., LTD for their free open samples.

## References

[1] S. Yang, A. Bryant, P. Mawby, D. Xiang, L. Ran, and P. Tavner, "An industry-based survey of reliability in power electronic converters," *IEEE Transactions on Industry Applications*, vol. 47, no. 3, pp. 1441–1451, 2011.

[2] S. Bernet, "Recent developments of high power converters for industry and traction applications," *IEEE Transactions on Power Electronics*, vol. 15, no. 6, pp. 1102–1117, 2000.

[3] Z. Chen, J. M. Guerrero, and F. Blaabjerg, "A review of the state of the art of power electronics for wind turbines," *IEEE Transactions on Power Electronics*, vol. 24, no. 8, pp. 1859–1875, 2009.

[4] S. Yang, D. Xiang, A. Bryant, P. Mawby, L. Ran, and P. Tavner, "Condition Monitoring for Device Reliability in Power Electronic Converters: A Review," *IEEE Transactions on Power Electronics*, vol. 25, pp. 2734–2752, 2011.

[5] A. Bryant, S. Y. Yang, P. Mawby et al., "Investigation into IGBT dV/dt during turn-off and its temperature dependence," *IEEE Transactions on Power Electronics*, vol. 26, pp. 3019–3031, 2011.

[6] D. Xiang, L. Ran, P. Tavner, A. Bryant, S. Yang, and P. Mawby, "Monitoring solder fatigue in a power module using case-above-ambient temperature rise," *IEEE Transactions on Industry Applications*, vol. 47, pp. 2578–2259, 2011.

[7] M. G. Pecht, *Prognostics and Health Management of Electronics*, John Wiley & Sons, Hoboken, NJ, USA, 2008.

[8] N. Patil, D. Das, and M. Pecht, "A prognostic approach for non-punch through and field stop IGBTs," *Microelectronics Reliability*, vol. 52, pp. 482–488, 2011.

[9] Y. Xiong, X. Cheng, Z. J. Shen, C. Mi, H. Wu, and V. K. Garg, "Prognostic and warning system for power-electronic modules in electric, hybrid electric, and fuel-cell vehicles," *IEEE Transactions on Industrial Electronics*, vol. 55, no. 6, pp. 2268–2276, 2008.

[10] D. W. Xiang, L. Ran, P. Tavner, S. Y. Yang, A. Bryant, and P. Mawby, "Condition monitoring power module solder fatigue using inverter harmonic identification," *IEEE Transactions on Power Electronics*, vol. 27, pp. 235–247, 2012.

[11] M. Musallam, C. Buttay, M. Whitehead, and C. M. Johnson, "Real-time compact electronic thermal modelling for health monitoring," in *Proceedings of the European Conference on Power Electronics and Applications (EPE '07)*, September 2007.

[12] N. Patil, S. Menon, D. Das, and M. Pecht, "Anomaly detection of non punch through insulated gate bipolar transistors (IGBT) by robust covariance estimation techniques," in *Proceedings of the 2nd International Conference on Reliability, Safety and Hazard: Risk-Based Technology and Physics-of-Failure Methods (ICRESH '10)*, pp. 68–72, December 2010.

[13] E. Sutrisno, Q. Fan, D. Das, and M. Pecht, "Anomaly Detection for Insulated Gate Bipolar Transistor (IGBT) under power cycling using a K–Nearest neighbor technique," in *Proceedings of the Prognostics and Health Management Solutions Conference (MFPT '12)*, Dayton, Ohio, USA, 2012.

[14] N. Patil, D. Das, C. Yin, H. Lub, C. Bailey, and M. Pecht, "A fusion approach to IGBT power module prognostics," in *Proceedings of the 10th International Conference on Thermal, Mechanical and Multi-Physics Simulation and Experiments in Microelectronics and Microsystems (EuroSimE '09)*, pp. 1–5, Delft, The Netherlands, April 2009.

[15] M. Musallam, C. M. Johnson, C. Yin, H. Lu, and C. Bailey, "In-service life consumption estimation in power modules," in *Proceedings of the 13th International Power Electronics and Motion Control Conference (EPE-PEMC '08)*, pp. 76–83, New York, NY, USA, September 2008.

[16] M. Ciappa, "Selected failure mechanisms of modern power modules," *Microelectronics Reliability*, vol. 42, no. 4-5, pp. 653–667, 2002.

[17] B. Lu and S. K. Sharma, "A literature review of IGBT fault diagnostic and protection methods for power inverters," *IEEE Transactions on Industry Applications*, vol. 45, no. 5, pp. 1770–1777, 2009.

[18] A. Qun-Tao, S. Li-Zhi, Z. Ke, and S. Li, "Switching function model-based fast-diagnostic method of open-switch faults in inverters without sensors," *IEEE Transactions on Power Electronics*, vol. 26, no. 1, pp. 119–126, 2011.

[19] D. Pan, X. Kai-gui, Z. Li, and R. Xianliang, "Open-switch fault diagnosis and system reconfiguration of doubly fed wind power converter used in a microgrid," *IEEE Transactions on Power Electronics*, vol. 26, no. 3, pp. 816–821, 2011.

[20] S. M. Pincus, "Approximate entropy as a measure of system complexity," *Proceedings of the National Academy of Sciences of the United States of America*, vol. 88, no. 6, pp. 2297–2301, 1991.

[21] W. W. Sheng and R. P. Colino, *Power Electronic Modules: Design and Manufacture*, CRC Press, Boca Raton, Fla, USA, 2005.

[22] N. Patil, J. Celaya, D. Das, K. Goebel, and M. Pecht, "Precursor parameter identification for insulated gate bipolar transistor (IGBT) prognostics," *IEEE Transactions on Reliability*, vol. 58, no. 2, pp. 271–276, 2009.

[23] V. K. Khanna, *The Insulated Gate Bipolar Transistor: IGBT Theory and Design*, IEEE Press, Piscataway, NJ, USA, 2003.

[24] K. Xing, F. C. Lee, and D. Boroyevich, "Extraction of parasitics within wire-bond IGBT modules," in *Proceedings of the 13th Annual Applied Power Electronics Conference and Exposition (APEC '98)*, pp. 497–503, Anaheim, Calif, USA, February 1998.

[25] N. Mohan, T. M. Undeland, and W. P. Robbins, *Power Electronics: Converters, Applications, and Design*, John Wiley & Sons, New York, NY, USA, 3rd edition, 2003.

[26] A. R. Hefner and D. M. Diebolt, "Experimentally verified IGBT model implemented in the Saber circuit simulator," *IEEE Transactions on Power Electronics*, vol. 9, no. 5, pp. 532–542, 1994.

# 3

# Comparative Study of $SiO_2$, $Al_2O_3$, and BeO Ultrathin Interfacial Barrier Layers in Si Metal-Oxide-Semiconductor Devices

**J. H. Yum,[1,2] J. Oh,[3] Todd. W. Hudnall,[4] C. W. Bielawski,[5] G. Bersuker,[2] and S. K. Banerjee[1]**

[1] *Microelectronics Research Center, Department of Electrical and Computer Engineering, The University of Texas, Austin, TX 78758, USA*
[2] *SEMATECH, 2706 Montopolis Drive, Austin, TX 78741, USA*
[3] *School of Integrated Technology, College of Engineering, Yonsei University, 162-1 Songdo-dong, Incheon 406-840, Republic of Korea*
[4] *Department of Chemistry and Biochemistry, Texas State University, 601 University Drive, San Marcos, TX 78666, USA*
[5] *Department of Chemistry, UT, Austin, TX 78712, USA*

Correspondence should be addressed to J. H. Yum, redeyes78@mail.utexas.edu

Academic Editor: Edward Yi Chang

In a previous study, we have demonstrated that beryllium oxide (BeO) film grown by atomic layer deposition (ALD) on Si and III-V MOS devices has excellent electrical and physical characteristics. In this paper, we compare the electrical characteristics of inserting an ultrathin interfacial barrier layer such as $SiO_2$, $Al_2O_3$, or BeO between the $HfO_2$ gate dielectric and Si substrate in metal oxide semiconductor capacitors (MOSCAPs) and n-channel inversion type metal oxide semiconductor field effect transistors (MOSFETs). Si MOSCAPs and MOSFETs with a BeO/$HfO_2$ gate stack exhibited high performance and reliability characteristics, including a 34% improvement in drive current, slightly better reduction in subthreshold swing, 42% increase in effective electron mobility at an electric field of 1 MV/cm, slightly low equivalent oxide thickness, less stress-induced flat-band voltage shift, less stress induced leakage current, and less interface charge.

## 1. Introduction

The CMOS scaling is bringing the $SiO_2$ thickness below 1.5 nm. For these very thin oxides, the leakage current becomes unacceptably large. One way to reduce the leakage current is the substitution of the $SiO_2$ by a material with a higher dielectric constant. The main advantage of high-k dielectrics is the low gale leakage achieved due to its high physical thickness. That also makes it attractive for low power applications. Because of these requirements, over the past 10 years, hafnium oxide ($HfO_2$) has gained considerable interest as a high dielectric constant material for fabricating complementary metal oxide semiconductor (CMOS) devices. It has several attractive properties such as a high dielectric constant, good thermodynamic stability with Si, and good electrical properties [1]. Unfortunately, some of the other physical properties like mobility reduction, charge trapping, and threshold voltage ($V_{th}$) instability are a major drawback for the performance of metal oxide semiconductor field effect transistors (MOSFETs) [2]. Especially $HfO_2$ high-k dielectric stacked MOSFETs were reported with low carrier mobility [3]. The main cause for the low mobility is still unknown, but has been attributed to remote Coulomb scattering caused by charges in the high-k dielectric [4] or optical phonon scattering [5]. Many researchers have believed that it is inevitable for all high-k dielectrics to have low energy bandgap and high scattering, compared to $SiO_2$. Therefore, if high-k dielectric with high energy bandgap and low scattering can be found, it will be the true solution for the above problems.

An alternative promising high-k gate dielectric material is beryllium oxide (BeO), which has superior interface stability [6–10] and is already known as an excellent gas diffusion barrier. This makes it a potentially suitable diffusion barrier

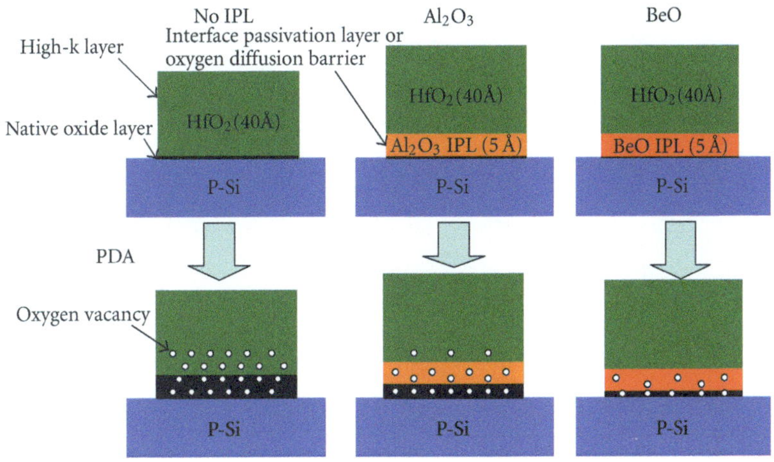


FIGURE 1: Cross-sectional MOS devices with various IL. The BeO interfacial layer is placed between $HfO_2$ and p-type Si substrate.

between $HfO_2$ and Si in CMOS processing. BeO also has metal-like thermal conductivity and a large energy bandgap (10.6 eV). These properties are indicative of low optical phonon and remote Coulomb scattering. Generally, a flow of phonons is responsible for heat conduction in dielectric materials. As the temperature increases, phonon density increases, but above 20 K, the phonon-phonon interaction becomes dominant and reduces the mean free path of the phonon drift, degrading thermal conductivity in the dielectrics [11]. BeO, however, has high thermal conductivity due to low phonon scattering because electrons in BeO are tightly and closely bound, so that the phonons in BeO are coupled to each other and have low energy and long wavelengths (or low phonon frequency). The high energy bandgap and band offset of BeO on Si makes intrinsic charge trapping difficult and results in a low trapped charge in the BeO dielectric (trapped charges in high-k dielectrics are the source of Coulomb scattering) [10]. Our previous studies have showed electrical and physical characteristics that BeO deposited with dimethylberyllium and water improves interface quality on III-V MOS devices by preventing sub-oxidation between high-k and III-V substrate during PDA [6]. In this paper, we compare the effect of interfacial barrier layer by inserting ultrathin $SiO_2$, $Al_2O_3$, or BeO barrier layer (IL) between the $HfO_2$ gate dielectric and Si substrate in metal oxide semiconductor capacitors (MOSCAPs) and NMOSFETs. The aim of using such a barrier layer was to improve the device performance and reliability while maintaining, as much as possible, the overall dielectric constant of the resulting film.

## 2. Fabrication Procedure

An ALD BeO IL was deposited on HF-last p-type Si substrates using dimethylberyllium precursors and water as an oxygen source. As a reference, ALD $Al_2O_3$ IL was deposited on the same cleaned substrate using trimethylaluminum and the same oxygen source. Samples with a BeO IL, $Al_2O_3$ IL, and without an IL were followed by ALD $HfO_2$. They were annealed for 3 min at 600°C in $N_2$ at atmospheric pressure. The physical thickness of the BeO and $Al_2O_3$ IL layers was controlled from the deposition rate which was measured on the bulk oxide using multiple-wavelength (200~900 nm) ellipsometry. The TaN electrode was deposited using reactive dc magnetron sputtering at 2000 Å followed by reactive ion etching (RIE) with Ar + $CF_4$ after electrode patterning of the gate. The source/drain (S/D) regions of NMOSFETs were implanted with phosphorus at 50 keV and a dose of $5 \times 10^{15}\,cm^{-2}$. High temperature (900°C, 1 min) annealing in $N_2$ ambient was used for S/D activation. E-beam evaporated Ni/AuGe/Au was used for both S/D and backside metallization. The final sintering was done at 400°C in forming gas for 30 min. For all MOSCAPs samples, PMA (500°C, 2 min) was done.

## 3. Results and Discussion

Figure 1 shows the cross-sectional MOS structure with various IL (or IPL). It is constructed based on the electrical and physical results in the previous experiments [10]. $SiO_2$, $Al_2O_3$, or BeO IL is placed between $HfO_2$ and the P-Si substrate. $Al_2O_3$ and BeO IL are intentionally inserted, but $SiO_2$ IL is thermally grown during post-deposition and S/D activation anneals. In Figure 2, the BeO(IL)/$HfO_2$ structures show the lowest leakage, comparable to those of $SiO_2$(IL)/$HfO_2$ and $Al_2O_3$(IL)/$HfO_2$ gate stacks. Insertion of BeO IL (5~10 Å) doesn't increase the EOT significantly after the post-deposition anneal (PDA) due to the efficient suppression of the oxygen diffusion during PDA. The effectiveness of oxygen diffusion barrier for BeO IL is more presented as the annealing temperature increases in Figure 3. BeO IL may have some advantage for EOT scaling and reliability improvement After S/D activation, around 15 Å $SiO_2$ is grown at the interface between $HfO_2$ and the Si substrate. The low EOT of BeO IL is an indication of efficient oxygen diffusion barrier. The similar results were presented using X-ray photoelectron spectroscopy (XPS) [10]. Oxygen diffusion through thin films is proportional to the number and size of pinholes in the respective film [12]. In general, smaller pinholes cause more collisions between the diffusing

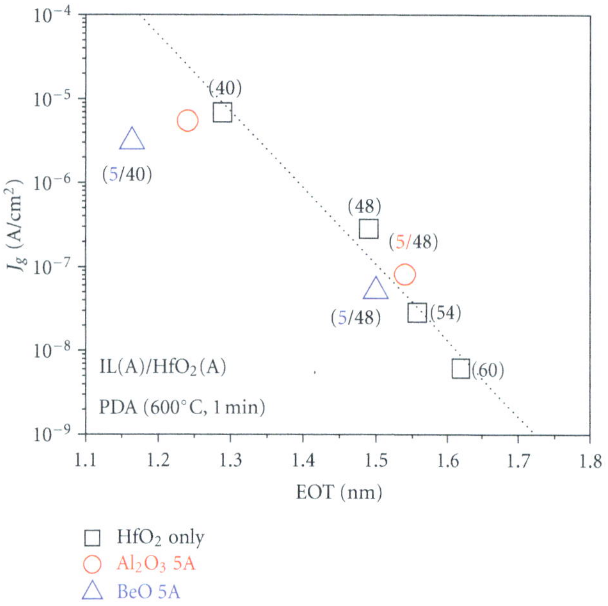


FIGURE 2: Gate leakage current versus EOT for $SiO_2/HfO_2$, $Al_2O_3$(IL)/$HfO_2$, and BeO(IL)/$HfO_2$ gate stacks.

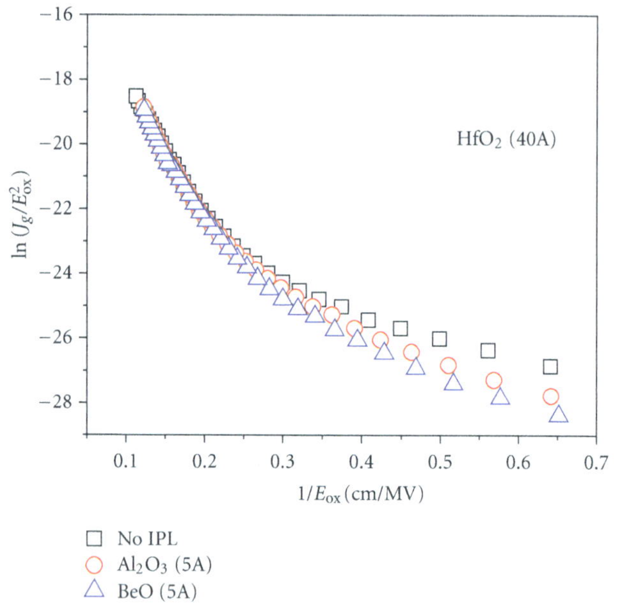


FIGURE 4: F-N plots to compare the effective potential barrier height for three different gate stacks.

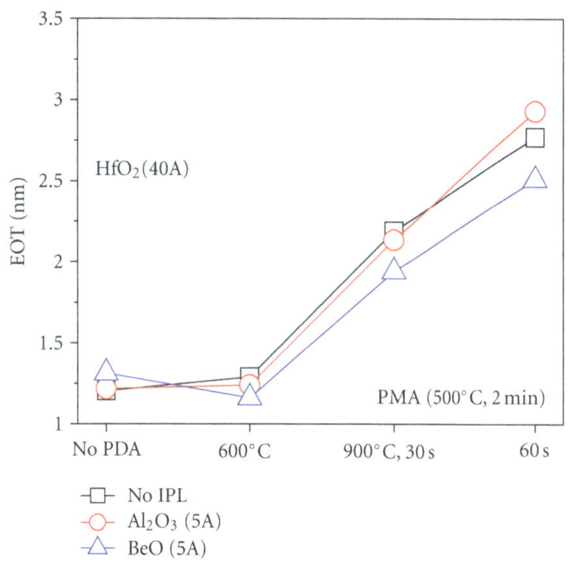


FIGURE 3: The change of EOT with the annealing temperature and duration for three different gate stacks.

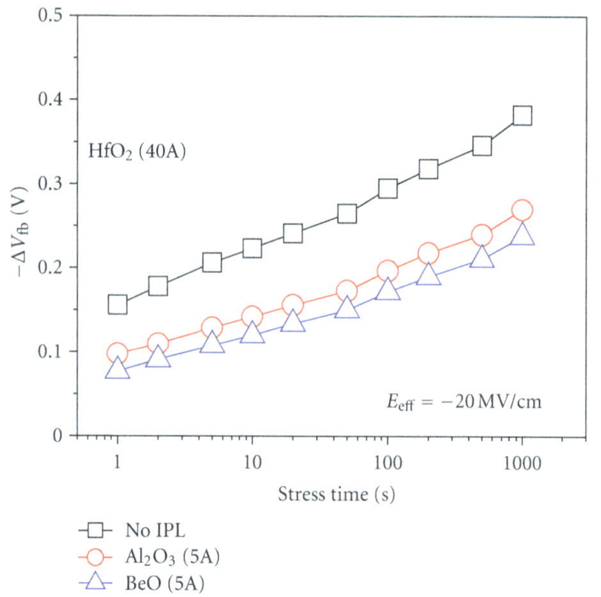


FIGURE 5: Stress-induced $V_{fb}$ shift ($\Delta V_{fb}$) versus stress time for three different gate stacks. $E_{eff} = (V_g - V_{fb})/EOT$.

molecules (e.g., oxygen) and the chemical groups present in the bulk film, reducing the rate of permeation. For reasons that are still under investigation, films of BeO, which have small molecular size, appear to exhibit relatively low oxygen diffusivity and are capable of effectively blocking the diffusion of impurities, such as Hf, thus minimizing defects in the substrate.

In general, the bandgap of the high-k material is inversely proportional to its permittivity, but BeO is an exception, having a very large energy bandgap (10.6 eV) combined with a still high dielectric constant of 6.8. As the bandgap, or correspondingly, band offset increases, a charge trapping in the dielectric decreases. The effective potential barrier heights for $SiO_2/HfO_2$, $Al_2O_3$(IL)/$HfO_2$, and BeO(IL)/$HfO_2$ gate stacks are compared using the Fowler-Nordheim plot in Figure 4. Due to bilayer gate structure, exact number of the effective barrier height is not extracted. But a higher barrier of the BeO IL stack is observed and it may results in the smaller electron tunneling currents, compared to other different gate stacks. Figures 5 and 6 are the representative results of reliability statistics characteristics. In Figure 5,

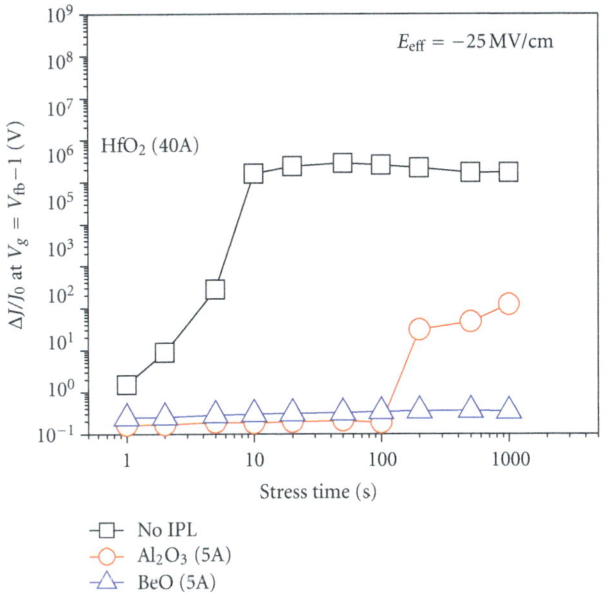


Figure 6: Stress-induced leakage current ($\Delta J_g/J_0$) versus stress time. $E_{eff} = (V_g - V_{fb})/EOT$.

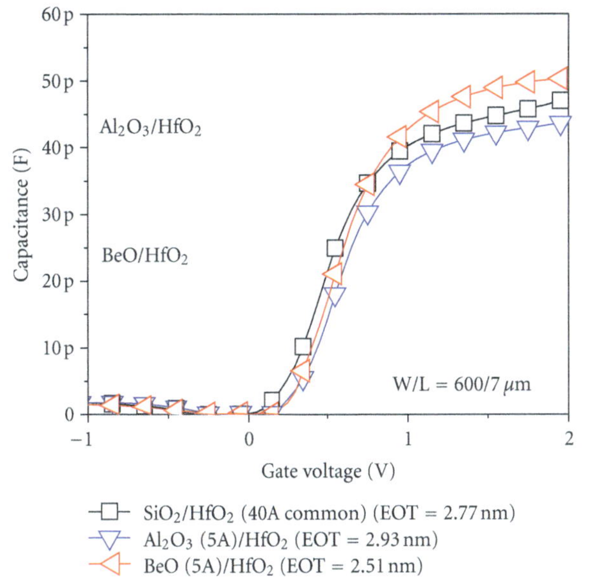


Figure 7: NMOSFETs inversion capacitance for three different gate stacks.

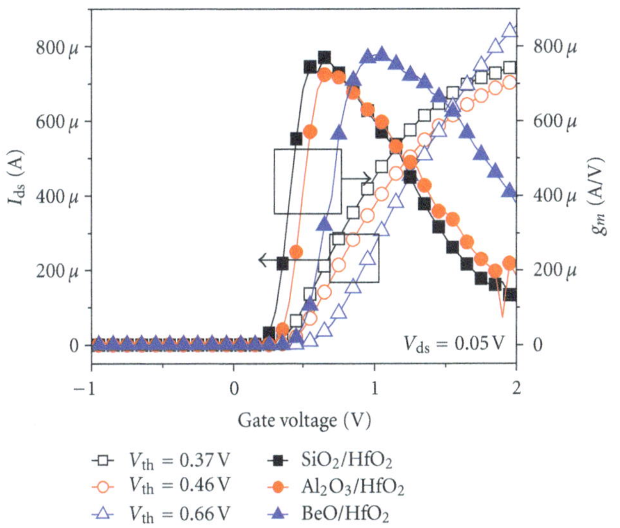


Figure 8: NMOSFETs $I_d - V_g$ characteristics of three gate stacks. BeO IL shows slightly higher $V_{th}$, $G_m$, and $I_d$.

the BeO(IL)/$HfO_2$ gate stack shows less initial $V_{fb}$ shift (after 1 sec stress) indicating fewer preexisting traps in the dielectric. A slightly smaller trap generation rate was also observed compared to other two gate stacks. In Figure 6, the BeO(IPL)/$HfO_2$ also shows the reduced stress-induced leakage current (SILC) degradation and no significant breakdown. But $SiO_2$/$HfO_2$ and $Al_2O_3$(IL)/$HfO_2$ show gradual breakdown with stress time. The lower trap generation rate and the reduced tunneling current of the BeO(IPL)/$HfO_2$ gate stack may improve the reliability characteristics and it may be the indication of the high structural stability. In the view point of thermodynamics of materials, the total entropy of a material consists of its thermal entropy, which is related to thermal conductivity, and configurational entropy, which is related to the crystallization (or crystallinity) of the material [13]. With high crystallinity and thermal conductivity, BeO may have high total entropy, and it means that BeO is more structurally stable, compared to other gate dielectrics, even though the direct correlation between thermodynamic stability and device performance is still questionable. For more details of BeO thermal stability, please see the reference [14]

Figure 7 is NMOSFET inversion capacitance for $SiO_2$/$HfO_2$ (40 Å), $Al_2O_3$(5 Å)/$HfO_2$ (40 Å), and BeO(5 Å)/$HfO_2$ (40 Å) gate stacks. The BeO/$HfO_2$ gate stack shows a slightly lower equivalent oxide thickness (EOT) (2.51 nm) than $SiO_2$/$HfO_2$ (2.77 nm) and $Al_2O_3$/$HfO_2$ (2.93 nm) even though the EOTs for all gate stacks significantly increased after S/D activation annealing (Figure 3). From the XPS analysis, EOT increase is mainly due to the oxygen in $HfO_2$ dielectric, instead of oxygen residue in anneal tool [10]. For $SiO_2$/$HfO_2$, $Al_2O_3$/$HfO_2$, and BeO/$HfO_2$ gate stacks, 1.7 nm, 1.5 nm, and 1.0 nm are expected for native oxide to be grown, respectively, based on Figure 3 results. Figure 8 shows NMOSFET drain current-gate voltage ($I_d - V_g$) characteristics of $SiO_2$/$HfO_2$, $Al_2O_3$/$HfO_2$, and BeO/$HfO_2$ gate stacks. With the slightly lower EOT, the BeO/$HfO_2$ stack exhibits more positive $V_{th}$ (0.66 V), higher drive current at $V_g$ = 2 V, and better subthreshold swing (69 mV/dec), compared to those of the $SiO_2$/$HfO_2$ stack ($V_{th}$ = 0.37 V, SS = 77 mV/dec) and $Al_2O_3$/$HfO_2$ stack ($V_{th}$ = 0.46 V, SS = 70 mV/dec). The threshold voltage equation obtained from an ideal MOS structure [15] is

$$V_{th} = \Phi_{ms} - \frac{Q_i}{C_i} - \frac{Q_d}{C_i} + 2\phi_F, \quad (1)$$

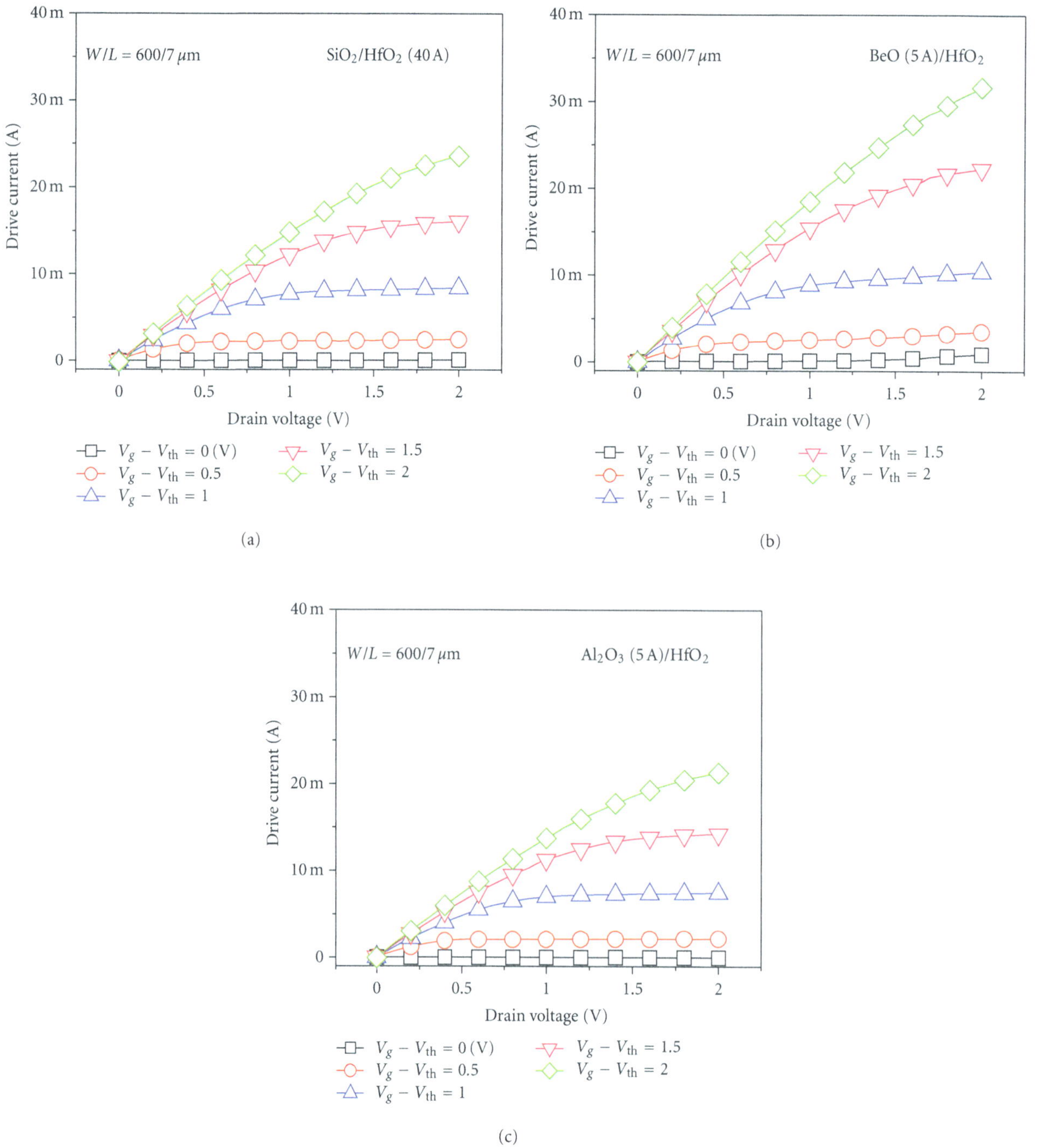


FIGURE 9: $I_d - V_d$ characteristics of three gate stacks. BeO IL shows significant increased drive current compared to $SiO_2$ and $Al_2O_3$ IL gate stacks.

where $\Phi_{ms}$, $Q_i$, $Q_d$, and $\phi_F$ are the work function differences between the metal and semiconductor ("−" value), interface charge ("+" value), depletion charge ("−" value) for the n-channel, and energy differences between the intrinsic energy level and Fermi energy level (+) for the n-channel, $\phi_F = (E_i - E_F)/q$. If we assume that $\Phi_{ms}$, $Q_d$, and $\phi_F$ are the same for all gate stacks because the only difference is interfacial layer, then the positive shift of $V_{th}$ of the BeO/$HfO_2$ stack is due to the less positive interface charges between BeO and the Si substrate. The fewer fixed charges in BeO layer may contribute to the fewer interface charges [10].

In Figure 9, the BeO/$HfO_2$ stack shows around 34% higher drive current (31.67 mA) at $V_d = 2$ V & $V_g - V_{th} = 2$ V than the $SiO_2$/$HfO_2$ stack (23.56 mA) and $Al_2O_3$/$HfO_2$ stack (21.28 mA). Only 5 Å BeO insertion between high-k and Si channel makes the drive current much improved. There is some reduction of drive current with the IL thickness increase for both the $Al_2O_3$/$HfO_2$ and BeO/$HfO_2$

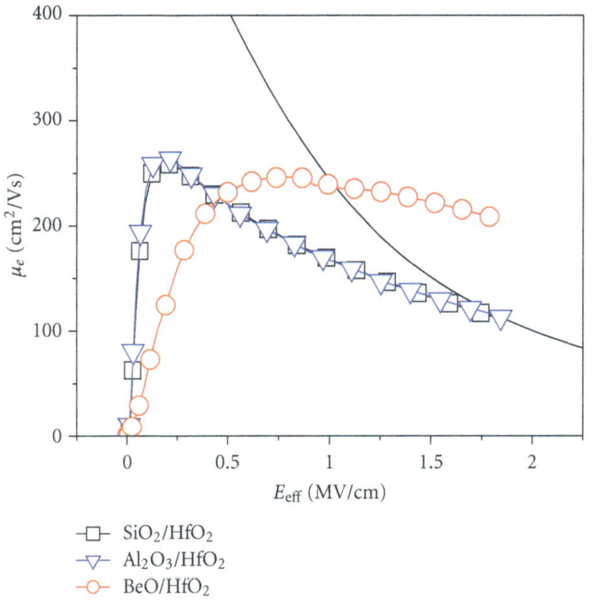


FIGURE 10: Effective channel mobility of NMOSFETs with three gate stacks.

gate stacks, but it is more significant on $Al_2O_3/HfO_2$ stack. It may be due to the less native interfacial oxide ($SiO_2$) growth for $Al_2O_3/HfO_2$, stack. Figure 10 illustrates the effective channel electron mobility using the split capacitance-voltage (C-V) method. The $BeO/HfO_2$ stack shows a 42% higher effective field ($E_{eff}$) mobility (238 cm²/Vs) than $SiO_2/HfO_2$ (167 cm²/Vs) and $Al_2O_3/HfO_2$ (166 cm²/Vs) at $E_{eff}$ = 1 MV/cm. It may require further investigation to confirm and explain these results. The electron mobility in $SiO_2/HfO_2$ and $Al_2O_3/HfO_2$ are fast-saturated to the universal trend, likely due to the thick $SiO_2$ interfacial layer grown during S/D activation. If the $SiO_2$ interfacial layer is thinner, the peak electron mobilities of the $HfO_2$ gate stack will decrease significantly [16]. In atomic configuration, physical roughness difference between amorphous $Al_2O_3$ and crystalline BeO is similar, but in electronic configuration, the electropotential roughness between them is quite different. In terms of electrostatic potential roughness, the two-dimensional ordered arrays of atoms on a crystalline surface generally give atomic scale surface height fluctuation, which exhibits low electrostatic potential roughness [17]. In a previous study, we demonstrated that ALD BeO on Si grows almost epitaxially [7], thereby may improve surface electro-potential roughness and high field electron mobility.

In this work, a BeO (IL)/$HfO_2$ gate stack was investigated and systematically compared to a $SiO_2/HfO_2$ gate stack. Inserting an ALD BeO IL between the Si channel and high-k gate dielectric enhances high field carrier mobility and improves MOSFET parameter and reliability characteristics while maintaining a similar EOT. Excellent BeO properties, such as a high energy bandgap, efficient oxygen diffusion barrier, and high crystallinity, improve the charge trapping, the suppression in EOT increase during S/D activation, and MOSFET performance, thus imparting significant advantages to MOS devices with a BeO IL.

## Acknowledgments

This work was supported in part by the Robert Welch Foundation (Grants F-1038 and F-1621) and NSF (Grant DMR-0706227).

## References

[1] K. J. Hubbard and D. G. Schlom, "Thermodynamic stability of binary oxides in contact with silicon," *Journal of Materials Research*, vol. 11, no. 11, pp. 2757–2776, 1996.

[2] A. Kerber, E. Cartier, R. Degraeve, L. Pantisano, P. Roussel, and G. Groeseneken, "Strong correlation between dielectric reliability and charge trapping in $SiO_2/Al_2O_3$ gate stacks with TiN electrodes," in *Proceedings of the Symposium on VLSI Technology Digest of Technical Papers*, pp. 76–77, June 2002.

[3] Y. Kim, G. Gebara, M. Freiler et al., "Conventional n-channel MOSFET devices using single layer $HfO_2$ and $ZrO_2$ as high-k gate dielectrics with polysilicon gate electrode," in *IEEE International Electron Devices Meeting (IEDM '01)*, pp. 455–458, Washington, DC, USA, December 2001.

[4] S. Saito, Y. Shimamoto, K. Torii et al., "The mechanism of mobility degradation in MISFETs with $Al_2O_3$ gate dielectric," in *Proceedings of the Symposium on VLSI Technology Digest of Technical Papers*, pp. 188–189, June 2002.

[5] M. V. Fischetti, D. A. Neumayer, and E. A. Cartier, "Effective electron mobility in Si inversion layers in metal-oxide-semiconductor systems with a high-$\kappa$ insulator: the role of remote phonon scattering," *Journal of Applied Physics*, vol. 90, no. 9, pp. 4587–4608, 2001.

[6] J. H. Yum, T. Akyol, M. Lei et al., "Atomic layer deposited beryllium oxide: effective passivation layer for III-V metal/oxide/semiconductor devices," *Journal of Applied Physics*, vol. 109, no. 6, Article ID 064101, 2011.

[7] J. H. Yum, T. Akyol, M. Lei et al., "A study of highly crystalline novel beryllium oxide film using atomic layer deposition," *Journal of Crystal Growth*, vol. 334, pp. 126–133, 2011.

[8] J. H. Yum, T. Akyol, M. Lei et al., "Inversion type InP metal oxide semiconductor field effect transistor using novel atomic layer deposited BeO gate dielectric," *Applied Physics Letters*, vol. 99, no. 3, Article ID 033502, 2011.

[9] J. H. Yum, T. Akyol, M. Lei et al., "Comparison of the self-cleaning effects and electrical characteristics of BeO and $Al_2O_3$ deposited as an interface passivation layer on GaAs MOS devices," *Journal of Vacuum Science and Technology A*, vol. 29, no. 6, Article ID 061501, 2011.

[10] J. H. Yum, T. Akyol, M. Lei et al., "Epitaxial ALD BeO: efficient oxygen diffusion barrier for EOT scaling and reliability improvement," *Electron Device Letters*, vol. 58, no. 12, pp. 4384–4392, 2011.

[11] R. E. Hummel, *Electronic Properties of Materials*, chapter 21, 3rd edition, 2001.

[12] A. A. Dameron, S. D. Davidson, B. B. Burton, P. F. Carcia, R. S. McLean, and S. M. George, "Gas diffusion barriers on polymers using multilayers fabricated by $Al_2O_3$ and rapid $SiO_2$ atomic layer deposition," *Journal of Physical Chemistry C*, vol. 112, no. 12, pp. 4573–4580, 2008.

[13] D. R. Gaskell, *Introduction to the Thermodynamics of Materials*, chapter 4, 5th edition, 2008.

[14] J. H. Yum, G. Bersuker, J. Oh, and S. K. Banerjee, "Theoretical approach evaluating beryllium oxide as A gate dielectric in the view points of electromagnetics and thermal stability," *Applied Physics Letters*, vol. 100, no. 5, Article ID 053501, 3 pages, 2012.
[15] B. G. Streetman and S. K. Banerjee, *Solid State Electronic Devices*, 6th edition, 2005.
[16] S. J. Rhee, C. Y. Kang, C. S. Kang et al., "Effects of varying interfacial oxide and high-k layer thicknesses for $HfO_2$ metal-oxide-semiconductor field effect transistor," *Applied Physics Letters*, vol. 85, no. 7, pp. 1286–1288, 2004.
[17] Y. Zhao, G. C. Wang, and T. M. Lu, *Characterization of Amorphous and Crystalline Rough Surface: Principles and Applications*, chapter 6, 7, and 16, 2000.

4

# Design Optimization of Transistors Used for Neural Recording

**Eric Basham and David Parent**

*Electrical Engineering Department, San Jose State University, San Jose, CA 95192-0084, USA*

Correspondence should be addressed to David Parent, david.parent@sjsu.edu

Academic Editor: Mingxiang Wang

Neurons cultured directly over open-gate field-effect transistors result in a hybrid device, the neuron-FET. Neuron-FET amplifier circuits reported in the literature employ the neuron-FET transducer as a current-mode device in conjunction with a transimpedance amplifier. In this configuration, the transducer does not provide any signal gain, and characterization of the transducer out of the amplification circuit is required. Furthermore, the circuit requires a complex biasing scheme that must be retuned to compensate for drift. Here we present an alternative strategy based on the $g_m/I_d$ design approach to optimize a single-stage common-source amplifier design. The $g_m/I_d$ design approach facilitates in circuit characterization of the neuron-FET and provides insight into approaches to improving the transistor process design for application as a neuron-FET transducer. Simulation data for a test case demonstrates optimization of the transistor design and significant increase in gain over a current mode implementation.

## 1. Introduction

A transistor represents a unique type of transducer because it can both convert one type of energy to an electrical signal and amplify the resulting signal simultaneously. To amplify the signal, the correct amplification circuit topology must be chosen. A clever designer will optimize both the amplifier circuit and the transistor as part of an iterative design process. To our knowledge, a systematic methodology has not been presented to optimize the transistor structure and the amplification circuitry of neuron-FETs. The approach presented here relies on a circuit design strategy using the level of inversion methodology, also referred to as the $g_m/I_d$ approach [1–3]. This technique has recently gained widespread attention in the integrated circuit design community because it offers techniques for reducing power consumption and maximizing bandwidth in circuits designed to perform analog signal processing.

The process of design may vary considerably as a function of the parameters under the designer's control. Printed circuit board-level engineering is done with discrete components. These system-level engineers select components from a range of available devices and then bias them appropriately. This is done by choosing the appropriate DC (static) operating points. Design with discrete components does not allow the designer control of intrinsic device parameters (i.e., width, length, oxide thickness, mobility, etc.). In contrast, integrated circuit engineers choose a process technology and then select dimensions (width and length) of individual transistors. However, integrated circuit designers rarely have control over the processes used to fabricate the transistors. The effects of process design on transistor performance can be significant. Any circuit will benefit from design optimization techniques applied at the process technology level. In digital circuits, transistors are typically optimized for low off-state leakage, high-drive current, and fast switching. For analog circuits, transistor optimization varies depending on the function of the circuit. For example, the same circuit topology can be optimized at the transistor design level for speed or low noise but not both.

Device engineers optimize the transistor fabrication process by tuning material parameters, oxide thickness, and transistor junctions. Design optimizations at this level may be extremely varied. Typical approaches include gate engineering, oxide engineering, source and drain engineering, and channel and substrate engineering. An excellent survey of the wide array of transistor topologies that have been developed is presented in [4]. Transistors employed as

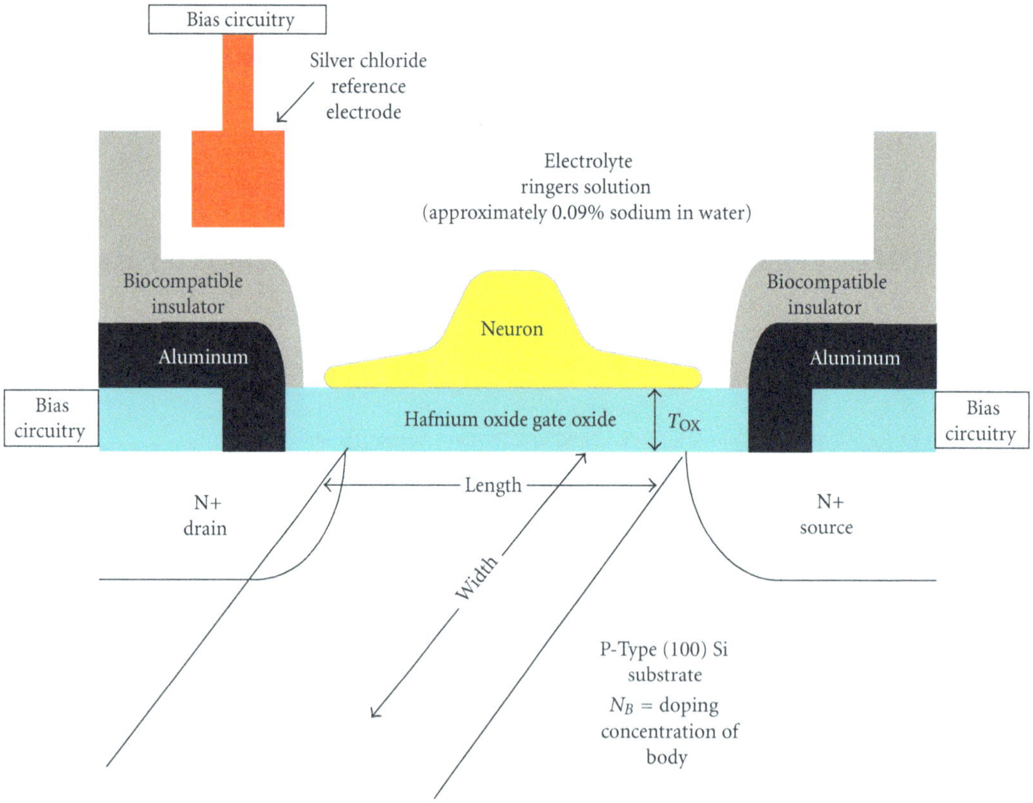


Figure 1: Neuron-FET cross-section. Modified from [27].

transducers represent an even more specialized case of design optimization. Transistor transducer elements are easily created by fabricating a transistor structure and modulating the electric field in the channel region via an energy conversion. For example, when a transistor is fabricated without a gate and submersed in a solution, the ionic charge in the solution serves to create a potential over the channel, modulating the current flow through the device [5, 6]. If neurons are cultured over the channel region where the gate would typically reside, as shown in Figure 1, the electrogenic activity of neurons may be directly detected. Using open-gate field-effect transistors in this manner was first proposed in [7, 8], and later a neuron was directly coupled to the transistor [9]. Excellent reviews that cover the development and application of the neuron-FET are [10–13].

The neuron-silicon hybrid transistor represents an interesting case for transistor and circuit design optimization. In addition to low-noise and high-gain requirements, transistor device sizing is constrained and not necessarily controllable. The boundary of the transistor sizing is set by the size of the neural cells under study and how well they adhere to the channel region. Shorter length and wider channel area result in greater intrinsic gain in transistors. However, for neuron-FETs, the actual width and length are set by the area of the cell membrane that adheres to the channel region. Thus, a large cell may have only a small patch of membrane adhered close enough to the substrate to induce a potential in the transistor channel region. There are several microstructures under study to increase and control the area of the cell that is adhered to substrate, but these will not be reviewed here.

There is a significant range of sizes of cultured neurons ranging from diameters as small as 2 $\mu$m for *Drosophila Melanogaster* (fruit fly) neurons to the large cells (300 $\mu$m) from *Aplysia Californica* (sea slug). Mammalian cells commonly used for culture include the *Rattus Norvegicus* (rat) Hippocampal neurons (20 $\mu$m). Invertebrate neurons grown in culture include retzius cells from *Hirudo Medicinalis* (leech) (60 $\mu$m), *Lymnaea Stagnalis* (pond snail) (30–100 um) [14], and *Helix Aspersa* (land snail) (100 $\mu$m). Thus, the effective width and length of the neuron-FET has an upper limit defined by the cell size, but is actually defined by the adherent membrane patch area. In each case, the length should be set to a minimum to ensure that a membrane patch is in place over the full length of the transistor or else flow of carriers from source to drain will not be modulated. The lower limit of the transistor length is set by the effect of channel length modulation on the intrinsic gain. The width is set to the approximate area where the neural cell will reside, in the hopes that a significant adherent patch will form across the width of the channel region.

Further design complication arises from the lack of access to the gate terminal. The gate area is covered by a neuron and the electrolyte solution and, thus, not readily available. The most common application of neuron-FETs is in an array topology sensing the response from many neurons simultaneously. The industry standard approach for sensing large numbers of neurons is the microelectrode array (MEA). Neuron-FETs offer the possibility of realizing significant advantages over MEAs in that the electrodes must be completely covered by the neuron for signal fidelity [15]. Since neurons rarely adhere to the substrate with perfectly circular footprints, shorting to the bath solution in low-density neural cultures is a common problem. Neuron-FETs may be able to record when part of the transducer is exposed to the bath potential, unlike MEAs, but in an array of neuron-FETs the gate DC potential is shorted across all terminals. Thus, separate DC gate biasing of individual transistors is not possible. This also means that decoupling capacitors on the gate or gate resistances cannot be part of the circuit design. The remaining design parameters are a relatively narrow range of lengths and access to the source and drain terminals.

The third major design challenge is the lack of a stable threshold voltage and an inherent DC gate bias. ISFETs are widely known to have threshold voltage drift [16–18]. Drift arises from several sources including hydration of the dielectric, ionic penetration, dielectric erosion under culturing conditions, and wear of the exposed dielectric due to cleaning and sterilization procedures. Drift is assumed to be part of the process of hydration of the dielectric layer over the channel region. Culturing neurons on top of silicon devices may result in mobile ions such as sodium and potassium, part of the required components of a culture media, penetrating into the gate dielectric. Penetration of ions into the dielectric may also cause a threshold voltage drift [19]. There has also been some evidence presented [14] that indicates that silicon dioxide is eroded in physiological saline at physiological temperatures. Gate oxide capacitance may also change through erosion due to cleaning processes if the neuron-FETs are reused. Finally, there is always a bias at the gate terminal because the sum of the metal-electrolyte voltage potential and the electrolyte-insulator dipole potential voltage creates an offset potential along the channel of the neuron-FET [20]. A typical value for a silver chloride electrode is about 250 mV.

Integrated circuit design is typically accomplished through adjusting the width, length, bias voltages and/or bias currents through the transistors. Transistors are assumed to have a stable threshold voltage and a fixed dielectric thickness. This is not the case with open-gate transistors used in biological applications. In amplifier designs using neuron-FETs dimensioning is limited to selection of the length, so optimization of the fabrication process and the circuitry providing biasing currents and amplification becomes paramount.

Neuron-FETs represent a special class of ion sensitive field effect transistors (ISFETs). ISFETs are commonly used as transducers for detection of ion concentrations in bulk solutions. In this application, ISFETs and the associated signal amplification circuitry are designed to operate on slower times scales (seconds) and used to measure ionic concentrations with precision and linearity. In looking to ISFET circuit topologies for design inspiration, there are three significant issues. First, ISFET topologies are optimized for signals that change very slowly (on the order of seconds) and not the 0.5 mS to 2 mS timescale of a neural action potential. Second, ISFETs also do not have constraints on the transistor dimensions, so often they are made large (long width and length) to improve flicker noise performance. Often a feedback system is used to maintain the potential at the reference electrode to improve linearity and enhance measurement of relative change. For this circuit implementation, the ISFET must be maintained in strong inversion to provide a linear output. A good review of topologies and limitations of each ISFET amplifiers topology can be found in [21]. We mention several cases here which are relevant to neuron-FET design. In several examples presented in the literature, the ISFET transducer has been used as a circuit element. In [22], a simple common-source amplifier with a tuned source resistance was introduced. The problem with this system is that the reference electrode supplying the gate bias system is controlled by an operational amplifier that directly controls the drain current potential using a negative-feedback loop. This architecture makes the ISFET less sensitive to changes in the material electrolyte interface and more sensitive to changes in the concentration. The former is the ideal case for neuron-FETs, the latter for ISFETs. A current mode technique is suggested in [23]. This approach uses a complex circuit and a reference FET (reFET) to maintain linearity. The reFET is a transistor designed to be much less sensitive to changes in ion concentration. Since it is impossible to place a reFET and a neuron-FET in near proximity and have the same adherent patch been in place across both transistor channels, approaches that use reFETs in their design are of limited utility.

In the first reports of ISFETs used for transduction of electrogenic cells, cross-coupled amplifiers were suggested as a way of amplifying signals. The issue with this architecture is that it does not allow biasing of the current through the neuron-FET and thus limits the design area. Weak inversion operation of ISFETs was first presented in [24, 25] and developed in subsequent work. The weak inversion regime circuits are designed for wide dynamic range as opposed to sensitivity in a narrow range. However, the weak inversion ISFETs and amplification circuitry were applied to the development of a circuit cell that was used as part of a larger circuit proposed for detecting action potential propagation along and axon, presented in [26]. This topology used pixels that are quite large, intended for recording from nerve bundles. The most common amplifier topology used for neuron-FET recording is the transimpedance amplifier with the transistor biased to produce gains below unity, shown in Figure 2, as presented in [9]. This circuit employs the neuron-FET as a current source with the output fed to a transimpedance amplifier. The problem with this is that the transducer operated in this manner has a less than unity gain. Since the transistor operates in the strong inversion region, it is a low-impedance source. The noise gain of a transimpedance amplifier driven by a low-impedance source is much higher in comparison to that of the noise gain of an equivalent voltage feedback amplifier. In addition, operation in this manner requires a voltage bias for the source and the drain, with the gate (bath) electrode held at ground. To characterize the transistor using a parameter extraction routine to measure changes in gain and threshold

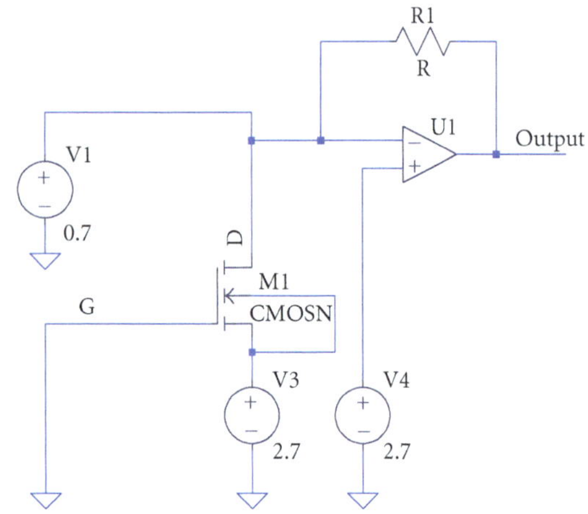


FIGURE 2: Neuron-FET transimpedance amplifier, as described in [9].

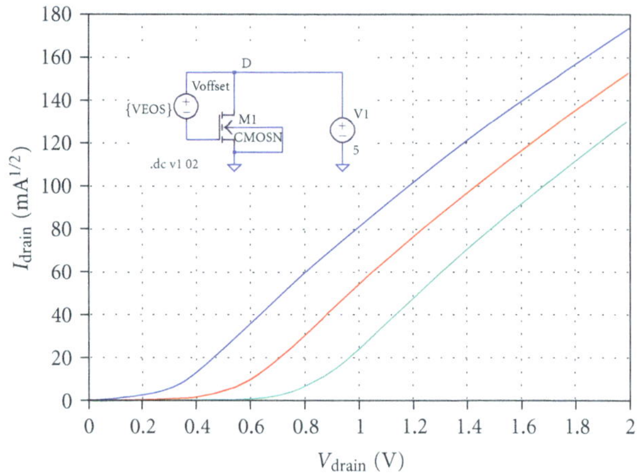


FIGURE 3: Extraction of $V_t$ and $K_n$ using a diode-connected transistor. Note that the offset potential produced by the reference electrode is incorporated into the estimate for threshold voltage used for design purposes. The offset voltages are −0.250 mV (blue line), 0 V (red line), and +0.250 mV (green line). The $V_t$ of the simulated device is 0.57 V.

voltage, this bias network must be removed. This is an issue because of the changes in the threshold voltage and dielectric properties that occur affecting transducer performance, as discussed earlier.

In the following section, a method to characterize the neuron-FET without direct access to the gate terminal is outlined. This characterization strategy must take into account the potential offsets that occur from the reference electrode and the dielectric layer. During this characterization process, offset bias from the reference electrode is included. The approach used differs from other $g_m/I_d$ design approaches in that transit frequency ($f_T$) is not considered. Neural signals are far below the upper frequency limits of most transistors and single-stage amplifiers. The characterization process drives the design of a gain stabilized common-source amplifier topology. The common-source degenerated amplifier is an excellent choice because of the gain stabilization it provides, simplicity that allows tuning of the circuit components, and biasing control. This design is compared to the current-transimpedance topology. A first-order evaluation of noise is included.

## 2. Methodology

Here, we outline a transistor characterization process that facilitates the circuit design process. The methodology provides insight into approaches to improve neuron-FET amplifier performance at both the transistor processing and circuit level. In the first section, a characterization process for the neuron-FET is described. The extracted parameters are required for design, and they provide insight into strategies to improve the operation of the open-gate neuron-FET. Biasing using the $g_m/I_d$ figure of merit is introduced to determine an ideal drain current and gate voltage pair for biasing for peak-gain and low-noise operation. In the second section, feedback is used to stabilize the AC and DC gain of a common-source amplifier. Verification of the design flow demonstrates gain stability in the presence of threshold voltage drift, sufficient bandwidth for amplification, and low harmonic distortion. Several opportunities for system improvement are outlined at the end of this section based upon insight gained from analysis of the design flow.

*2.1. Characterization of the Fabricated Device.* There can be significant variations in intrinsic transistor performance in neuron-FETs. Prototype transistors are often made using experimental processes and can exhibit wide variabilities when compared to transistors fabricated in a commercial production process. During operation, drift has an unknown effect on the threshold voltage, and the oxide capacitance may also be variable. Therefore a simplified strategy of extracting transistor performance curves is critical. Furthermore, the design strategy should be modified to use data from the simplified extraction routine. Only two parameters and a reference graph are required for the design method presented here. They are; the drain current at moderate inversion and the channel length modulation parameter, $\lambda$ [$V^{1/2}$], or output resistance, $r_o$ [$\Omega$]. The extracted transconductance efficiency curve then allows determination of a point for gate bias. It is prudent to extract the threshold voltage, $V_t$ [V], and the transconductance parameter, $K_n$ [$\mu$A/v$^2$], to record changes due to drift and dielectric changes. The transconductance parameter and threshold voltage are extracted by shorting the source terminal and applying a voltage source to the drain terminal and the reference electrode and sweeping voltage while measuring current. Data are plotted as $V_d$ [V] versus $I_d^{1/2}$ [A]. The test circuit and resulting curve are shown in Figure 3. $K_n$ is calculated as

$$K_n' = 2 * \text{slope}^2. \quad (1)$$

And the $V_t$ is estimated as the intersection of the line with the $x$ intercept, using the standard slope intercept form of $y = mx + b$; thus the threshold voltage may be evaluated as

$$V_t = \frac{-b}{m}. \quad (2)$$

This allows determination of the transistor performance parameters $V_t$ and $K_n$. Electrochemical potentials introduce an error as shift in the threshold voltage. Consequently, the offset becomes part of the system design. Alternatively, the voltage offset may be measured between the bulk and the reference electrode treating the device as a MEIS (metal electrolyte insulator semiconductor) capacitor. Since sensitive calibrated capacitance voltage meters may not be available and would require removing the transistor from the operating circuit, the voltage sweep method has more utility. In reality, the slope will vary significantly with higher drain currents. However, the extraction routine is performed in the operating area of interest. In this region, the values are more stable.

The other critical parameter to enable design is the output resistance, $r_o$. This is determined by measuring the channel length modulation effect. The test is conducted by measuring the drain current versus the drain voltage as the gate voltage is stepped. However, the output resistance changes as a function of both the gate voltage and the applied drain voltage at each fixed width and length. Thus, the biasing point for the transistor needs to be determined before the output resistance can be measured. Output conductance is a complex function of channel length, gate-to-source voltage, and drain-to-source voltage. The maximum is reached at the onset of moderate inversion and then stabilizes through the weak inversion regime.

Using the $g_m/I_d$ figure of merit, shown in Figure 5, the relationship between transconductance efficiency and drain current can be more easily understood. Transistors biased deeply into the weak inversion region have maximal gain per unit current, but very low drive currents. A minimum acceptable drive current may be determined from the current required to produce a full-swing output across a typical bypass capacitor, according to

$$I = C\frac{\partial V}{\partial t}. \tag{3}$$

Transistors in strong inversion have less gain per unit current, so while the gain increases, the noise, power consumption, and drive current requirements also increase. A good compromise is the transition between the two that is described as moderate inversion. Moderate inversion also describes a region of high-output conductance. Supplementing Figure 4, an alternate method to understand the relationship between intrinsic gain and output resistance is found by plotting the product of $g_m/g_{ds}$ versus $I_d$ (not shown). Both of these methods show that the maximum gain at a fixed width and length is reached at the onset of weak inversion, in the moderate inversion region.

The inversion coefficient can be measured directly by further characterizing the transistor and extracting a process-related value called specific current. Instead of determining an exact value for the inversion coefficient the drain voltage of the diode-connected transistor can be swept and the plot can be inspected for the transition between weak and strong inversion, shown in Figure 5. Measurement is made by diode connecting the transistor and sweeping the drain voltage while the drain current is monitored, the same data as is

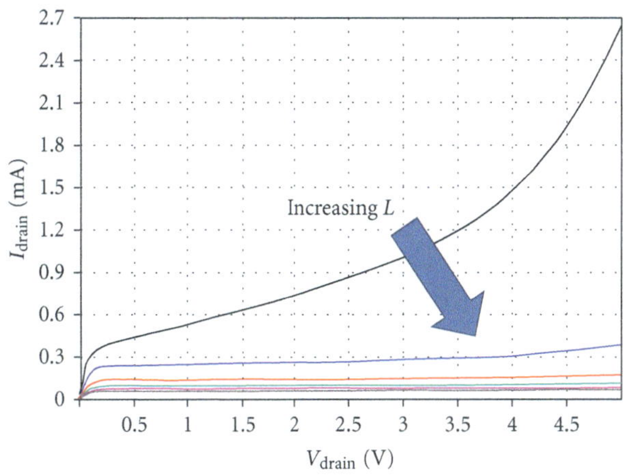


Figure 4: Channel length modulation effects on output resistance as an effect of increasing drain voltage and increasing length. The upper line (black) is a 100 $\mu$m transistor with a 0.5 $\mu$m length, with each line describing a transistor of increasing length by 0.5 $\mu$m.

collected for estimation of the output resistance. The rate of change of the drain current with respect to the drain current ($\partial I_d/I_d$ versus $I_d$) is plotted versus the drain current on a log-log plot, shown in Figure 6. This approach is derived from

$$\frac{g_m}{I_d} = \frac{\partial I_d/V_d}{I_d} = \frac{\partial I_d}{I_d}(V_d). \tag{4}$$

Since the procedure for extracting the bias point is so streamlined, it can be routinely performed as part of a measurement process to account for significant changes in threshold voltage and offset potentials.

*2.2. Selection of the Amplifier Topology.* Typically, in $g_m/I_d$ design voltages are not used to control the level of inversion, but instead drain currents are precisely controlled. In neuron-FET designs, the drain current is difficult to control due to leakage current and inherent variation. The feedback network, part of the amplifier design, can then be used to mitigate variation in the transistor operation and hold the transistor in a moderate inversion regime. Using this technique compensates for the offset produced by the electrochemical potential of the reference electrode, any offset generated at the electrolyte-semiconductor interface and any effects of drift. The drain current bias point determined from Figure 7 allows the correct DC gate voltage biasing for the transistor to maintain operation in the moderate inversion regime and subsequently drives the design of the biasing network for the amplifier circuitry. This figure of merit is generated by shorting the gate and drain and measuring the drain voltage while sweeping the drain current. The drain voltage and $g_m/I_d$ are plotted versus the swept drain current. Notice that the moderate inversion region remains stable across different voltage offset values. The weak inversion region is unavailable if the voltage offset is biased positively.

There are three basic single-transistor amplifiers. They are the common-source, common-drain, and common-gate topologies. Since neurons source very small currents

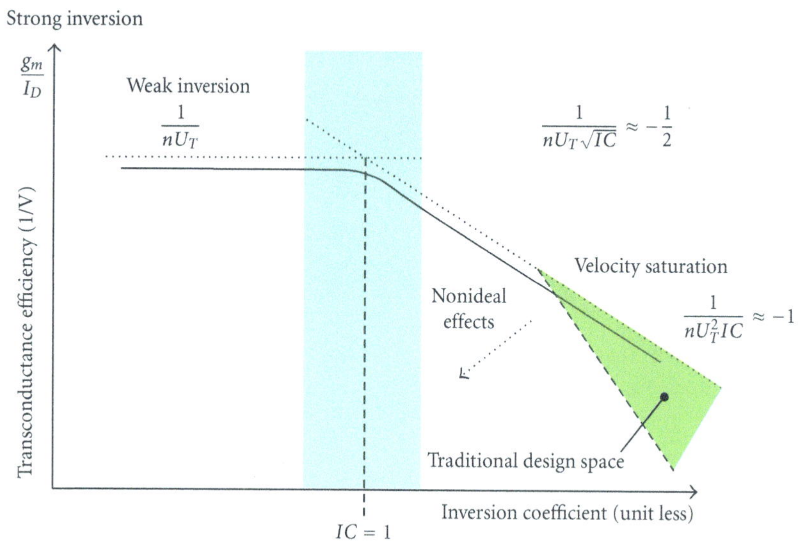


Figure 5: Theoretical transconductance efficiency figure of merit graph. $U_T$ is thermal potential KT and n is the slope factor or body effect factor.

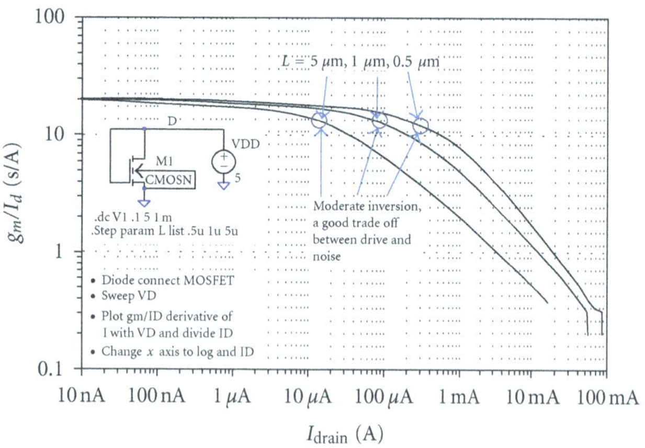


Figure 6: Transconductance figure of merit extracted from simulation of a diode-connected FET. The effect of length on the transition from weak to strong inversion is clearly seen. Longer transistor channel lengths have larger output resistance because they are less affected by channel length modulation, but have lower drive currents in moderate inversion.

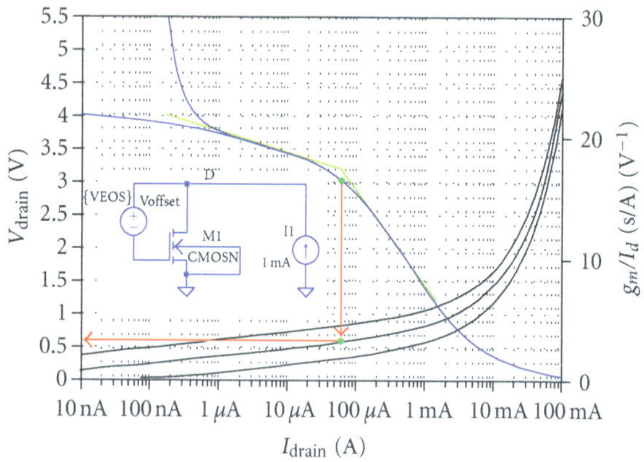


Figure 7: Determination of the DC gate bias as a function of the moderate inversion transition. The blue line is the transconductance efficiency curve (left axis), and the black line is the drain (gate) voltage (right axis). Each of these are stepped a −0.250 V, 0 V and +0.250 V, voltage offset values. The selection of the knee in the transconductance efficiency curve leads to the selection of a 2 $\mu$A $I_d$. The point at which this intersects the gate voltage to provide the DC gate voltage bias point. This pair of values is used to design the single-stage amplifier circuitry.

(nA to pA), they are an inappropriate source for low-input-impedance amplifiers such as the common-gate amplifier. The common-drain amplifier is not appropriate for design because gains are typically less than one and the output resistance is low. Only the common-source amplifier (Figure 8) provides gain from a high-impedance signal source with a high impedance output. Using a source degeneration resistor provides additional benefits because the offset voltage becomes less important, so small variations are mitigated.

The bypass capacitors are selected to pass signals in the bandwidth covering typical neural signals (approximately 500 Hz–10 kHz). Using standard-size components this is a capacitor size of 1 $\mu$F and is selected for the source capacitance. Basically, the impedance of the bypass capacitor at the lowest operation frequency (100 Hz) should be much smaller than the resistance of the source resistor (generally 10 times smaller). After a source resistance is selected, the capacitance may be calculated using

$$Z = \frac{R_S}{10} = \frac{1}{2\pi C f}. \quad (5)$$

This in turn drives the current requirements for the device. Feedback is used to reduce the gain and stabilize the circuit over a wider range. The estimate used is to set $R_d$ to a value equal to the output resistance divided by 10. This makes the drain resistor the dominant effect over the output resistance, $r_o$, and thus small variations in output resistance are mitigated. $R_s$ is then selected by using a value that distributes equal voltages across $R_d$, $R_s$, and the neuron-FET. This results in the following design relationship:

$$V_{DS} = \frac{1}{3}V_{DD}, \tag{6}$$

then,

$$V_{DD} = I_D R_D + \frac{1}{3}V_{DD} + I_D R_S \tag{7}$$

and, by simplification,

$$\begin{aligned} \frac{2}{3}V_{DD} &= I_D R_D + I_D R_S, \\ I_D R_S &= \frac{2}{3}V_{DD} - I_D R_D, \\ R_S &= \frac{2}{3}\frac{V_{DD}}{I_D} - R_D. \end{aligned} \tag{8}$$

The outcome is that values of $R_s$ are typically slightly larger than $R_d$. Since the DC intrinsic transistor gain, $g_m$, approaches $1/R_s$ as $R_s$ becomes large with respect to $1/g_m$ the total DC gain can be estimated as $A_v = -(R_d/R_s)$ and will typically be less than 1. AC gain, in contrast, will approach $g_m R_d$ as the bypass capacitor shorts out the AC signal. If the transistor length is long enough that $r_o \gg Rd$, then bigger $r_o$ means a larger $R_d$ can be used in the circuit and the overall circuit gain can be larger. Also, if $r_o \gg R_d$ then $r_o$ drops out of the gain equation and thus makes the system even less sensitive to $r_o$ variation due to process variation (i.e., variations in length and CLM effects). Overall the resistor source degeneration makes the entire system more stable by providing feedback. The constraint is that the transistor length must be short enough to provide enough drive current to drive bypass capacitors, but long enough to mitigate channel length effects. $R_s$ also has the effect of mitigating the effects of an unstable $V_t$. As $R_s$ goes to infinity, $V_{gs}$ becomes equal to $V_t$. This means that $K_n$ becomes dominant and improvements in the process to improve $K_n$ have significant impact on the overall performance.

## 3. Results and Analysis

Determination of gain and distortion provide valuable metrics in the evaluation of the design stability.

*3.1. Gain.* Figure 9 clearly shows that a stable gain of approximately 10 is achieved. In addition, the gain varies less than 3 dB across a threshold variation of ±0.5 V.

*3.2. Distortion.* The distortion is shown in Figure 10 to be even throughout the signal bandwidth of interest. The peak at 1 kHz is due to the input test vector. The first harmonic appears at 3 kHz.

*3.3. Noise Effects.* Noise analysis reveals interesting characteristic of this amplifier topology. A common source transistor has a voltage thermal noise of

$$\overline{\mathbf{V}^2_{n,th}} = 4kT\left(\frac{2}{3}\frac{1}{g_m}\right) \tag{9}$$

and a current thermal noise of

$$\overline{\mathbf{I}^2_{n,th}} = 4kT\left(\frac{2}{3}\right)g_m \tag{10}$$

that implies that the transistor transconductance must be maximized if the transistor is to amplify a voltage signal at the gate and minimized if the transistor operates as a current source input. The source-degenerated common-source amplifier has a thermal noise of

$$\overline{\mathbf{V}^2_{n,in}} = 4kT\left(\frac{2}{3}\frac{1}{g_m} + R_s + R_D\left(\frac{1+g_m R_S}{g_m R_D}\right)^2\right) \tag{11}$$

excluding the bypass capacitors. This means that at DC, the noise of the source-degenerated common-source amplifier is larger than the simple common-source stage. However, the bypass capacitor at the output blocks DC signals, and at AC the source bypass capacitor shorts $R_s$. The effect is that no DC noise is passed to the amplifier and the AC thermal noise is equivalent to the common-source amplifier without source degeneration. Neuron-FETs are typically used as current source inputs to a transimpedance amplifier, but this configuration does not provide for amplification or DC blocking. By maximizing the gain and using the neuron-FET as a voltage amplifier, an improvement may be realized.

Noise is an important issue with small signals, and this is especially true in neuron-FETs where the signals are so highly attenuated. The issues are nicely reviewed in [28]. $1/f$ noise dominates the operation of ISFETs, and the same is true of Neuron-FETs. Nyquist noise of adhesion between the neuron and the transistor is another important source of noise [29]. An alternate strategy to mitigate noise effects based on chopper amplification was first presented in [30], but measurement results were inconclusive. Gate-related spectral noise density is given by

$$S_{V_{gs}}(f) \approx \frac{4kT}{c_{ox}^2}\frac{n_{ot}}{WL}\frac{1}{f}, \tag{12}$$

where $c_{ox}$ is the oxide capacitance in $F$, $n_{ot}$ is the number of oxide traps, $W$ is the transistor width in $\mu$m, and $L$ is the transistor length in $\mu$m, and $f$ is frequency in Hz. It is already known that increased width and length result in a lower $1/f$ noise; however, these variables are constrained by the use of a transistor as a neuron-FET. Evaluation of the formula shows that significant increases in the oxide capacitance have significant effect on the gate-related spectral noise density when all other variables are held the same. In [28], the authors also suggest that increasing the oxide capacitance will lead to lower $1/f$ noise. In our lab, we have begun to explore high-k films for neuron-FETs [31–34].

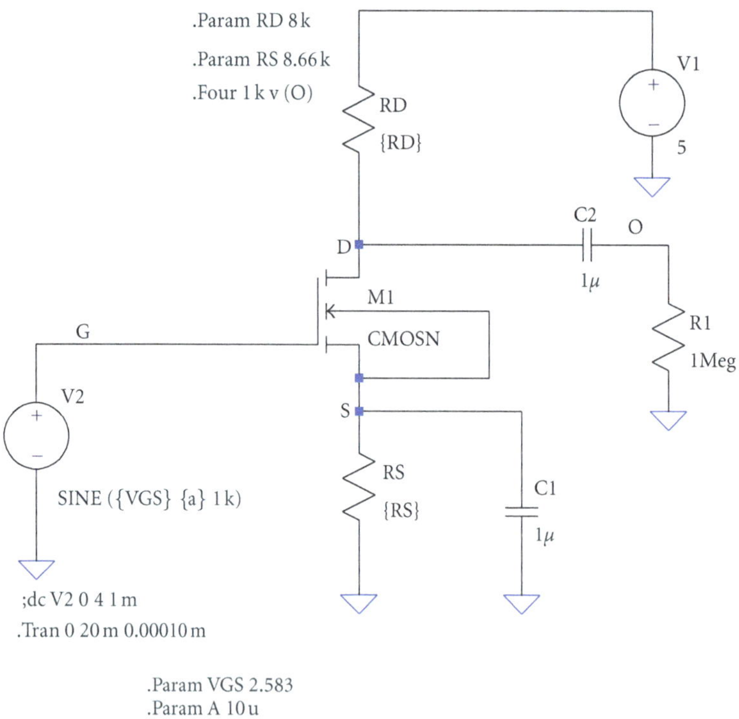


FIGURE 8: Common source with source degeneration, bypass capacitors, and load resistances.

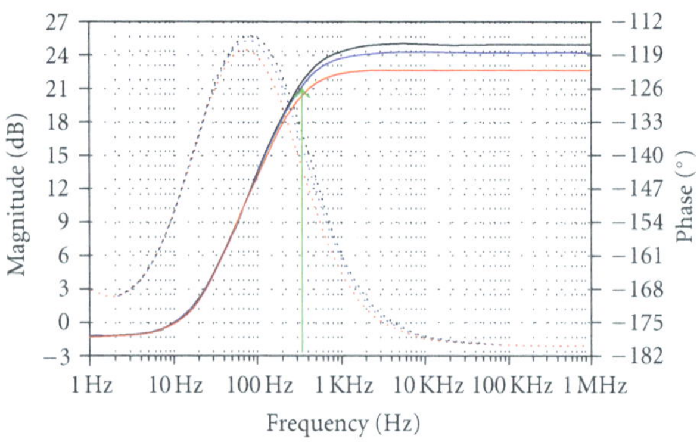


FIGURE 9: Gain verification. Output is measured from a 10 $\mu$V input source as approximately 11, missing the design estimate of 12.5 by about 15%. The green arrow shows the 350 Hz −3 dB frequency. The upper 3 dB (not shown) is much greater than the signal range of interest (approximately 10 kHz). The curves show performance of variable $V_t$ in this configuration. Notice the gain has been stabilized to less than 3 dB across a threshold variation of 1volt.

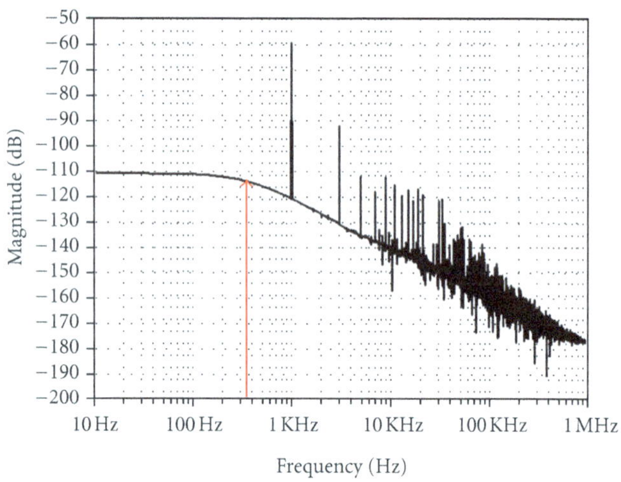


FIGURE 10: Distortion analysis using the fast Fourier transform. The worst case total harmonic distortion, measured at the lower −3 dB limit of 350 Hz is −114.116 dB or 0.0001969%, shown as a red arrow.

## 4. High-k Gate Dielectrics

Beyond the circuit topology the transistor fabrication can be optimized. Several labs have reported examples of transistor-level design optimizations. These include the use of high-k dielectrics for the gate. In addition to predicted noise enhancement high-k dielectrics may enhance the transconductance efficiency as well. This in turn may lead to higher gains per unit area, an important design boundary for neuron-FETs. In both cases, the transconductance figure of merit is a valuable design and diagnosis tool for evaluating process technologies. An example of the evaluation of thin and thick oxides and short and long devices using the approach presented in this paper is shown in Figure 11.

*4.1. Effects of Ionic Penetration into the Gate Dielectric.* To reduce the drift of the threshold voltage of neural sensing transistors, Wallrapp and Fromherz [35] proposed to use high-k dielectrics such as $HfO_2$ and $TiO_2$ as the dielectric (oxide) layer to prevent ion penetration of

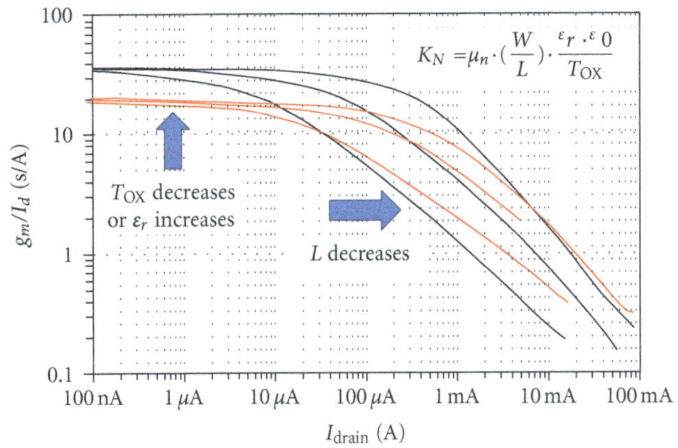


FIGURE 11: Use of the transconductance figure of merit to evaluate process changes. Effect of increasing oxide dielectric constant, increasing length or decreasing oxide thickness. The red line is a thick-oxide transistor; the black lines are from a thin oxide transistor. Both sets of curves are from transistors of a width of 100 μm and lengths $L = 5$ μm, 1 μm, and 0.5 μm from left to right.

the oxide layer. However, there is some evidence that ISFETs with high-k dielectrics still demonstrate drift [36]. Capacitance measurements were presented for $HfO_2$ and $TiO_2$ films with thicknesses ranging from 50 to 190 Å that were measured with electrolyte-oxide-semiconductor (EOS) structures. These showed no ionic penetration from the electrolyte. In 2006 Meyburg et al. [37, 38] recorded extracellular signals from embryonic-rat cardiac myocytes using n-channel and p-channel MOSFETs with an extended-gate sensing area similar to that used by Cohen [39, 40]. The sensing oxide was 70 Å thick $SiO_2$. Sensors did not show the effect of light metallic ion penetration for up to two hours of exposure. In 2007, Meyburg et al. [41, 42] extended his previous work with a full CMOS process that had 2-, 5-bit decoders that were able to address an array of 32 × 32 sensors. In 2004, Cohen et al. [39] was able to measure extracellular voltage signals from *Aplysia* by using a floating-gate depletion mode MOS transistor. The sensing area was comprised of a 420 Å thick $SiO_2$ that was extended from the polysilicon gate of the transistor. The separation of the sensing area and the active channel reduced the effects of ionic contamination from the tissue culture and allowed the design of the sensing dimensions channel to be separated from the design of the channel dimensions. The depletion mode transistor (normally on) did not require a DC bias to record neural signals. In 2006, Cohen et al. [40] used this recording system to show that the morphology of the neuron controls the shape the field potentials measured from *Aplysia* neurons. This system was used again in 2008 [43] to show that applying mechanical pressure to a neuron can increase the signal-to-noise ratio of the recording. These examples represent important avenues for further investigation.

## 5. Conclusion

Neurons cultured directly on top of open-gate field-effect transistors result in a hybrid device, the neuron-FET. These represent a special class of an ion-sensitive field-effect transistor (ISFET). Neuron-FETs have significant differences in their operating regime in comparison to ISFETs that result in different design boundaries. These differences call for amplification circuitry specifically designed to amplify signals from neuron-FETs.

Therefore, we have presented a method to optimize circuits that include a transistor used as a transducer to record neural activity. This procedure is not dependent upon the transistor and will benefit from design optimization of the physical transistor device. Using the $g_m/I_d$ technique the circuit designer can optimize the transistor design as part of the design flow. Using this technique we were able to find a transistor length long enough to provide a reasonable output resistance so that it can be made independent of device threshold voltage variations and limitations inherent in the test setup and short enough to drive the bypass capacitors. Simulation data for a test case demonstrates tolerance to variation and a significant increase in gain over typical implementations.

## References


[1] P. Jespers, *The GM/ID Methodology, A Sizing Tool for Low-Voltage Analog Cmos Circuits*, Springer, New York, NY, USA, 2009.

[2] D. Foty, D. Binkley, M. Bucher et al., "Starting over: gm/ID-based MOSFET modeling as a basis for modernized analog design methodologies," in *Proceedings of the Technical International Conference on Modeling and Simulation of Microsystems (Nanotech '02)*, pp. 682–685, 2002.

[3] D. M. Binkley, *Tradeoffs and Optimization in Analog CMOS Design*, John Wiley & Sons, Chichester, UK, 2007.

[4] K. K. Ng, *Complete Guide to Semiconductor Devices*, IEEE Press, John Wiley & Sons, New York, NY, USA, 2nd edition, 2002.

[5] M. J. Madou and S. R. Morrison, *Chemical Sensing with Solid State Devices*, Academic Press, Boston, Mass, USA, 1989.

[6] P. Bergveld, "Thirty years of ISFETOLOGY: what happened in the past 30 years and what may happen in the next 30 years," *Sensors and Actuators B*, vol. 88, no. 1, pp. 1–20, 2003.

[7] P. Bergveld, "Development, operation, and application of the ion-sensitive field-effect transistor as a tool for electrophysiology," *IEEE Transactions on Biomedical Engineering*, vol. 19, no. 5, pp. 342–351, 1972.

[8] P. Bergveld, J. Wiersma, and H. Meertens, "Extracellular potential recordings by means of a field effect transistor without gate metal, called OSFET," *IEEE Transactions on Biomedical Engineering*, vol. 23, no. 2, pp. 136–144, 1976.

[9] P. Fromherz, A. Offenhäusser, T. Vetter, and J. Weis, "A neuron-silicon junction: a retzius cell of the leech on an insulated-gate field-effect transistor," *Science*, vol. 252, no. 5010, pp. 1290–1293, 1991.

[10] M. J. Schöning and A. Poghossian, "Recent advances in biologically sensitive field-effect transistors (BioFETs)," *Analyst*, vol. 127, no. 9, pp. 1137–1151, 2002.

[11] C. S. Lee, S. Kyu Kim, and M. Kim, "Ion-sensitive field-effect transistor for biological sensing," *Sensors*, vol. 9, no. 9, pp. 7111–7131, 2009.

[12] A. Poghossian, S. Ingebrandt, A. Offenhäusser, and M. J. Schöning, "Field-effect devices for detecting cellular signals,"

*Seminars in Cell and Developmental Biology*, vol. 20, no. 1, pp. 41–48, 2009.
[13] P. Fromherz, "Semiconductor chips with ion channels, nerve cells and brain," *Physica E*, vol. 16, no. 1, pp. 24–34, 2003.
[14] M. Merz and P. Fromherz, "Silicon chip interfaced with a geometrically defined net of snail neurons," *Advanced Functional Materials*, vol. 15, no. 5, pp. 739–744, 2005.
[15] D. A. Stenger and T. M. McKenna, *Enabling Technologies for Cultured Neural Networks*, Academic Press, San Diego, Calif, USA, 1994.
[16] J.-C. Chou and C.-N. Hsiao, "Drift behavior of ISFETs with a-Si:H-$SiO_2$ gate insulator," *Materials Chemistry and Physics*, vol. 63, no. 3, pp. 270–273, 2000.
[17] M. Klein, "Time effects of ion-sensitive field-effect transistors," *Sensors and Actuators*, vol. 17, no. 1-2, pp. 203–208, 1989.
[18] C. Jakobson, M. Feinsod, and Y. Nemirovsky, "Low frequency noise and drift in Ion sensitive field effect transistors," *Sensors and Actuators B*, vol. 68, no. 1, pp. 134–139, 2000.
[19] T. Prodromakis, P. Georgiou, K. Michelakis, and C. Toumazou, "Effect of mobile ionic-charge on CMOS based Ion-sensitive field-effect transistors (ISFETs)," in *Proceedings of the IEEE International Symposium on Circuits and Systems (ISCAS '09)*, pp. 2165–2168, May 2009.
[20] S. Martinoia and P. Massobrio, "ISFET-neuron junction: circuit models and extracellular signal simulations," *Biosensors and Bioelectronics*, vol. 19, no. 11, pp. 1487–1496, 2004.
[21] A. Morgenshtein, *Design and methodology of ISFET (Ion sensitive field effect transistor) microsystems for bio-telemetry*, M.S. thesis, Senate of The Technion—Israel Institute of Technology, Haifa, Israel, 2003.
[22] A. K. Covington and P. D. Whalley, "Recent advances in microelectronic ion-sensitive devices (ISFETs). The operational transducer," *Journal of the Chemical Society, Faraday Transactions*, vol. 82, no. 4, pp. 1209–1215, 1986.
[23] C. Wang and Y. Zhao, "A novel current-mode readout circuit for ISFET sensor," in *Proceedings of the IEEE Asia Pacific Conference on Circuits and Systems*, pp. 407–410, December 2008.
[24] L. Shepherd and C. Toumazou, "Towards direct biochemical analysis with weak inversion ISFETS," in *Proceedings of the IEEE International Workshop on Biomedical Circuits and Systems*, pp. S1.5–S1.5, December 2004.
[25] L. Shepherd and C. Toumazou, "A biochemical translinear principle with weak inversion ISFETs," *IEEE Transactions on Circuits and Systems I*, vol. 52, no. 12, pp. 2614–2619, 2005.
[26] P. Georgiou, I. F. Triantis, T. G. Constandinou, and C. Toumazou, "Spiking chemical sensor (SCS): a new platform for neuro-chemical sensing," in *Proceedings of the 3rd International IEEE EMBS Conference on Neural Engineering (CNE '07)*, pp. 126–129, May 2007.
[27] D. W. Parent and E. J. Basham, "Hafnium transistor process design for neural interfacing," in *Proceedings of the Annual International Conference of the IEEE Engineering in Medicine and Biology Society: Engineering the Future of Biomedicine (EMBC '09)*, pp. 5875–5878, September 2009.
[28] C. Jakobson and Y. Nemirovsky, "I/f noise in ion sensitive field effect transistors from subthreshold to saturation," *IEEE Transactions on Electron Devices*, vol. 46, no. 1, pp. 259–261, 1999.
[29] M. Voelker and P. Fromherz, "Nyquist noise of cell adhesion detected in a neuron-silicon transistor," *Physical Review Letters*, vol. 96, no. 22, Article ID 228102, 2006.
[30] M. Dagtekin, *A chopper modulated amplifier system design for in vitro neural recording*, Ph.D.dissertation, North Carolina State University, Raleigh, North Carolina, 2006.
[31] D. Parent and E. Basham, "Hafnium transistor design for neural interfacing," in *Proceedings of the 30th Annual International Conference of the IEEE Engineering in Medicine and Biology Society (EMBS '08)*, pp. 3356–3359, August 2008.
[32] D. Parent and E. Basham, "A course for designing transistors for high gain analog applications," in *Proceedings of the 17th Biennial University/Government/Industry Micro-Nano Symposium—UGIM*, pp. 75–78, July 2008.
[33] D. W. Parent and E. J. Basham, "Hafnium transistor process design for neural interfacing," in *Proceedings of the Annual Engineering in Medicine and Biology Society (EMBC '09)*, pp. 5875–5878, Minneapolis, Minn, USA, September 2009.
[34] D. W. Parent, J. Davis, and E. J. Basham, "High-k dielectric fabrication process to minimize mobile ionic penetration," in *Proceedings of the Annual Engineering in Medicine and Biology Society Conference (EMBC '10)*, Buenos Aires, Argentina, 2010.
[35] F. Wallrapp and P. Fromherz, "$TiO_2$ and $HfO_2$ in electrolyte-oxide-silicon configuration for applications in bioelectronics," *Journal of Applied Physics*, vol. 99, no. 11, Article ID 114103, 2006.
[36] H. Hara and T. Ohta, "Dynamic response of a $Ta_2O_5$-gate pH-sensitive field-effect transistor," *Sensors and Actuators B*, vol. 32, no. 2, pp. 115–119, 1996.
[37] S. Meyburg, M. Goryll, J. Moers et al., "N-Channel field-effect transistors with floating gates for extracellular recordings," *Biosensors and Bioelectronics*, vol. 21, no. 7, pp. 1037–1044, 2006.
[38] S. Meyburg, G. Wrobel, R. Stockmann, J. Moers, S. Ingebrandt, and A. Offenhäusser, "Single cell recordings with pairs of complementary transistors," *Applied Physics Letters*, vol. 89, no. 1, Article ID 013901, 2006.
[39] A. Cohen, M. Spira, S. Yitshaik, G. Borghs, O. Shwartzglass, and J. Shappir, "Depletion type floating gate p-channel MOS transistor for recording action potentials generated by cultured neurons," *Biosensors and Bioelectronics*, vol. 19, no. 12, pp. 1703–1709, 2004.
[40] A. Cohen, J. Shappir, S. Yitzchaik, and M. E. Spira, "Experimental and theoretical analysis of neuron-transistor hybrid electrical coupling: the relationships between the electro-anatomy of cultured Aplysia neurons and the recorded field potentials," *Biosensors and Bioelectronics*, vol. 22, no. 5, pp. 656–663, 2006.
[41] S. Meyburg, R. Stockmann, J. Moers, A. Offenhäusser, and S. Ingebrandt, "Advanced CMOS process for floating gate field-effect transistors in bioelectronic applications," *Sensors and Actuators B*, vol. 128, no. 1, pp. 208–217, 2007.
[42] M. Schindler, S. Ingebrandt, S. Meyburg, and A. Offenhäusser, "Design and function principle of a large scale sensor array for the bi-directional coupling to electrogenic cells," in *Proceedings of the 4th International Conference on Solid-State Sensors, Actuators and Microsystems*, pp. 1243–1246, June 2007.
[43] A. Cohen, J. Shappir, S. Yitzchaik, and M. E. Spira, "Reversible transition of extracellular field potential recordings to intracellular recordings of action potentials generated by neurons grown on transistors," *Biosensors and Bioelectronics*, vol. 23, no. 6, pp. 811–819, 2008.

# 5

# Logic Gates and Ring Oscillators Based on Ambipolar Nanocrystalline-Silicon TFTs

**Anand Subramaniam,[1] Kurtis D. Cantley,[2] and Eric M. Vogel[2,3]**

[1] *Department of Electrical Engineering, University of Texas at Dallas, Richardson, TX 75080, USA*
[2] *Department of Materials Science and Engineering, University of Texas at Dallas, Richardson, TX 75080, USA*
[3] *School of Materials Science and Engineering, Georgia Institute of Technology, Atlanta, GA 30332, USA*

Correspondence should be addressed to Anand Subramaniam; asubram3@utdallas.edu

Academic Editor: Olga Korostynsk

Nanocrystalline silicon (nc-Si) thin film transistors (TFTs) are well suited for circuit applications that require moderate device performance and low-temperature CMOS-compatible processing below 250°C. Basic logic gate circuits fabricated using ambipolar nc-Si TFTs alone are presented and shown to operate with correct outputs at frequencies of up to 100 kHz. Ring oscillators consisting of nc-Si TFT-based inverters are also shown to operate at above 20 kHz with a supply voltage of 5 V, corresponding to a propagation delay of <10 $\mu$s/stage. These are the fastest circuits formed out of nanocrystalline silicon TFTs to date. The effect of bias stress degradation of TFTs on oscillation frequency is also explored, and relatively stable operation is shown with supply voltages >5 V for several hours.

## 1. Introduction

Nanocrystalline silicon (nc-Si) has attracted much interest for use in circuit applications that require low temperature, large-area fabrication, and alternative substrates. Examples of these applications may include neuromorphic architectures for vision sensing [1, 2], flexible and flat-panel displays [3–5], radio frequency identification (RFID) tags [6], solar cells [7], and many more. Many of these applications do not require high performance devices, but rather the viability of a simple and inexpensive fabrication process. Neuromorphic circuits are also defect tolerant to a certain extent, allowing low-yield devices to be employed. Nc-Si thin film transistors (TFTs) can be fabricated using a CMOS-compatible fabrication process [8, 9], which also facilitates three-dimensional integration with high-performance CMOS structures [5, 10, 11]. In such a case, TFT circuitry such as biosensor arrays or neural networks could be fabricated around a CMOS core to improve the overall functionality.

The limited use of noncrystalline silicon devices in circuits was historically due to low carrier mobility [12], device degradation under bias stress [13, 14], and the lack of p-channel operation [15]. Use of nc-Si instead of amorphous silicon (a-Si) improves the electron mobility and device degradation to a certain extent. The above drawbacks have been further addressed in the recent years by high purity nc-Si deposition which reduces the incorporation of impurity oxygen in the channel layer and ensures ambipolar operation of the devices [8, 15]. In this context, the term "ambipolar" refers to either n- or p-channel operation depending on the bias conditions. The maximum mobility observed in nc-Si TFTs is 150 cm$^2$/V·s for electrons and 25 cm$^2$/V·s in case of holes [15]. The performance of these devices has also been considerably improved by utilizing thin high-$\kappa$ gate dielectrics that reduce the subthreshold swing and threshold voltages ($V_T$) of these devices [16]. For instance, the switching power consumption of an ambipolar nc-Si TFT-based inverter operating at 100 Hz reduced from 5 pJ per switch to 350 fJ per switch when the gate dielectric was changed from 200 nm $SiO_2$ to 20 nm $HfO_2$. Fabrication of nanoscale TFTs has been demonstrated using electron beam lithography (EBL) [9]. Submicron device dimensions are essential for use in high-density architectures such as neuromorphic systems [17]. In addition, nc-Si TFTs can be fabricated at temperatures of 250°C and below, allowing the incorporation of alternative substrates including polyimide [18] and steel foil [19]. The feasibility of fabricating circuits on such flexible substrates provides TFT technology with a distinct advantage over conventional CMOS technology.

The ambipolar operation of nc-Si TFTs also allows a single device type to be used to form a complementary inverter [9]. Such an arrangement ensures better inverter output range and a higher voltage gain than a source-follower inverter formed using an n-channel device. In contrast, many other low-temperature fabrication devices such as amorphous oxide TFTs operate solely in the n-channel mode. Therefore, inverters formed from such devices require either the integration of other devices such as resistors or use of a source follower configuration that prevents rail-to-rail output [20–22]. Both *n*- and *p*-channel organic TFTs have been recently developed and are the only other materials capable of producing complementary TFT inverters at low temperatures [23] for many of the applications discussed above. However, such devices typically have low mobilities, and therefore the circuits operate at very low switching speeds [24, 25]. Furthermore, rapid atmospheric degradation is observed in organic semiconductors [26], limiting their use.

In this paper, we present logic gates and ring oscillators that operate at low supply voltages formed using ambipolar nc-Si TFTs alone. The fabrication of these circuits is described in detail, followed by their electrical characterization. The operation of ring oscillators at different supply voltages is analyzed. The performance of these circuits is compared with that of similar circuits formed using alternate low-temperature materials. Finally, the circuits are electrically stressed for several hours to examine the effect of device degradation on their operation.

## 2. TFT Fabrication

The fabrication of nc-Si TFT-based circuits employs a four-level lithography process. The circuits contain 100 $\mu$m square pads at input and output nodes and were laid out for ease of electrical probing rather than the minimization of circuit area. A p-type Si wafer with moderate doping ($10^{17}$ cm$^{-3}$) is used as the substrate on which 300 nm $SiO_2$ was grown with thermal oxidation. Initially, 80 nm Cr metal is deposited and patterned. This layer acts as source/drain of the individual TFTs as well as the bottom level of metal lines. All metal lines are 6 $\mu$m wide. Next, an 80 nm thick nc-Si layer is deposited using plasma-enhanced chemical vapor deposition (PECVD) at 13.56 MHz, 125 W RF power, 0.9 Torr, and 250°C deposition temperature in a 1:100 mixture of $SiH_4$ and $H_2$. A detailed description of the deposition condition is contained in [8]. The channel layer is then patterned with Cl chemistry-based plasma etching.

Following this, the $HfO_2$ gate dielectric with equivalent oxide thicknesses (EOTs) ranging from 7 nm to 2.5 nm was deposited using atomic layer deposition (ALD). EOTs were calculated assuming dielectric constant of 17 for $HfO_2$. This number is lower than the standard accepted value of 24 because the process temperature is 200°C. The third level of lithography was then used to pattern contact via holes through the gate dielectric, and 100 nm Au was used as contact metal. The final metal layer was Al (thickness 100 nm), used as gate metal as well as the top level of metal lines. A 2-hour long, 250°C anneal in forming gas completed the process. Individual transistors, inverters, logic gates, and the ring oscillators were fabricated using the same mask set. Figure 1 shows a schematic of the nc-Si TFT.

## 3. Electrical Characteristics of TFT and Inverter

Current-voltage (I-V) characteristics (both transfer and output) when a 20 $\mu$m × 10 $\mu$m nc-Si TFT with 20 nm HfO2 gate dielectric is operated with top-gate bias (grounding the substrate) are shown in Figures 2(a) and 2(b). The device exhibits ambipolar operation, that is, it operates in both the electron and hole regimes. The direction of current flow depends solely on the drain voltage. The applied gate voltage determines which charge carriers dominate in the channel: electrons constitute the charge flow when the gate voltage is positive and holes at negative bias.

Low hysteresis of less than 50 mV/dec. was observed in the *I-V* measurements (also Figure 2(a)), and the $I_{ON}/I_{OFF}$ ratio when operated at a drain voltage of |1 V| is >$10^5$. At higher drain voltage values, the on/off ratio reduces because both hole and electron conduction occur for a larger range of gate voltages. Mobility values were extracted from the *I-V* characteristics. The maximum electron and hole mobilities observed are 20 cm$^2$/V·s and 5 cm$^2$/V·s, respectively. The corresponding subthreshold swing values are 0.6 V/decade and 0.65 V/decade.

Figure 3(a) shows the voltage transfer characteristics of a nc-Si TFT-based inverter ($W_{pull\text{-}up}/W_{pull\text{-}down} = 4$) at different operating voltages. The static CMOS topology was selected for the fabricated inverter since it offers the best output voltage range and voltage gain for ambipolar TFTs. Since the TFTs are ambipolar, no need exists for separate *n*- and *p*-channel devices in the inverter. The inverter exhibits close to peak-to-peak operation and high voltage gain (in the range 12–15). Slight deviation from peak-to-peak exists due to the ambipolar nature of the TFTs: the drive TFT does not completely turn off at low input and vice versa. The high gain can be attributed to the good saturation characteristics of the component devices. The dynamic characteristics of the inverter were also investigated by applying square voltage pulses at different frequencies. The inverter was able to respond closely to the input beyond 2 MHz [9], as shown in Figure 3(b). The switching speed could be minimized further by reducing the gate-to-source/drain overlap capacitance, which was not optimized in the current process.

## 4. Logic Gates

Logic gate circuits including NOR, NAND, OR, and AND were fabricated. OR and AND gates were realized by adding

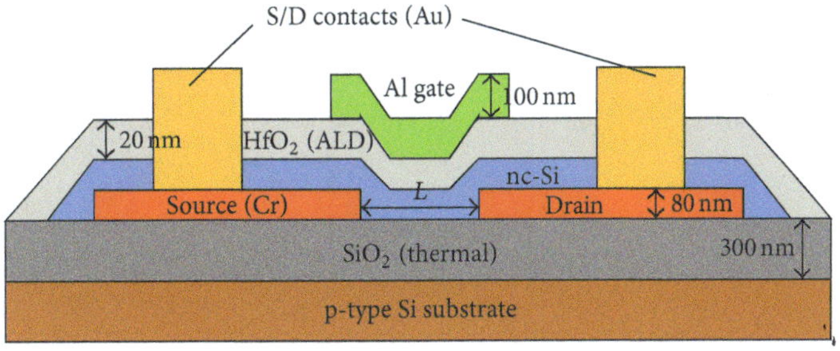


FIGURE 1: Schematic cross-section of a staggered top-gated nc-Si TFT.

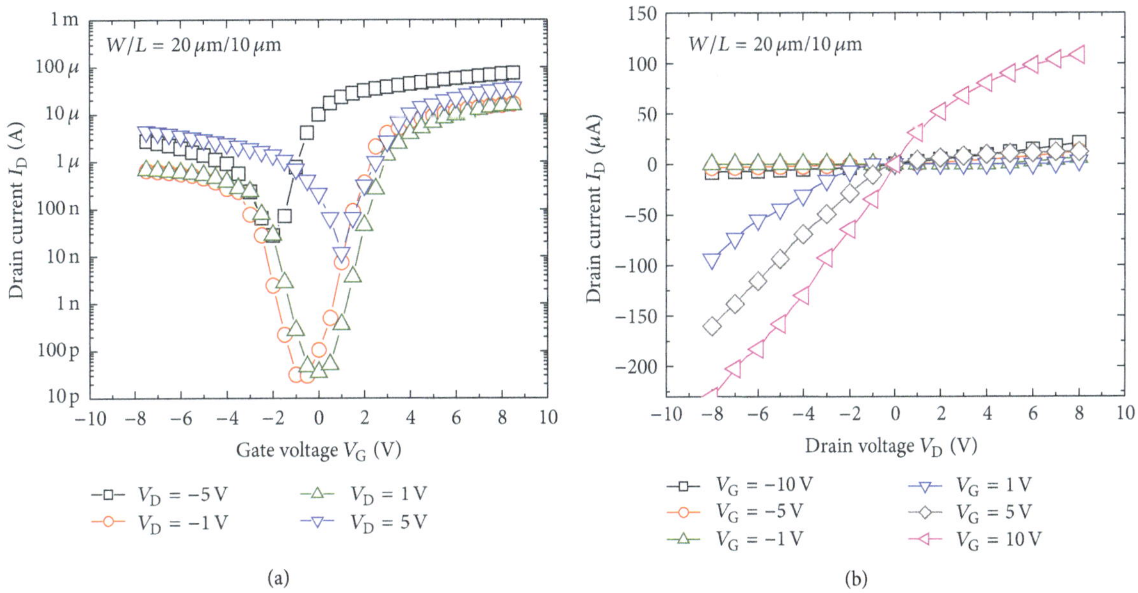


FIGURE 2: Current-voltage ($I$-$V$) characteristics of a $20\,\mu m \times 10\,\mu m$ ambipolar nc-Si TFT with 20 nm $HfO_2$ gate dielectric. (a) Transfer characteristics ($I_D - V_G$) in both forward (symbols) and reverse (lines) directions illustrating the low hysteresis in the device and (b) output characteristics ($I_D - V_D$).

an inverter at the output of the NOR and NAND gates, respectively, and were composed of six ambipolar TFTs each (Figure 4). A single device type is sufficient for each gate, but TFTs were sized for minimum delay according to the circuit configuration, as shown in the figure.

Figure 5 shows the operation of the different logic gates at $V_{DD} = 5$ V. The input waveforms A and B have amplitude 5 V and frequency 100 kHz. The transition between input states occurs with rise and fall times of 10 ns. All possible combinations of inputs are shown. The outputs respond with no perceptible delay to each input change. The logic output values are correct in all cases, although the outputs deviated from the supply voltage or ground in some cases by up to 10% of $V_{DD}$ due to the ambipolar nature of the TFTs. The nc-Si TFT-based logic gates operate at a similar frequency to high-quality organic transistor-based circuits at a much lower supply voltage [24].

## 5. Ring Oscillators

Three-, five-, or seven-stage ring oscillators were fabricated by connecting nc-Si TFT inverters in series, with the output of the last stage feeding back to the first. Figure 6 shows such a fabricated seven-stage ring oscillator. Each inverter was similar to the one in Section 3 and had a geometry ratio of four. To start the oscillations, the output of one of the stages was pulsed with a short square pulse. This set off the oscillations which stabilized at a set frequency depending on the number of stages and the supply voltage $V_{DD}$. Figure 7 shows the buildup of oscillations in three- and five-stage ring oscillators. At a supply voltage of 5 V, the oscillation frequencies are 21 kHz and 12 kHz, respectively. The total number of ring oscillators tested for each number of stages was approximately 15; the observed frequency variation was below 10%. The yield of operating circuits was >97%, which

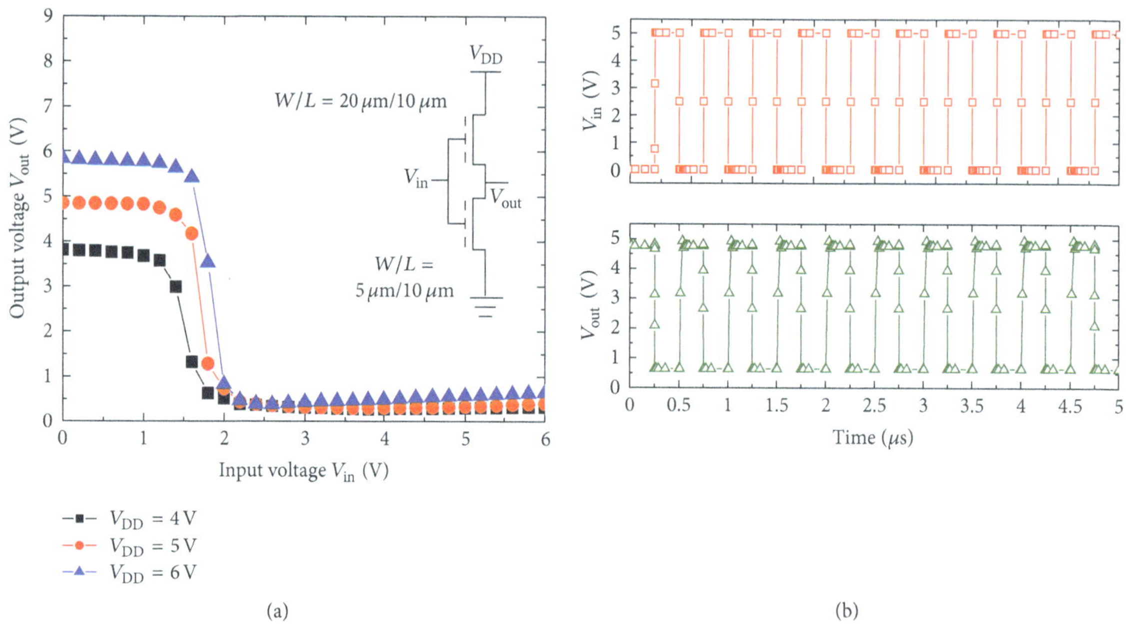


FIGURE 3: (a) Voltage transfer characteristics of an inverter formed using two ambipolar nc-Si TFTs with 20 nm $HfO_2$ gate dielectric measured at different operating voltages. (b) Operation of the inverter at 2 MHz input frequency and $V_{DD}$ = 5 V.

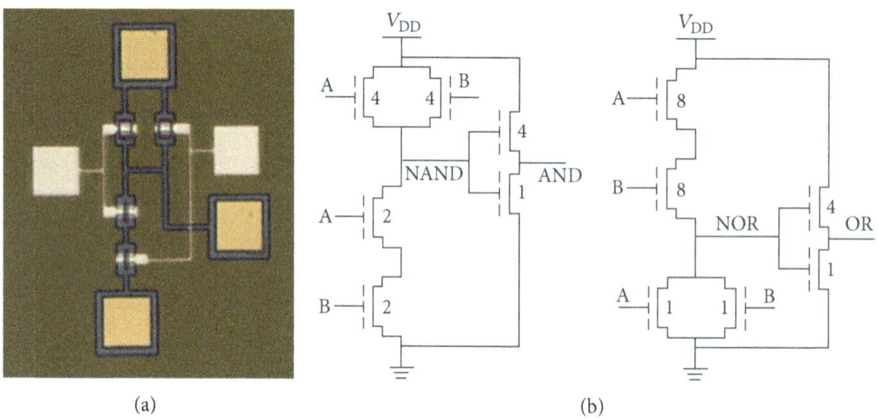


FIGURE 4: (a) Optical microscope image of a 2-input NAND gate fabricated using nc-Si TFTs. (b) Circuit diagram of the NAND, AND, NOR, and OR logic gates. The AND and OR gates are built by adding an inverter at the out of the NAND and NOR gates, respectively.

is reasonable for a first demonstration. A five-stage ring oscillator operating at 20 kHz ($V_{DD}$ = 5 V) consumes approximately 0.5 μW.

The supply voltage was varied while testing the oscillators. An increase in voltage swing and a reduction in the propagation delay were observed with increase in $V_{DD}$. For instance, the circuit starts oscillating with $V_{DD}$ as low as 3 V with propagation delay of ~14 μs/stage. At $V_{DD}$ = 8 V, the frequency increases to 38 kHz, corresponding to a delay of 4 μs/stage. This trend is illustrated in Figure 8. The propagation delays compare well with other ring oscillator circuits fabricated using low-temperature technology [20, 22–28]. This is also the first instance of noncrystalline silicon TFT-based ring oscillators operating at low-(<5 V)-supply voltages. The operating frequency of these circuits would also benefit from reduced parasitic capacitance in the devices.

In addition to varying the supply voltage, ring oscillators with different channel lengths were also tested (retaining the same channel widths). Figure 9 shows the relationship between channel length and propagation delay for three different values of $V_{DD}$. The delay is minimum for $L$ = 2 μm and increases with channel length, following a similar trend

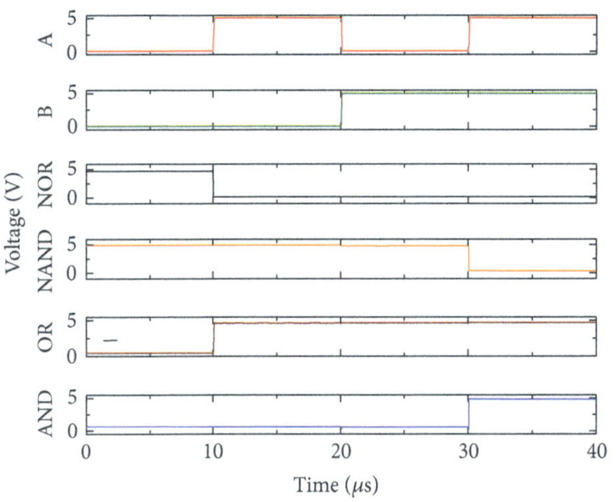


FIGURE 5: Operation of nc-Si TFT-based logic gates: (a) and (b) show input waveforms, and (c)–(f) show the outputs of OR, AND, NOR, and NAND gates, respectively.

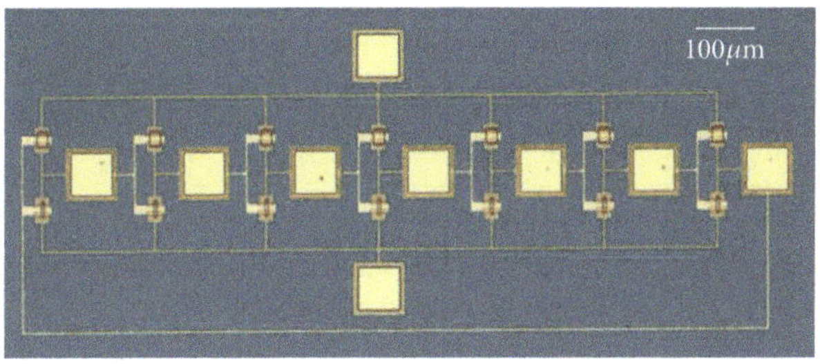


FIGURE 6: A fabricated seven-stage ring oscillator consisting of ambipolar nc-Si TFTs.

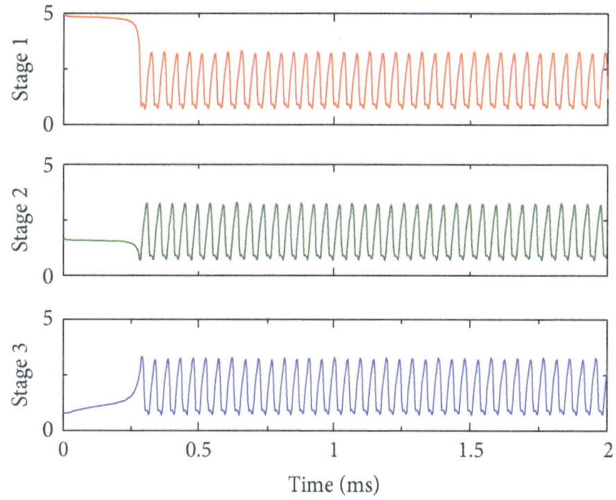


(a)

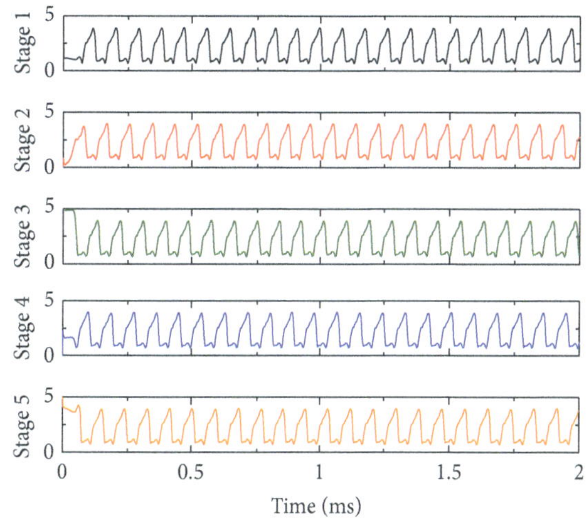


(b)

FIGURE 7: (a) Operation of a three-stage ring oscillator at $V_{DD}$ = 5 V. Waveforms at the output of each stage are shown. The oscillation frequency is approximately 21 kHz. (b) Operation of a five-stage ring oscillator at $V_{DD}$ = 5 V. The oscillation frequency is approximately 12 kHz.

as observed in [25]. This trend can be attributed to higher drive currents in TFTs with higher $W/L$ values.

It is known that nc-Si TFTs suffer from threshold voltage degradation upon bias stressing [9, 13, 14, 29]. The dominant mechanism for this is believed to be defect state creation in the channel layer [5, 13]. To quantify the effect of device degradation on ring oscillator operation, a three-stage ring oscillator was operated at different supply voltages for increasing time periods. The oscillation frequency was measured after each stress period. The results (Figure 10) show that after long periods of continuous operation the oscillation frequency decreased for all supply voltages. The maximum reduction occurred for $V_{DD}$ = 7 V in which case the frequency reduced by 43% after a stress period of $10^4$ s. Similar trends were observed for each $V_{DD}$ value. Although any device degradation is adverse to the circuit, the observed frequency reduction is a large improvement over other low-temperature fabricated ring oscillators [22]. This can be further improved by performing nc-Si depositions at slightly higher temperatures or potentially by further optimization of the deposition conditions.

## 6. Conclusion

In this work, logic gates and ring oscillators fabricated using ambipolar nc-Si TFTs have been demonstrated. Use of nc-Si as channel material enables low-cost large-area fabrication of circuits using these TFTs. The fabrication process employed uses low temperature (maximum of 250°C) and is CMOS compatible. The devices do not degrade upon exposure to the atmosphere and are robust. The logic gates were shown to operate with correct outputs at frequencies up to 100 kHz and rapid switching between states. Ring oscillators consisting of up to seven stages were fabricated and characterized. The oscillation frequencies varied from 8 kHz to 55 kHz depending on the supply voltage and channel length, corresponding to delays of approximately 3 μs to 20 μs per

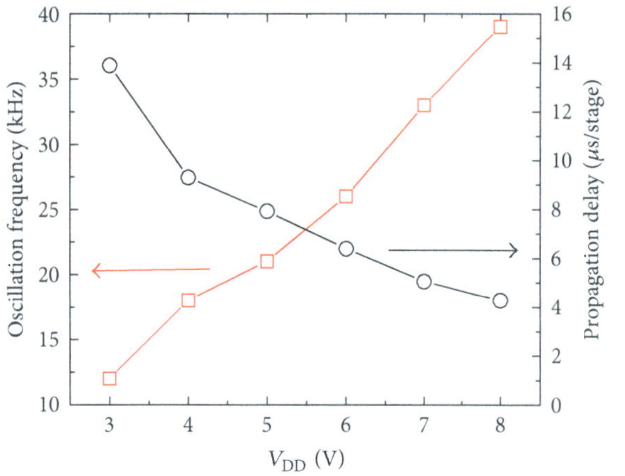


FIGURE 8: Variation of frequency of oscillation and propagation delay with supply voltage in a three-stage ring oscillator.

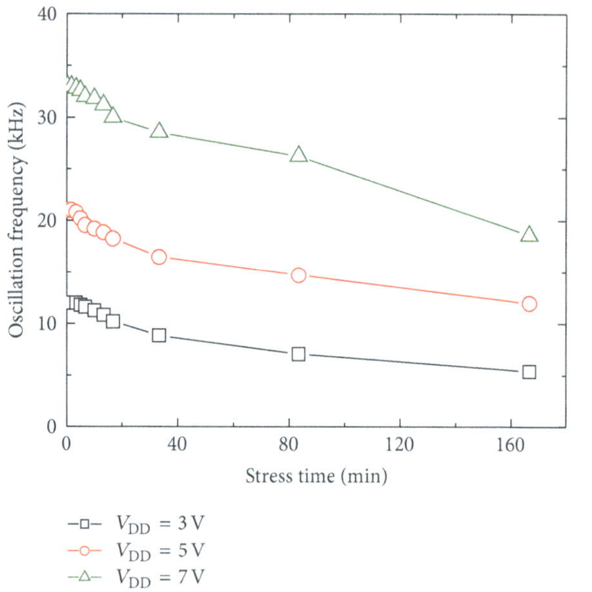


FIGURE 10: Effect of bias stress degradation on frequency of oscillation of a three-stage ring oscillator at various supply voltages.

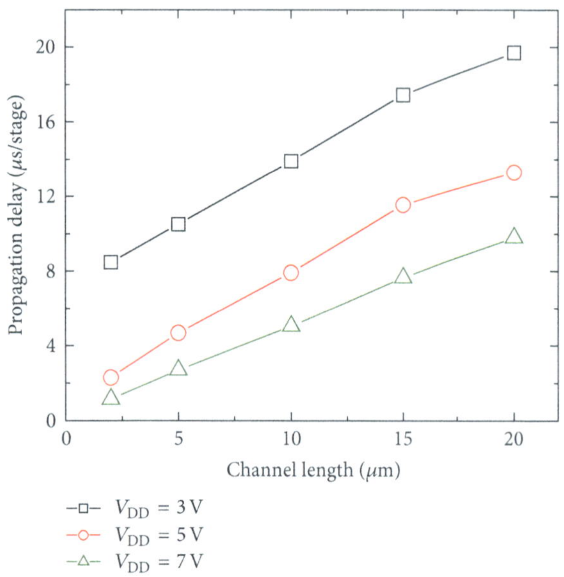


FIGURE 9: Variation of propagation delay with channel length in three-stage ring oscillators for different values of supply voltage.

stage. The yield and reproducibility of either type of circuit were >97%, which is reasonable for a first demonstration. Finally, the ring oscillator circuits were electrically stressed by continuous operation beyond $10^4$ s to examine the effect of bias stress degradation of nc-Si TFTs. The oscillation frequencies reduced upon long periods of stress, but this reduction was lower than that observed in other circuits with low-temperature fabrication processes.

Although the circuits were fabricated as large area for ease of probing, nanoscale circuits using nc-Si TFTs are certainly feasible. Submicron nc-Si TFTs fabricated using electron beam lithography have been demonstrated elsewhere, and the same process could be extended to produce compact circuits.

## References

[1] K. D. Cantley, A. Subramaniam, H. Stiegler, R. Chapman, and E. Vogel, "Hebbian learning in spiking neural networks with nano-crystalline silicon TFTs and memristive synapses," *IEEE Transactions on Nanotechnol*, vol. 10, no. 5, pp. 1066–1073, 2011.

[2] C. Koch, "Neuromorphic vision chips," *IEEE Spectrum*, vol. 33, no. 5, pp. 38–46, 1996.

[3] A. J. Snell, K. D. Mackenzie, W. E. Spear, P. G. LeComber, and A. J. Hughes, "Application of amorphous silicon field effect transistors in addressable liquid crystal display panels," *Applied Physics*, vol. 24, no. 4, pp. 357–362, 1981.

[4] T. Tsukuda, *Technology and Applications of Amorphous Silicon, Springer Series in Material Science*, vol. 37, Springer, Berlin, Germany, 2000.

[5] D. Knipp, R. A. Street, H. Stiebig et al., "Vertically integrated amorphous silicon color sensor arrays," *IEEE Transactions on Electron Devices*, vol. 53, no. 7, pp. 1551–1558, 2006.

[6] I. C. Y. Kwong, H. J. Lee, and A. Sazonov, "Nanocrystalline silicon diodes for rectifiers on RFID tags," *MRS Proceedings*, vol. 1066, 2008.

[7] C. Min, Z. Weijia, W. Tianmin, J. Fei, L. Guohua, and D. Kun, "Nanocrystalline silicon films with high conductivity and the application for PIN solar cells," *Vacuum*, vol. 81, no. 1, pp. 126–128, 2006.

[8] A. Subramaniam, K. D. Cantley, R. A. Chapman, B. Chakrabarty, and E. M. Vogel, "Ambipolar non-crystalline-silicon TFTs with submicron dimensions and reduced threshold voltage shift," in *Proceeding of the 69th Annual Device Research Conference (DRC '11)*, June 2011.

[9] A. Subramaniam, K. D. Cantley, H. J. Stiegler, R. A. Chapman, and E. M. Vogel, "Submicron ambipolar nanocrystalline silicon

thin-film transistors and inverters," *IEEE Transactions on Electron Devices*, vol. 59, no. 2, pp. 359–366, 2012.

[10] K. K. Likharev, "Neuromorphic CMOL circuits," in *Proceedings of the 3rd IEEE Conference on Nanotechnology Conference Digest (IEEE Nano '03)*, vol. 2, pp. 339–342, August 2003.

[11] Y. Nara, Y. Kudou, and M. Matsumura, "Application of amorphous-silicon field-effect transistors in three-dimensional integrated circuits," *Japanese Journal of Applied Physics*, vol. 22, no. 6, pp. L370–L372, 1983.

[12] B. Stannowski, R. E. I. Schropp, R. B. Wehrspohn, and M. J. Powell, "Amorphous-silicon thin-film transistors deposited by VHF-PECVD and hot-wire CVD," *Journal of Non-Crystalline Solids*, vol. 299–302, no. 2, pp. 1340–1344, 2002.

[13] M. J. Powell, C. Van Berkel, I. D. French, and D. H. Nicholls, "Bias dependence of instability mechanisms in amorphous silicon thin-film transistors," *Applied Physics Letters*, vol. 51, no. 16, pp. 1242–1244, 1987.

[14] A. R. Hepburn, J. M. Marshall, C. Main, M. J. Powell, and C. Van Berkel, "Metastable defects in amorphous-silicon thin-film transistors," *Physical Review Letters*, vol. 56, no. 20, pp. 2215–2218, 1986.

[15] C. H. Lee, A. Sazonov, and A. Nathan, "High hole and electron mobilities in nanocrystalline silicon thin-film transistors," *Journal of Non-Crystalline Solids*, vol. 352, no. 9-20, pp. 1732–1736, 2006.

[16] A. Subramaniam, K. D. Cantley, R. A. Chapman, H. J. Stiegler, and E. M. Vogel, "Submicron ambipolar nanocrystalline-silicon TFTs with high-$\kappa$ gate dielectrics," in *Proceedings of the International Semiconductor Device Research Symposium (ISDRS '11)*, December 2011.

[17] K. D. Cantley, A. Subramaniam, H. J. Stiegler, R. A. Chapman, and E. M. Vogel, "SPICE simulation of nanoscale non-crystalline silicon TFTs in spiking neuron circuits," in *Proceedings of the 53rd IEEE International Midwest Symposium on Circuits and Systems (MWSCAS '10)*, pp. 1202–1205, August 2010.

[18] H. Gleskova, S. Wagner, V. Gašparík, and P. Kováč, "150∘C amorphous silicon thin-film transistor technology for polyimide substrates," *Journal of the Electrochemical Society*, vol. 148, no. 7, pp. G370–G374, 2001.

[19] E. Y. Ma and S. Wagner, "Amorphous silicon transistors on ultrathin steel foil substrates," *Applied Physics Letters*, vol. 74, no. 18, pp. 2661–2662, 1999.

[20] C. G. Lee, T. Joshi, K. Divakar, and A. Dodabalapur, "Circuit applications based on solution-processed zinc-oxide TFTs," in *Proceedings of the 69th Device Research Conference, Conference Digest (DRC '11)*, June 2011.

[21] J. Sun, D. A. Mourey, D. Zhao et al., "ZnO thin-film transistor ring oscillators with 31-ns propagation delay," *IEEE Electron Device Letters*, vol. 29, no. 7, pp. 721–723, 2008.

[22] K. H. Kim, Y. H. Kim, H. J. Kim, J. I. Han, and S. K. Park, "Fast and stable solution-processed transparent oxide thin-film transistor circuits," *IEEE Electron Device Letters*, vol. 32, no. 4, pp. 524–526, 2011.

[23] T. D. Anthopoulos, D. M. De Leeuw, E. Cantatore et al., "Organic complementary-like inverters employing methanofullerene-based ambipolar field-effect transistors," *Applied Physics Letters*, vol. 85, no. 18, pp. 4205–4207, 2004.

[24] B. Crone, A. Dodabalapur, Y. Y. Lin et al., "Large-scale complementary integrated circuits based on organic transistors," *Nature*, vol. 403, no. 6769, pp. 521–523, 2000.

[25] H. Klauk, M. Halik, U. Zschieschang et al., "Flexible organic complementary circuits," *IEEE Transactions on Electron Devices*, vol. 52, no. 4, pp. 618–622, 2005.

[26] E. J. Meijer, D. M. De Leeuw, S. Setayesh et al., "Solution-processed ambipolar organic field-effect transistors and inverters," *Nature Materials*, vol. 2, no. 12, pp. 678–682, 2003.

[27] D. Zhao, D. A. Mourey, and T. N. Jackson, "Fast flexible plastic substrate ZnO circuits," *IEEE Electron Device Letters*, vol. 31, no. 4, pp. 323–325, 2010.

[28] K. Hiranaka, T. Yamaguchi, and S. Yanagisawa, "Self-alignment processed amorphous silicon ring oscillators," *IEEE Electron Device Letters*, vol. 5, no. 7, pp. 224–225, 1984.

[29] D. R. Allee, L. T. Clark, B. D. Vogt et al., "Circuit-level impact of a-Si: H thin-film-transistor degradation effects," *IEEE Transactions on Electron Devices*, vol. 56, no. 6, pp. 1166–1176, 2009.

# 6

# Single-Input Four-Output Current Mode Filter Using Operational Floating Current Conveyor

**Neeta Pandey, Deva Nand, and Zubair Khan**

*Department of Electronics and Communications, Delhi Technological University, Delhi, India*

Correspondence should be addressed to Neeta Pandey; n66pandey@rediffmail.com

Academic Editor: Sudhanshu Maheshwari

This paper presents operational floating current conveyor (OFCC) based single input four output current mode filter. It employs only three OFCCs and two grounded capacitors and resistors each. The MOS based grounded resistors implementation is used, which adds feature of electronic tunability to the filter parameters. The filter also enjoys low component spread and low sensitivity performance. The effect of finite transimpedance and parasites of OFCC on the proposed circuit is also analyzed. The functionality of the proposed circuit is demonstrated through SPICE simulations using 0.5 $\mu$m CMOS process model provided by MOSIS (AGILENT).

## 1. Introduction

The development of current mode continuous time filters has received growing interest due to its applications in sampled data systems, communications, and control systems [1, 2]. This is also in consonance with the trend of operating circuits at lower supply voltages where the current mode concept is more useful. The circuit operation in current mode circuits is decided by current which enables the system design to have wider dynamic range [3]. These circuits are also simple and compact as current addition/subtraction does not require additional circuit elements. Owing to the advantages of current mode filters, considerable research has directed towards the realization of single-input multiple output filters. A variety of current mode building blocks [4–10], namely, current follower transconductance amplifier [4], current conveyor [5–7], operational transconductance amplifier [8], $Z$ copy current follower transconductance amplifier [9], and $Z$-Copy Current Inverter Transconductance amplifier [10], have been used to implement these filters [1, 4–22]. The structures—use excessive number of elements [6, 7, 13, 15] some of the references cited in [17], do not present low input impedance [1, 8, 13–16, 19] which is necessary for a current mode filter, employ multiple inputs to provide output responses [15, 16], put matching constraint [7, 10, 15, 16] to obtain all five responses of universal filter, has less than three simultaneous output responses [4, 8, 15, 16, 21], employ different type of active blocks [12, 19, 20]. The operational floating current conveyor (OFCC) [23, 24], which was introduced as operational floating conveyor (OFC) [25], combines the features of current conveyor and the current feedback op-amp along with additional current outputs that add flexibility in the circuit design. Literature survey on OFCC based circuits [26–32] shows that voltage mode filter [26–28], variable gain amplifier [29], wheatsone bridge [30], and instrumentation amplifier [31] and read out circuit [32] are available. The study of current mode filters [1, 4–22] shows that no OFCC based current mode filter is available in the literature. Therefore, this paper aims at presenting a single-input four-output OFCC based current mode filter. The proposed filter uses three OFCCs, two grounded capacitors and two grounded resistors. The resistors are implemented using MOSFETs so as to achieve electronic tunability of filter parameters. The filter enjoys low component spread and low sensitivity performance. The proposed circuit is also analyzed to take finite transimpedance and parasites of OFCC into account. The functionality of the proposed circuit

TABLE 1: Transistors aspect ratios of the circuit shown in Figure 2.

| Transistor | $W$ ($\mu$m)/$L$ ($\mu$m) |
|---|---|
| M1, M2 | 50/1 |
| M3, M4, M11, M12, M14<br>M16, M18, M20, M22, M24 | 50/2.5 |
| M5, M7, M10, M15, M17, M19<br>M21, M23, M25 | 20/2.5 |
| M6, M8 | 40/2.5 |
| M9, M13 | 100/2.5 |

is demonstrated through SPICE simulations using 0.5 $\mu$m CMOS process model provided by MOSIS (AGILENT).

## 2. Circuit Description

*2.1. OFCC.* The circuit symbol of operational floating current conveyor (OFCC) [23, 24] is shown in Figure 1. It has a low impedance current input port $X$ and a high impedance voltage input port $Y$. It also has a low impedance voltage output port $W$ and high impedance current output ports $Z1$, $Z2$, $Z3$, and $Z4$. The output voltage at port $W$ is multiplication of input current at port $X$ and the open loop transimpedance gain $Z_t$. The port relationships of the OFCC is characterized by the following matrix:

$$\begin{bmatrix} I_Y \\ V_X \\ V_W \\ I_{Z1} \\ I_{Z2} \\ I_{Z3} \\ I_{Z4} \end{bmatrix} = \begin{bmatrix} 0 & 0 & 0 & 0 & 0 & 0 & 0 \\ 1 & 0 & 0 & 0 & 0 & 0 & 0 \\ 0 & Z_t & 0 & 0 & 0 & 0 & 0 \\ 0 & 0 & 1 & 0 & 0 & 0 & 0 \\ 0 & 0 & 1 & 0 & 0 & 0 & 0 \\ 0 & 0 & -1 & 0 & 0 & 0 & 0 \\ 0 & 0 & -1 & 0 & 0 & 0 & 0 \end{bmatrix} \begin{bmatrix} V_Y \\ I_X \\ I_W \\ V_{Z1} \\ V_{Z2} \\ V_{Z3} \\ V_{Z4} \end{bmatrix}. \quad (1)$$

It may be noted that the voltage at port $X$ is the same as input voltage at port $Y$, so voltage tracking action is available at input port. The output current flowing through port $W$ is copied to ports $Z1$ and $Z2$ in phase and to ports $Z3$ and $Z4$ out of phase, thereby offering current tracking at the output ports. Figure 2 shows the CMOS based schematic of OFCC which is based on topology proposed in [29].

*2.2. Proposed Filter.* In this section the implementation of OFCC based current mode single-input four-output filter, as shown in Figure 3, is proposed. It employs three OFCCs, and two grounded capacitors and resistors each. The analysis of the circuit gives the following transfer functions:

$$T_{\text{LPF1}} = T_{\text{LPF2}} = \frac{I_{\text{LPF1}}}{I_{\text{in}}} = \frac{I_{\text{LPF2}}}{I_{\text{in}}} = \frac{-1}{D(s)}, \quad (2)$$

$$T_{\text{BPF}} = \frac{I_{\text{BPF}}}{I_{\text{in}}} = \frac{-sC_2R_2}{D(s)},$$

$$T_{\text{HPF}} = \frac{I_{\text{HPF}}}{I_{\text{in}}} = \frac{-s^2C_1C_2R_1R_2}{D(s)}, \quad (3)$$

$$T_{\text{NOTCH}} = \frac{I_{\text{NOTCH}}}{I_{\text{in}}} = \frac{s^2C_1C_2R_1R_2 + 1}{D(s)},$$

where

$$D(s) = s^2C_1C_2R_1R_2 + sC_2R_2 + 1. \quad (4)$$

Thus the proposed circuit provides low pass, band pass, and notch (band stop) response simultaneously at high impedance. It may be noted that the current $I_{\text{HPF}}$ is not explicitly available at high impedance and therefore cannot be used directly. However, by connecting $I_{\text{NOTCH}}$ and $I_{\text{LPF1}}$, the high-pass response can be made available at high output impedance. This arrangement makes low pass, band pass, and high-pass responses available simultaneously. Similarly, the all pass function can easily be obtained by connecting band pass and notch output currents, that is, $I_{\text{AP}} = I_{\text{BP}} + I_{\text{NOTCH}}$, and the corresponding transfer function is obtained as

$$T_{\text{AP}} = \frac{I_{\text{AP}}}{I_{\text{in}}} = \frac{s^2C_1C_2R_1R_2 - sC_2R_2 + 1}{D(s)}. \quad (5)$$

It may be noted that all the five filter responses can be made simultaneously available by adding $Z$ stages in 2nd OFCC (Figure 3). Further, there is no matching constraint for realization of filter responses.

All the responses are characterized by pole frequency ($\omega_0$), bandwidth ($\omega_0/Q_0$), and quality factor ($Q_0$) given as

$$\omega_0 = \sqrt{\frac{1}{R_1R_2C_1C_2}},$$

$$\text{BW} = \frac{\omega_0}{Q_0} = \frac{1}{R_1C_1}, \quad (6)$$

$$Q_0 = \sqrt{\frac{R_1C_1}{R_2C_2}}.$$

Equation (6) reveals that the pole frequency ($\omega_0$) and quality factor ($Q_0$) can be adjusted by $R_2$, without disturbing $\omega_0/Q_0$. The orthogonal adjustability of $\omega_0$ and $Q_0$ can be achieved by simultaneous adjustment of $R_1$ and $R_2$ such that the product $R_1R_2$ remains constant and the quotient $R_1/R_2$ varies and vice versa. With moderate values of ratios of component values $((R_1/R_2) = (C_1/C_2) = Q_0)$, that is, from low component spread [34], high values of Q-factor can be obtained. Hence the component spread is of the order of $\sqrt{Q_0}$.

The proposed filter uses grounded resistors which can easily be implemented using the MOS based structure given in Figure 4 [33]. It uses two diode connected matched transistors, operating in saturation region. Assuming $V_1 = -V_2$, the value of resistor is given by

$$R = \frac{L}{2\mu C_{\text{ox}}W(V_1 - V_T)}, \quad (7)$$

where $\mu$ is carrier mobility, $C_{\text{ox}}$ is gate capacitance per unit area, $V_T$ is threshold voltage, and $W$, $L$ are the channel width and length respectively.

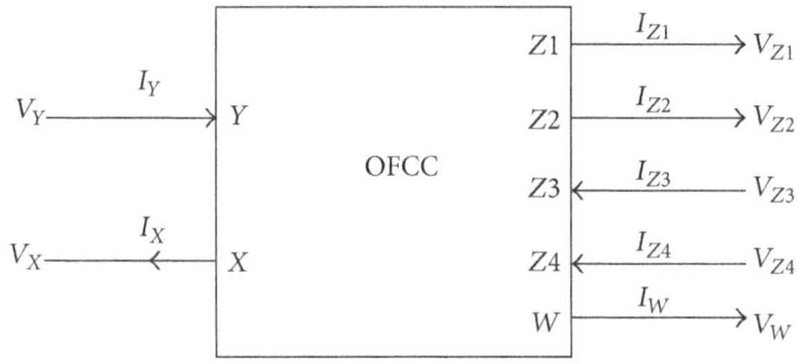


FIGURE 1: Circuit symbol of OFCC.

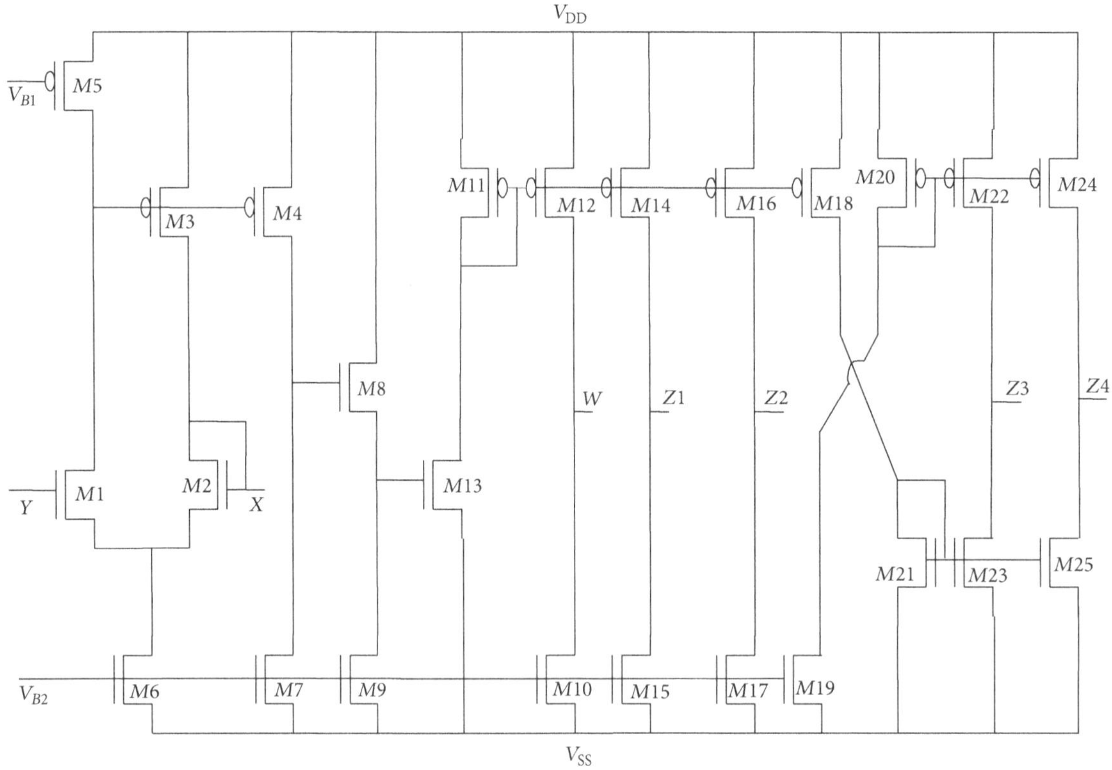


FIGURE 2: CMOS schematic of OFCC [29].

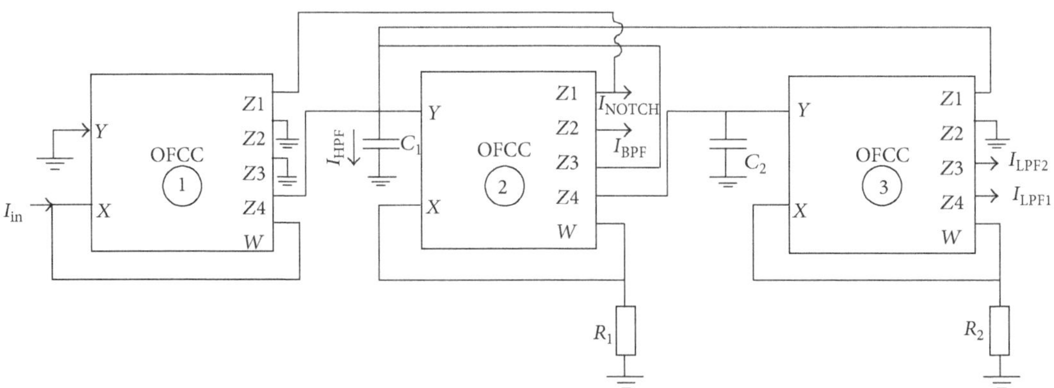


FIGURE 3: Proposed OFCC based single-input four-output current mode filter.

TABLE 2: Bias voltages/resistor values for orthogonal adjustment of $f_0$ and $Q_0$.

| $V_{11}$ *(V) | $V_{21}$ *(V) | $R_1$ (KΩ) | $V_{12}^*$(V) | $V_{22}^*$(V) | $R_2$ (KΩ) | Q | F (KHz) |
|---|---|---|---|---|---|---|---|
| 1.310 | −1.310 | 1 | 1.310 | −1.310 | 1 | 1 | 1590 |
| 0.869 | −0.869 | 2 | 0.869 | −0.869 | 2 | 1 | 800 |
| 0.726 | −0.726 | 5 | 0.726 | −0.726 | 5 | 1 | 320 |
| 0.711 | −0.711 | 10 | 0.711 | −0.711 | 10 | 1 | 160 |

*$V_{1i}$ and $V_{2i}$ refer to bias voltages corresponding to resistance $R_i$.

TABLE 3: Bias voltages/resistor values for orthogonal adjustment of $Q_0$ with $f_0$.

| $V_{11}$ *(V) | $V_{21}$ *(V) | $R_1$ (KΩ) | $V_{12}$ *(V) | $V_{22}$ *(V) | $R_2$ (KΩ) | $C_1$ (nF) | $C_2$ (nF) | Q | F (KHz) |
|---|---|---|---|---|---|---|---|---|---|
| 1.310 | −1.310 | 1 | 0.869 | −0.869 | 2 | 0.1 | 0.2 | 0.5 | 800 |
| 0.869 | −0.869 | 2 | 0.869 | −0.869 | 2 | 0.1 | 0.1 | 1 | 800 |
| 0.869 | −0.869 | 2 | 1.310 | −1.310 | 1 | 0.2 | 0.1 | 2 | 800 |

*$V_{1i}$ and $V_{2i}$ refer to bias voltages corresponding to resistance $R_i$.

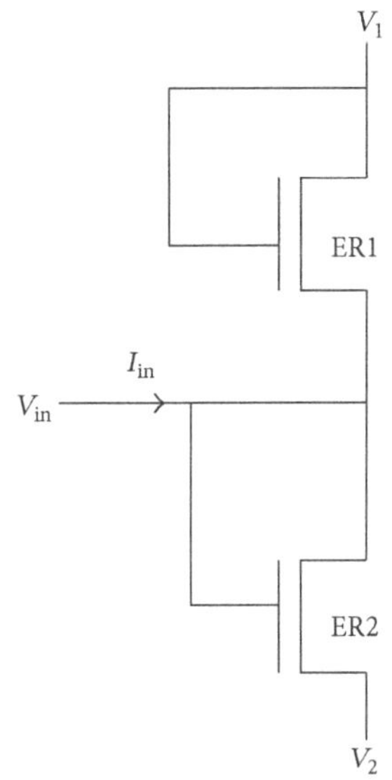


FIGURE 4: MOS realization of a resistor [33].

The sensitivity analysis of the proposed circuit is as follows:

$$S_{R_1}^{\omega_0} = S_{R_2}^{\omega_0} = S_{C_1}^{\omega_0} = S_{C_2}^{\omega_0} = S_{R_2}^{Q} = S_{C_2}^{Q} = -\frac{1}{2},$$
$$S_{R_1}^{Q} = S_{C_1}^{Q} = -S_{R_2}^{Q} = -S_{C_2}^{Q} = \frac{1}{2}. \quad (8)$$

Thus the all passive sensitivities are not more than unity in magnitude. So the proposed filter circuit can be classified as insensitive.

*2.3. Effect of Finite Transimpedance Gain.* In this section the effect of finite transimpedance gain of OFCC is considered and compensation is employed for high frequency applications. Ideally the transimpedance gain $Z_t$ is assumed to approach infinity. However, in practice, $Z_t$ is a frequency dependent finite value. Using single pole model for transimpedance gain, $Z_t(s)$ is expressed as [29]

$$Z_t(s) = \frac{Z_{to}}{1 + (s/\omega_{tc})}. \quad (9)$$

The parameter $Z_{to}$ is the DC open loop transimpedance gain and $\omega_{tc}$ is the transimpedance cutoff frequency. For high frequency applications, the transimpedance gain, $Z_t(s)$, is approximated as

$$Z_t(s) \cong \frac{1}{sC_p}, \quad (10)$$

where

$$C_p = \frac{1}{Z_{to}\omega_{tc}}. \quad (11)$$

Taking single pole model of $Z_t(s)$ into account, the low pass transfer function of (2) modifies to

$$T_{LPF1} = T_{LPF2} = \frac{I_{LPF1}}{I_{in}} = \frac{I_{LPF2}}{I_{in}} = \frac{-1}{D_n(s)}, \quad (12)$$

where

$$D_n(s) = s^2 R_1 \varepsilon_1(s) R_2 \varepsilon_2(s) C_1 C_2 + s R_2 \varepsilon_2(s) C_2 + 1,$$
$$\varepsilon_1(s) = \frac{1}{1 + sC_p R_1}, \qquad \varepsilon_2(s) = \frac{1}{1 + sC_p R_2}. \quad (13)$$

The inclusion of a capacitor of value $C_p$ connected between the terminal $W$ and ground in second and third OFCCs cancels the coefficient of "$s$" thereby making both $\varepsilon_1(s)$ and $\varepsilon_2(s)$ equal to unity. Thus the transfer function of low pass filter after compensation reduces to (2). Similar analysis is valid for other responses as well.

*2.4. Effect of Other Nonidealities.* The second group of nonidealities comes from parasites of OFCC comprising of resistances and capacitances connected in parallel at terminals

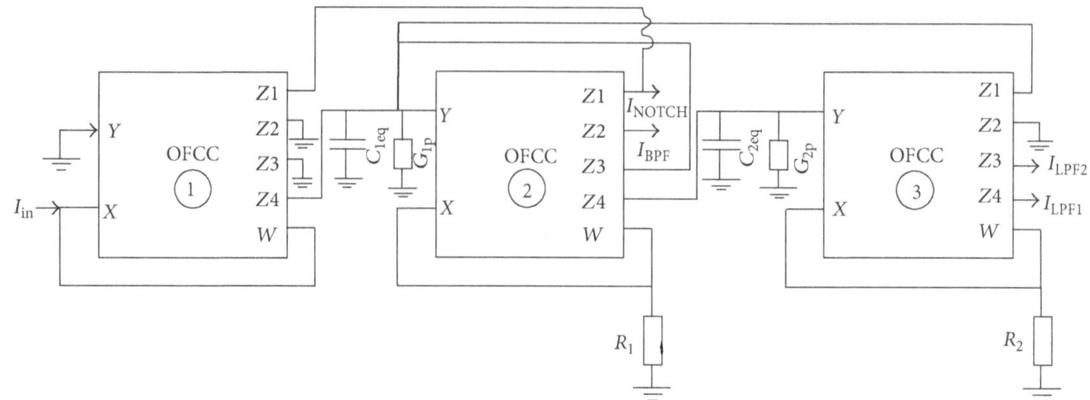


FIGURE 5: Proposed OFCC based current mode filter with parasites.

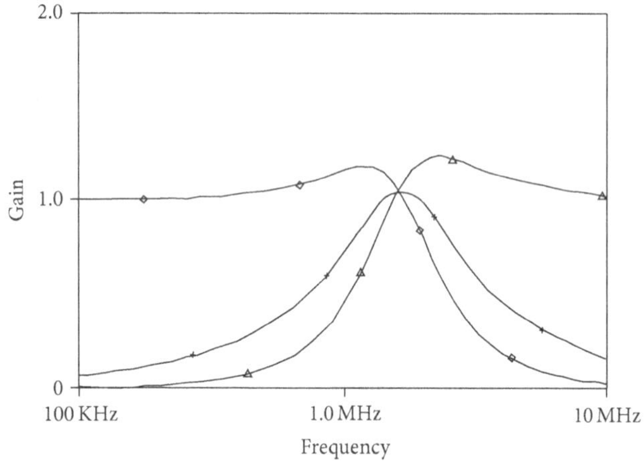


FIGURE 6: Simulated low pass, band pass, and high-pass responses.

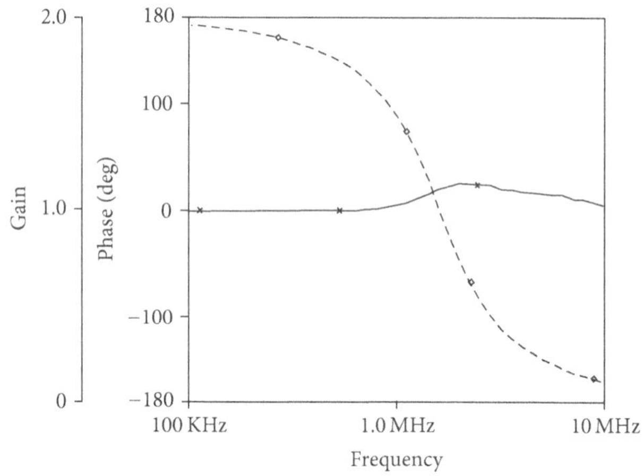


FIGURE 8: Magnitude and phase plots for all-pass response.

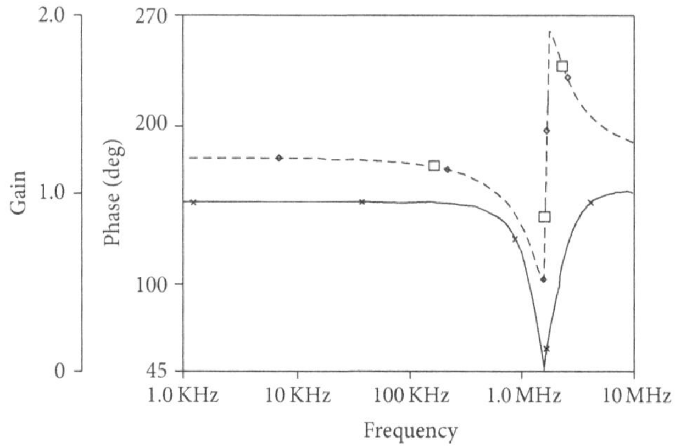


FIGURE 7: Magnitude and phase plot for notch response.

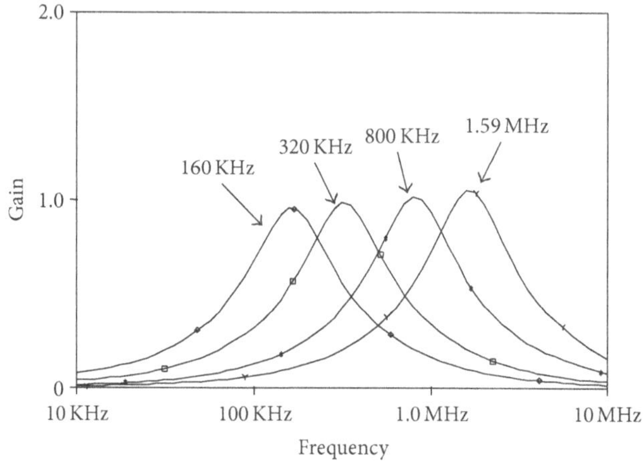


FIGURE 9: Tuning of $f_0$.

$Y$, $Z1$, $Z2$, $Z3$, and $Z4$ (i.e., $R_Y$, $C_Y$, $R_z$, $C_Z$). The effects of these parasites on filter response depend strongly on circuit topology. The proposed filter topology of Figure 3 in the presence of these parasites modifies to Figure 5 where $C_{1eq} = C_1//C_{Y1}//C_{Z41}//C_{Z31}//C_{Z13}$, $C_{2eq} = C_2//C_{Y3}//C_{Z42}$, $G_{1p} = 1/(R_{Z41}//R_{Z31}//R_{Z13})$, and $G_{2p} = 1/(R_{Y3}//R_{Z42})$; the subscript $i$ (=1, 2, 3) refers to parasites of $i$th OFCC and $j$ denotes $j$th $Z$ terminal. It may be noted that effect of parasitic capacitances can be accommodated by pre adjusting the value of external capacitor. The parasitics introduce poles at $G_{1P}/C_{1eq}$ and $G_{2P}/C_{2eq}$. So by choosing operating frequencies

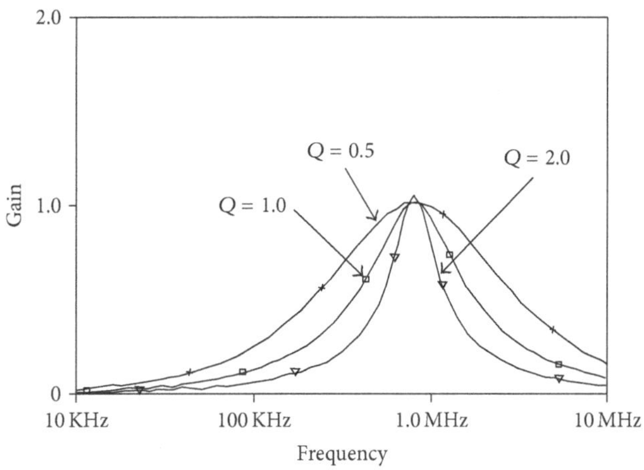


FIGURE 10: Tuning of $Q_0$.

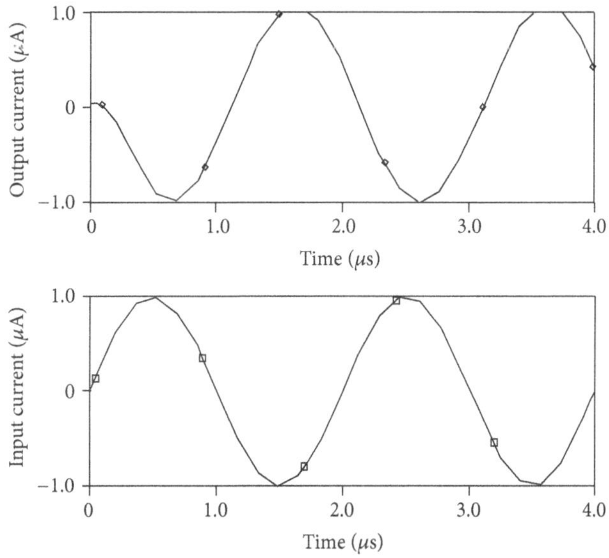


FIGURE 11: Transient response for low pass output.

sufficiently below these frequencies the effect of parasitic impedance can be practically eliminated and hence filter may approach towards ideal response.

## 3. Simulation Results

To verify the functionality of the proposed OFCC based current mode filter, SPICE simulations have been carried out using 0.5 $\mu$m CMOS process model provided by MOSIS (AGILENT) and CMOS schematic of Figure 2 with power supply voltage of $V_{DD} = -V_{SS} = 1.5$ V and $V_{B1} = -V_{B2} = 0.8$ V. The aspect ratios of the transistor are reported in Table 1 [29].

The simulations have been performed for a pole frequency of 1.59 MHz with component values as $C_1 = C_2 =$ 100 pF and MOS based resistors of value 1 $k\Omega$ by selecting bias voltages as $\pm$1.310 V. Figure 6 shows the simulation results for low pass, band pass, and high-pass responses. The phase and magnitude plots for notch and all pass responses are depicted in Figures 7 and 8, respectively. The responses confirm the theoretical predictions.

The orthogonal adjustment of $f_0$ with $Q_0$ is depicted in Figure 9 where the value of $Q_0 = 1$ is considered. The capacitors $C_1$ and $C_2$ are taken as 100 pF and the bias voltages along with the resistor values for different values of $f_0$ as given in Table 2. Figure 10 shows orthogonal adjustment of $Q_0$ with $f_0 = 800$ kHz. The values of $Q_0$ for constant value of $f_0$ as obtained with $C_1 = C_2 = 100$ pF and bias voltages and resulting resistor values are listed in Table 3.

To study the time domain behavior of the proposed filter, an input sinusoidal signal of 500 KHz frequency and amplitude 2 $\mu$A is applied. The transient response for low pass output is shown in Figure 11. To show the effectiveness of proposed filter a mixed sinusoidal signal of frequencies of 50 KHz, 500 KHz, and 4 MHz having amplitude of 2 $\mu$A each is applied at the input of the filter. The transient response with its spectrum for input and output signals is shown in Figures

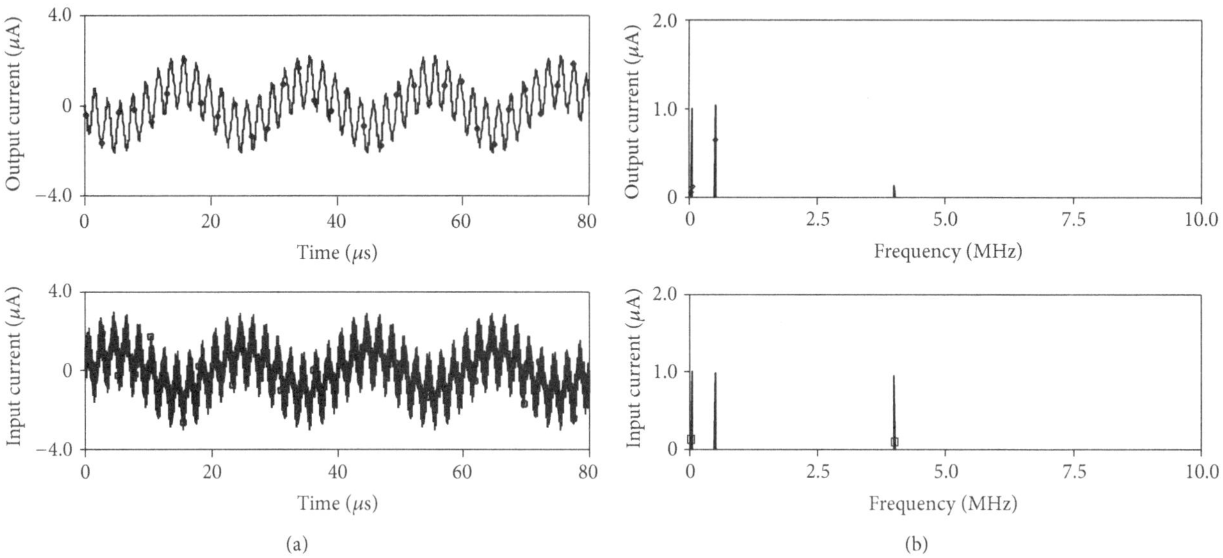


FIGURE 12: (a) Transient response of input and output signals (b) Spectrum of input and output signals.

12(a) and 12(b). It is clear that the 50 KHz, and 500 KHz signals are passed without attenuated and 4 MHz signal is significantly attenuated.

## 4. Conclusion

In this paper, a single-input four-output current mode filter based on OFCC is presented. The circuit uses only three OFCCs, two grounded resistors, and two grounded capacitors. The low pass, band pass and notch responses are available at high impedance whereas high-pass and all pass responses can be obtained by adding $Z$ stages in the second OFCC. The filter parameters are tuned electronically via MOS based grounded resistor. The effect of finite transimpedance and parasites on filter parameters is also included.

## References


[1] M. A. Ibrahim, S. Minaei, and H. Kuntman, "A 22.5 MHz current-mode KHN-biquad using differential voltage current conveyor and grounded passive elements," *AEU—International Journal of Electronics and Communications*, vol. 59, no. 5, pp. 311–318, 2005.

[2] A. S. Sedra and K. C. Smith, *Microelectronic Circuits*, Holt, Rinehart and Winston, Orlando, Fla, USA, 3rd edition, 1991.

[3] S. S. Rajput and S. S. Jamuar, "Advanced applications of current conveyors: a tutorial," *Journal of Active and Passive Electronic Devices*, vol. 2, pp. 143–164, 2007.

[4] W. Tangsrirat, "Single-input three-output electronically tunable universal current-mode filter using current follower transconductance amplifiers," *AEU—International Journal of Electronics and Communications*, vol. 65, no. 10, pp. 783–787, 2011.

[5] C. Wang, J. Xu, A. Ü. Keskin, S. Du, and Q. Zhang, "A new current-mode current-controlled SIMO-type universal filter," *AEU—International Journal of Electronics and Communications*, vol. 65, no. 3, pp. 231–234, 2011.

[6] J.-W. Horng, "Current-mode and transimpedance-mode universal biquadratic filter using multiple outputs cciis," *Indian Journal of Engineering and Materials Sciences*, vol. 17, no. 3, pp. 169–174, 2010.

[7] K. K. Abdalla, D. R. Bhaskar, and R. Senani, "Configuration for realising a current-mode universal filter and dual-mode quadrature single resistor controlled oscillator," *Institution of Engineering and Technology Circuits, Devices and Systems*, vol. 6, no. 3, pp. 159–167, 2012.

[8] A. Qadir and T. Altaf, "Current mode canonic OTA-C universal filter with single input and multiple outputs," in *Proceedings of International Conference on Electronic Computer Technology (ICECT '10)*, pp. 32–34, May 2010.

[9] J. Satansup and W. Tangsrirat, "Single-input five-output electronically tunable current-mode biquad consisting of only ZC-CFTAs and grounded capacitors," *Radioengineering*, vol. 20, no. 3, pp. 650–655, 2011.

[10] D. Biolek, V. Biolkova, Z. Kolka, and J. Bajer, "Single-input multi-output resistorless current-mode biquad," in *Proceedings of European Conference on Circuit Theory and Design Conference Program (ECCTD '09)*, pp. 225–228, August 2009.

[11] R. Senani and A. K. Singh, "A new universal current-mode biquad filter," *Frequenz*, vol. 56, no. 1-2, pp. 55–58, 2002.

[12] J.-W. Horng, "Current conveyors based current-mode universal biquadratic filter," *Journal of the Chinese Institute of Electrical Engineering*, vol. 9, no. 2, pp. 147–150, 2002.

[13] R. Senani, A. K. Singh, and V. K. Singh, "New tunable SIMO-type current mode universal biquad using only three MOCCs and all grounded passive elements," *Frequenz*, vol. 57, no. 7-8, pp. 160–161, 2003.

[14] R. Senani, V. K. Singh, A. K. Singh, and D. R. Bhaskar, "Novel electronically controllable current-mode universal biquad filter," *IEICE Electronics Express (Japan)*, vol. 1, pp. 410–415, 2004.

[15] R. Senani, "New universal current-mode biquad employing all grounded passive components but only two DOCCs," *Journal of Active and Passive Electronic Devices (USA)*, vol. 1, pp. 281–288, 2006.

[16] T. Tsukutani, Y. Sumi, and N. Yabuki, "Novel current-mode biquadratic circuit using only plus type DO-DVCCs and grounded passive components," *International Journal of Electronics*, vol. 94, no. 12, pp. 1137–1146, 2007.

[17] A. M. Soliman, "Current-mode universal filters using current conveyors: classification and review," *Circuits, Systems, and Signal Processing*, vol. 27, no. 3, pp. 405–427, 2008.

[18] M. Sagbas and M. Koksal, "Current-mode state-variable filter," *Frequenz*, vol. 62, no. 1-2, pp. 37–42, 2008.

[19] W. Chunhua, A. U. Keskin, L. Yang, Z. Qiujing, and D. Sichun, "Minimum configuration insensitive multifunctional current-mode biquad using current conveyors and all-grounded passive components," *Radioengineering*, vol. 19, no. 1, pp. 178–184, 2010.

[20] K. K. Abdalla, "Universal current-mode biquad employing dual output current conveyors and MO-CCCA with grounded passive elements," *Circuits and Systems*, vol. 4, pp. 83–88, 2013.

[21] J. Jerabek, R. Sotner, and K. Vrba, "Comparison of the SITO current-mode universal filters using multiple-output current followers," in *Proceedings of the 35th International Conference on Telecommunications and Signal Processing (TSP '12)*, pp. 406–410, 2012.

[22] E. Yuce, "Current-mode electronically tunable biquadratic filters consisting of only CCCIIs and grounded capacitors," *Microelectronics Journal*, vol. 40, no. 12, pp. 1719–1725, 2009.

[23] A. A. Khan, M. A. Al-Turaigi, and M. Abou El-Ela, "Operational floating current conveyor: characteristics, modelling and applications," in *Proceedings of the IEEE Instrumentation and Measurement Technology Conference*, pp. 788–791, Hamamtsu, Japan, May 1994.

[24] Y. H. Ghallab, M. A. El-Ela, and M. Elsaid, "Operational floating current conveyor: characteristics, modeling and experimental results," in *Proceedings of International Conference on Microelectronics*, pp. 59–62, 1999.

[25] C. Toumazou, A. Payne, and F. J. Lidgey, "Operational floating conveyor," *Electronics Letters*, vol. 27, no. 8, pp. 651–652, 1991.

[26] Y. H. Ghallab, W. Badawy, K. V. I. S. Kaler, M. A. El-Ela, and M. H. El-Said, "A new second-order active universal filter with single input and three outputs using operational floating current conveyor," in *Proceedings of IEEE International Conference on Microelectronics*, pp. 42–45, 2002.

[27] Y. H. Ghallab, M. A. El-Ela, and M. Elsaid, "A novel universal voltage mode filter with three inputs and single output using only operational floating current conveyor," in *Proceedings of International Conference on Microelectronics*, pp. 95–98, 2000.

[28] Y. H. Ghallab, W. Badawy, M. A. El-Ela, and M. H. El-Said, "The operational floating current conveyor and its applications," *Journal of Circuits, Systems and Computers*, vol. 15, no. 3, pp. 351–372, 2006.

[29] H. M. Hassan and A. M. Soliman, "Novel CMOS realizations of the operational floating conveyor and applications," *Journal of Circuits, Systems and Computers*, vol. 14, no. 6, pp. 1113–1143, 2005.

[30] Y. H. Ghallab and W. Badawy, "A new design of a current-mode wheatstone bridge using operational floating current conveyor," in *Proceedings of International Conference on MEMS, NANO and Smart Systems (ICMENS '06)*, pp. 41–44, December 2006.

[31] Y. H. Ghallab, W. Badawy, K. V. I. S. Kaler, and B. J. Maundy, "A novel current-mode instrumentation amplifier based on operational floating current conveyor," *IEEE Transactions on Instrumentation and Measurement*, vol. 54, no. 5, pp. 1941–1949, 2005.

[32] Y. H. Ghallab and W. Badawy, "A new differential PH sensor current mode readout circuit using only two operational floating current conveyor," in *Proceedings of IEEE International Workshop on Biomedical Circuits and Systems*, pp. 13–16, December 2004.

[33] Z. Wang, "2-MOSFET transresistor with extremely low distortion for output reaching supply voltages," *Electronics Letters*, vol. 26, no. 13, pp. 951–952, 1990.

[34] S.-I. Liu, "High input impedance filters with low component spread using current-feedback amplifiers," *Electronics Letters*, vol. 31, no. 13, pp. 1042–1043, 1995.

# 7

# Design and Analysis of Wideband Ladder-Type Film Bulk Acoustic Wave Resonator Filters in Ku-Band

**N. Izza M. Nor,[1,2] K. Shah,[1] J. Singh,[1] and Z. Sauli[2]**

[1] *Centre for Technology Infusion, La Trobe University, Bundoora, VIC 3086, Australia*
[2] *School of Microelectronic Engineering, Universiti Malaysia Perlis, 02600 Perlis, Malaysia*

Correspondence should be addressed to N. Izza M. Nor; nimohdnor@students.latrobe.edu.au

Academic Editor: Egidio Ragonese

This paper presents the design of ladder-type filters based on film bulk acoustic wave resonator (FBAR) in Ku-band. The proposed FBAR filter has an insertion loss of −3 dB, out-of-band rejection of −12 dB and 3 dB bandwidth of 1.0 GHz from 15 GHz to 16 GHz. Based on the characteristics of the FBAR filter, the expected characteristics of FBAR resonators are determined by using the 1D numerical analysis. This design proves that it is possible to design a wide-bandwidth FBAR filter in Ku-band.

## 1. Introduction

Transceiver systems working in Ku-band frequency range of 12–18 GHz are primarily used for satellite communications and radars. Cost, area, and power consumption are the key figures of merit for such transceiver system. However, the literature shows that most of these transceivers are designed using low-temperature cofire ceramic (LTCC) technology and are relatively large in size and heavy due to the use of discrete components such as filter and separately located modules. The multilayer LTCC and the systems-on-package (SOP) implementations are capable of overcoming these issues by integrating active and passive components on one board. Various Ku-band filters have been reported in the literature using different designs and manufacturing methods such as defect ground structure (DGS), interdigital structure, coupled line filters, and couple strip line filters have been integrated using LTCC technology [1–3]. However, improvement in filter performance and better integration methods with microwave monolithic integrated circuit (MMIC) and radio frequency (RF) microelectromechanical systems (MEMS) technology as used in WiFi and WiMAX applications [4] can be used to improve integration and reduce power consumption.

Film bulk acoustic wave resonator (FBAR) filter and FBAR diplexer designed using RF MEMS technology have been developed for WiFi and WiMAX applications [4]. Such MEMS components have shown better performance and higher integration level which can also be achieved in Ku-band transceivers using MEMS-based FBAR filters. Thin membrane-type resonator and solidly mounted resonator (SMR) are two types of FBARs based on acoustical isolation from the substrate [5]. Acoustical isolation achieved by air-gap is the preferred method to achieve high-quality (Q) factor using simpler fabrication methods [6]. The literature shows that AlN is the preferred material due to its moderate mechanical coupling factor, higher acoustic velocity, and higher *Q* value [7] at operating frequencies higher than 10 GHz. Surface acoustic wave (SAW) resonators whose resonance frequency depends on the pitch of integrated device technology (IDT) electrode patterns [8] are limited to less than 15 GHz [9]. However, FBARs operating in frequency range of 5 GHz to 20 GHz have been reported in, the literature [8, 10]. Due to high *Q* factor of FBARs, low insertion loss filters can be designed in Ku-band which will improve system performance and allow higher integration resulting in lost cost and low-power consumption [11].

There are several topologies of FBAR filters including ladder-type, lattice-type, or the combination of both [12]. FBAR filters have very high *Q* and small size. In addition, FBAR can be manufactured on a flat silicon substrate using surface micromachining and befits mass production, and

TABLE 1: Specifications of Ku-band transceiver front-end [3].

| | Input frequency (GHz) | Output frequency (GHz) | Input power (dBm) | Output power (dBm) | Gain (dB) | NF (dB) | Reject ion (dBc) |
|---|---|---|---|---|---|---|---|
| Receiver | 15.5 ± 0.5 | 0.8 ± 0.5 | ≤ −10 | — | ≥25 | ≤5.5 | — |
| Transmitter | 0.8 ± 0.5 | 15.5 ± 0.5 | 0 | ≥27 | ≥27 | — | ≥35 |
| IF | LO frequency: 14.7 GHz<br>Output power: 19 dBm | | | | | | |

this allows the filter components to become compact and cheap [13]. Furthermore, FBAR is compatible MMIC and MEMS technology. FBAR filter has been reported operating in frequency range of X-band, K-band, and Ka-band [13–15].

The $Q$ factor and electromechanical coupling coefficient ($k_{\text{eff}}^2$) of the piezoelectric thin film are two figures of merit for an FBAR. Lower value of $k_{\text{eff}}^2$ leads to a high insertion loss and higher value of $Q$ leads to lower insertion loss [16]. It is important to notice that improving $k_{\text{eff}}^2$ results in decreased $Q$ factor, and so optimization of both parameters is defined by one figure of merit (FOM), which is a product of $k_{\text{eff}}^2 \times Q$ [16]. In filter applications, it has been shown that this FOM parameter is inversely proportional to filter insertion loss [16].

The literature shows that traditionally FBARs with high $Q$ factors are designed and used in FBAR filter circuits to evaluate their performance. The filters are manufactured and their performance is measured. However, in this research collections of closed-form expressions from [17] are used to analyse the filter performance based on FBAR design variables like series and parallel resonance frequencies, $Q$ factor, coupling coefficient along with filter design variables like filter order, insertion loss, and out-of-band rejection. A typical Ku-band transceiver specification shows that Ku-band FBAR filters are expected to have insertion loss less than 3 dB, out-of-band rejection smaller than 12 dB, and 3 dB bandwidth of 1.2 GHz [3]. Then, based on the FBAR filter characteristics modelled using Butterworth Van Dyke (BVD) circuit [18], the expected FBAR characteristics are determined.

This paper is organised as follows: Section 2 presents the introduction of Ku-band transceiver front-end. Section 3 describes the theory of FBAR filter design. Section 4 presents Ku-band FBAR filter. Section 5 presents results and Section 6 presents conclusion.

## 2. Ku-Band Transceiver Front-End

A typical Ku-band transceiver consisting of low-noise amplifier (LNA), power amplifier (PA), voltage-controlled oscillator (VCO), local oscillator (LO), upconveter, downconverter, and filters is presented in Figure 1 as in [3]. In this transceiver, two types of Ku-band filter are used which have centre frequency of 15.5 GHz and 14.7 GHz. The filter operating at the centre frequency of 15.5 GHz is used to select the frequency band while the filter operating at the centre frequency of 14.7 GHz is used for filtering LO signal. The specifications of this transceiver are summarised in Table 1. Many different Ku-band transceivers exist [3, 19–21], but for this research this typical transceiver is selected to obtain performance parameters for a typical Ku-band filter working at centre frequency of 15.5 GHz. In this study, the ladder-type FBAR filter is designed based on typical Ku-band filter specifications and the characteristics of the FBAR filter and the striplines interdigital filter are compared. The next section discusses design process of FBAR filter based on closed-form equations.

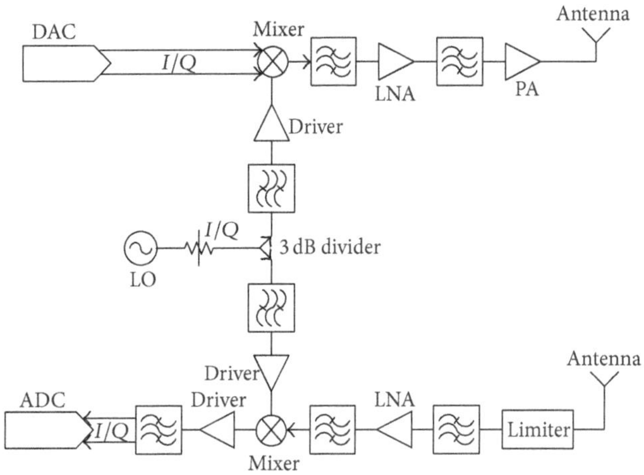


FIGURE 1: Front-end of Ku-band transceiver [3].

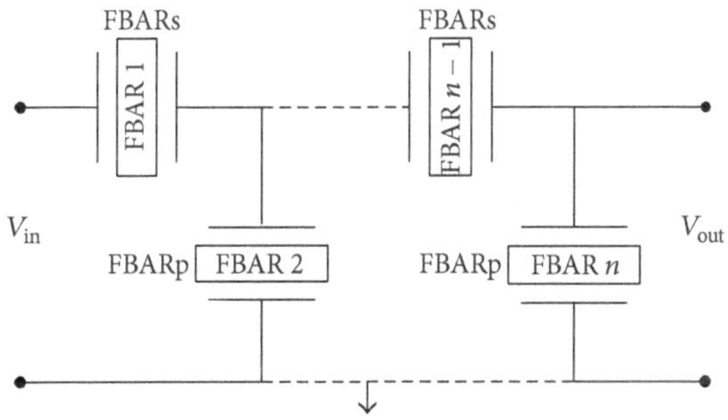


FIGURE 2: $n$th order of ladder filter.

## 3. FBAR Filter Design

The ladder-type FBAR filter consists of a number of L-sections comprising two resonators, one connected in series and the other connected in shunt as shown in Figure 2. The order of the filter is equal to the number of FBARs used in ladder connection. The resonator has two resonant frequencies which are series resonant frequency, $f_s$, and parallel resonant

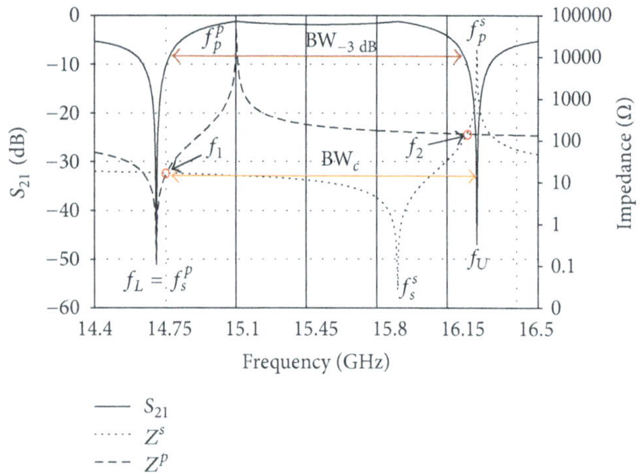


FIGURE 3: Working principle and design parameters of a ladder-type FBAR filter.

frequency, $f_p$. At $f_s$, the electrical impedance is minimum, $Z_{\min}$, and at $f_p$ the electrical impedance is maximum, $Z_{\max}$. At frequency far from $f_s$ or $f_p$, the impedance has the characteristics of static capacitance, $C_o$ [11].

Figure 3 presents the working principle of a second-order ladder-type FBAR filter. The continuous line represents the typical transmission response of ladder filter, while the dotted and dashed lines characterise the electrical impedance of series ($Z^s$) and parallel ($Z^p$) resonators, respectively. $f_s^s$ is the series frequency of the series resonator and $f_p^s$ is the parallel frequency of the series resonator. Meanwhile, $f_s^p$ is the series frequency of the shunt resonator and $f_p^p$ is the parallel frequency of the shunt resonator. The $f_s^p$ is tuned to be lower than the $f_s^s$. The ladder-type filter gives a steep roll-off but a poor out-of-band (OoB) rejection characteristic. A better OoB can be achieved by cascading more L-section to the filter; however this will have to be traded off with the insertion loss.

*3.1. Filter Design Methodology.* Figure 3 described the design parameters of FBAR filter. A set of closed-form expressions for designing FBAR filters are given in [17]. The design of ladder-type FBAR filter consists of determining the order of the filter. The specifications such as bandwidth, OoB rejection, and the frequency allocation of the upper $f_U$ and lower $f_L$ transmission zeros will be used as design variables. The design procedure is as follows.

*Step 1* (determine $f_s^p$ and $f_p^s$). The value of $f_s^p$ will be located at the lower transmission zero, $f_L$, and $f_p^s$ will be located at the upper transmission zero, $f_U$, as shown in Figure 3.

*Step 2* (determine $f_p^p$ and $f_s^s$). The value of $f_p^p$ and $f_s^s$ can be calculated using [14]

$$k_{\text{eff}}^2 = \left(\frac{\pi^2}{4}\right)\left(\frac{f_p - f_s}{f_p}\right). \tag{1}$$

This value can be set as required, but two conditions must be accomplished: the value of $k_{\text{eff}}^2$ must be lower than the maximum achievable value with the used materials, and the design is restricted to those filters whose upper and lower transmission zeroes fulfil [17]

$$f_U \geq \left(\frac{\pi^2}{4k_{\text{eff}}^2}\right)\left(\frac{\pi - \sqrt{\pi^2 - 16k_{\text{eff}}^2}}{\pi + \sqrt{\pi^2 - 16k_{\text{eff}}^2}}\right) f_L. \tag{2}$$

*Step 3* (determine $f_1$ and $f_2$). The approximation of bandwidth ($\text{BW}_c$) is $\text{BW}_c = f_2 - f_1$, where $f_1$ and $f_2$ are the frequencies where $|Z^s(f_i)| = |Z^p(f_i)|$ and $i = 1, 2$ is made. The $\text{BW}_c$ will be smaller than the 3 dB bandwidth ($\text{BW}_{-3\,\text{dB}}$) and it is independent on the filter order, thus simplifying significantly the design equations.

*Step 4* (determine $\psi = C_o^s/C_o^p$). The $\text{BW}_c$ depends on the value of $f_s^p$, $f_p^p$, $f_s^s$, $f_p^s$, and $\psi$. Since all the resonance frequencies have been determined, the $\psi$ will be calculated to achieve the specified $\text{BW}_c$. A simplified approach of electrical impedance of the FBAR is as given in [22]:

$$Z_{\text{in}} = \frac{1}{j2\pi f C_o}\left(\frac{f^2 - f_s^2}{f^2 - f_p^2}\right) \tag{3}$$

resulting as given in [17]:

$$\psi = \frac{-f_i^4 + f_i^2 f_p^{p2} + f_i^2 f_s^{s2} - f_s^{s2} f_p^{p2}}{f_i^4 - f_i^2 f_p^{s2} - f_i^2 f_s^{p2} + f_p^{s2} f_s^{p2}}, \tag{4}$$

where $f_i$ can be either $f_1$ or $f_2$. In order to obtain a closed-form expression of $\psi$ as a function of $\text{BW}_c$, an approximation of $f_1$ and $f_2$ that are symmetrically spaced around the centre frequency, $f_c$, is made, where $f_1 = f_c - \text{BW}_c/2$ and $f_2 = f_c + \text{BW}_c/2$. It can be seen from (4) that $\psi$ increases with $\text{BW}_c$.

*Step 5* (determine $\Theta = C_o^s C_o^p$). One of the conditions to find the static capacitance that the filter must accomplish is the image impedance matching condition. This condition is define as [17]

$$C_o^s C_o^p = \frac{1}{(2\pi f_c Z_o)^2}, \tag{5}$$

where $Z_o = 50\,\Omega$ is the impedance of the source and the load.

*Step 6* (compute $C_o^s$ and $C_o^p$ and determine $N$). $C_o^s$ and $C_o^p$ can now be easily computed using (4) and (5). Since the OoB rejection level depends only on $N$ and $\psi$, the order $N$ of the filter will be chosen to achieve the required OoB rejection.

In designing a wideband FBAR ladder filter at high frequencies, many considerations are required as discussed in the next section.

## 4. Ku-Band FBAR Filter Design

In this study, the ladder-type FBAR filter is designed to achieve $\text{BW}_{-3\,\text{dB}}$ of 1.2 GHz with centre frequency of

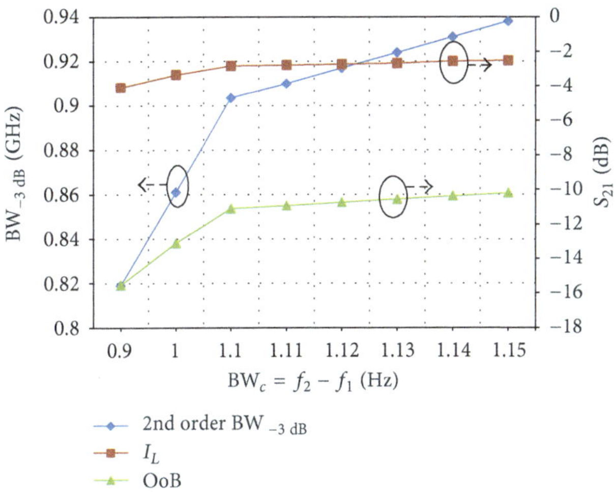


FIGURE 4: Effect of varying the $f_1$ and $f_2$.

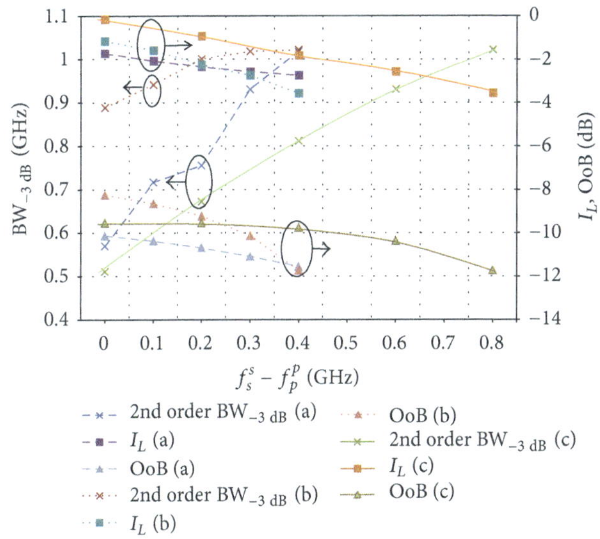


FIGURE 5: Effect of varying $f_s^s$, $f_s^p$, $f_p^s$, and $f_p^p$.

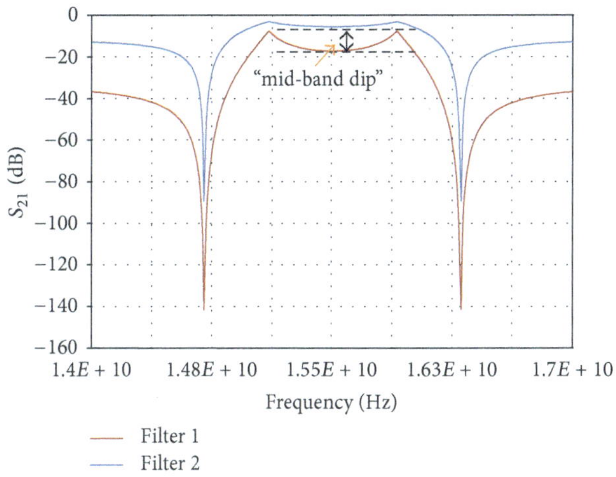


FIGURE 6: Comparison of transmission response ($S_{21}$) of 2nd order FBAR filter of Filter 1 and Filter 2.

15.5 GHz, insertion loss of 3 dB, and out-of-band rejection of 12 dB. In order to achieve the required specifications of FBAR filter, an analysis of the parameters that affect the characteristics of the filters has been carried out. The effects of varying $f_1$ and $f_2$ followed by the effect of varying $f_s^s$, $f_s^p$, $f_p^s$, and $f_p^p$ are presented in Sections 4.1 and 4.2. Finally, the effects of filter order $N$ on OoB rejection and filter characteristics are discussed in Section 4.3.

*4.1. Effect of $f_1$ and $f_2$.* The effect of $f_1$ and $f_2$ is shown in Figure 4, where BW$_c$ is equal to $f_2 - f_1$. Eight different filters are designed to investigate the effect. The bandwidth of each FBAR is set to 500 MHz and the difference between $f_s^s$ and $f_p^p$ is set to 600 MHz. Second-order filter characteristics are used for comparison. From Figure 4, it is observed that as the BW$_c$ increases, the BW$_{-3\text{ dB}}$ also increases. It is also observed that the insertion loss ($I_L$) has no significant change. Meanwhile the OoB rejection decreases significantly after BW$_c$ reaches 1.0 GHz and then remains constant. From this observation, we can conclude that to get wider BW$_{-3\text{ dB}}$, the $f_2 - f_1$(BW$_c$) should be increased to the required value.

*4.2. Effect of Varying $f_s^s$, $f_s^p$, $f_p^s$, and $f_p^p$.* Five different filters are designed to investigate the effect of varying $f_s^s$ while other parameters are constant and the obtained results are presented as dashed line in Figure 5. Firstly, the value of $f_p^p$ is kept constant to 15.2 GHz while $f_s^s$ varies from 15.5 GHz to 15.9 GHz with 100 MHz steps. The results in Figure 5 indicate that as the $f_s^s$–$f_p^p$ decreases, the 2nd order BW$_{-3\text{ dB}}$ also decreases.However, it is observed that the $I_L$ and the OoB are inversely proportional to $f_s^s$–$f_p^p$.

Secondly, the value of $f_s^s$ is kept constant to 15.2 GHz, while $f_p^p$ varies from 15.5 GHz to 15.9 GHz with 100 MHz steps. The results achieved are depicted as dotted line in Figure 5. It can be seen that as the $f_s^s$–$f_p^p$ decreases, the BW$_{-3\text{ dB}}$ decreases. It is also observed that $I_L$ improves, while the OoB degrades as the BW$_{-3\text{ dB}}$ decreases.

The effect of varying both $f_s^s$ and $f_p^p$ while other parameters are constant is shown as continuous line in Figure 5. It shows that as the $f_s^s$–$f_p^p$ decreases BW$_{-3\text{ dB}}$ also decreases. $I_L$ improves significantly as BW$_{-3\text{ dB}}$ decreases. On the other hand, the OoB degrades as BW$_{-3\text{ dB}}$ decreases.

Based on the analysis described in this section, varying the values of $f_s^s$, $f_s^p$, $f_p^s$, and $f_p^p$ will directly affect the filter characteristics. From Figure 5, it is observed that the BW$_{-3\text{ dB}}$ and OoB gradually decrease as the difference between $f_s^s$ and $f_p^p$ decreases. On the other hand, $I_L$ improves as $f_s^s$ and $f_p^p$ decrease.

Figure 6 presents the comparison of transmission response ($S_{21}$) of the filters designed in this analysis. It is observed that some of the filters have the "mid-band dip" at the centre due to the increasing gap between $f_s^s$ and $f_p^p$ of the FBARs. Hence, one of the purposes of this analysis is to minimise the "mid-band dip." Both filters have same values

Table 2: FBAR filters design parameters.

| Parameter | $f_1$ (GHz) | $f_2$ (GHz) | $f_s^s$ (GHz) | $f_p^s$ (GHz) | $f_s^p$ (GHz) | $f_p^p$ (GHz) | $BW_c$ (GHz) | $\psi$ |
|---|---|---|---|---|---|---|---|---|
| Filter A | 14.725 | 16.275 | 15.90 | 16.30 | 14.70 | 15.10 | 1.55 | 5.2 |
| Filter B | 14.75 | 16.25 | 15.90 | 16.30 | 14.70 | 15.10 | 1.50 | 11.2 |

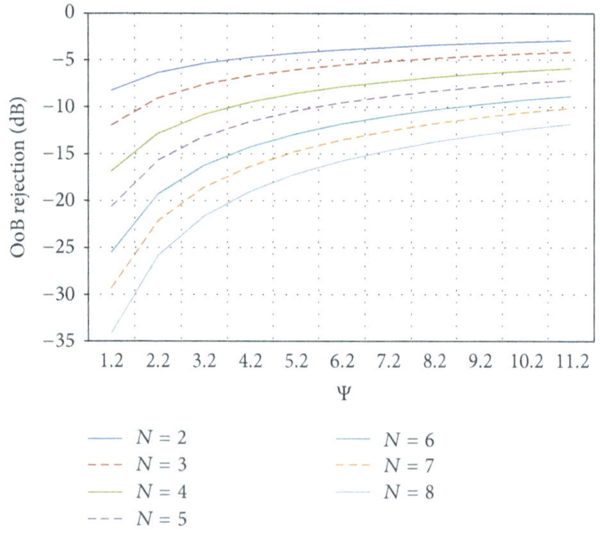


Figure 7: Out-of-band rejection (OoB) as a function of Ψ and $N$.

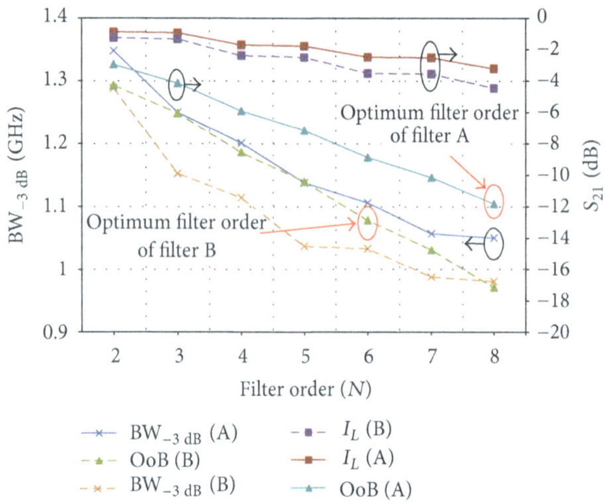


Figure 8: Characteristics of Ku-band FBAR filters.

of $f_s^s$, $f_s^p$, $f_p^s$, and $f_p^p$ but different values of $f_1$ and $f_2$. As mentioned previously, $BW_c = f_2 - f_1$. Filter 2 has the $BW_c$ of 1.55 GHz while Filter 1 has the $BW_c$ of 1.14 GHz. It can be seen that for Filter 2 the "mid-band dip" is less than 3 dB and the achieved $BW_{-3\,dB}$ for these filters is 1.0 GHz. The variation of $f_1$ and $f_2$ values has significant effect on $\psi$. The relationship of OoB, $\psi$, and $N$ depicted in Figure 7. From Figure 7 it can be observed that an increase in $\psi$ will degrade the OoB. As a result of this analysis, the FBAR filter design parameters listed in Table 2 have been chosen.

Next we will discuss optimisation of filter order for these filters.

*4.3. Effect of N.* The OoB of the FBAR filter only depends on $\psi$ and $N$. As $\psi$ is already determined previously, the $N$ is chosen to achieve the specified $BW_{-3\,dB}$. The effect of $N$ on filter characteristics of Filters A and B is presented in Figure 8. It is noticed that the $BW_{-3\,dB}$ and $I_L$ decrease gradually as $N$ increases. At the same time OoB steadily improves as $N$ increases. From this study, it is noticeably seen that even though increasing the filter order will improve the OoB, it will degrade $I_L$ and $BW_{-3\,dB}$ of the filter. The higher filter order requires more resonators which will result in increased size and cost. Hence, optimum value of filter order is selected to be 8th and 5th for filter A and filter B, respectively.

## 5. Results

From the analysis described above, it is possible to design FBAR filters with 1.0 GHz bandwidth operating at 15.5 GHz. The parameters used in designing the filter are presented in Table 2. The characteristics of each filter are summarised and compared in Figure 8. It is observed that Filter A has the advantage of wider $BW_{-3\,dB}$ and better $I_L$ compared to Filter B. However, the drawback of Filter A is the OoB rejection which is slightly higher than that of Filter B. Filter A has achieved the $BW_{-3\,dB}$ up to 1.35 GHz and extensive value of insertion loss, but the OoB rejection is too high. From all the analysis, it can be seen that FBAR filter with wider $BW_{-3\,dB}$ will degrade the OoB rejection of the filter, even though the $I_L$ is improved.

Table 3 shows a comparison of FBAR filters operating in X-band to Ka-band. All these filters were designed at 4-stage ladder filter which is equal to 8th order filter. The 7th and 8th order FBAR filter A and the 5th and 6th order FBAR filter B are chosen as the best filter characteristics in this work. It is shown that FBAR filters in this work have better performance than reported in [14], where the insertion loss is too high and the fractional bandwidth is only 2%. To date, this is the first FBAR filter operating in Ku-band frequency range ever reported.

Based on the characteristics of the designed FBAR filters, two sets of FBARs were designed using the BVD equivalent circuit. The derived equations for equivalent circuit elements of FBAR are as presented in [23]. The FBAR parameters extracted are listed in Table 4.

In order to further analyse and optimise the FBAR designs evaluated using numerical analysis, 3D Finite Element Method (FEM) was used. Tables 5 and 6 show the comparisons of the extracted FBAR parameters between the 1-D and 3-D FEM analyses for Filter A and Filter B. It can be seen that the parameter values are highly similar.

Figure 9 shows the transmission response ($S_{21}$) of the 3rd stage FBAR filter A designed using the 1-D modelling and 3-D FEM. The results from both analyses match well

TABLE 3: Comparison of FBAR filter characteristics.

| Reference | Centre frequency (GHz) | Insertion loss (dB) | Out-of-band (GHz) | Fractional bandwidth (%) |
|---|---|---|---|---|
| [13] | 29.2 | 3.8 | 11 | 3.4 |
| | 23.8 | 3.8 | 13 | 3.4 |
| [14] | 19.8 | 4.1 | 18 | 2.0 |
| [15] | 9.08 | 1.7 | 21 | 3.1 |
| Filter A* | 15.5 | 3.2 | 12 | 6.45 |
| Filter B* | 15.5 | 3.5 | 13 | 6.45 |

*indicates the FBAR filter characteristics in this work.

TABLE 4: Extracted FBAR parameters.

| Parameter | Filter A | | Filter B | |
|---|---|---|---|---|
| FBAR type | FBAR 1 (series) | FBAR 2 (shunt) | FBAR 1 (series) | FBAR 2 (shunt) |
| $C_o$ (pF) | 0.6872 | 0.0614 | 0.4681 | 0.0901 |
| $C_m$ (pF) | 0.0350 | 0.0379 | 0.0239 | 0.0258 |
| $L_m$ (nH) | 2.8613 | 34.616 | 4.2 | 23.582 |
| $k_{\text{eff}}^2$ (%) | 6.06 | 6.54 | 6.06 | 6.54 |
| $f_s$ (GHz) | 15.904 | 14.707 | 15.904 | 14.707 |
| $f_p$ (GHz) | 16.313 | 15.12 | 16.313 | 15.12 |

TABLE 5: Comparisons of 1D and 3D FEM of Filter A.

| Parameters | Series FBAR | | Shunt FBAR | |
|---|---|---|---|---|
| | 1D | 3D FEM | 1D | 3D FEM |
| $C_o$ (pF) | 0.6872 | 0.6837 | 0.0614 | 0.0651 |
| $C_m$ (pF) | 0.0351 | 0.0373 | 0.0379 | 0.0357 |
| $L_m$ (nH) | 2.86 | 2.69 | 34.616 | 32.897 |
| $k_{\text{eff}}^2$ (%) | 6.06 | 6.36 | 6.54 | 6.51 |
| $f_s$ (GHz) | 15.90 | 15.87 | 14.70 | 14.67 |
| $f_p$ (GHz) | 16.30 | 16.30 | 15.10 | 15.12 |

TABLE 6: Comparisons of 1D and 3D FEM of Filter B.

| Parameters | Series FBAR | | Shunt FBAR | |
|---|---|---|---|---|
| | 1D | 3D FEM | 1D | 3D FEM |
| $C_o$ (pF) | 0.468 | 0.471 | 0.0901 | 0.107 |
| $C_m$ (pF) | 0.024 | 0.025 | 0.058 | 0.058 |
| $L_m$ (nH) | 4.20 | 3.97 | 23.58 | 19.96 |
| $k_{\text{eff}}^2$ (%) | 6.06 | 6.36 | 6.54 | 6.51 |
| $f_s$ (GHz) | 15.90 | 15.87 | 14.70 | 14.67 |
| $f_p$ (GHz) | 16.30 | 16.29 | 15.10 | 15.12 |

within the pass-band. The attenuation poles of both filters also agree well. The disagreement on the suppression is due to piezoelectric material losses that were introduced in 3-D FEM. It can be concluded that the 1-D modelling used can be applied to predict the characteristics of the FBAR filter.

## 6. Conclusion

FBAR filter with insertion loss of 3.5 dB, out-of-band rejection of 12 dB, centre frequency of 15.5 GHz, and fractional

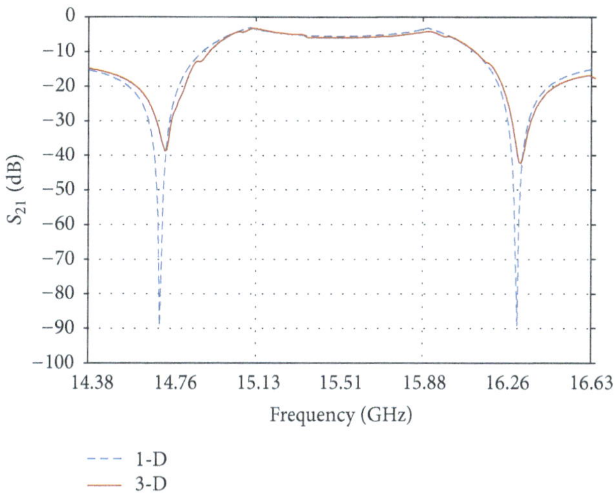


FIGURE 9: Comparison of FBAR filter transmission response ($S_{21}$) between 1D and 3D FEM.

bandwidth of 6.45% was designed using closed-form expressions. To best of the author's knowledge this analysis has for the first time shown that the closed-form expressions and 1-D models can be used to characterise wideband FBAR filters. From the analysis, we conclude that increasing the order of the FBAR filter will improve the out-of-band rejection; however the insertion loss and the 3 dB bandwidth will degrade. The results show that the proposed FBAR filter is a good candidate for implementation in a Ku-band transceiver.

## References

[1] M. F. Islam, M. A. Mohd. Ali, and B. Y. Majlis, "Ku-band band-pass filter using surface micromachined process," in *Proceedings of the Asia Pacific Microwave Conference (APMC '09)*, pp. 547–550, December 2009.

[2] L. Chang-Ho, A. Sutono, H. Sangwoo et al., "A compact LTCC-based Ku-band transmitter module," *IEEE Transactions on Advanced Packaging*, vol. 25, no. 3, pp. 374–384, 2002.

[3] H. Su, M. Xing, Y. Li, and W. Li, "Design of a front-end of a Ku-band transceiver based on LTCC technology," in *Proceedings of the 11th International Conference on Electronic Packaging Technology and High Density Packaging (ICEPT-HDP '10)*, pp. 851–855, August 2010.

[4] D. Kim, D. H. Kim, G. H. Baek et al., "An FBAR-and LTCC-based RF front-end module for Wi-Fi and WiMAX dual-mode

systems," *Microwave and Optical Technology Letters*, vol. 52, no. 3, pp. 753–757, 2010.

[5] G. G. Fattinger, J. Kaitila, R. Aigner, and W. Nessler, "Thin film bulk acoustic wave devices for applications at 5.2 GHz," in *Proceedings of the IEEE Ultrasonics Symposium*, pp. 174–177, October 2003.

[6] H. C. Pineda, *Thin Film bilk acoustic wave resonator-FBAR fabrication, heterogeneous integration with CMOS technologies, and sensor applications [Ph.D. thesis]*, Universite de Montpellier II, Montpellier, France; Universitat Autonoma de Barcelona, Barcelona,Spain, 2007.

[7] K.-W. Kim, G.-Y. Kim, J.-G. Yook, and H.-K. Park, "Air-gap-type TFBAR-based filter topologies," *Microwave and Optical Technology Letters*, vol. 34, no. 5, pp. 386–387, 2002.

[8] K. Umeda, H. Kawamura, M. Takeuchi, and Y. Yoshino, "Characteristics of an AlN-based bulk acoustic wave resonator in the super high frequency range," *Vacuum*, vol. 83, no. 3, pp. 672–674, 2008.

[9] M. Hara, T. Yokoyama, T. Sakashita et al., "Super-high-frequency band filters configured with air-gap-type thin-film bulk acoustic resonators," *Japanese Journal of Applied Physics*, vol. 49, no. 7, Article ID 07HD13, 2010.

[10] N. I. M. Nor, K. Shah, J. Singh, N. Khalid, and Z. Sauli, "Design and analysis of film bulk acoustic wave resonator in Ku-band frequency for wireless communication," in *Proceeding of SPIE, Active and Passive Smart Structures and Integrated Systems 2012, 83411R*, vol. 8341, 2012.

[11] F. Z. Bi and B. P. Barber, "Bulk acoustic wave RF technology," *IEEE Microwave Magazine*, vol. 9, no. 5, pp. 65–80, 2008.

[12] A. A. Shirakawa, J.-M. Pham, P. Jarry, and E. Kerhervé, "Design of FBAR filters at high frequency bands," *International Journal of RF and Microwave Computer-Aided Engineering*, vol. 17, no. 1, pp. 115–122, 2007.

[13] M. Hara, T. Yokoyama, T. Sakashita, M. Ueda, and Y. Satoh, "A study of the thin film bulk acoustic resonator filters in several ten GHz band," in *Proceedings of the IEEE International Ultrasonics Symposium (IUS '09)*, pp. 851–854, September 2009.

[14] M. Hara, T. Yokoyama, M. Ueda, and Y. Satoh, "X-band filters utilizing AlN thin film bulk acoustic resonators," in *Proceedings of the IEEE Ultrasonics Symposium (IUS '07)*, pp. 1152–1155, October 2007.

[15] T. Yokoyama, M. Hara, M. Ueda, and Y. Satoh, "K-band ladder filters employing air-gap type thin film bulk acoustic resonators," in *Proceedings of the IEEE International Ultrasonics Symposium (IUS '08)*, pp. 598–601, November 2008.

[16] G. F. Perez-Sanchez and A. Morales-Acevedo, "Design of bulk acoustic wave resonators based on ZnO for filter applications," in *Proceedings of the 6th International Conference on Electrical Engineering, Computing Science and Automatic Control (CCE '09)*, pp. 1–6, November 2009.

[17] O. Menéndez, P. De Paco, R. Villarino, and J. Parrón, "Closed-form expressions for the design of ladder-type FBAR filters," *IEEE Microwave and Wireless Components Letters*, vol. 16, no. 12, pp. 657–659, 2006.

[18] C. Min-Chiang, H. Zi-Neng, P. Shih-Yung, Z. Wang, and C. S. Lam, "Modified BVD-equivalent circuit of FBAR by taking electrodes into account," in *Proceedings of the IEEE Symposium on Ultrasonics*, pp. 973–976, October 2002.

[19] J. Jianhua, L. Yinqiao, H. Sha, J. Zhou, Y. Fei, and F. Tan, "A compact LTCC Transmit Receive module at Ku-band," in *Proceedings of the International Conference on Microwave and Millimeter Wave Technology (ICMMT '10)*, pp. 1239–1241, May 2010.

[20] R. Gilmore, C. Bruckner, and D. Dunn, "Ku-band MMIC transceiver for mobile satellite communications," in *Proceedings of the 21st European Microwave Conference*, pp. 1261–1265, September 1991.

[21] Z. Zhang, G. Zhao, X. Zeng, H. Sun, and X. Lv, "A transceiver for Ku band digital monopulse radar system," in *Proceedings of the International Conference on Microwave and Millimeter Wave Technology (ICMMT '10)*, pp. 979–982, May 2010.

[22] A. Shirakawa, J. M. Pham, P. Jarry, E. Kerherve, and E. Hanna, "Ladder-type FBAR filter synthesis methodology," in *Proceedings of the 19th Conference on Design of Circuits and Integrated Systems (DCIS '04)*, pp. 519–523.

[23] K. Kun-Wook, G. Myeong-Gweon, Y. Jong-Gwan, and P. Han-Kyu, "Resonator size effects on the TFBAR ladder filter performance," *IEEE Microwave and Wireless Components Letters*, vol. 13, no. 8, pp. 335–337, 2003.

# 8

# A 12 GHz 30 mW 130 nm CMOS Rotary Travelling Wave Voltage Controlled Oscillator

G. Jacquemod,[1] F. Ben Abdeljelil,[2] L. Carpineto,[3] W. Tatinian,[2] and M. Borgarino[4]

[1] *University Nice Sophia Antipolis, IM2NP UMR CNRS 7334, 06410 Biot, France*
[2] *University Nice Sophia Antipolis, LEAT UMR CNRS 6071, 06540 Valbonne, France*
[3] *SDRF, Sophia Antipolis, 06410 Biot, France*
[4] *University of Modena and Reggio Emilia, 41100 Modena, Italy*

Correspondence should be addressed to G. Jacquemod, gilles.jacquemod@unice.fr

Academic Editor: Ulrich L. Rohde

This paper reports a 12 GHz rotary travelling wave (RTW) voltage controlled oscillator designed in a 130 nm CMOS technology. The phase noise and power consumption performances were compared with the literature and with telecommunication standards for broadcast satellite applications. The RTW VCO exhibits a −106 dBc/Hz at 1 MHz and a 30 mW power consumption with a sensibility of 400 MHz/V. Finally, requirements are given for a PLL implementation of the RTW VCO and simulated results are presented.

## 1. Introduction

In a Ku-band satellite receiver, usually the signal picked up by the dish antenna is amplified by a low noise block (LNB) designed with compound semiconductor HEMT (high electron mobility transistor) and the local oscillator is a dielectric resonator oscillator (DRO) [1]. Recently, efforts have been made to design a satellite receiver using CMOS technologies [2, 3]. In [2] the design was focused on the receiver chain only (LNA and mixer) while in [3] the whole LNB was addressed, where the receiver exhibits a traditional superheterodyne and the local oscillator is a traditional differential VCO. Because of the image frequency issue, the RF designer should introduce image rejection filters, facing therefore all the related troubles.

Solutions alternative to the traditional approach (superheterodyne architecture plus image rejection filter) can be envisaged at architectural level in two ways: an image rejection architecture or a direct conversion architecture. In the former solution the architecture is an interferometric structure where the image frequency signal destructively interferes with itself while the desired RF signal constructively interferes with itself. In the latter solution image and RF signal coincide. Both above cited solutions require a demodulator configuration where a couple of mixers are driven by differential in quadrature signals. Their generation is usually obtained using quadrature VCOs, as preferred solution with respect to polyphase filters, ring oscillators, or frequency dividers.

In particular, in the present paper, a rotary travelling wave voltage controlled oscillator (RTW VCO) in a 130 nm CMOS technology is investigated having in mind the idea of improving the Ku-band satellite receiver architecture by replacing the previously sketched out traditional superheterodyne architecture with an image rejection architecture, so that the image frequency rejection filters can be avoided in the receiver design.

In the design of the VCO attention should be paid to the phase noise, which is a very stringent specification for the satellite broadcasting, because of the use of amplitude and phase shift keying (APSK) modulation schemes that make the constellation round and therefore prone to suffer from cycle slips if the phase noise of the local oscillator is too high.

In order to optimize the phase noise performances of the RTW VCO, an impulse sensitivity function (ISF) based model and the design guidelines for required transmission line are proposed.

Table 1: PLL specifications.

| Characteristics | Specifications |
|---|---|
| Frequency band of the input signal | 12.2–12.7 GHz |
| Output frequency (IF) | 950–1450 MHz |
| Local oscillator frequency | 11.25 GHz |
| Local oscillator stability | <1.5 MHZ (30°C–60°C) |
| Phase noise | < −95 dBc/Hz at 100 kHz<br>< −115 dBc/Hz at 1 MHz |

The paper is organized as follows. In Section 2, the PLL specifications, its architecture, and its different blocks are presented. Section 3 deals with the RTW oscillator architecture and the corresponding line theory is proposed. Phase noise considerations are presented in Section 4. Section 5 is dedicated to the measurement results and Section 6 is devoted to the design and simulation of the PLL performance, in terms of transient response and phase noise error. Finally, the paper ends with drawing some conclusions in Section 7.

## 2. PLL Specifications and Design

Figure 1 depicts a Ku-band satellite heterodyne receiver. In this paper, we suggest replacing the DRO of local oscillator (LO), which has a very good stability but a very high cost, with a CMOS PLL. The first generation architecture for satellite receiver was not image rejection mixer. Nevertheless, a QVCO (RTW) is chosen to realize this PLL in order to implement image rejection architecture for the mixer, easing the requirement for the input filter and it is not presented in this paper. The specifications of this PLL are summarized in Table 1. The second oscillator (IF to base band, BB) is a classical PLL, with a tuning range from 1 to 1.5 GHz, without any specific design challenges.

For the band translation switch satellite application, settling time is not a mandatory requirement as in GSM or any standard using hopping (as Bluetooth). Nevertheless, the PLL itself would lock within five times the loop time constant. The loop bandwidth was set for best noise performances to be around 550 to 600 kHz.

The chosen PLL architecture is a classical analog PLL topology, as depicted in Figure 2. The reference frequency is generated by a 50 MHz quartz oscillator. The satellite application forces the output frequency to be around 12 GHz [4].

With the exception of the RTW VCO (described in Section 3), we present in the following paragraphs the various blocks of the PLL.

To divide by $N$ the output frequency, a series of three dividers are used due to the relatively high value of the rand division ($N = 12\,\text{GHz}/50\,\text{MHz}$). Indeed, depending of the working frequency, different divider topologies can be used to get better efficiency [5]. As a consequence, the first divider is a fixed division (by 4 in our case) [6], the second one is a CML [7] divider by 8, and the last one is a CML dual modulus 7/8 frequency divider.

The phase/frequency detector (PFD) is a conventional one. It is based on D-type registers, customized to limit the dead zone. This PFD reduces the nonlinear transfer characteristic. It produces a narrow pulse during the time difference between the rising edge of the reference signal and the signal from the frequency divider in the PLL feedback loop. The use of a charge pump naturally adds a pole at the origin in the loop transfer function of the PLL, since the charge pump current is driven into a filter to generate a voltage. This additional pole integrates the error signal causes the system to track the input with one more order. The charge pump consists of pull-up and pull-down transistors, driving a 1 mA charge pump current.

The third order loop filter has been chosen external (see Figure 3). The resistance and capacitance values are computed according to the following equations developed in [8]:

$$T_1 = \frac{\cos^{-1}(\text{PM}) - \tan(\text{PM})}{\text{BW}}, \quad (1)$$

$$T_2 = \frac{1}{\text{BW}^2 * T_1}, \quad (2)$$

$$C_1 = \frac{T_1 * K_{\text{VCO}} * K_\phi * \sqrt{\left(1 + (\text{BW}*T_2)^2\right)/\left(1 + (\text{BW}*T_1)^2\right)}}{T_2 * \text{BW}^2 * N}, \quad (3)$$

$$C_2 = C_1 \frac{T_2}{T_1 - 1}, \quad (4)$$

$$R_2 = \frac{T_2}{C_2}, \quad (5)$$

$$C_3 = \frac{C_1}{2}, \quad (6)$$

$$R_3 = \frac{R_2}{2}, \quad (7)$$

where PM is the phase margin, BW is the open loop bandwidth, $K_\phi$ is the charge pump current/PFD gain, $K_{\text{VCO}}$ is the VCO gain, and $N$ is the rand division.

## 3. RTW VCO Architecture and Design

The RTW architecture is a reasonable alternative compared to the LC tank classical topology at frequencies above 10 GHz. Most particularly, it provides a more compact design and a lower phase noise at the cost of a higher consumption [9]. The principle of such a VCO is based on a distributed amplifier, with wave amplification over a transmission line (constructive amplification) where the output (2) is connected to the input (1), as shown in Figure 4. In this way, the oscillating frequency is given by

$$f_{\text{osc}} = \frac{v_{\text{phase}}}{2NL} = \frac{1}{\sqrt{L_T C_T}}, \quad (8)$$

where $v_{\text{phase}}$ is the phase velocity, $L = l_g = l_d$ the distance of the transmission line between two amplification stages, and

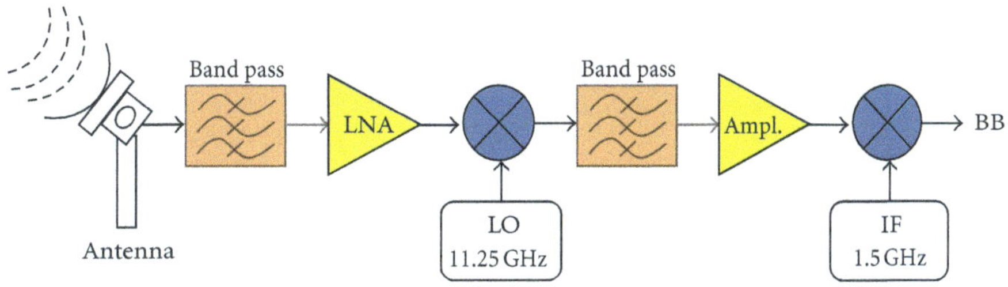


FIGURE 1: Typical satellite receiver block diagram.

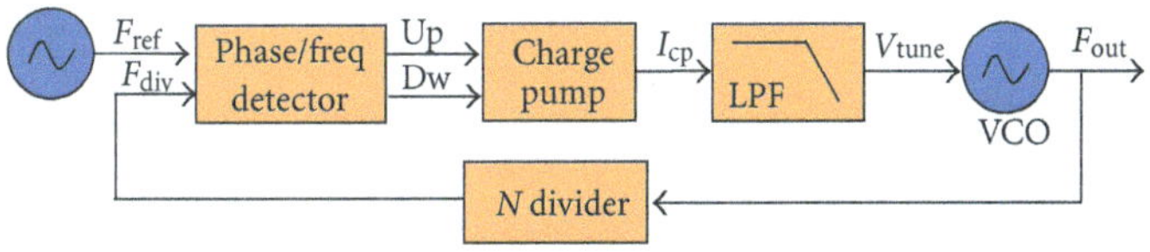


FIGURE 2: PLL architecture.

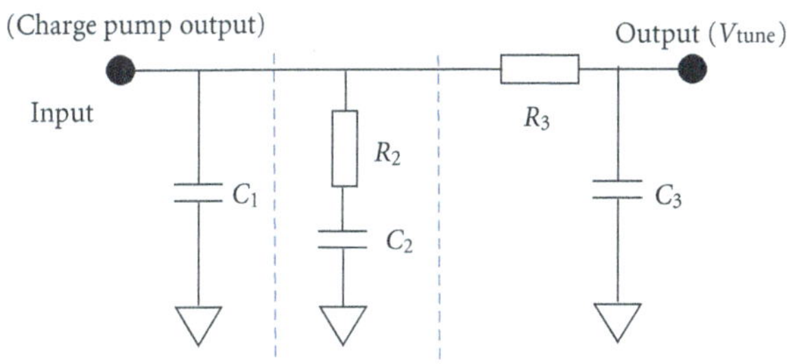


FIGURE 3: Loop filter schematic.

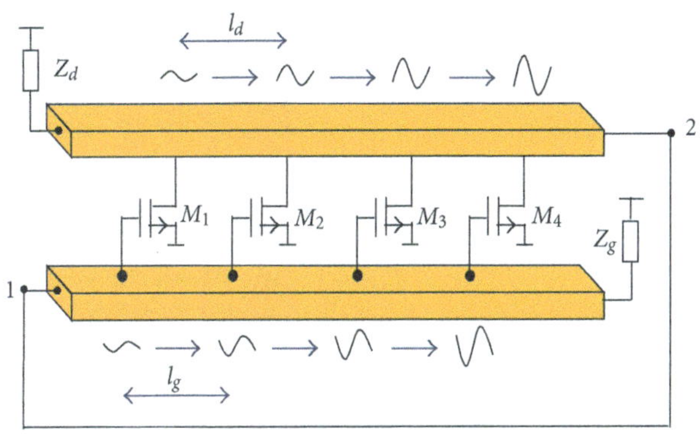


FIGURE 4: Basic RTW oscillator.

$N$ the number of stages. $L_T$ represents the total inductance of the line, and $C_T$ the capacitance.

In order to avoid the dissipation of half power in terminal $Z_g$ and $Z_d$, a double crossing line transmission is realized. The corresponding topology is proposed in Figure 10. The amplifiers are realized by CMOS inverters loaded by varactors and switched capacitors in order to reach the desired VCO gain, $K_{VCO}$. Four amplification stages are used in order to realize a QVCO (Figure 5).

To investigate more in detail the proposed architecture and provide an accurate design method, it is essential to gain a good understanding of the transmission lines theory. As discussed in [9], the oscillating frequency is determined by the odd mode propagation along the line. The characteristic impedance and propagation constant have to be replaced by the ones calculated in differential mode, $Z_{0diff}$ and $\gamma_{diff}$, respectively. Neglecting the losses in the metal and dielectric layers, $Z_{0diff}$ can be written as

$$Z_{0diff} = \sqrt{\frac{L_{diff}}{C_{diff}}} = \sqrt{\frac{L - M}{C + 2C_f}}, \tag{9}$$

where $L_{diff}$ and $C_{diff}$ are the inductance and capacitance per unit length of the line in differential mode and $L$, $M$, $C$ and $C_f$, are defined as in Figure 6.

In the case of MOS technologies the bottom metal is used to fabricate the ground plane and the top metal or a metal stack is used for the transmission line [10]. This configuration reduces the coupling between lines and ground allowing higher oscillating frequencies, as a consequence. Moreover, as the top metal is usually the thicker one, its use reduces the series resistance providing the advantage of a higher quality factor [11].

In a first approximation, the inductances and capacitances can be computed using Greenhouse [12] and Sakurai and Tamaru [13] formulas. The self- and mutual inductances can be approximated by

$$\begin{aligned} L &= \frac{\mu_0}{2\pi} l \left[ \ln\left(\frac{1}{w+t}\right) + 1.193 + \frac{w+t}{3l} \right], \\ M &= 2l \ln\left( \frac{1}{\text{GMD}} + \sqrt{1 + \left(\frac{l}{\text{GMD}}\right)^2} \right) \\ &\quad - \sqrt{1 + \left(\frac{\text{GMD}}{l}\right)^2} + \frac{\text{GMD}}{l}, \end{aligned} \tag{10}$$

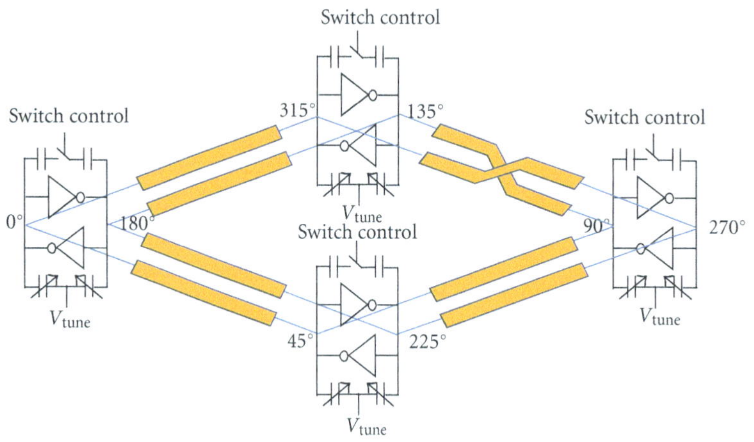


FIGURE 5: RTW VCO topology.

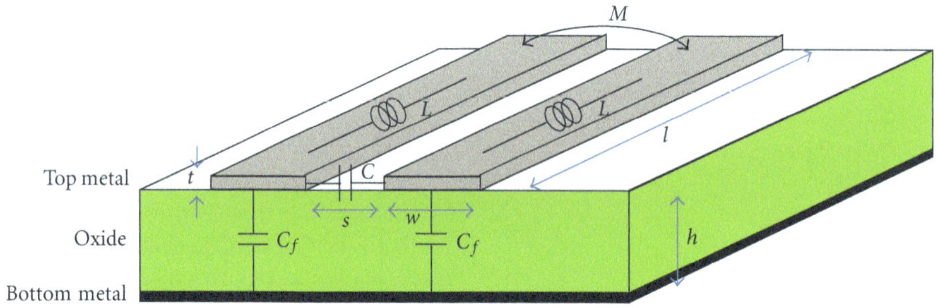


FIGURE 6: Metal cross-section.

where $l$, $t$, and $w$ are the line length, metal thickness, and width, respectively, and GMD is the geometrical mean distance that can be approximated by

$$\mathrm{GMD} = \exp\left[\log(l) - \left[\frac{1}{12}\left(\frac{w}{l}\right)^2 + \frac{1}{60}\left(\frac{w}{l}\right)^4 + \frac{1}{168}\left(\frac{w}{l}\right)^6\right]\right]. \tag{11}$$

The capacitances per unit length are estimated through the following formulas:

$$C = \varepsilon_0\varepsilon_r\left[0.03\left(\frac{w}{h}\right) + 0.83\left(\frac{t}{h}\right) - 0.07\left(\frac{t}{h}\right)^{0.222}\right]\left(\frac{s}{h}\right)^{-1.34},$$
$$C_f = \frac{\varepsilon_0\varepsilon_r}{4}M\sqrt{1-\left(1+\frac{2w}{s}\right)^{-2}}, \tag{12}$$

where $\varepsilon_0$, $\varepsilon_r$, $s$, and $h$ are the vacuum permittivity, the dielectric constant, the spacing between the lines, and the dielectric thickness (see Figure 6), respectively, and $M$ is an elliptic integral of the second kind.

The above equations are guaranteed with an accuracy of about 5 to 10%. If a more accurate calculation is required, one should carry out electromagnetic simulations and then extract an RLGC (or simply LC) compact model compatible with SPICE-like simulators. This entire procedure can become very time consuming.

As the RLC values of the line simultaneously affect the quality factor ($Q$), which is proportional to the inductance and degraded by the series resistance, and the characteristic impedance, the first step is to choose the line shape that reduces the silicon area and respects symmetry considerations. The second step consists in choosing the spacing, width, and length in order to optimize $Q$ and $Z_c$.

It has been demonstrated that the characteristics impedance increases as a function of the line spacing and decreases as a function of the width. On the contrary, the quality factor reaches an optimal value for given spacing and width. So, the width has to be optimized to get $Q$ as high as possible; using a 130 nm standard CMOS technology, the optimized width is 5 $\mu$m [4]. The drawback is that the characteristic impedance is not very close to the optimal value. This trouble can be fixed by enlarging the spacing to maximize the characteristic impedance almost without impact on $Q$. As a best trade-off with the line size, we chose a spacing of 14 $\mu$m [4].

Finally, the line length was chosen equal to 1500 $\mu$m, to get the right amount of total line inductance and capacitance for the desired oscillation frequency.

For simulation purpose, as previously cited, one should extract a RLCG compact model of the line suitable to be adopted in SPICE-like simulator as done by Hsieh et al. [14]. In our case, the line has been optimized to reduce the total space on silicon. The corresponding layout makes it difficult to use theoretical equations accurately.

Practical experience and mismatch between theoretical and simulated scattering parameters clearly suggest the interest of performing electromagnetic simulations (using HFSS) [4].

Once achieved the target oscillation frequency $f_{OSC}$, varactors, and capacitances can be designed. Moving from the frequency tuning range, the highest allowed $K_{VCO}$, and the available tuning voltage (1.2 V), the desired frequency tuning range was divided into 500 MHz subbands.

The choice of the switched capacitor value ($C_{on}$) is carried out according to quality factor $Q$ considerations. For a switch transistor and a capacitor connected in series, $Q$ is equal to $(R_{on}C_{on}\omega)^{-1}$, where $R_{on}$ is the transistor series resistance. $C_{on}$ has therefore to be not too high to have a good quality factor and not too low to limit the effect of the switch parasitic capacitors ($C_{off}$). In the same way, the switch width has to be chosen to get a good trade-off between $R_{on}$ and parasitic capacitor. In practice, the quality factor of the network should be set to a value higher than the one imposed by the resonator, which is the bottleneck for the quality factor in the RTW oscillator. Eventually, the $C_{on}/C_{off}$ ratio is chosen close to 2. A 100 MHz step between each characteristic leads to 16-switched-capacitor network.

## 4. Phase Noise Analysis

The phase noise in LC oscillators has been studied for many years [15–17]. However, for the RTW architecture, the theory is slightly different. Legrand de Mercey proposed an approach to compute the phase noise based on the approximation that the output signal is a pure square wave [9]. In our case, the coupled lines act as a band-pass filter and the signal is more sinusoidal. This is mostly the case when the number of inverters becomes low enough and the cut-off frequency of the line decreases.

The theory we propose here is a generalization of Mercey's work to any shape of output signal and is based on the computation of the impulse sensitivity function (ISF) introduced by Hajimiri and Lee as a general theory of phase noise in electrical oscillators [15]. In this kind of approach to the phase noise analysis, the ISF has to be exactly known and thus a transient simulation has to be run prior to the phase noise computation. The ISF, in the following indicated as $\Gamma$, describes the phase deviation in the periodic signal due to a current impulse $i_n(t)$. The phase deviation is maximum (minimum) if $i_n(t)$ occurs close to the zero-crossing (peak) instant of the periodic signal. The phase shift $\varphi_{out}$ due to $i_n(t)$ is described by the following formula:

$$\varphi_{out}(t) = \frac{1}{q_{max}} \int_{-\infty}^{t} \Gamma(2\pi f_0 \tau) \cdot i_n(\tau) d\tau, \quad (13)$$

with $q_{max}$ being the maximal charge stocked by the oscillator. The corresponding phase noise is

$$L(\Delta f) = 10 \log \left( \frac{\Gamma_{RMS}^2 \overline{i_{nw}^2(\Delta f)}}{2 q_{max}^2 (2\pi \Delta f)^2} \right), \quad (14)$$

where $\Delta f$ is the frequency offset, $\Gamma_{RMS}$ is the root-mean-squared value of the ISF function and $i_{nw}$ is the power spectral density of the current impulses assumed to be white noise current.

Assuming the oscillator as a second order system, the ISF is

$$\Gamma = \frac{g'}{g'^2 + g''^2}, \quad (15)$$

where $g'$ and $g''$ are the first and second derivatives of the normalized waveform of the signal generated by the oscillator.

Moving from these equations, White and Hajimiri [18] proposed to recompute the characteristic impedance including also the discrete capacitances contributed by the varactors and the switched capacitors. It is here worth noticing that this is valid as long as the distributed condition is respected (e.g., for high number of inverters).

The total capacitance $C_T$ is then given by

$$C_T = C_{diff} + C_{active}, \quad (16)$$

where $C_{diff}$ is the line capacitance in differential mode and $C_{active}$ is the sum of the input and output capacitances of the inverters.

This value of $C_T$, once inserted into (9) in the place of $C_{diff}$, leads to a new expression of the adjusted characteristic impedance ($Z_{0T}$).

In order to compute the total phase noise of the RTW oscillator, $q_{max}$ is expressed as a function of the output signal amplitude ($A$) and $C_T$:

$$q_{max} = A C_T. \quad (17)$$

The obtained phase noise expression is given by

$$L(\Delta f) = 10 \log \left( \frac{\Gamma_{RMS}^2 Z_{0T}^2 f_0^2 \overline{i_n^2(\Delta f)}}{2 q_{max}^2 (2\pi \Delta f)^2} \right). \quad (18)$$

Finally, the evaluation of the phase noise requires to know the noise spectral density of the noise sources introduced by the active devices while the thermal noise contribution of the line can be calculated using

$$\overline{i_{line}^2(\Delta f)} = \frac{4kT}{\mathrm{Re}(Z_{0T})}. \quad (19)$$

Concerning the contribution of active devices, assuming all the transistors exhibit the same transconductance ($g_m$), the noise contribution of each transistor is provided by the following expression:

$$\overline{i_{transistor}^2(\Delta f)} = 4kT\gamma g_m. \quad (20)$$

So, the total contribution from the $N$ inverters is given by

$$\overline{i_{active}^2(\Delta f)} = 2N \overline{i_{transistor}^2(\Delta f)}. \quad (21)$$

Adding (19) and (21) one obtains the total white noise power spectral density:

$$\overline{i_{nw}^2(\Delta f)} = 4kT \left( N\gamma g_m + \frac{1}{Z_{0T}} \right). \quad (22)$$

The substitution of this expression into (18) leads to the following expression of the phase noise:

$$L(\Delta f) = 10\log\left(\frac{\Gamma_{\mathrm{RMS}}^2 Z_{0T} f_0^2 4kT(1 + N\gamma g_m Z_{0T})}{2q_{\max}^2 (2\pi\Delta f)^2}\right). \quad (23)$$

In practice, it is worth noticing that a first transient simulation is required to determine the signal amplitude and the RMS value of the impulse sensitivity function while all the other parameters are known prior to the design. We can notice that this formula (23) does not take into account the flicker noise, which is due to the nonlinear capacitors of the circuit. The frequency transition between $1/f^3$ and $1/f^2$ (the flicker knee is mainly fixed by the capacity of the transmission line) is made having relatively low frequency offset below 100 kHz. In fact, the nonlinear capacitors in the circuit are mainly varactors and parasitic capacitors of the transistors (inverters). During the design of the circuit, we have try to reduce at most the varactors (and thus the nonlinear capacitors), for another reason than the flicker noise, to reduce the $K_{\mathrm{VCO}}$ with the switched capacitors. By having lowering the flicker noise knee, and by considering the high-pass filtering of the PLL on the VCO noise, the integrated phase noise can be considered as independent from the flicker noise in our case.

In (23), the $\Gamma_{\mathrm{RMS}}$ decreases with the number of invertersis as shown previously, while the term $N \cdot g_m$ represents the total required transconductance and is constant whatever the number of invertersis. As a consequence, the total phase noise will decrease when $N$ increases. This is verified in Figure 7(a) where the ISF function of the output signal for VCOs with 2, 4, 8, or 16 inverters pairs is shown. The corresponding phase noise simulations are given in Figure 7(b).

We can observe that when $N$ increases from 8 to 16 the phase noise variation is lower than expected, which is due to the inverters slew rate that prevents the output signal from being a square wave, so that the $\Gamma_{\mathrm{RMS}}$ tends to reach a minimal value.

## 5. Implementation and Measured Results

A RTW VCO has been designed and fabricated in a bulk 130 nm CMOS technology using the previously described phase noise modeling and line optimization as a design guideline. The microphotography of the prototype is shown in Figure 8 where $V_{\mathrm{tune}}$ is the varactor tuning voltage and $V_{\mathrm{comp}}$ is injected in an analog to digital converter to generate the switch word SW. This latter can be checked thank to the SW test outputs. The total size is $300 \times 300\,\mu\mathrm{m}^2$. With all the switches set on, that is, with the digital word controlling the switched capacitor network set to "0000", the RTW VCO exhibited a $f_{\mathrm{OSC}} = 11.24$ GHz and delivered about −30.7 dBm (see Figure 9) to a load of 50 Ω by drawing 25 mA from a 1.2V bias.

The phase noise measurements were carried out with an Agilent E5500 phase noise meter. Under the same switch configuration as before, the measured phase noise is shown in Figure 10. At 1 MHz frequency offset of the central frequency $f_{\mathrm{OSC}} = 11.24$ GHz, the measured phase noise is −105 dBm/Hz instead of the −107 dBm/Hz value from the simulation (−109 dBm/Hz value computed from the theory developed in Section 4). Because of the intrinsic hard nature of both the nonlinear phenomena involved in the generation of the phase noise and of the experimental difficulties affecting the phase noise measurement, it is the authors' opinion that the measured value has to be considered in well agreement with the numerical simulation and especially with the analytical computation. It is worth pointing out that the achieved phase noise value sounds very promising when compared with the phase noise specifications required by satellite applications for the low noise block oscillator. On the basis of the electromagnetic simulations carried out in the present work, it is the authors' opinion that the lacking 10 dB, or at least a large fraction of them, can be achieved by making thicker the top metal layer used to design the line. In particular, it can be estimated that a thickness of 4 $\mu$m would be enough to completely fulfill the phase noise specifications of −115 dBc/Hz at 1 MHz offset from the carrier, leading to a 1.2° RMS integrated phase noise which has been commonly admitted by the design community as a reasonable specification.

Figure 11 depicts the experimental dependence of the output frequency on the tuning voltage measured for all the possible configuration of the switched capacitor network. The lowest (highest) frequency band is obtained by setting the digital word controlling the switched capacitor network equal to "1111" ("0000"). The total frequency coverage is 1.2 GHz divided into 16 parts and the maximal value for the $K_{\mathrm{VCO}}$ is 400 MHz/V.

Figure 12 compares through the following Figure Of Merit (FOM) the RTW-VCO reported in the present paper with the state-of-the-art of distributed oscillators [9, 14, 19–23]:

$$\mathrm{FOM} = 10 \cdot \log\left(\left(\frac{f_{\mathrm{osc}}}{f_m}\right)^2 \frac{f_{\mathrm{osc}}}{L(\Delta f) \cdot P_{\mathrm{DC}}}\right) + 10\log(T_{\%}). \quad (24)$$

In the frequency range of interest, the oscillator reported in the present paper well compares with the performances claimed by other authors in the literature. In particular, it is worth pointing out that the FOM achieved in the present paper with a VCO fabricated in 0.13 $\mu$m CMOS technology is fairly close to the FOM achieved by a VCO fabricated in the more expensive 65 nm technology. In addition, in Figure 12, it is also reported an estimation of the FOM achievable if a top metal layer of 4 $\mu$m (RF CMOS process) was used. Measured characteristics of the RTW VCO are listed in Table 2.

## 6. PLL Implementation

Since the VCO is used in a PLL, the previous measured phase noise spectrum depicted has been adopted in a PLL linear Matlab model [24]. The cut-off frequency of the loop filter has been designed having in mind the minimization of the PLL integrated phase noise. Figure 13 plots the integrated

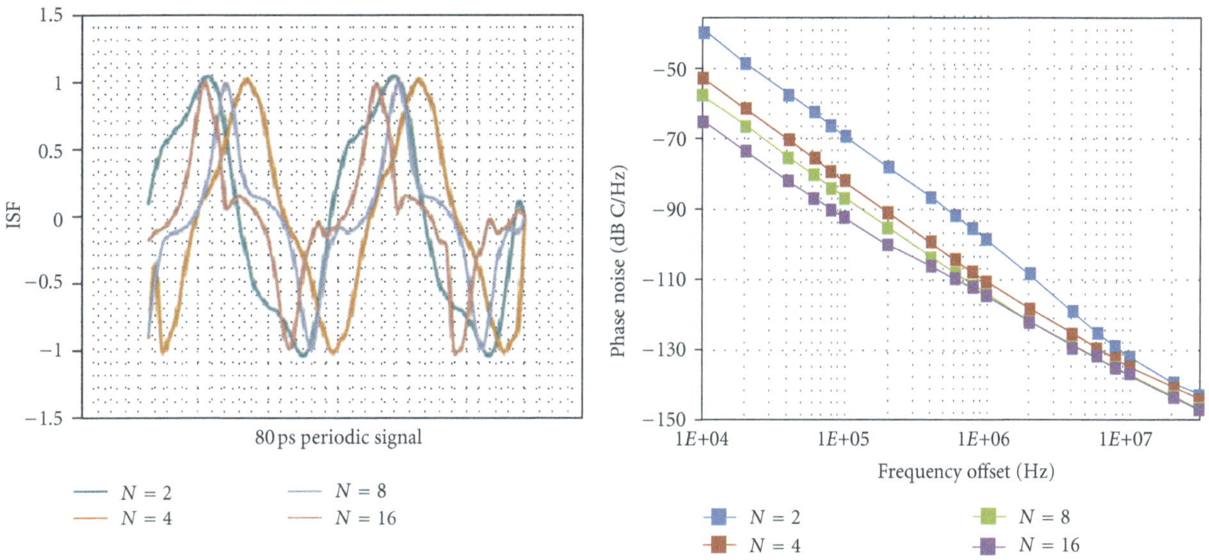


(a) ISF function

(b) Simulated phase noise

Figure 7: ISF functions and phase noise for $N = 2, 4, 8$, or 16 amplifiers.

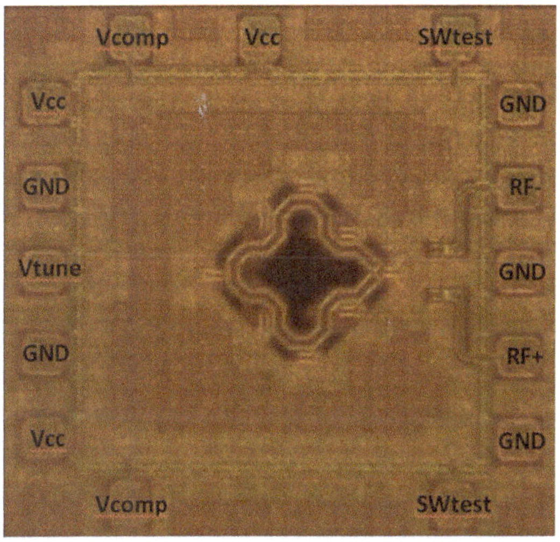


Figure 8: RTW VCO microphotography.

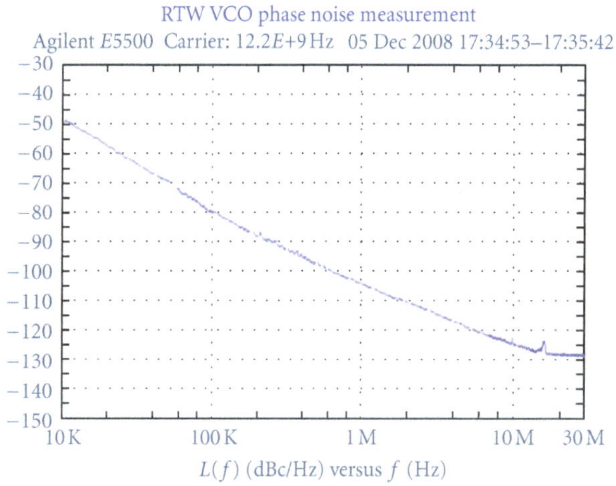


Figure 10: RTW VCO phase noise measurement.

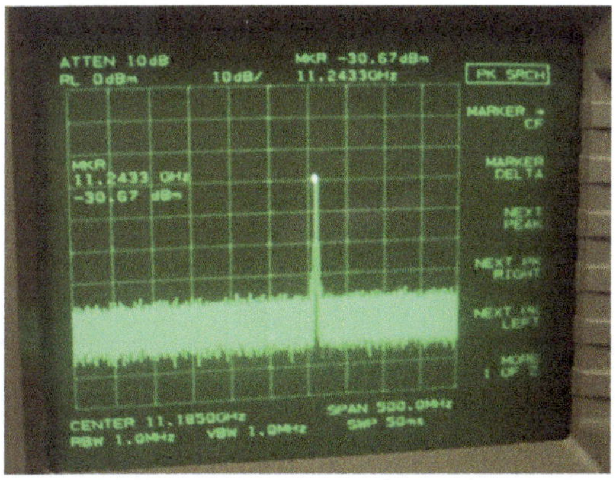


Figure 9: Frequency measurement.

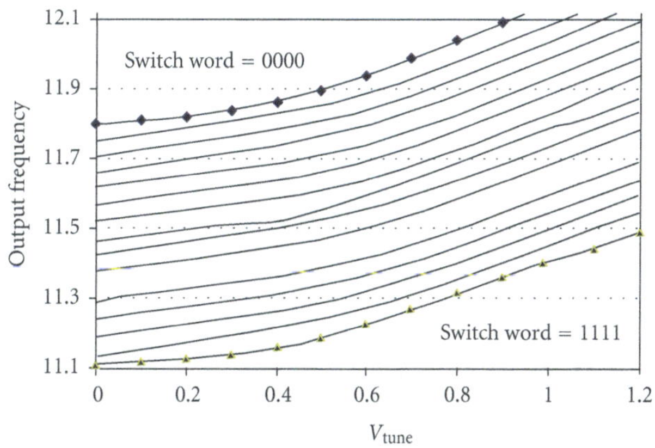


Figure 11: Output frequency versus $V_{\text{tune}}$.

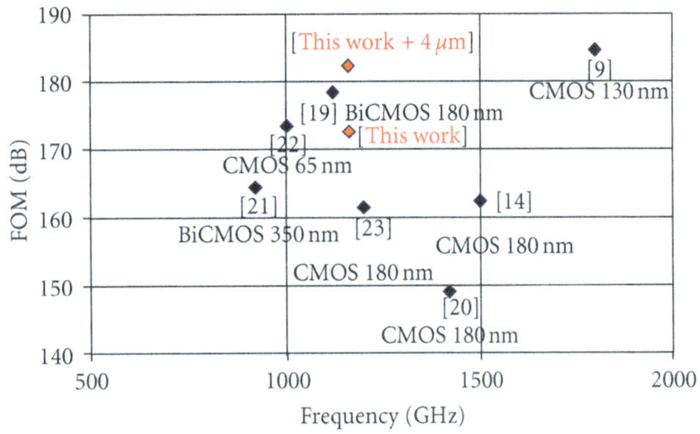


FIGURE 12: RTW VCO FOM comparison.

TABLE 2: Measured VCO characteristics comparison.

| | |
|---|---|
| Central frequency | 11.25 GHz |
| $K_{VCO}$ | 400 MHz |
| Tuning range | 1200 MHz |
| Power consumption | 30 mW |
| Size | 0.105 mm$^2$ |
| PN at 1 MHz | −105 dBc/Hz |
| FOM | 174 dB |

TABLE 3: Simulated PLL characteristics.

| | |
|---|---|
| Central frequency | 11.5 GHz |
| Power consumption | 39 mW |
| Bandwidth | 550 kHz |
| Size | 0.105 mm$^2$ |
| PN at 1 MHz | −102 dBc/Hz |
| Settling time | 11 $\mu$s |

phase error versus the open loop bandwidth as computed by the linear model.

The minimum phase error is 1.6° RMS with a 550 kHz optimal bandwidth. From the cut-off frequency of the loop filter, the $R$ and $C$ values of the loop filter were calculated: $C_1 = 15$ pF, $C_2 = 1.9$ nF, $C_3 = 7.5$ pF, $R_2 = 1.7$ kΩ, $R_3 = 1.5$ kΩ. In addition to the previously cited Matlab model, the PLL design has been also addressed through SPICE simulations to gain more details about the noise contribution of each block (PFD/CP, prescaler, divider, and reference), as depicted in Figure 14, and an estimation of the PLL settling time from transient simulation (cf. Figure 15).

The performances of the designed PLL are summarized in Table 3. The phase noise specifications at a 1 MHz offset are not met. Different solutions can be used to solve this problem. For the RTW structure, we can use a thicker and higher quality metal to realize the transmission line [4], further increase the discrete capacitive steps to lower the sensibility (i.e., $K_{VCO}$), and use an internal regulator to provide a supply that is very clean; the $K_{VCO}$ versus supply is actually one of the biggest problem of this structure.

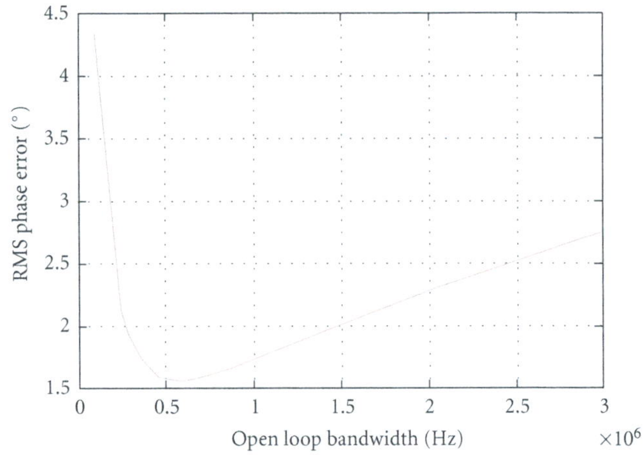


FIGURE 13: Integrated RMS phase error.

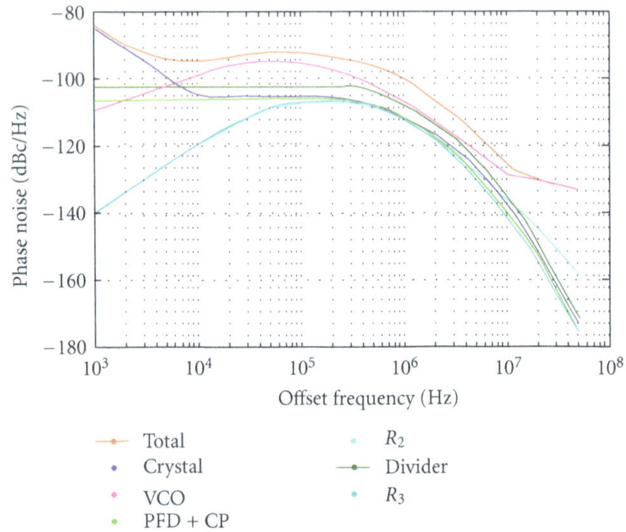


FIGURE 14: Simulated RTW VCO PLL phase noise.

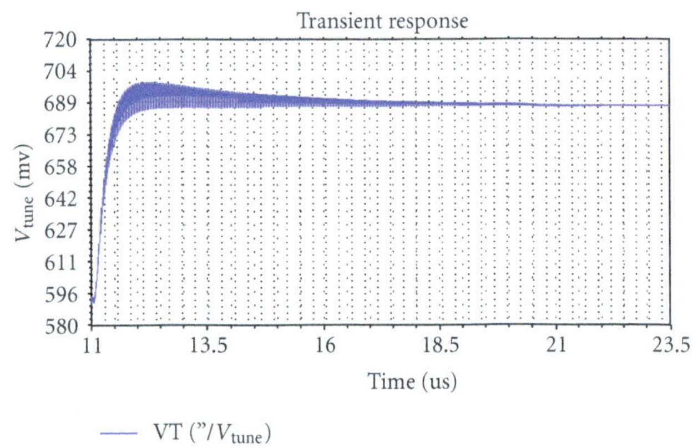


FIGURE 15: Simulated settling time.

## 7. Conclusion

In the present paper a RTW VCO designed in a 130 nm CMOS technology has been reported. The VCO was designed featuring two tune modes: a broadband tune mode obtained through switched capacitors and a fine tune mode realized with traditional varactors. This approach allows reducing the VCO gain VCO with improvements for the phase noise and the stability of the PLL. The VCO design was optimized through the use of a phase noise modelling and a line optimization procedure presented in the paper.

In particular the phase noise model was based on the use of the impulse sensitivity function (ISF) which is based on a linear-time-varying (LTV) model of the oscillator. The theory developed in Section 4 estimates a phase noise of −109 dBc/Hz at an offset frequency of 1 MHz, a value that fairly well compares with the measured value of −105 dBc/Hz as well as the simulated value of −107 dBc/Hz. In spite of this fairly well agreement between the proposed analytical model, simulations, and measurements, it is mandatory to pointing out that the theory provides an analytical expression of the phase noise (see (18)) where the impact of the quality factor ($Q$) on the phase noise is not clear, even if one can attend, as a rule of thumb, that the phase noise decreases with increasing $Q$. This limitation stems from the fact that an ISF description of the phase noise is based on a linear-time-varying (LTV) model where the computation of the quality factor remains in some way obscure, because of the lack of a unified getting [25–27]. In this perspective, under a conceptual point of view physically based approaches perform better. In [28] the authors report on a physically based theory of the phase noise in a rotary travelling wave oscillator. The theory leads to a fairly well agreement with the measurements for an offset frequency of 1 MHz, as in the case of the present paper. Again, the physically based theory proposed in [28] makes provision for a decrease of the phase noise in the $1/f$ frequency range with increasing the number of amplifiers constituting the oscillator as the ISF theory proposed in the present paper (see Figure 7(b)). Moving closer to the carrier, differences appear between the two approaches. At an offset frequency of 10 kHz the physically based theory fails to predict the phase noise for more than 10 dB while the ISF based model proposed in the present paper matches the experimental value within a 10 dB of tolerance. The authors in [25] state indeed that a physically based modelling of the flicker-noise up-conversion in the rotary travelling wave oscillator is still an open area of investigation. In short, even if the phase noise modelling proposed in the present paper suffers from limited insights into the physical mechanisms responsible for the phase noise, on the other hand, it actually exhibits a prediction capability of the phase noise on a larger offset frequency range.

When compared with other distributed oscillators reported in the literature, the RTW VCO reported in the present work exhibits a figure of merit close to that recently reported for a 65 nm CMOS distributed VCO. On the basis of the experience carried out during this work, the authors believe that the achieved phase noise can be reduced down to −115 dBc/Hz at 1 MHz by using a technology offering a thicker top metal layer.

The obtained results suggest not only that a RTW VCO designed using a CMSO technology can replace the DRO as a local oscillator, with a reduction in size and costs of the satellite receiver, as a consequence, but that it can be also usefully applied for the design of Ku-band satellite receiver with an image frequency rejection or direct conversion architecture, offering the advantage of eliminating the troubles related to the design of the image frequency rejection filters.

## Acknowledgments

The authors wish to thank the CIM-PACA Design Platform for its support. The circuits were fabricated through the CMP consortium with PACA region grants.

## References

[1] NXP, "Integrated mixer oscillator PLL for satellite LNB," *TFF1004HN/N1 Datasheet*, 2008.

[2] Z. Deng, J. Chen, J. Tsai, and A. M. Niknejad, "A CMOS Ku-band single-conversion low-noise block front-end for satellite receivers," in *Proceedings of the IEEE Radio Frequency Integrated Circuits Symposium, (RFIC '09)*, pp. 135–138, Boston, Mass, USA, June 2009.

[3] K. Miyashita, "A plastic packaged ku-band LNB with very high susceptibility to supply PLL in 0.18 um CMOS," in *Proceedings of the 10th Topical Meeting on Silicon Monolithic Integrated Circuits in RF Systems, (SiRF '10)*, pp. 188–191, New Orleans, La, USA, January 2010.

[4] F. Ben Abdeljelil, W. Tatinian, L. Carpineto, and G. Jacquemod, "Design of a CMOS 12 GHz rotary travelling wave oscillator with switched capacitor tuning," in *Proceedings of the IEEE Radio Frequency Integrated Circuits Symposium, (RFIC '09)*, pp. 579–582, Boston, Mass, USA, June 2009.

[5] B. Razavi, K. F. Lee, and R. H. Yan, "Design of high-speed, low-power frequency dividers and phase-locked loops in deep submicron CMOS," *IEEE Journal of Solid-State Circuits*, vol. 30, no. 2, pp. 101–111, 1995.

[6] D. Dermit, F. Ducati, D. Balsamo, P. Lucchi, M. Borgarino, and G. Jacquemod, "A 130 nm CMOS tunable digital frequency divider for dual-band microwave radiometer," in *Proceedings of the 16th IEEE International Conference on Electronics, Circuits and Systems, (ICECS '09)*, pp. 203–206, Hammamet, Tunisia, December 2009.

[7] U. Singh and M. M. Green, "High-frequency CML clock dividers in 0.13-$\mu$m CMOS operating Up to 38 GHz," *IEEE Journal of Solid-State Circuits*, vol. 40, no. 8, pp. 1658–1660, 2005.

[8] C. D. Hedayat, A. Hachem, Y. Leduc, and G. Benbassat, "Modeling and characterization of the 3rd order charge-pump PLL: a fully event-driven approach," *Analog Integrated Circuits and Signal Processing*, vol. 19, no. 1, pp. 24–45, 1999.

[9] G. Le Grand de Mercey, "A 18 GHz rotary traveling wave VCO in CMOS with I/Q outputs," in *Proceedings of the 29th European Solid-State Circuits Conference (ESSCIRC '03)*, pp. 489–492, Lisboa, Portugal, 2003.

[10] B. Razavi, "The role of monolithic transmission lines in high-speed integrated circuits," in *Proceedings of the IEEE Custom Integrated Circuits Conference (CICC '02)*, pp. 367–374, May 2002.

[11] R. A. Pucel, "Losses in Microstrip," *IEEE Transaction on Microwave Theory and Techniques*, vol. 16, pp. 342–350, 1968.

[12] H. M. Greenhouse, "Design of planar rectangular microelectronic inductors," *IEEE Trans Parts Hybrids Packag*, vol. 10, no. 2, pp. 101–109, 1974.

[13] T. Sakurai and K. Tamaru, "Simple formulas for two- and three-dimensions capacitance," *IEEE Transactions on Electron Devices*, vol. 30, no. 2, pp. 183–185, 1983.

[14] H. H. Hsieh, Y. C. Hsu, and L. H. Lu, "A 15/30-GHz dual-band multiphase voltage-controlled oscillator in 0.18-$\mu$m CMOS," *IEEE Transactions on Microwave Theory and Techniques*, vol. 55, no. 3, pp. 474–483, 2007.

[15] A. Hajimiri and T. H. Lee, "A general theory of phase noise in electrical oscillators," *IEEE Journal of Solid-State Circuits*, vol. 33, no. 2, pp. 179–194, 1998.

[16] T. H. Lee, *The Design of CMOS Radio Frequency Integrated Circuits*, Cambridge, 1998.

[17] D. B. Leeson, "A simple model of feed-back oscillator noise spectrum," *Proceedings of the IEEE*, vol. 54, pp. 329–330, 1885.

[18] C. J. White and A. Hajimiri, "Phase noise in distributed oscillators," *Electronics Letters*, vol. 38, no. 23, pp. 1453–1454, 2002.

[19] N. Seller, A. Cathelin, H. Lapuyade, J. B. Bégueret, E. Chataigner, and D. Belo, "A 10 GHz distributed voltage controlled oscillator for WLAN application in a VLSI 65 nm CMOS process," in *Proceedings of the IEEE Radio Frequency Integrated Circuits Symposium, (RFIC '07)*, pp. 115–118, June 2007.

[20] G. P. Bilionis, A. N. Birbas, and M. K. Birbas, "Fully integrated differential distributed VCO in 0.35-$\mu$m SiGe BiCMOS technology," *IEEE Transactions on Microwave Theory and Techniques*, vol. 55, no. 1, pp. 13–22, 2007.

[21] D. Guckenberger and K. T. Kornegay, "Differential distributed amplifier and oscillator in SiGe BiCMOS using close-packed interleaved on-chip transmission lines," in *Proceedings of the IEEE Bipolar/BiCMOS Circuits and Technology Meeting*, pp. 68–71, September 2004.

[22] H. Wu and A. Hajimiri, "10 GHz CMOS distributed voltage controlled oscillator," in *Proceedings of the 22nd Annual Custom Integrated Circuits Conference (CICC '00)*, pp. 581–584, May 2000.

[23] K. Bhattacharyya and T. H. Szymanski, "Performance of a 12 GHz monolithic microwave distributed oscillator in 1.2 V 0.18$\mu$m CMOS with a new simple design technique for frequency changing," in *Proceedings of the IEEE Annual Conference on Wireless and Microwave Technology, (WAMICON '05)*, pp. 174–177, April 2005.

[24] G. Jacquemod, L. Geynet, B. Nicolle, E. D. Foucauld, W. Tatinian, and P. Vincent, "Design and modelling of a multi-standard fractional PLL in CMOS/SOI technology," *Microelectronics Journal*, vol. 39, no. 9, pp. 1130–1139, 2008.

[25] G. J. Coram, "A simple 2-D oscillator to determine the correct decomposition of perturbations into amplitude and phase noise," *IEEE Transactions on Circuits and Systems I: Fundamental Theory and Applications*, vol. 48, no. 7, pp. 896–898, 2001.

[26] U. L. Rohde, A. K. Poddar, and G. Boeck, *Modern Microwave Oscillators For Wireless Applications: Theory and Optimization*, John Wiley & Sons, 2005.

[27] R. Undheim, *Design of a linear FMCW radar synthesizer with focus on phase noise [M.S. thesis]*, Department of Electronics and Telecommunications, Norwegian University of Science and Technology, 2012.

[28] K. Takinami, R. Walsworth, S. Osman, and S. Beccue, "Phase-noise analysis in rotary traveling-wave oscillators using simple physical model," *IEEE Transactions on Microwave Theory and Techniques*, vol. 58, no. 6, pp. 1465–1474, 2010.

# 9

# Comprehension of Postmetallization Annealed MOCVD-$TiO_2$ on $(NH_4)_2S$ Treated III-V Semiconductors

**Ming-Kwei Lee and Chih-Feng Yen**

*Department of Electronic Engineering, Chung Yuan Christian University, Chung Li City 32023, Taiwan*

Correspondence should be addressed to Ming-Kwei Lee, mklee@cycu.edu.tw

Academic Editor: Kuan-Wei Lee

The electrical characteristics of $TiO_2$ films grown on III-V semiconductors (e.g., p-type InP and GaAs) by metal-organic chemical vapor deposition were studied. With $(NH_4)_2S$ treatment, the electrical characteristics of MOS capacitors are improved due to the reduction of native oxides. The electrical characteristics can be further improved by the postmetallization annealing, which causes hydrogen atomic ion to passivate defects and the grain boundary of polycrystalline $TiO_2$ films. For postmetallization annealed $TiO_2$ on $(NH_4)_2S$ treated InP MOS, the leakage current densities can reach $2.7 \times 10^{-7}$ and $2.3 \times 10^{-7}$ A/cm$^2$ at $\pm 1$ MV/cm, respectively. The dielectric constant and effective oxide charges are 46 and $1.96 \times 10^{12}$ C/cm$^2$, respectively. The interface state density is $7.13 \times 10^{11}$ cm$^{-2}$ eV$^{-1}$ at the energy of 0.67 eV from the edge of valence band. For postmetallization annealed $TiO_2$ on $(NH_4)_2S$ treated GaAs MOS, The leakage current densities can reach $9.7 \times 10^{-8}$ and $1.4 \times 10^{-7}$ at $\pm 1$ MV/cm, respectively. The dielectric constant and effective oxide charges are 66 and $1.86 \times 10^{12}$ C/cm$^2$, respectively. The interface state density is $5.96 \times 10^{11}$ cm$^{-2}$ eV$^{-1}$ at the energy of 0.7 eV from the edge of valence band.

## 1. Introduction

Due to its high electron mobility and direct energy band gap compared with Si, much attention has been focused on III-V compound semiconductor (e.g., InP and GaAs) high-speed devices. Usually, the metal-semiconductor field-effect transistor (MESFET) is one of III-V main high-speed devices due to the lack of high quality of oxide on it. The main disadvantage of MESFET is the high gate currents of Schottky contact under the positive bias of several tenths of a volt, which severely limits the maximum drain currents, the lower noise margin, and the less flexibility of the circuit design. Metal-oxide-semiconductor field-effect transistor (MOSFET) can alleviate these problems. Many high-*k* dielectrics, such as $Al_2O_3$ [1], $TiO_2$ [2], $Ga_2O_3$ [3], HfAlO [4], and $HfO_2$ [5], are currently being explored on III-V substrates. For high-*k* dielectrics, the same gate capacitance per unit area can be realized using a much thicker gate materials and results in less tunneling leakage current. Of various high-*k* materials, $TiO_2$ has generated much interest as it offers a large dielectric constant (*k* value 4–86) [6] and a higher transconductance of MOSFET is expected [7].

High dielectric constant polycrystalline $TiO_2$ films were prepared by metal-organic chemical vapor deposition (MOCVD) [8], sol-gel [9], and sputtering [10]. MOCVD-$TiO_2$ was used in this study because of its simple process and higher quality. Usually, the leakage current of MOCVD-$TiO_2$ on III-V is high from the high interface state ($D_{it}$) [11, 12] and the polycrystalline grain boundary [13, 14]. From previous studies [13, 14], the high $D_{it}$ from native oxides on III-V surface can be removed by $(NH_4)_2S$ treatment. It can also passivate the surface dangling bonds of III-V surface and prevent it from oxidizing.

The low temperature postmetallization annealing (PMA) is an effective process to reduce the oxide charge density and the $D_{it}$ in $SiO_2$/Si metal-oxide-semiconductor (MOS) technology [15, 16]. The mechanism of PMA process is from the reaction between the aluminum contact and hydroxyl groups existed on $SiO_2$ films surface resulting in hydrogen atomic ions diffusing through the oxide and passivate the

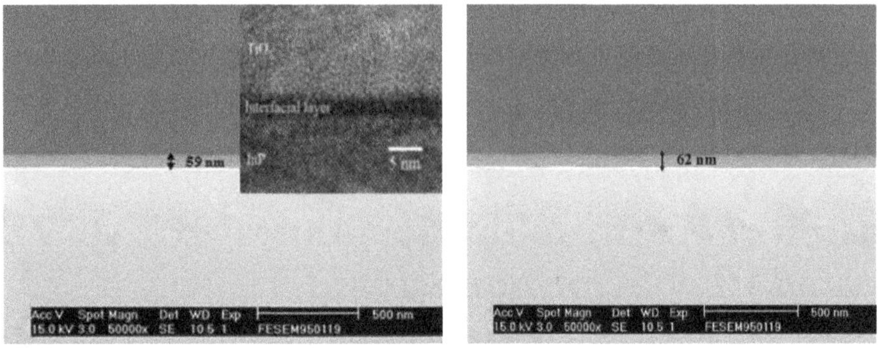


(a) (b)

Figure 1: (a) Images of SEM and HRTEM of $TiO_2$/S-InP and (b) SEM cross section of $TiO_2$/S-GaAs substrate.

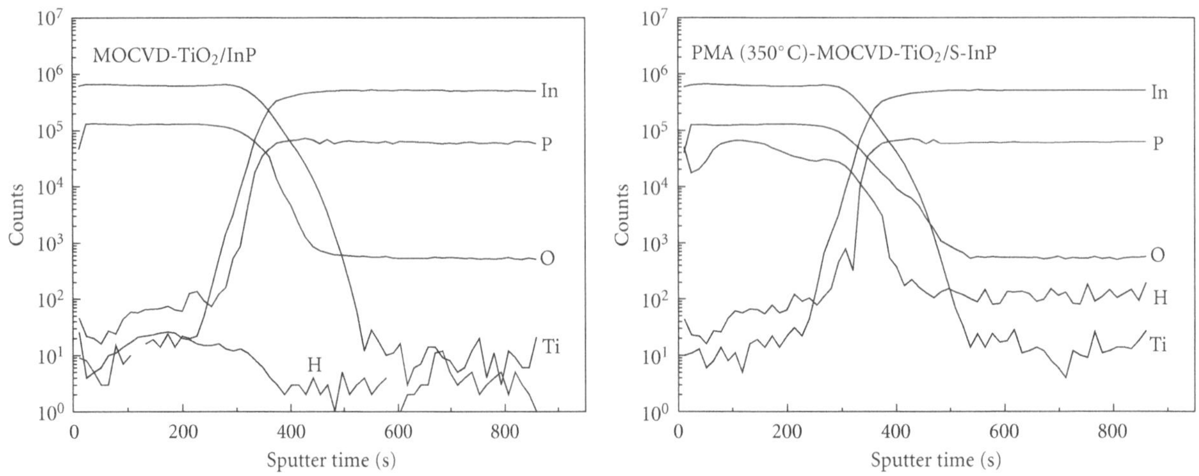


(a) (b)

Figure 2: SIMS depth profiles for (a) $TiO_2$/S-InP and (b) PMA (350°C)-$TiO_2$/S-InP.

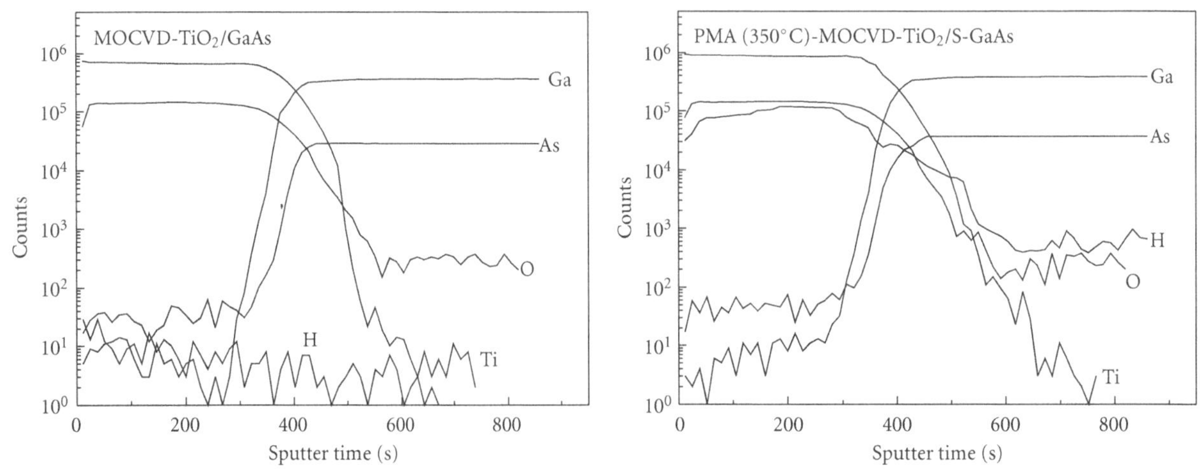


(a) (b)

Figure 3: SIMS depth profiles for (a) $TiO_2$/S-GaAs and (b) PMA (350°C)-$TiO_2$/S-GaAs.

TABLE 1: Electrical characteristics by PMA(350°C)-$TiO_2$/InP and PMA(350°C)-$TiO_2$/GaAs.

| MOS structures | Dielectric constant | Leakage current at 1 MV/cm | Interface state density | $\Delta V_{FB}$ of hysteresis loop |
|---|---|---|---|---|
| PMA(350°C)-$TiO_2$/S-InP | 44 | $2.7 \times 10^{-7}$ and $2.3 \times 10^{-7}$ A/cm$^2$ | $7.13 \times 10^{11}$ cm$^{-2}$ eV$^{-1}$ | 17 mV |
| PMA(350°C)-$TiO_2$/S-GaAs | 66 | $9.7 \times 10^{-8}$ and $1.4 \times 10^{-7}$ | $5.96 \times 10^{11}$ cm$^{-2}$ eV$^{-1}$ | 9 mV |

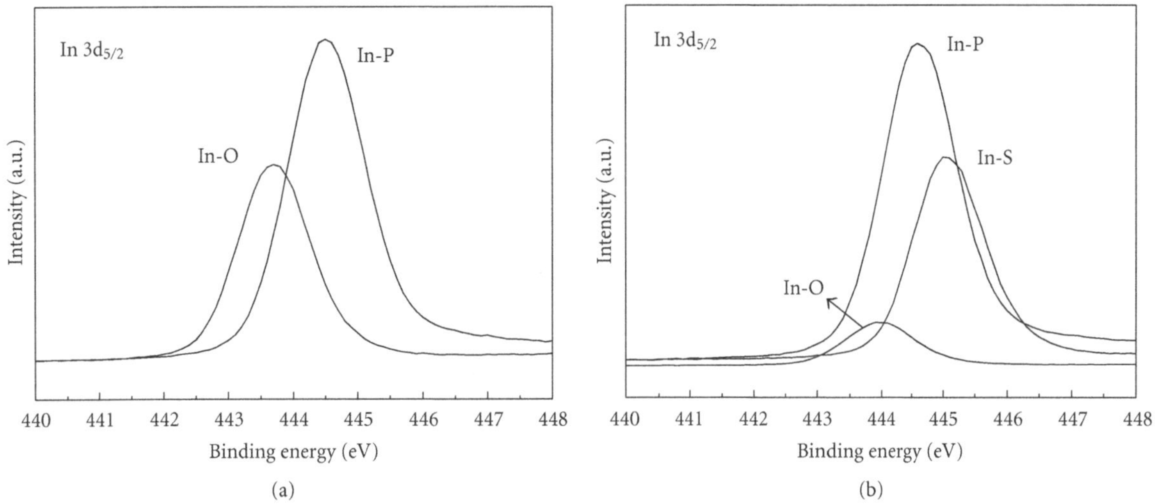


FIGURE 4: XPS spectra of In $3d_{5/2}$ core level from (a) InP without $(NH_4)_2S$ treatment and (b) InP with $(NH_4)_2S$ treatment.

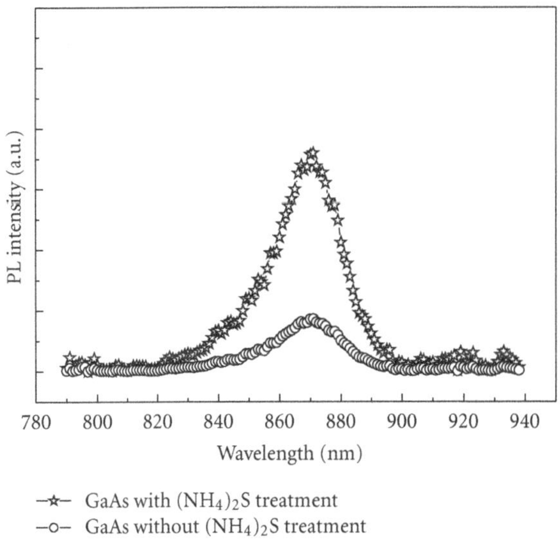


FIGURE 5: PL spectra of GaAs with and without $(NH_4)_2S$ treatment.

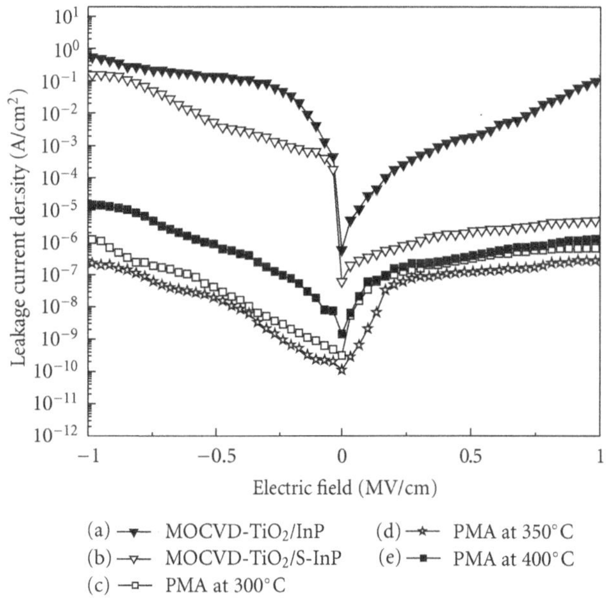


FIGURE 6: Leakage current densities of $TiO_2$/InP with and without $(NH_4)_2S$ treatments and PMA-$TiO_2$/S-InP at different PMA temperatures.

oxide traps [15–17]. From our previous study [18], the PMA was used to reduce the leakage current from the defects and grain boundary of polycrystalline $TiO_2$ films grown on silicon. Both treatments also show the same function on high-*k*/III-V. In this study, we try to review the improvement of electrical characteristics of $TiO_2$/InP and $TiO_2$/GaAs by the combination of $(NH_4)_2S$ and PMA treatments (PMA-$TiO_2$/S-InP (GaAs)).

## 2. Experimental

Zn doped p-type (100) InP and GaAs with carrier concentration of $5 \times 10^{16}$ and $6 \times 10^{16}$ cm$^{-3}$ were used as the substrates. These substrates were degreased in solvent and followed by chemical etching in a solution ($H_2SO_4 : H_2O_2 : H_2O$ = 5 : 1 : 1) for 3 min and then rinsed in deionized water.

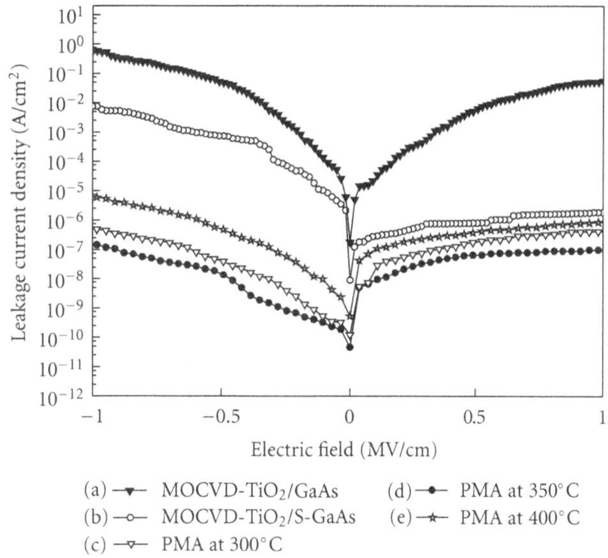


Figure 7: Leakage current densities of $TiO_2$/GaAs with and without $(NH_4)_2S$ treatments and without $(NH_4)_2S$ treatment and PMA-$TiO_2$/S- GaAs at different PMA tempreatures.

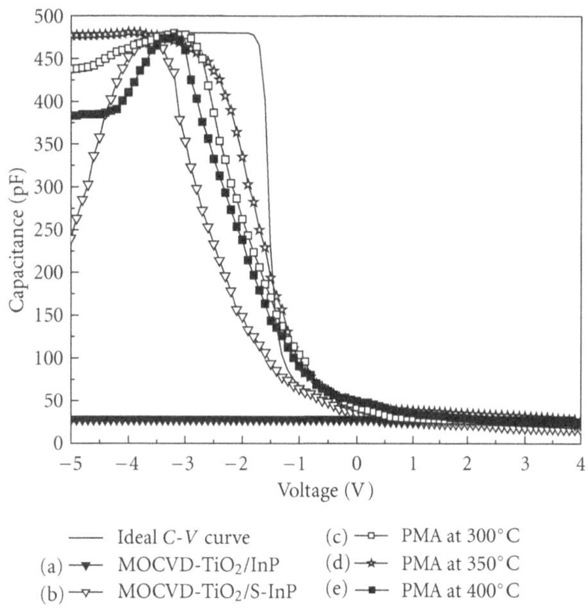


Figure 8: *C-V* characteristics of $TiO_2$/InP with and without $(NH_4)_2S$ treatments and PMA-$TiO_2$/S-InP at different PMA temperatures.

After cleaning, substrates were immediately dipped into $(NH_4)_2S$ solution for 40 min at 50°C and then blow-dried with nitrogen gas. After $(NH_4)_2S$ treatment, substrates were thermally treated at 220°C in a nitrogen atmosphere for 10 min in order to desorb the excess of weakly bonded sulfur and were ready for MOCVD-$TiO_2$ growth.

Polycrystalline $TiO_2$ thin films were grown on substrates by a horizontal cold-wall MOCVD system. Tetraisopropoxytitanium ($Ti(i\text{-}OC_3H_7)_4$) was used as a Ti precursor and kept at 24°C. Nitrogen was used as the carrier gas and its flow rate was 10 sccm. Nitrous oxide gas ($N_2O$) was used as an oxidizing agent and its flow rate was 100 sccm. Molybdenum was used as the oxidation-resist susceptor. The reactor pressure was kept at 5 Torr during the growth. The growth temperature was kept at 400°C for 5 min. The chemical reaction steps during the deposition of $TiO_2$ on substrate in MOCVD system are as follows.

In PMA procedure, aluminum (Al) was deposited upon the $TiO_2$ films as the cap layer. Then, these films were annealed in nitrogen ambient for 10 min at the temperature of 300, 350, and 400°C, respectively. Finally, the Al was etched away with an etching solution ($H_3PO_4 : HNO_3 : CH_3COOH : H_2O = 73 : 4 : 3.5 : 19.5$).

A metal-oxide-semiconductor (MOS) structure was used to examine the electrical characteristics. In-Zn alloy (In 90% and Zn 10%) was evaporated on the III-V back side for ohmic contact and then thermally annealed at 400°C for 3 min in nitrogen atmosphere. The ohmic contact was confirmed by the current-voltage characteristics. Then, Al was evaporated on $TiO_2$ films as the top contact with the area of $7.07 \times 10^{-4}\,cm^2$. Scanning electron microscopy (SEM) was used to examine the thickness of $TiO_2$ film. A HP4145B semiconductor-parameter analyzer was used for current-voltage (*I-V*) characterization. A high frequency (1 MHz) HP4280A capacitance-voltage (*C-V*) meter was used for *C-V* characterization scanned from accumulation region to inversion region. The DC bias was swept by 1/30 V/sec. The $D_{it}$ was derived from *C-V* curves by Terman method [19], which can provide a good evaluation [20] of the $D_{it}$ higher than $10^{10}\,cm^{-2}\,eV^{-1}$ with 10% error [21, 22].

## 3. Results and Discussion

The SEM cross section of $TiO_2$/S-InP is shown in Figure 1(a) and the thickness of $TiO_2$ film is 59 nm. The SEM picture shows that there is an interfacial layer. From the image of high-resolution transmission electron microscopy (HRTEM) shown in the inset in Figure 1(a), the interfacial layer is 2.5 nm. It is from the interdiffusion between $TiO_2$ and InP examined by SIMS depth profile shown in Figure 2(a). The SEM cross section of $TiO_2$/S-GaAs is shown in Figure 1(b) and the thickness of $TiO_2$ film is 62 nm. The SEM picture also shows an interfacial layer. The mechanism is the same as $TiO_2$/S-InP and examined by the SIMS depth profile shown in Figure 3(a). It indicates that a low temperature growth process is essential for decreasing the interfacial layer for ultrathin $TiO_2$ film. Atomic layer deposition (ALD) may be the candidate.

The X-ray photoelectron spectroscopy (XPS) core level spectra of In $3d_{5/2}$ of InP without and with $(NH_4)_2S$ treatments are shown in Figures 4(a) and 4(b), respectively. For InP without $(NH_4)_2S$ treatment, the strong In $3d_{5/2}$ XPS peak at 444.7 eV can be attributed to In-P bond [23] and the peak at 443.7 eV is from In-O bond [24]. For InP with $(NH_4)_2S$ treatment, a new strong peak at 444.9 eV [25] and a small satellite peak at 443.9 eV are from In-S and In-O, respectively. The strong In-S bond shows that In empty dangling bond is passivated by S. The much weaker

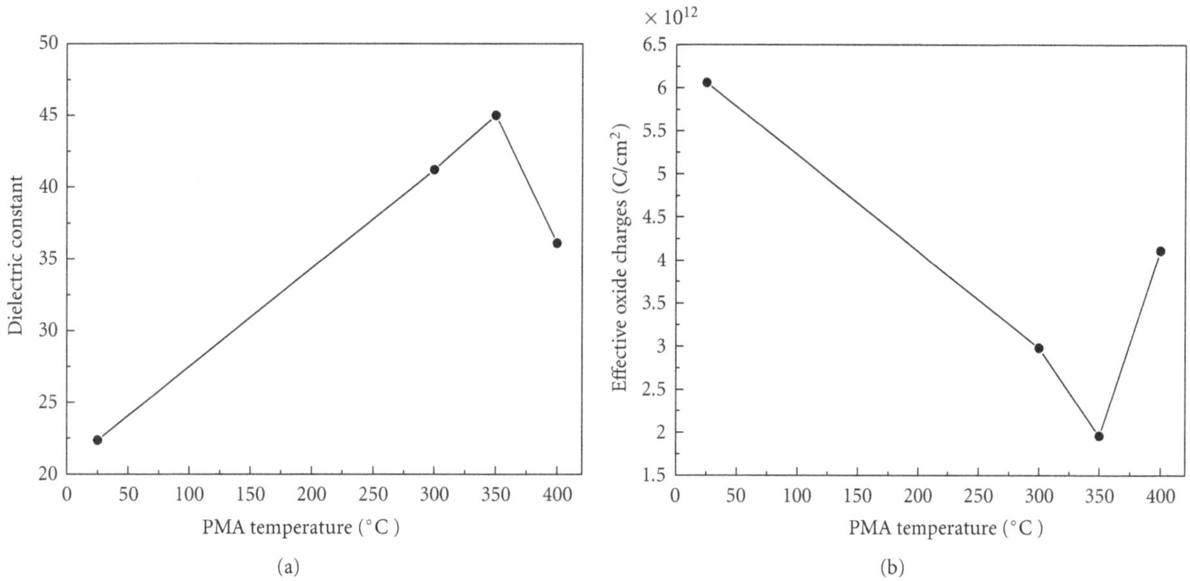


Figure 9: (a) Dielectric constant and (b) effective oxide charges of $TiO_2$/S-InP and PMA-$TiO_2$/S-InP at different PMA temperatures.

In-O peak indicates that native oxides are significantly reduced after $(NH_4)_2S$ treatment. It suggests the $(NH_4)_2S$ treatment can not completely remove native oxides on III-V semiconductors. It needs a further investigation.

In order to double check the function of $(NH_4)_2S$ passivation, the photoluminescence (PL) spectra of GaAs and S-GaAs are shown in the inset of Figure 5. The lower PL intensity for GaAs without $(NH_4)_2S$ treatment shows a high surface recombination velocity [26, 27]. The PL intensity is much improved for GaAs with $(NH_4)_2S$ treatment. It indicates that the $(NH_4)_2S$ passivation is an effective way to reduce the surface recombination velocity [28]. It supports that $(NH_4)_2S$ treatment can remove native oxides and prevent III-V surface from oxidizing.

The leakage current densities of $TiO_2$ film deposited on InP substrate with and without $(NH_4)_2S$ treatments are shown in Figure 6. It also shows the leakage current densities of PMA-$TiO_2$/S-InP treated at the PMA temperatures of 300, 350, and 400°C. The leakage current densities of $TiO_2$/InP are 0.1 and 0.57 A/cm$^2$ at ±1 MV/cm as shown in Figure 6(a). The high leakage currents are mainly from the high density of defects in the grain boundary of polycrystalline $TiO_2$ film [11, 12] and the high interface states at $TiO_2$/InP interface due to InP native oxide [13, 14]. For $TiO_2$/S-InP as shown in Figure 6(b), the leakage current densities are $4.56 \times 10^{-6}$ and 0.16 A/cm$^2$ at ±1 MV/cm, respectively. The leakage currents are mainly from $D_{it}$ and grain boundary of $TiO_2$ film. After $(NH_4)_2S$ treatment, higher quality $TiO_2$ film can be deposited on reconstructed InP surface [14]. The leakage current is much improved under positive bias from the reduction of $D_{it}$. However, only one order improvement of the leakage current under negative bias is from grain boundary.

The leakage current density can be further improved by PMA treatment as shown in Figures 6(c)–6(e). The lowest leakage current densities of PMA-$TiO_2$/S-InP can reach $2.7 \times 10^{-7}$ and $2.3 \times 10^{-7}$ A/cm$^2$ at ±1 MV/cm at the PMA treatment of 350°C as shown in Figure 6(d). After PMA process, the thickness of the interfacial layer does not change and is examined from SIMS depth profiles as shown in Figures 2(a) and 2(b). Therefore, the improvement of leakage current is not from the increase of interfacial layer thickness after PMA. Figure 2(b) shows that H atoms are uniformly distributed in the whole $TiO_2$ film due to H fast diffusion after PMA treatment. It would diffuse along and passivate the grain boundary. At the PMA treatment of 300°C, the leakage current densities are $6.58 \times 10^{-7}$ and $1.2 \times 10^{-6}$ A/cm$^2$ at ±1 MV/cm as shown in Figure 6(c). The slight increase of leakage current compared with Figure 6(d) could be from the lower PMA temperature, which cannot provide sufficient energy for H atoms for passivation. For the PMA temperature at 400°C as shown in Figure 6(e), the leakage currents are higher than that of 300 and 350°C. It is that the higher PMA temperature would destroy the H passivation and is examined by *C-V* characteristics.

For $TiO_2$ on GaAs substrate with and without $(NH_4)_2S$ treatments, the leakage current densities are shown in Figure 7. It also exhibits the leakage current densities of PMA-$TiO_2$/S-GaAs treated at the PMA temperatures of 300, 350, and 400°C. The leakage current densities of $TiO_2$/GaAs without $(NH_4)_2S$ treatment are $5 \times 10^{-2}$ and $5.9 \times 10^{-1}$ A/cm$^2$ at ±1 MV/cm as shown in Figure 7(a). For $TiO_2$/S-GaAs as shown in Figure 7(b), the leakage current densities are $2.1 \times 10^{-6}$ and $6 \times 10^{-3}$ A/cm$^2$ at ±1 MV/cm, respectively. The leakage current density can be further improved by PMA treatment as shown in Figures 7(c)–7(e). The leakage current densities are $4.5 \times 10^{-7}$ and $4.9 \times 10^{-7}$ A/cm$^2$ at ±1 MV/cm at the PMA treatment of 300°C as shown in Figure 7(c). The lowest leakage current densities of PMA-$TiO_2$/S-GaAs can reach $9.7 \times 10^{-8}$ and $1.4 \times 10^{-7}$ A/cm$^2$ at ±1 MV/cm at the PMA treatment of 350°C as shown in Figure 7(d). For the PMA temperature at 400°C,

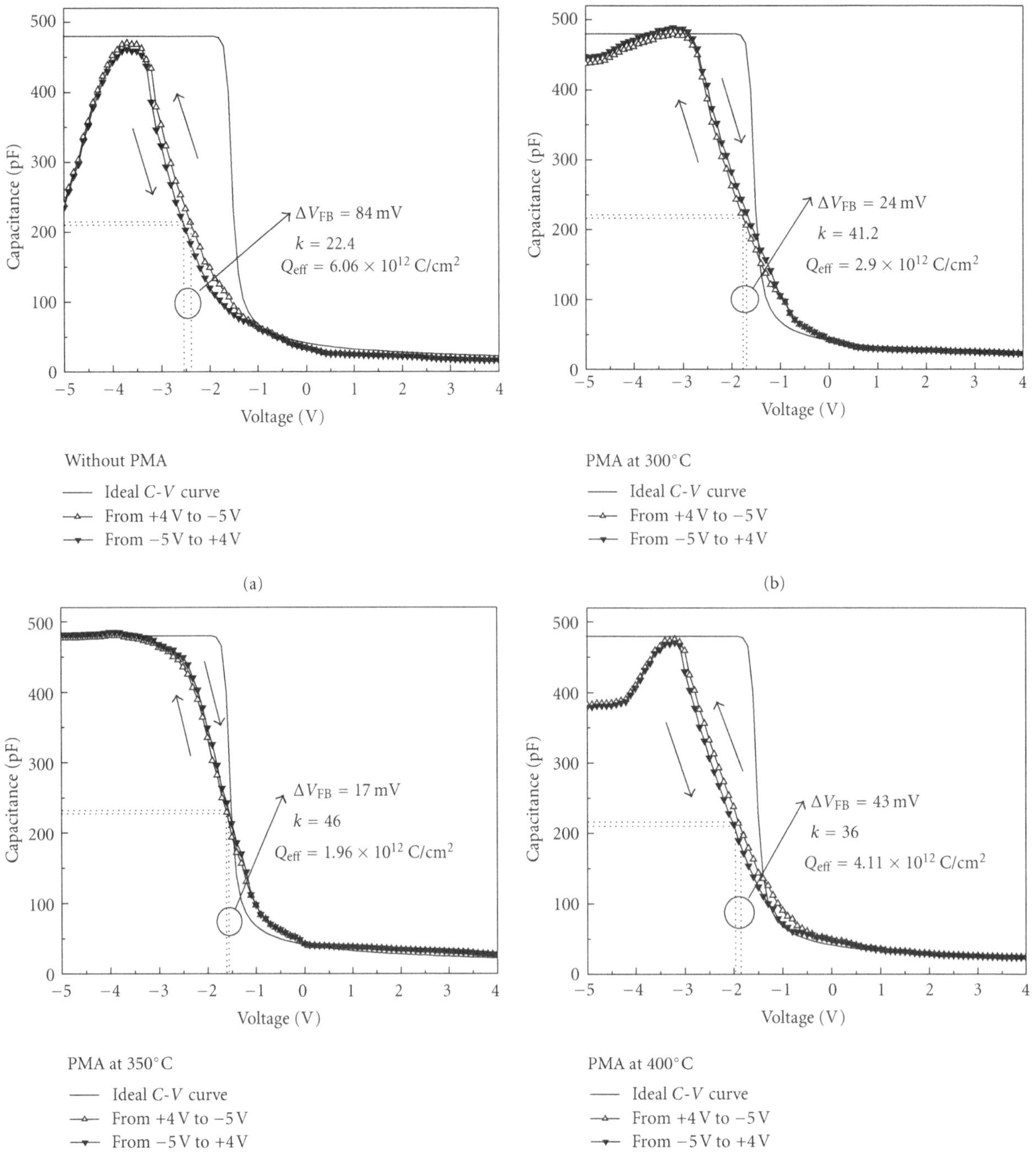


FIGURE 10: *C-V* hysteresis loops of (a) $TiO_2$/S-InP, (b) PMA at 300°C, (c) PMA at 350°C, and (d) PMA at 400°C.

the leakage currents are $9.5 \times 10^{-7}$ and $6 \times 10^{-6}$ A/cm$^2$ at ±1 MV/cm as shown in Figure 3(e), which are higher than that of 300 and 350°C. The mechanisms are similar to InP as mentioned in the previous paragraph.

The *C-V* characteristics of $TiO_2$/InP, $TiO_2$/S-InP, and PMA-$TiO_2$/S-InP are shown in Figure 8. The *C-V* characteristics of $TiO_2$/InP show a flat curve as in Figure 8(a). It is from the high density of interface states due to the existence of native oxides on InP surface, which causes the pinning of the surface Fermi level near the middle of the band gap [29]. Figure 8(b) shows the *C-V* characteristics of $TiO_2$/S-InP. The capacitance in the accumulation region is high due to the improved interface quality. The capacitance decay at higher negative bias is due to the high leakage current, which comes from the defects and the grain boundary of polycrystalline $TiO_2$ film. Sharp *C-V* curves PMA-$TiO_2$/S-InP can be obtained after PMA treatments at 300, 350, and 400°C as shown in Figures 8(c)–8(e), respectively. The ideal

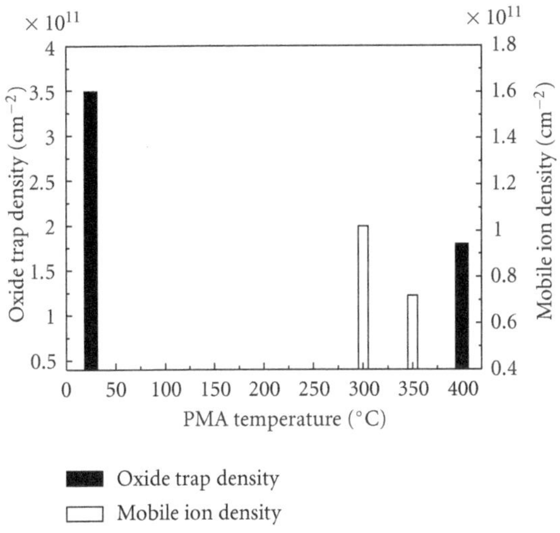


Figure 11: Oxide trap density and mobile ion density of $TiO_2$/S-InP MOS structures as a function of PMA temperature.

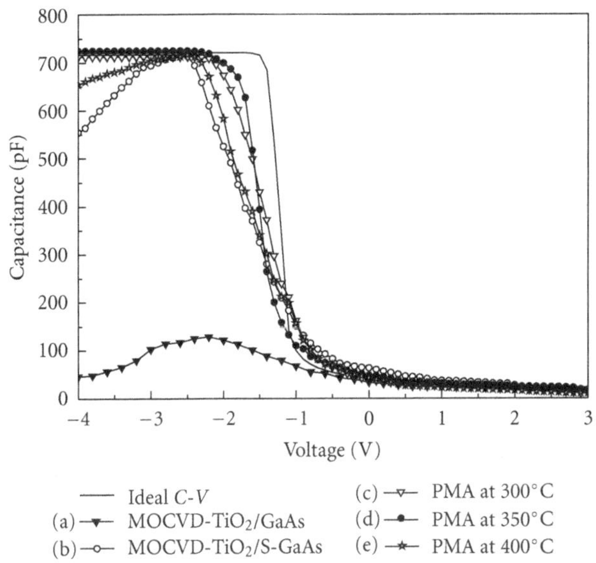


Figure 12: *C-V* characteristics of $TiO_2$/GaAs with and without $(NH_4)_2S$ treatments and PMA-$TiO_2$/S-GaAs at different PMA temperatures.

C-V curve is also shown in the figure as a reference. It is derived from the neglect of the effective oxide charges and the interface states, but the work function difference ($\Phi_{ms}$ = $-1.51$ V) of metal (Al) and semiconductor (InP) is taken into account. The optimized PMA temperature is 350°C as shown in Figure 8(d), in which the stretch-out phenomenon and the flat-band voltage shift are minimized. The dielectric constants and effective oxide charges of PMA-$TiO_2$/S-InP films as functions of PMA temperature are shown in Figures 9(a) and 9(b), respectively. The dielectric constant increases with the PMA temperature due to the improvements of interface and film qualities. But the value decreases at PMA temperature high than 350°C. The higher PMA temperature will break the H bonds and lose the passivation function [30, 31] and results in the increase of leakage current as shown in Figure 6. The dielectric constant and the effective oxide charges can reach 46 and 1.96 $\times$ $10^{12}$ C/cm$^2$ at the PMA temperature of 350°C. The thickness of $TiO_2$ film is 59 nm. The interfacial layer is very thin, which has minor effect during the extraction of dielectric constants.

Moreover, the *C-V* hysteresis loops as a function of PMA temperature are shown in Figure 10. The *C-V* hysteresis loop of $TiO_2$/S-InP without PMA treatment is counterclockwise as shown in Figure 10(a), which is from high density of oxide trapped charges [18, 20] in $TiO_2$/S-InP film without H passivation. The *C-V* hysteresis loops of PMA-$TiO_2$/S-InP film are clockwise at 300 and 350°C as shown in Figures 10(b) and 10(c). The mobile ions are responsible for the *C-V* clockwise hysteresis loop due to the decrease of oxide trapped charges from film quality improvement. The *C-V* hysteresis loop changes back to counterclockwise at 400°C as shown in Figure 10(d). It is dominated by oxide trapped charges due to the break of H bonds and hence the loss of the passivation function at higher PMA temperature [30, 31]. The sum ($N_{ot}$) of oxide trapped density and mobile ion density can be derived from the difference of flat-band voltage ($\Delta V_{FB}$) of the *C-V* hysteresis loops measured at high frequency [20]. The formula is as follows:

$$N_{ot} = \frac{-\Delta V_{FB} C_{ox}}{Aq}, \quad (1)$$

where $C_{ox}$ is the oxide capacitance, $A$ is the contact area ($7.07 \times 10^{-4}$ cm$^{-2}$), and $q$ is magnitude of an electron charge. In *C-V* measurement, the bias scans first from accumulation region to inversion region (forward scan) and then back to accumulation region (backward scan). $\Delta V_{FB}$ is defined as the difference of $V_{FB}$ between the forward scan and the backward scan. So, the polarity of oxide trapped charge is negative and that of mobile ion charge is positive. The $N_{ot}$ of PMA-$TiO_2$/S-InP as a function of PMA temperature is shown in Figure 11. The lowest $N_{ot}$ is $7.18 \times 10^{10}$ C/cm$^2$ at the PMA temperature of 350°C.

The *C-V* characteristics of $TiO_2$/GaAs, $TiO_2$/S-GaAs and PMA-$TiO_2$/S-GaAs are shown in Figure 12. The *C-V* characteristics of $TiO_2$/GaAs show a stretch-out phenomenon under negative bias as shown in Figure 12(a). It is from the high $D_{it}$ due to the existence of native oxides on GaAs surface. The breakdown at higher negative bias is from the higher leakage current resulted in the grain boundary of $TiO_2$ polycrystalline structure. Figure 12(b) shows the *C-V* characteristics of $TiO_2$/S-GaAs. The capacitance in the accumulation region is high due to the improved interface quality and the capacitance decay at higher negative bias is due to the high leakage current. Sharp *C-V* curves PMA-$TiO_2$/S-GaAs can be obtained after PMA treatments at 300, 350, and 400°C as shown in Figures 12(c), 12(d), and 12(e), respectively. The ideal C-V curve is also shown in the figure as a reference. The work function difference ($\Phi_{ms}$ = $-1.31$ V) of metal (Al) and semiconductor (GaAs) is taken into account. The optimized PMA temperature is 350°C as shown in Figure 12(d), in which the stretch-out phenomenon and the flat-band voltage shift are minimized. The dielectric constants and effective oxide charges of PMA-$TiO_2$/S-GaAs

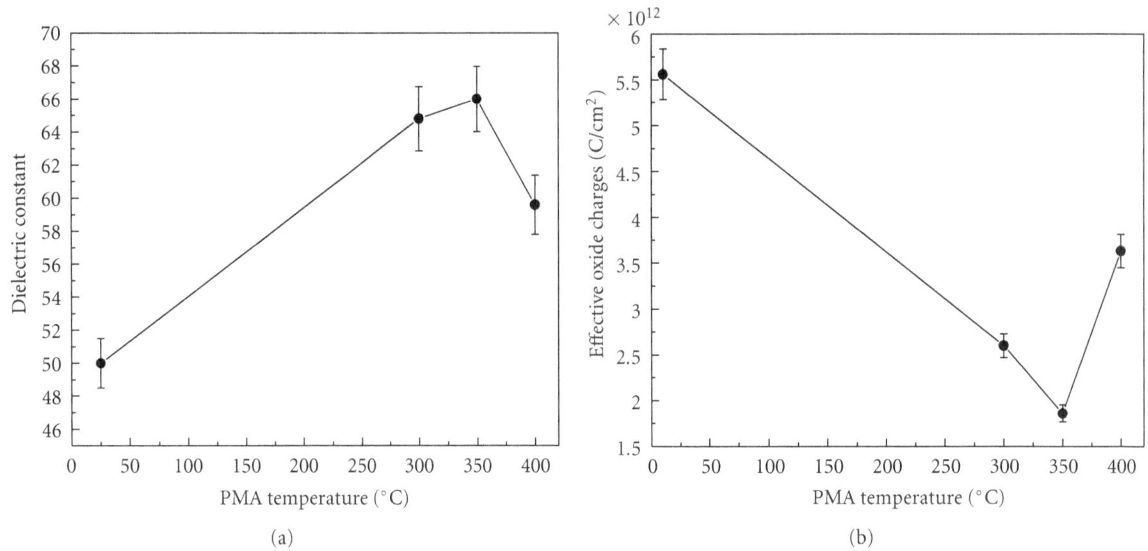


Figure 13: (a) Dielectric constant and (b) effective oxide charges of $TiO_2$/S-GaAs and PMA-$TiO_2$/S-GaAs at different PMA temperatures.

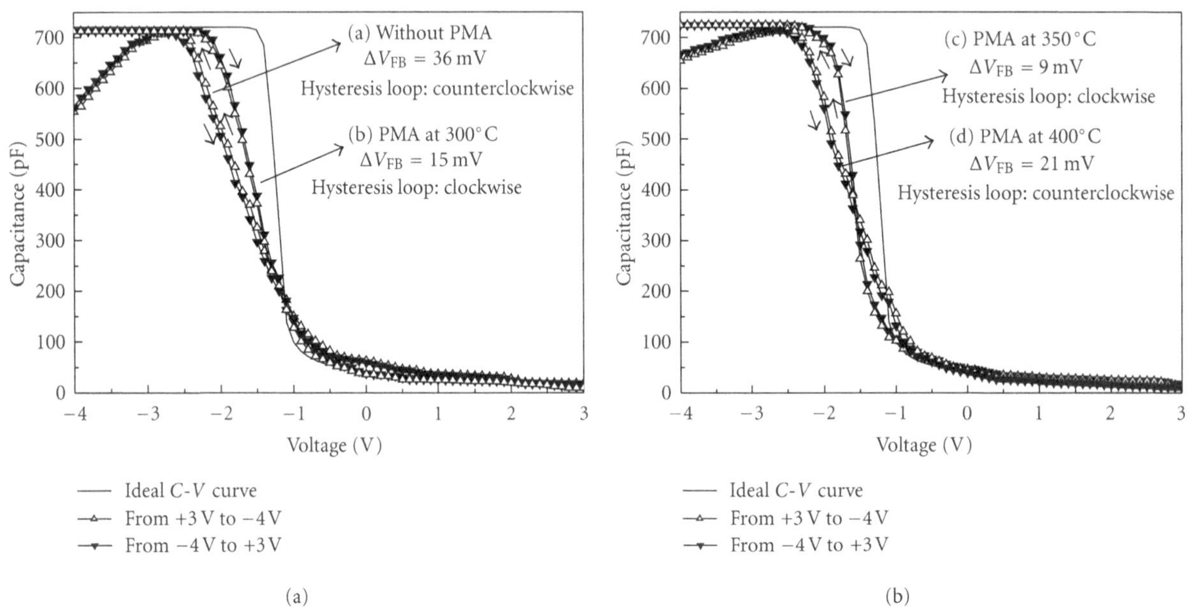


Figure 14: *C-V* hysteresis loops of (a) $TiO_2$/S-GaAs, (b) PMA at 300°C, (c) PMA at 350°C, and (d) PMA at 400°C.

films as functions of PMA temperature are shown in Figures 13(a) and 13(b), respectively. The dielectric constant and the effective oxide charges can reach 66 and $1.86 \times 10^{12}$ C/cm$^2$ at the PMA temperature of 350°C.

Moreover, the *C-V* hysteresis loops as a function of PMA temperature are shown in Figure 14. The *C-V* hysteresis loop of $TiO_2$/S-GaAs without PMA treatment is counterclockwise as shown in Figure 14(a). The *C-V* hysteresis loops of PMA-$TiO_2$/S-GaAs film are clockwise at 300 and 350°C as shown in Figures 14(b) and 14(c). The *C-V* hysteresis loop changes back to counterclockwise at 400°C as shown in Figure 14(d). These *C-V* behaviors are similar to InP. The $N_{ot}$ of PMA-$TiO_2$/S-GaAs as a function of PMA temperature is shown in Figure 15. The lowest $N_{ot}$ is $5.7 \times 10^{10}$ C/cm$^2$ at the PMA temperature of 350°C.

The $D_{it}$ of $TiO_2$/S-InP and PMA-$TiO_2$/S-InP at different PMA temperatures is shown in Figure 16. The lowest $D_{it}$ is $7.13 \times 10^{11}$ cm$^{-2}$ eV$^{-1}$ at the energy of 0.67 eV from the edge of valence band. The $D_{it}$ of $TiO_2$/S-GaAs and PMA-$TiO_2$/S-GaAs at different PMA temperatures is shown in Figure 17. The lowest $D_{it}$ is $5.96 \times 10^{11}$ cm$^{-2}$ eV$^{-1}$ at the energy of 0.7 eV from the edge of valence band. The PMA temperature of two samples was fixed at 350°C.

Table 1 shows the comparisons of electrical characteristics by PMA(350°C)-$TiO_2$/InP and PMA(350°C)-$TiO_2$/GaAs. Form this table we can clearly recognize that electrical

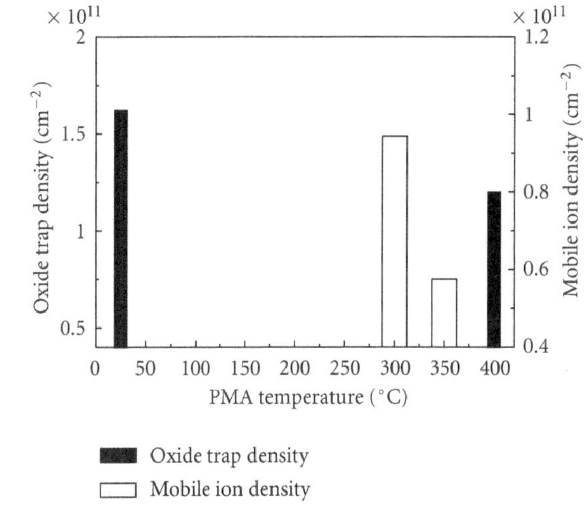


FIGURE 15: Oxide trap density and mobile ion density of $TiO_2$/S-GaAs MOS structures as a function of PMA temperature.

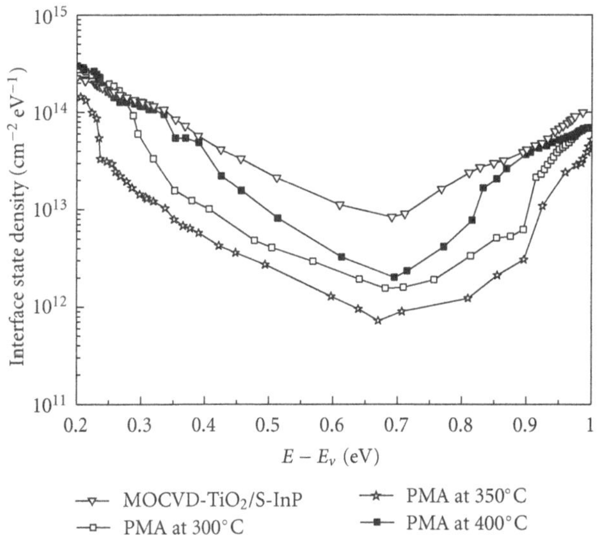


FIGURE 16: Interface state densities of $TiO_2$/S-InP and PMA-$TiO_2$/S-InP at different PMA temperatures.

characteristics of GaAs are superior to InP. It would be from the fact that the $P$ outgas is more serious than that of As due to the higher vapor pressure. It slightly degrades the interface and film quality of InP MOS structure. PMA and $(NH_4)_2S$ treatments highly improve the electrical properties of III-V MOS structures, and dielectric films prepared by lower growth temperature can give more benefits, such as liquid phase deposition and ALD.

## 4. Conclusions

$TiO_2$ films grown on III-V semiconductors with $(NH_4)_2S$ treatments were investigated. With $(NH_4)_2S$ treatment, the interface quality of $TiO_2$/III-V is much improved. PMA treatment further improves the electrical characteristics. The electrical characteristics of GaAs MOS is better than that of

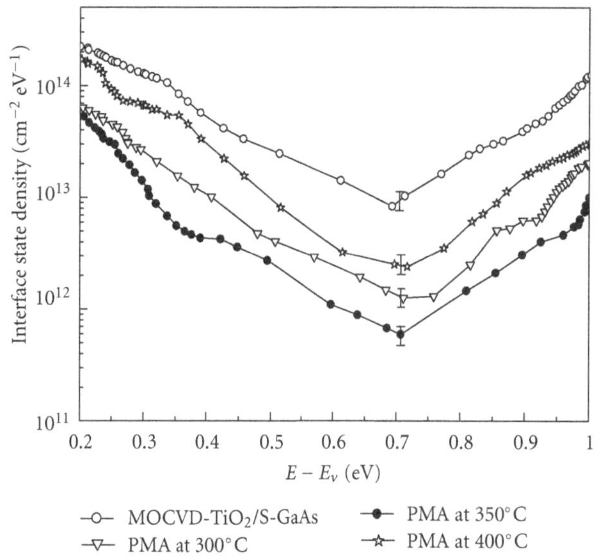


FIGURE 17: Interface state densities of $TiO_2$/S-GaAs and PMA-$TiO_2$/S-GaAs at different PMA temperatures.

InP, which is from the innate character of higher $P$ vapor pressure. There is an interface layer from the inter-diffusion between $TiO_2$ and substrate. Dielectric films prepared by lower growth temperature for III-V MOS structures can give more benefits.

## Acknowledgment

The authors would like to thank the National Science Council, Taiwan, for their support under Contract no. 98-2221-E110-073-MY3.

## References

[1] H. C. . Lin, P. D. Ye, and G. D. Wilk, "Leakage current and breakdown electric-field studies on ultrathin atomic-layer-deposited $Al_2O_3$ on GaAs," *Applied Physics Letters*, vol. 87, Article ID 182904, 3 pages, 2005.

[2] M. K. Lee, C. F. Yen, and J. J. Huang, "Electrical characteristics of liquid-phase-deposited $TiO_2$ films on GaAs substrate with $(NH_4)_2$ Sx treatment," *Journal of the Electrochemical Society*, vol. 153, no. 5, pp. F77–F80, 2006.

[3] C. C. Cheng, C. H. Chien, G. L. Luo et al., "Improved electrical properties of Gd2 $O_3$ GaAs capacitor with modified wet-chemical clean and sulfidization procedures," *Journal of the Electrochemical Society*, vol. 155, no. 3, pp. G56–G60, 2008.

[4] H. C. Chin, M. Zhu, G. S. Samudra, and Y. C. Yeo, "N-xhannel GaAs MOSFET with TaNHfAlO gate stack formed using in situ vacuum anneal and silane passivation," *Journal of the Electrochemical Society*, vol. 155, no. 7, pp. H464–H468, 2008.

[5] G. He, L. D. Zhang, M. Liu, and Z. Q. Sun, "$HfO_2$-GaAs metal-oxide-semiconductor capacitor using dimethylaluminumhydride-derived aluminum oxynitride interfacial passivation layer," *Applied Physics Letters*, vol. 97, Article ID 062908, 3 pages, 2010.

[6] R. Paily, A. DasGupta, N. DasGupta et al., "Pulsed laser deposition of $TiO_2$ for MOS gate dielectric," *Applied Surface Science*, vol. 187, no. 3-4, pp. 297–304, 2002.

[7] S. M. Sze, *Physics of Semiconductor Devices*, chapter 8, John Wiley & Sons, New York, NY, USA, 2nd edition, 1981.

[8] S. A. Campbell, D. C. Gilmer, X. C. Wang et al., "MOSFET transistors fabricated with high permitivity $TiO_2$ dielectrics," *IEEE Transactions on Electron Devices*, vol. 44, no. 1, pp. 104–109, 1997.

[9] R. S. Sonawane, S. G. Hegde, and M. K. Dongare, "Preparation of titanium(IV) oxide thin film photocatalyst by sol-gel dip coating," *Materials Chemistry and Physics*, vol. 77, pp. 744–750, 2003.

[10] P. Zeman and S. Takabayashi, "Effect of total and oxygen partial pressures on structure of photocatalytic $TiO_2$ films sputtered on unheated substrate," *Surface and Coatings Technology*, vol. 153, no. 1, pp. 93–99, 2002.

[11] D. M. Shang and W. Y. Ching, "Electronic and optical properties of three phases of titanium dioxide: rutile, anatase, and brookite," *Physical Review B*, vol. 51, pp. 13023–13032, 1995.

[12] D. C. Gilmer, X. C. Wang, M. T. Hsieh, H. S. Kim, W. L. Glasfelter, and J. Yan, "MOSFET transistors fabricated with high permitivity $TiO_2$ dielectrics," *IEEE Transactions on Electron Devices*, vol. 44, pp. 104–109, 1997.

[13] R. Lyer, R. R. Chang, A. Dubey, and D. L. Lile, "The effect of phosphorous and sulfur treatment on the surface properties of InP," *Journal of Vacuum Science & Technology B*, vol. 6, p. 1174, 1988.

[14] R. W. M. Kwok, L. J. Huang, W. M. Lau et al., "X-ray absorption near edge structures of sulfur on gas-phase polysulfide treated InP surfaces and at $SiN_x$/InP interfaces," *Journal of Vacuum Science & Technology A*, vol. 12, p. 2701, 1994.

[15] E. H. Nicollian and J. R. Brews, *MOS (Metal Oxide Semiconductor) Physics and Technology*, chapter 15, John Wiley & Sons, New York, NY, USA, 2003.

[16] E. K. Badih and J. B. Richard, *Introduction to VLSI Silicon Device Physics, Technology and Characterization*, Kluwer Academic, 1986.

[17] M. L . Reed and J. D. Plummer, "Chemistry of Si-$SiO_2$ interface trap annealing," *Journal of Applied Physics*, vol. 63, p. 5776, 1988.

[18] M. K. Lee, J. J. Huang, and Y. H. Hung, "Variation of electrical characteristics of metallorganic chemical vapor deposited $TiO_2$ films by postmetallization annealing," *Journal of the Electrochemical Society*, vol. 152, no. 11, pp. F190–F193, 2005.

[19] L. M. Terman, "An investigation of surface states at a silicon/ silicon oxide interface employing metal-oxide-silicon diodes," *Solid State Electronics*, vol. 5, no. 5, pp. 285–299, 1962.

[20] D. K. Schroder, *Semiconductor Material and Device Characterization*, chapter 5 and 6, John Wiley & Sons, New York, NY, USA, 1998.

[21] E. H. Nicollian and J. R. Brews, *MOS (Metal Oxide Semiconductor) Physics and Technology*, chapter 8 and 9, John Wiley & Sons, New York, NY, USA, 2003.

[22] C. T. Sah, A. B. Tole, and R. F. Pierret, "Error analysis of surface state density determination using the MOS capacitance method," *Solid State Electronics*, vol. 12, no. 9, pp. 689–709, 1969.

[23] Y. Tao, A. Yelon, E. Sacher, Z. H. Lu, and M. J. Graham, "S-passivated InP (100)-(1×1) surface prepared by a wet chemical process," *Applied Physics Letters*, vol. 60, p. 2669, 1992.

[24] H. P. Song, A. L. Yang, H. Y. Wei et al., "Determination of wurtzite InN/cubic $In_2O_3$ heterojunction band offset by x-ray photoelectron spectroscopy," *Applied Physics Letters*, vol. 94, Article ID 222114, 3 pages, 2009.

[25] T. K. Oh, C. H. Baek, and B. K. Kang, "Surface treatment for enhancing current gain of AlGaAs/GaAs heterojunction bipolar transistor," *Solid-State Electronics*, vol. 48, no. 9, pp. 1549–1553, 2004.

[26] M. Passlack, M. Hong, J. P. Mannaerts, J. R. Kwo, and L. W. Tu, "Recombination velocity at oxide-GaAs interfaces fabricated by in situ molecular beam epitaxy," *Applied Physics Letters*, vol. 68, no. 25, pp. 3605–3607, 1996.

[27] M. Passlack, M. Hong, J. P. Mannaerts, R. L. Opila, and F. Ren, "Thermodynamic and photochemical stability of low interface state density $Ga_2O_3$-GaAs structures fabricated by in situ molecular beam epitaxy," *Applied Physics Letters*, vol. 69, no. 3, pp. 302–304, 1996.

[28] R. S. Besser and C. R. Helms, "Comparison of surface properties of sodium sulfide and ammonium sulfide passivation of GaAs," *Journal of Applied Physics*, vol. 65, p. 4306, 1989.

[29] Y. Ishikawa, T. Fujui, and H. Hasegawa, "Kink defects and Fermi level pinning on (2×4) reconstructed molecular beam epitaxially grown surfaces of GaAs and InP studied by ultrahigh-vacuum scanning tunneling microscopy and x-ray photoelectron spectroscopy," *Journal of Vacuum Science & Technology B*, vol. 15, pp. 1163–1172, 1997.

[30] C. K. Jung, D. C. Lim, H. G. Jee et al., "Hydrogenated amorphous and crystalline SiC thin films grown by RF-PECVD and thermal MOCVD; comparative study of structural and optical properties," *Surface and Coatings Technology*, vol. 171, pp. 46–50, 2003.

[31] H. D. Fuchs, M. Stutzman, M. S. Brandt et al., "Porous silicon and siloxene: vibrational and structural properties," *Physical Review B*, vol. 48, pp. 8172–8189, 1993.

# Modeling of Current-Voltage Characteristics of the Photoactivated Device Based on SOI Technology

**Doron Abraham, Avraham Chelly, David Elbaz, Shimron Schiff, Michah Nabozny, and Zeev Zalevsky**

*Faculty of Engineering, Bar Ilan University, 52900 Ramat Gan, Israel*

Correspondence should be addressed to Zeev Zalevsky, zalevsz@biu.ac.il

Academic Editor: Abdelkarim Mercha

An analytical model of the silicon on insulator photoactivated modulator (SOI-PAM) device is presented in order to describe the concept of this novel device in which the information is electronic while the modulation command is optical. The model, relying on the classic Shockley's analysis, is simple and useful for analyzing and synthesizing the voltage-current relations of the device at low drain voltage. Analytical expressions were derived for the output current as function of the input drain and gate voltages with a parameterization of the physical values such as the doping concentrations, channel and oxide thicknesses, and the optical control energy. A prototype SOI-PAM device having an area of $4\,\mu m \times 3\,\mu m$ with known parameters is used to experimentally validate and support the model. Finally, the model allows the understanding of the physical mechanisms inside the device for both dark and under illumination conditions, and it will be used to optimize and to find the performance limits of the device.

## 1. Introduction

Silicon-based nanophotonic devices have attracted appreciable attention and effort because of their potential applications in silicon-integrated circuits. Although silicon may be not the ideal platform for optical devices, the usage of silicon is mandatory for the generation of optoelectronic integrated circuitry. The quality of commercial silicon on insulator (SOI) wafers driven by the microelectronics industry still continues to improve while the cost continues to decrease. Moreover, as previously mentioned the compatibility with silicon integrated circuits manufacturing is an important reason for this interest in silicon photonics [1].

The optical capabilities are used to overcome some limits of the non optical devices. Those capabilities are high propagation velocity, reduced cross talks and absence of noise coupled by electrical inductance and capacitance [2–4]. Such devices can operate at very low operational power and have parallel processing capabilities [5, 6].

In this paper, we present the first analytic model of the previously presented hybrid optical-electrical device, the *silicon on insulator photo-activated modulator* (SOI-PAM) [7]. As deduced from its name, the device is controlled by optical command. like a junction field effect transistor (JFET), the information current flows between two terminals (source and drain) in a silicon channel. The device concept relays on the fact that the thickness of depletion layer in the channel, which affects the information current, can be controlled through external illumination and not only by applying a gate voltage.

The model was developed for the linear domain of the channel (drain to source) current voltage characteristic and adapted to experimental measurements.

In Section 2, we describe the operation principle of the SOIPAM device. In Section 3, we present the general assumptions for the model. In Section 4, we derive the analytical model in the depletion mode and in Section 5, we develop unified lineal model and extract its parameters. Section 6 concludes the paper.

## 2. SOI-PAM Device

The device is based on a structure consisting of a thin n-type silicon layer (30 nm) on insulator above p-type substrate (SOI-bonded wafer) [8]. The n-type layer was obtained by compensation of the original low-doped p-type silicon layer using an ion implantation of phosphorus.

As depicted in Figure 1, the channel is delimited by source and drain ohmic contacts made of $n^+$ doped polysilicon. The length of the channel is about few microns as fabricated in the prototype version of the device [8]. The information channel is insulated from the p-type substrate by a buried oxide layer having a thickness $t_{ox}$ of 150 nm (SOI wafer). The device is insulated from the surrounding devices by a thicker thermal field oxide.

In principle, when a given negative gate potential $V_G$ is applied to the bottom side of the p-type substrate, a negatively charged depletion layer appears under the buried oxide layer. As a result, a positively charged depletion layer appears inside the n-type channel. Since the n-type doping level (donors concentration of $N_d$ of about $10^{17}$ cm$^{-3}$) of the channel is designed to be larger than the p-type substrate (acceptors concentration $N_a$ of $10^{15}$ cm$^{-3}$), the thickness of the depletion layer in the n-type area is narrower than the depletion layer in the p-type region. Moreover, the device is designed in such a way that, in dark conditions, the depletion layer in the channel is smaller than the channel thickness so the device is partially depleted and can be considered as normally on, that is, moderately conducting under dark condition and zero gate voltage. Note that the gate voltage should be applied with a sufficiently high frequency (few kHz) or swept at a sufficient rate in order to bring the substrate into the "deep" depletion mode and consequently to avoid the inversion of the channel.

If a short pulse of a visible laser illuminates the area under the oxide layer (oblique illumination), the photo generated electrons will accumulate at the bottom silicon-oxide interface and consequently increase the positively charged depletion layer in the channel. With a pulse of sufficient energy, the channel will be cut off. In a future version of the device, a sweeping potential will be added so that the modulation will be limited by the drift time of the photogenerated carriers in a micrometric region rather than by the time consuming recombination process. Assuming a saturation velocity of $10^7$ cm/s and a one micron sweeping length, the drift time is expected to be as small as 10 ps, that is, achievable modulation rates can be as high as 100 GHz.

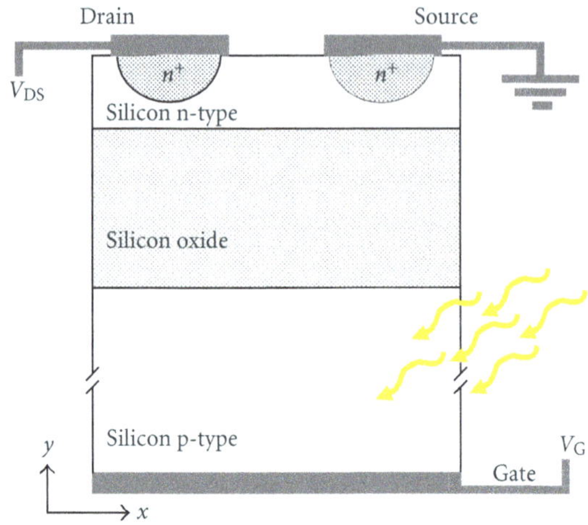


Figure 1: Schematic cross section of the SOI-PAM device. Like in a field effect device the information current flows through silicon channel that is limited by two terminals (source and drain). The source is connected to ground. Gate voltage $V_G$ is applied at sufficient rate to bring the substrate into its deep depletion mode. An oblique illumination generates charges in the p-type substrate.

## 3. General Assumptions

In a similar way used for other field effect devices, the 2D-Poisson's equation will be solved in one dimension (which is the $y$ axis that is being perpendicular to the plane of the Si-SiO$_2$-Si interface) by assuming the Shockley's Gradual Channel Approximation (GCA) [9]. For small drain voltage values, it is reasonable to assume that the electric field varies more rapidly in the $y$ direction than the variations of the lateral field along the channel direction ($x$ axis) due to the large ratio between the channel length and the buried oxide thickness.

The geometry of the device is sketched in Figure 1. The channel, having a length $L$ of 3 $\mu$m, is assumed to be along the $x$ direction. The thickness of the channel $t_n$ is 30 nm and is measured in the $y$ direction. The width of the device $W$ is 4 $\mu$m measured in the $z$ direction. The channel potential $V(x)$ is a function of the distance $x$ along the channel.

In order to derive an expression for the drain current, it is useful to make some simplifying assumptions.

(1) All the regions of the device are assumed to be uniformly doped.

(2) The current flow occurs only within the nondepleted (neutral) portion of the channel.

(3) The mobility of the carriers is not drain voltage dependent (low field mobility is mainly gate voltage dependent).

(4) The oxide is an ideal insulator such the oxide and interface charges are negligible.

(5) The contact series resistances are negligible.

(6) The generation current is negligible.

## 4. Derivation of the Analytic Model for the Depletion Mode

The Poisson's equation is solved separately for each region of the SOI-PAM device structure composed by a p-n depleted junction separated by an oxide layer. As supposed in the GCA approximation, the channel voltage $V(x)$ induced by the drain-source bias $V_{DS}$ is added to the applied negative gate voltage $V_{GS}$ according to the following equation:

$$V_n + V_p + V_{ox} = V_{FB} - V_{GS} + V(x), \quad (1)$$

where $V_n$, $V_p$, $V_{ox}$ are the potentials in n-type depleted region, p-type depleted region and the oxide layer respectively. $V_{FB}$ is the Flat Band (FB) bias of the SOI-PAM device that includes the built-in bias of the p-n junction and the work function differences between the contact metals and the silicon regions (substrate and channel, resp.). Oxide and interface charges are considered to be negligible. Finally, we get from (1) that:

$$\frac{qN_d d_n^2}{2\varepsilon_s} + \frac{qN_a d_p^2}{2\varepsilon_s} + \frac{qN_a d_p t_{ox}}{\varepsilon_{ox}} = V_{FB} - V_{GS} + V(x), \quad (2)$$

where $d_n$ and $d_p$ are the depletion thicknesses of the n and p-type areas respectively. $\varepsilon_{ox}$, $\varepsilon_s$ are the dielectric permittivities of silicon oxide and silicon, respectively. Then, the charge neutrality of the device is expressed in the following equation:

$$N_d d_n = N_a d_p. \quad (3)$$

The channel depletion layer thickness, $d_n(x)$, can be solved accordingly:

$$d_n(x) = \alpha_0 t_n\left(\sqrt{1 + \frac{V_{FB} - V_{GS} + V(x)}{\phi_0}} - 1\right), \quad (4)$$

where $\alpha_0$ is defined as the oxide coupling factor (dimensionless):

$$\alpha_0 = \left(\frac{N_a}{N_a + N_d}\right)\frac{\varepsilon_s t_{ox}}{\varepsilon_{ox} t_n} = 0.148 \quad (5)$$

and $\phi_0$ is the oxide coupling voltage defined as:

$$\phi_0 = \left(\frac{N_a N_d}{N_a + N_d}\right)\left(\frac{\varepsilon_s t_{ox}}{\varepsilon_{ox}}\right)^2 \frac{q}{2\varepsilon_s} = 0.155\,\text{V}. \quad (6)$$

We can easily check that the depleted $d_n$ is zero for $V_{GS}$ equals $V_{FB}$, that is, the flat band voltage is the transition between accumulation and depletion modes as for MOS capacitor.

The maximum value of $d_n(x)$ equals to $t_n$ when the corresponding gate voltage $V_G$, equals to the threshold (or cut off) voltage $V_T$ when taken without drain voltage ($V(x) = 0$). Then from (4) we obtain:

$$\begin{aligned} V_T &= V_G(d_n(x) = t_n)|_{V(x)=0} = V_{FB} - \frac{\phi_o}{\alpha_0^2}(1 + 2\alpha_0) \\ &= V_{FB} - \varphi_{PO}(1 + 2\alpha_0) = V_{FB} - 9.170\,\text{V}, \end{aligned} \quad (7)$$

where $\varphi_{PO}$ is the internal pinch off voltage for the device without oxide and at zero drain voltage as following:

$$\varphi_{PO} \equiv \frac{\phi_o}{\alpha_0^2} = N_d\left(1 + \frac{N_d}{N_a}\right)\frac{q}{2\varepsilon_s}t_n^2 = 7.076\,\text{V}. \quad (8)$$

Assuming that the drain current $I_{DS}$, flowing in the neutral upper part of the depleted channel, is constant along the $x$ axis, we can write that:

$$I_{DS} = qN_D W\mu_n[t_n - d_n(x)]\frac{dV}{dx}. \quad (9)$$

By integrating the expression for $V(x)$ between 0 and $V_{DS}$ we get the oxide-modified Shockley's equation for drain current (as far as $V_{DS} < V_{GS} - V_T$):

$$I_{DS} = g_{max}\left[\begin{array}{c} V_{DS}(1 + \alpha_0) \\ + \dfrac{2}{3\sqrt{\varphi_{PO}}}\left\{\begin{array}{c} [\phi_0 + V_{FB} - V_{GS}]^{3/2} \\ -[\phi_0 + V_{FB} - V_{GS} + V_{DS}]^{3/2} \end{array}\right\} \end{array}\right] \quad (10)$$

where $g_{max}$ is the maximum value of the conductance (for undepleted channel).

$$g_{max} = \frac{qN_d W\mu_n}{L}t_n. \quad (11)$$

One can then derive the channel conductance $g_d$ of the SOI-PAM device from (10):

$$g_d = \frac{\partial I_{DS}}{\partial V_{DS}} = g_{max}\left[1 + \alpha_0 - \sqrt{\frac{V_{FB} + \phi_0 - V_{GS} + V_{DS}}{\varphi_{PO}}}\right]. \quad (12)$$

In the depleted case (where $V_{DS} \ll V_{FB} + \phi_0 - V_{GS}$, that is, $V_{GS} \ll V_{FB} + \phi_0 - V_{DS}$) the channel conductance does not longer depend on $V_{DS}$ but on $V_{GS}$. So, according to (12) and (4), the conductance can be simply related to the closure of the channel by the depletion region entirely controlled by $V_{GS}$ as following:

$$g_d = g_{max}\left[\frac{t_n - d_n(V_{GS})}{t_n}\right]. \quad (13)$$

At the saturation point, for a given $V_{GS}$, the drain current reaches its maximum value, so that the saturation voltage $V_{DS,sat}$ is obtained by zeroing the channel conductance $g_d$, in (12), yielding:

$$V_{DS,sat} = \phi_{PO}(1 + 2\alpha_0) - V_{FB} + V_G = V_G - V_T \equiv V_{GT}. \quad (14)$$

From (7) the threshold voltage $V_T$ is expected to be high in the device, so the saturation will not occurred for the relatively small values of $V_{DS}$ considered in the depletion mode.

According the above equations, we can derive some qualitative expectations on the influence of material parameters on the conductance $g_d$. According to (11) and (12) $g_d$ is increased by increasing: the channel doping concentration $N_d$ or the channel thickness $t_n$ or also the buried oxide thickness $t_{ox}$. Also, gd is increased by decreasing the substrate concentration $N_a$. Finally, we can notice that for the limit case where the oxide layer vanishes ($t_{ox} \rightarrow 0$) both the silicon oxide coupling factor in (5) and the silicon oxide coupling voltage in (6) vanish while $V_{FB}$ equals to the junction's built-in voltage $\varphi_i$. So, the above expressions (12)–(14) are reduced to those describing a classic JFET n-channel device (for which $N_a \gg N_d$).

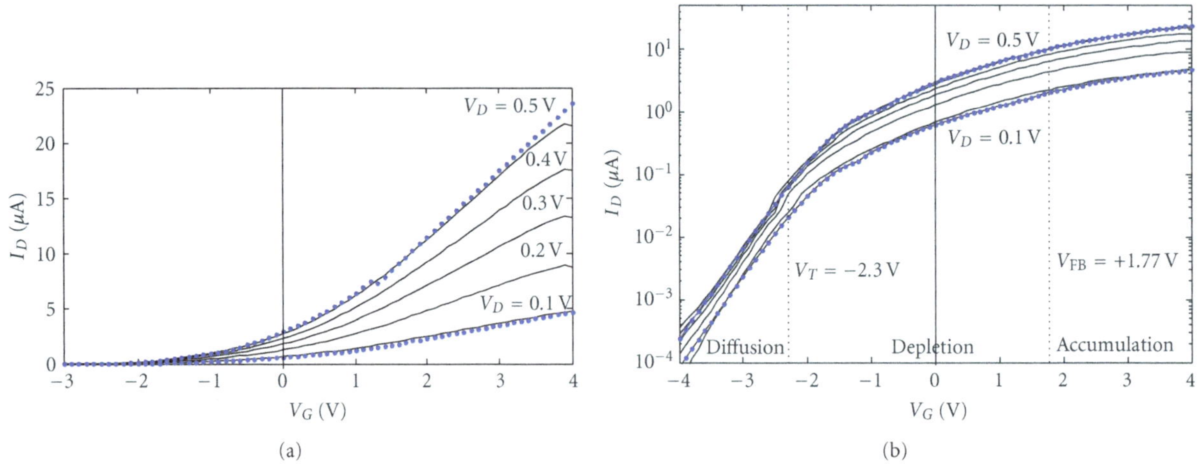


FIGURE 2: Experimental $I_{DS}$-$V_{GS}$ characteristics: linear scale in (a) and semi-log in (b). The measurements are taken in dark conditions (solid lines) for several values of $V_{DS}$ between 0.1 V and 0.5 V. The dotted lines result from the calculated current according to the proposed unified model. For clarity, only results obtained for $V_{DS}$ = 0.1 V and $V_{DS}$ = 0.5 V are presented. The $V_{GS}$ domains of the operation modes are specified in (b).

## 5. The Unified Linear Model and Parameters Extraction

*5.1. $I_{DS}$-$V_{GS}$ Characteristics and Operation Modes.* A prototype of the constructed SOI-PAM device was experimentally characterized using the Keithley SCS-4200 parameter analyzer. As seen in Figure 2, the $I_{DS}$-$V_{GS}$ characteristics are measured over [−4 V, 4 V] $V_{GS}$ range for several low $V_{DS}$ values (0.1–0.5 V) in order to get a linear relation between $I_{DS}$ and $V_{DS}$.

In this section we develop a unified model of the $I_{DS}$-$V_{GS}$ characteristics based on the MESFET unified model [9] available for the three operation modes of the device and at low $V_{DS}$ values. A similar approach was used by Colinge for a SOI p-channel MOSFET [10].

*(1) Accumulation Mode ($V_{GS} > V_{FB} > 0$).* The n-channel is conducting through the accumulation layer of majority carriers located at the top of the silicon-buried oxide interface. The electron concentration is determined by $V_{GS}$-$V_{FB}$.

*(2) Depletion Mode ($V_T < V_{GS} < V_{FB}$).* The n-channel is partially depleted but is still conducting through the squeezed neutral region of the channel above the depletion region.

*(3) Diffusion Mode ($V_{GS} < V_T$).* The n-channel is fully depleted; there is a small diffusion current flowing from the $n^+$ source region through the $n$ depleted channel. The current is controlled exponentially by the gate voltage as observed in MESFET devices [9]. The $V_{GS}$ sweep is performed quickly enough such the channel inversion is not achieved so the channel is considered in a "deep depletion" state.

We will now express the drain current model for each one of those mode as following:

*(1) Accumulation Mode ($V_{GS} > V_{FB} > 0$).* The linear drain current can be described by the following relation as observed in the pseudo-MOSFET device [11]:

$$I_{DS} = \frac{W}{L}\mu_n C_{ox}(V_{GS} - V_{FB})V_{DS}, \tag{15}$$

where $C_{ox}$ is the buried oxide capacitance per unit area, given by:

$$C_{ox} = \frac{\varepsilon_{ox}}{t_{ox}} = 23\,\text{nF/cm}^2. \tag{16}$$

From the linear extrapolation at positive values of $V_{GS}$, we can extract the values of $V_{FB}$ (~0.1 V) and $\mu_n$ (~700 cm$^2$/Vs) which are roughly constant with $V_{DS}$. The extracted mobility value is found consistent with the similar thin pseudo SOI-MOSFET [11] having the same silicon film thickness and the doping concentration range of the n-channel ($10^{17}$ cm$^{-3}$). However due to the non uniformity of the ion implant process, the doping value is not precisely known for each device. Moreover due to the rapid variation of the mobility with doping concentration in this range ($10^{17}$ cm$^{-3}$), the doping concentration cannot be extracted accurately from the mobility.

*(2) Depletion Mode ($V_T < V_{GS} < V_{FB}$).* The drain current in the partially depleted channel can be rewritten by the following relation [9]:

$$I_{DSa} = q n_{sa} \mu_n \frac{W}{L} V_{DS} \tag{17}$$

with $n_{sa}$ being the electron sheet density in the channel above the threshold voltage $V_T$. This density is simply given by:

$$n_{sa} = N_d(t_n - d_n) \tag{18}$$

while $d_n$ is the thickness of the depleted region in the channel at low $V_{DS}$ so it is mainly a square root function of $V_{GS}$ as given by (4).

*(3) Diffusion Mode ($V_{GS} < V_T$).* For $V_{DS}$ positive values, the saturation condition ($V_{DS} > V_{GS} - V_T$) is always satisfied in this mode, so the drain current in the fully depleted channel is being described as a saturated diffusion current (independent of $V_{DS}$) according to [9]:

$$I_{DSb} = qn_{sb}D_n\frac{W}{L} = qn_{sb}\mu_n\frac{W}{L}V_{th}, \tag{19}$$

where $V_{th}$ is the thermal voltage (25.6 mV at room temperature) and $n_{sb}$ is the electron sheet density in the channel below the threshold voltage $V_T$. For $V_{DS}$ larger than $V_{th}$, $n_{sb}$ is defined by a Boltzmann's distribution like:

$$n_{sb} = n_0 \exp\left(\frac{V_{GS} - V_T}{\eta V_{th}}\right) < n_0 \tag{20}$$

with $n_0$ being the electron sheet density at $V_T$ [9]:

$$n_0 = \left(\frac{C_{si}C_{ox}}{C_{si} + C_{ox}}\right)\frac{\eta V_{th}}{q}. \tag{21}$$

And $C_{Si}$ is the fully depleted silicon channel capacitance per unit area at $V_T$, given by:

$$C_{si} = \frac{\varepsilon_{si}}{t_n} = 345\,\text{nF/cm}^2 = 15C_{ox}, \tag{22}$$

$\eta$ is the ideality factor extracted from the slope of the subthreshold exponential fit and is evaluated to be 11.5 as the mean value over several $V_{DS}$ values as seen in Figure 2(b). This value is comparable to the $C_{Si}/C_{ox}$ ratio as also reported in the subthreshold current model for MOSFET devices [9].

According to (19) and (20), the threshold voltage $V_T$ can be extracted as the departure point of the exponential fit [9] and is found to be −2.3 V as shown in Figure 2(b) for $V_{DS}$ equals 0.1 V. The corresponding threshold current $I_{DSb,T}$ is 25 nA which is the maximum diffusion current allowed in the channel.

However, according to (7), the expected value of $V_T$ is −9 V. Actually, the measured current is not vanishing at −9 V. Moreover, the depletion current $I_{DSa}$ is a power law function of $V_{GS}$ while the diffusion current $I_{DSb}$ is an exponential decay with $V_{GS}$ below −2.3 V. So the diffusion is the limiting process in the subthreshold mode. Consequently, the measured drain current can be modeled by the following expression [9]:

$$I_{DS,depl} = \frac{I_{DSa}I_{DSb}}{I_{DSa} + I_{DSb}}. \tag{23}$$

Experimentally (see Figure 2(b)), $I_{DSb}$ is found to be slightly $V_{DS}$ dependent in spite of the saturation condition assumed above. In fact, we can link this dependence to those of $V_T$ with $V_{DS}$ at low $V_{DS}$ values. This is similar to the DIBL effect found in MOSFET's devices [9]. The $V_T$ values are extracted from the threshold current $I_{DSb,T}$ fixed at 25 nA for each $I_{DS}$-$V_{GS}$ from Figure 2(b). For instance, for $V_{DS}$ equals 0.5 V, the extracted $V_T$ value is −2.6 V. These values are used in the model.

*5.2. Fine Tuning of the Doping Concentration.* Due to the lack of accuracy of the doping concentration, this latter can be refined in order that (23) will best fit the measured dain current in Figure 2(b) in particular at $V_{DS} = 0.5$ V. The matching value for $N_d$ is found as $8 \times 10^{16}\,\text{cm}^{-3}$, instead of the initial designed value of $10^{17}\,\text{cm}^{-3}$, but still reasonable in regard to the doping distribution. This tuning is performed according the mobility voltage-dependent model as presented below.

*5.3. Mobility Voltage Dependence.* The electron mobility can be evaluated from the extracted threshold voltage $V_T$ (−2.3 V) using (19) and (20) as follows:

$$\mu_n(V_T) = \left(\frac{qn_0}{I_{DSb,T}}\frac{W}{L}V_{th}\right) = 200\,\text{cm}^2/\text{Vs}. \tag{24}$$

This mobility value is found to be much lower than the value measured in the accumulation mode (700 cm²/Vs). This is not surprising since the low field mobility in the depletion mode is reported to increase rapidly for gate voltages above $V_T$ in MESFET devices [12]. This should be explained by the decreasing of the depletion layer while increasing $V_{GS}$ above $V_T$. According to this reference, we can expect a linear relation of the mobility with $V_{GS}$ in order to match the experimental $I_{DS}$-$V_{GS}$ characteristics in the depletion mode like

$$\mu_n(V_{GS}) = \mu_0 + \theta(V_{GS} - V_T). \tag{25}$$

Another assumption is necessary for the physical coherence of the proposed model. On the one hand, the mobility cannot vanish with $V_{GS}$ so it should be pinned to the lower threshold limit value of 200 cm²/Vs even for $V_{GS}$ values below $V_T$. On the other hand, for the continuity sake, the depletion current above $V_T$, $I_{DSa}$, should match the accumulation current voltage [11]. So the current matching between the accumulation and the depletion current (for which $d_n \sim 0$) may occur at a specific "effective" flat band bias $\underline{V}_{FB}$ defined as follows:

$$I_{DS,acc} = \frac{W}{L}\mu_n C_{ox}(\underline{V}_{FB} - V_{FB})V_{DS} = I_{DSa} = qN_d t_n \mu_n \frac{W}{L}V_{DS}, \tag{26}$$

the expression of $\underline{V}_{FB}$ is consequently given by

$$\underline{V}_{FB} = V_{FB} + \frac{qN_D}{C_{ox}}t_n = 1.77\,\text{V}, \tag{27}$$

with: $N_D = 8 \times 10^{16}\,\text{cm}^{-3}$.

This "effective" flat band bias $\underline{V}_{FB}$ value actually defines the lower limit of $V_{GS}$ for which the accumulation current may be described by (15). Also, $\underline{V}_{FB}$ defines the upper limit of $V_{GS}$ for which the depletion current may be described by (10).

The mobility variation with $V_{GS}$ can finally be modeled by using the following step function:

$$\mu_n(V_{GS}) = \begin{cases} \mu_0 \approx 200\,\text{cm}^2/V_S, & V_{GS} \le V_T, \\ \theta(V_{GS} - V_T), & \theta \approx 175\,\text{cm}^2/V_S, \\ & V_T \le V_{GS} \le \underline{V}_{FB}, \\ \mu_n \approx 700\,\text{cm}^2/V_S, & V_{GS} \ge \underline{V}_{FB}. \end{cases} \quad (28)$$

The comparison between the experimental characterization and the unified model as described in this section shows a satisfying matching for $V_{DS}$ values of 0.1 V and 0.5 V as can be seen from Figures 2(a) and 2(b).

*5.4. $I_{DS}$-$V_{DS}$ Characteristics.* In order to confirm and complete the current-voltage model of the SOI-PAM device, the measured $I_{DS}$-$V_{DS}$ characteristics are presented in Figure 3. The unified model is used to simulate these characteristics up to the saturation region for $V_{GS}$ equals −1, 0, 1, 2, and 3 V, respectively, and for positive $V_{DS}$ values up to 3 V. The linear behavior of the drain current in the saturation region can be interpreted as a drain-induced channel length modulation [9].

For $V_{GS}$ equals 3 V, the device is in the accumulation mode so the corresponding $I_{DS}$-$V_{DS}$ characteristic is well fitted by (15) by adding the parabolic term in $V_{DS}$ up to the saturation region as used for MOSFET device:

$$I_{DS} = \frac{W}{L}\mu_n C_{ox}\left[(V_{GS} - V_{FB})V_{DS} - \frac{V_{DS}^2}{2}\right]. \quad (29)$$

The saturation voltage is then defined as

$$V_{DS,sat} = V_{GS} - V_{FB} \quad (30)$$

with $V_{FB}$ equals to 0.1 V.

The measured saturation voltage (Figure 3) matches the expected value (−2.9 V).

For $V_{GS}$ equals 2 V, the device is still in the accumulation mode, so the previous equation is valid till the $V_{GS}$ equals $\underline{V}_{FB}$, that is, 1.77 V as extracted from (26). However, a small discrepancy is observed above 1 V near the saturation region. This may happen due to an enhancement of the channel length modulation effect as mentioned above [9].

For $V_{GS}$ equals 1 V, the device is now in the depletion mode, so (17) can be used to fit the corresponding characteristic by taking $\mu_n$ value of 577 cm²/Vs as estimated from (28). The saturation voltage $V_{DS,sat}$ measured as 2.3 V on Figure 3, is overestimated by the modeled $V_{GS}$-$V_T$ value (3.3 V from (14)) may be due to the velocity saturation [9] that was not included in the present model.

For $V_{GS}$ equals 0 V, the contribution of the diffusion current is negligible and (17) is still sufficient to fit the corresponding characteristic by taking $\mu_n$ value of 402 cm²/Vs as estimated from (28) but with slight overestimation of the current value and of $V_{DS,sat}$ as noticed above.

For $V_{GS}$ equals −1 V, the contribution of the diffusion current is now not negligible and (23) is used to fit the corresponding characteristic by taking $\mu_n$ value of 227 cm²/Vs as estimated from (28) but again with slight overestimation of the current value and of $V_{DS,sat}$ as noticed above.

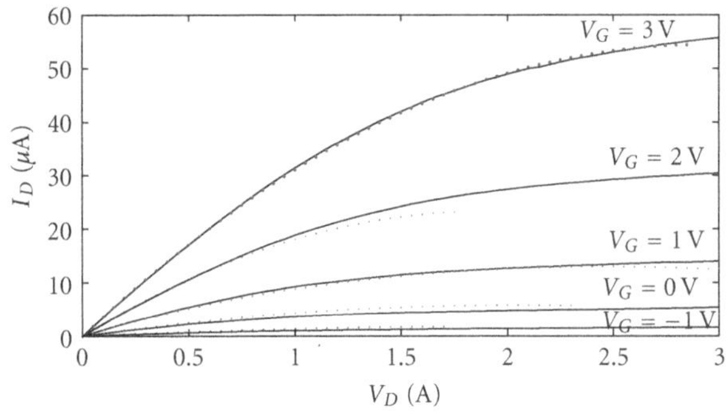


FIGURE 3: Experimental $I_{DS}$-$V_{DS}$ characteristics: the measurements are taken in dark conditions (solid lines) for several values of $V_{GS}$ between 3 V and −1 V. The dotted lines result from the calculated current according to the proposed unified model.

*5.5. Simulation of the Illumination Effect.* By synchronously and directionally illuminating the p-type substrate, the free photogenerated electrons concentration can be significantly and instantly increased to build the inversion (negatively charged) layer under the silicon oxide. Consequently, the positively charged depletion layer in the n-type channel is widening and could fully deplete the channel.

We can express the n-type drop voltage dependency by the amount of electrons that are generated by the applied illumination. The number of incident photons $N_{ph}$ is related to the total energy $E_{ph}$ as:

$$E_{ph} = N_{ph}\frac{hc}{\lambda}, \quad (31)$$

where $h$ is Planck's constant, $c$ is the speed of light and $\lambda$ is the photon's wavelength. The capacitances of the partially depleted silicon layer $C_{Si}$ and the buried oxide $C_{ox}$ are associated in series, so the total capacitance per unit area is

$$C = \frac{C_{ox}C_{Si}}{C_{ox} + C_{Si}}\,[\text{F/cm}^2]. \quad (32)$$

Then, channel's voltage drop, that is, the "illumination induced voltage" can be expressed by

$$V_{ph} = \frac{Q_{ph}}{CWL} = \frac{q\eta_\lambda N_{ph}}{CWL}, \quad (33)$$

where $\eta_\lambda$ is the quantum efficiency of silicon for the given $\lambda$. So, for relatively low illumination levels, the thickness of the channel depletion layer $d_n(x)$ can be increased by the illumination-induced voltage $V_{ph}$ according to (4)

$$d_n(x) = \alpha_0 t_n\left(\sqrt{1 + \frac{V_{FB} - V_{GS} + V(x) + V_{ph}}{\phi_0}} - 1\right). \quad (34)$$

Consequently, the depleted channel region will grow and the depletion mode current $I_{DSa}$ will decrease when

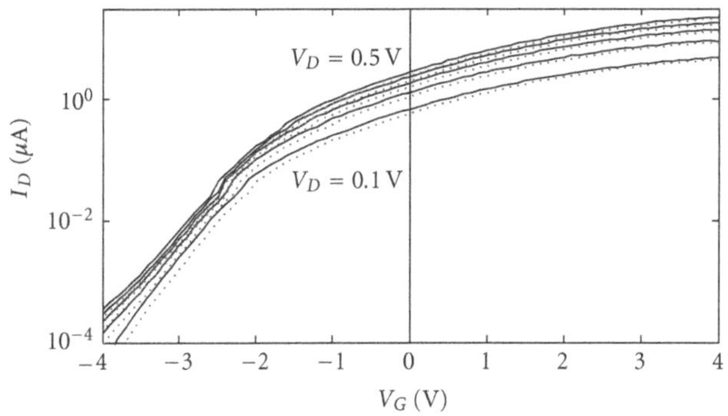


FIGURE 4: Measured $I_{DS}$-$V_{GS}$ characteristics (semilog) taken at dark (solid line) and under room light illumination (dashed line). The dashed line is slightly shifted to more positive $V_{GS}$ values especially in the depletion and the diffusion modes ($V_{GS} < 1.77$ V).

the illumination energy is increased, as expected from the operation principle of the device. For the actual device parameters, we can evaluate from (34) that the channel will be fully closed (for small $V_{DS}$) by taking $V_{ph}$ equals 4.5 V, that is, by supplying 75,000 photons (assuming an ideal unity quantum efficiency). In terms of light power, taking for instance, a green diode laser ($\lambda = 532$ nm), it would be estimated to 2.8 mW per device for a 10 ps pulse. This value can be minimized by decreasing the size of the device and increasing the oxide thickness.

Experimentally, by simply exposing the device to room (fluorescent) light, we can notice a slight positive shift of the whole $I_{DS}$-$V_{GS}$ characteristics as shown in Figure 4. This evidence roughly the light-induced decreasing of the drain current as expected from the present model.

Of course, a more in depth measurements of this photoelectric modulation should be performed to validate the model more accurately and will be the object of further publication.

## 6. Conclusions

In this paper, we have derived a unified analytical model of the $I_{DS}$-$V_{GS}$ and $I_{DS}$-$V_{DS}$ characteristics for whole the operation modes of the SOI-PAM device matching the linear $I_{DS}$-$V_{GS}$ domain.

The analysis of the depletion mode is based on the classical Shockley's model but originally adapted to the presence of buried oxide thickness between the n-type channel and the p-type substrate.

The effect of the light on the channel current modulation could be simply expressed as a positive shift of the gate voltage, and a qualitative behavior of the modulation has been demonstrated.

## References

[1] A. Rudnitsky, A. Shahmoon, M. Nathan et al., "All-optical integrated micro logic gate," *Microelectronics Journal*, vol. 42, no. 2, pp. 472–476, 2011.

[2] G. Steinmeyer, "A review of ultrafast optics and optoelectronics," *Journal of Optics A*, vol. 5, no. 1, pp. R1–R15, 2003.

[3] A. Rostami, "Low threshold and tunable all-optical switch using two-photon absorption in array of nonlinear ring resonators coupled to MZI," *Microelectronics Journal*, vol. 37, no. 9, pp. 976–981, 2006.

[4] H. K. Tsang, C. S. Wong, T. K. Liang et al., "Optical dispersion, two-photon absorption and self-phase modulation in silicon waveguides at 1.5 $\mu$m wavelength," *Applied Physics Letters*, vol. 80, no. 3, p. 416, 2002.

[5] J. Faist, "Silicon shines on," *Nature*, vol. 433, no. 7027, pp. 691–692, 2005.

[6] Q. Xu, B. Schmidt, S. Pradhan, and M. Lipson, "Micrometre-scale silicon electro-optic modulator," *Nature*, vol. 435, no. 7040, pp. 325–327, 2005.

[7] D. Abraham, Z. Zalevsky, A. Chelly, J. Shappir, and M. Rosenbluh, "Silicon on insulator photo-activated modulator," *Microelectronics Journal*, vol. 39, no. 12, pp. 1429–1432, 2008.

[8] D. Abraham, Z. Zalevsky, A. Chelly, and J. Shappir, "Fabrication of vertically positioned silicon on insulator photo-activated modulator," *Photonics and Nanostructures*, vol. 7, no. 4, pp. 190–197, 2009.

[9] K. Lee, M. Shur, T. A. Fjeldly, and T. Ytterdal, *Semiconductor Device Modeling for VLSI*, Prentice Hall Series in Electronics and VLSI, chapter 4, 1993.

[10] J. P. Colinge, "Conduction mechanisms in thin-film accumulation-mode SOI p-channel MOSFET's," *IEEE Transactions on Electron Devices*, vol. 37, no. 3, pp. 718–723, 1990.

[11] S. Cristoloveanu, D. Munteanu, and M. S. T. Liu, "A review of the pseudo-MOS transistor in SOI wafers: operation, parameter extraction, and applications," *IEEE Transactions on Electron Devices*, vol. 47, no. 5, pp. 1018–1027, 2000.

[12] B. J. Moon, M. J. Helix, and S. Lee, "A new technique to determine the average low-field electron mobility in MESFET using C-V measurement," *IEEE Transactions on Electron Devices*, vol. 39, no. 9, pp. 1982–1986, 1992.

11

# A Very Robust AlGaN/GaN HEMT Technology to High Forward Gate Bias and Current

**Bradley D. Christiansen,[1] Eric R. Heller,[2] Ronald A. Coutu Jr.,[1] Ramakrishna Vetury,[3] and Jeffrey B. Shealy[3]**

[1] *Department of Electrical and Computer Engineering, Air Force Institute of Technology, Wright-Patterson Air Force Base, OH 45433, USA*
[2] *Materials and Manufacturing Directorate, Air Force Research Laboratory, Wright-Patterson Air Force Base, OH 45433, USA*
[3] *Defense and Power Business Unit, RF Micro Devices, Inc., Charlotte, NC 28269, USA*

Correspondence should be addressed to Bradley D. Christiansen, bradley.christiansen@wpafb.af.mil

Academic Editor: Jung-Hui Tsai

Reports to date of GaN HEMTs subjected to forward gate bias stress include varied extents of degradation. We report an extremely robust GaN HEMT technology that survived—contrary to conventional wisdom—high forward gate bias (+6 V) and current (>1.8 A/mm) for >17.5 hours exhibiting only a slight change in gate diode characteristic, little decrease in maximum drain current, with only a 0.1 V positive threshold voltage shift, and, remarkably, a persisting breakdown voltage exceeding 200 V.

## 1. Introduction

There are recent reports stating that high negative gate bias causes the gates of GaN HEMTs to degrade. The signature of this degradation mechanism is an increase in gate leakage current [1–3].

Other reports state that forward gate current limits the survival times of GaN HEMTs, especially during RF operation [4, 5]. GaN HEMTs ($L_g = 0.7\,\mu$m, $W_g = 2 \times 100\,\mu$m) in [6] reached about 400 mA/mm forward gate current before burning out. The semi-insulating Fe-doped GaN of [6] was grown by MOCVD on sapphire substrates.

Reference [7] specifically considered the effects of high positive gate bias (up to +6 V) on GaN HEMTs with gate-integrated field plates, $L_g = 0.25\,\mu$m, and $W_g = 2 \times 25\,\mu$m. By stepping $V_G$ from +0.5 V to +6 V in 0.5 V steps for 30 minutes per step, a reduction in $V_{\text{Gon}}$ ($V_G$ at a normalized gate current of 1 mA/mm) after 360 minutes was observed, accompanied by about a $10^4$ increase in gate leakage current and strong Ohmic gate behavior. The authors concluded that the large forward gate current and high temperature degraded the Schottky contact. Despite the reduction in $V_{\text{Gon}}$, [7] showed that there was little degradation in drain and source resistances and maximum drain current.

We show the results of stressing GaN HEMTs at extremely high gate current densities but for longer times and with more positive results. Despite the extremely high biases, the devices we tested survived well past the current density in [6] and were less conductive and with less $V_{\text{Gon}}$ degradation than those in [7]. Remarkably, it is noted that the devices we tested survived the stresses—contrary to conventional wisdom—with very little degradation in device drain current and voltage capability.

## 2. Experimental

We stressed four nominally identical AlGaN/GaN HEMTs. All four devices had the same structure which consisted of a semi-insulating SiC substrate [8], a 0.5 $\mu$m length optically defined gate with a gate-integrated field plate [8], and a source-connected field plate [9]. Gate width was $2 \times 50\,\mu$m. The gate contained a nickel Schottky barrier and thick gold overlay for low gate resistance. The highly resistive GaN buffer was grown by OMVPE [8]. The AlGaN barrier was undoped [8]. SiN, grown by PECVD, was used for passivation [8]. The gate-to-drain gap was greater than the gate-to-source gap [8].

The devices were tested on a Peltier thermal base plate in air. The power supplies used were Agilent models E5280A High-Power Source/Monitor Unit (SMU) Module (for the drain) and E5281A Medium-Power SMU Module (for the gate) in a model E5273A 2-Channel SMU.

We stressed the devices to define a safe operating area. An initially tested part (not shown) was stressed to $I_G \approx 260\,\text{mA/mm}$ at a base plate temperature of $T_{\text{bp}} = 35°\text{C}$ without any discernable degradation, which provided the reference and motivation for this study. Based on this observation, a detailed study of gate robustness was conducted as described next. For this detailed study, the source and drain were wire-bonded and the gate was contacted by a needle probe. The following sequence was used at $T_{\text{bp}} = 45°\text{C}$. The voltage-sweep and transfer-curve voltage ranges were divided into 201 linear, ~35 ms dwell steps (~7 seconds total sweep time).

(1) Characterize with a transfer curve ($V_G = -6$ V to 1 V and $V_D = 10$ V) before any stress and after each stress sweep (steps 2–6).

(2) Sweep $V_G$ three times from 0 V to +2.5 V with $V_S = V_D = 0$ V.

(3) Sweep $V_G$ four times from 0 V to +2.5 V with $V_S = V_D = 0$ V and hold $V_G$ at +2.5 V for 1 minute.

(4) Sweep $V_G$ from 0 V to +3.0 V with $V_S = V_D = 0$ V and hold $V_G$ at +3.0 V for 1 minute. Repeat at +0.5 V increments to $V_G = +6.5$ V ($I_G = 1.89$ A/mm).

(5) Sweep $V_G$ from 0 V to +6.0 V with $V_S = V_D = 0$ V and hold $V_G$ at +6.0 V for 30 minutes.

(6) Repeat step 5 with 30-minute, 150-minute, 120-minute, and >12-hour holding times, respectively, at $V_G = +6.0$ V ($I_G \approx 1.82$ A/mm). After the last hold, the device was in a small-bias state ($I_G \approx 300\,\mu\text{A/mm}$, $V_S = V_D = 0$ V) for two days due to the gate contact needle probe coming loose.

In summary, the test lasted more than 17.5 hours at $V_G = +6.0$ V in addition to 1 minute at $V_G = +6.5$ V.

## 3. Results and Discussion

Results from one of the tested devices are described in the following. In summary, the device survived for >17.5 hours the stress of biasing at $V_G = +6.0$ V and $I_G \geq 1820$ mA/mm forward gate current (see Figure 2), which is a current density of >360 kA/cm$^2$ and >10.9 W/mm power *through the gate*. $I_{D\max}$ (defined as drain current at $V_G = 1$ V and $V_D = 10$ V) degraded slightly and saturated over stress duration (see Figure 1). After 210 minutes of stress, degradation in the ideality of $I_G$-$V_G$ was observed at low currents (see the upper left inset of Figure 2), although it is noted that this ideality degradation did not significantly impact the current handling capability and breakdown voltage capability of the device.

The forward gate current values from our tested devices were far more than the value of 400 mA/mm (57 kA/cm$^2$) reported in [6] that destroyed the particular GaN HEMTs tested in that report. The devices tested in this report were also less conductive at high gate bias (1.63 A/mm, 326 kA/cm$^2$ at $V_G = +5$ V) compared to those of [7], wherein the forward gate current was reported to be 2 A/mm (800 kA/cm$^2$) at $V_G = +5$ V. We note that these comparisons are to devices with different gate lengths, contact and sheet resistances, and source-to-gate-to-drain gaps, although the point made strictly regards the robustness of the Schottky gate which is less sensitive to these differences.

Figure 1 shows the transfer curves and associated gate current in absolute value and transconductance of the device throughout stressing. Black lines show step 1 (initial) and the last repeat of step 2; red lines show the last repeat of step 3 and results of step 4; finally, green lines show steps 5-6. Initial slight improvement in gate current (trapping or burn-in behavior likely) gives way to degradation. The transfer curves exhibit a degradation trend with a *saturation* apparently after the first 30-minute holding time, much like a transient burn-in effect. Two separate causes—resulting in quick degradation in the short term and slow degradation in the long term—appear responsible for the electrical changes observed during exposure to bias. The gate current increases in a different fashion, with a decreased rate of change—leading to possible saturation—after the first several hours of stress. There was little drain current degradation at the tested current density—a 6.1% reduction, comparing the prestress $I_{D\max}$ of 787 mA/mm to the poststress $I_{D\max}$ of 739 mA/mm. In addition, there was only a 0.1 V positive shift in threshold voltage.

Figure 2 shows the device's gate diode curves during the stress sweep from $V_G = 0\text{ V to } +2.5\text{ V} \leq V_G \leq +6.5$ V (steps 2–6). During the first few sweeps, the gate current improved. After this, excess leakage at low gate bias appeared (see upper left inset), then the gate current increased with stress time and saturated. Despite the ideality degradation, the breakdown voltage remained above 200 V (based on satisfying the nominal 1 mA/mm industry criterion for breakdown). If there is any change in $V_{\text{Gon}}$ it is slight and masked by other effects. In contrast, [7] observed a noticeable change in $V_{\text{Gon}}$ of ~0.5 V in significantly shorter stress time.

There are insufficient details in [6, 7] to adequately compare and contrast those structures and the present structure. However, technology maturity, gate metal stack differences, and the source-connected field plate in the present technology are possible factors for the observed improvements. Reference [10] provides a physical explanation why positive gate bias and current may not be as damaging to GaN HEMTs as conventional wisdom dictated. Reference [10] also provides an example of observations that do not conform to expectations for thermally induced degradation: transconductance degraded more for a semi-ON state than for a higher-power ON state.

The high forward gate current seen in this testing would have caused significant degradation or failure in earlier vintage GaN HEMTs. The tested GaN HEMTs exhibited the ability to withstand nearly constant high forward gate stress and

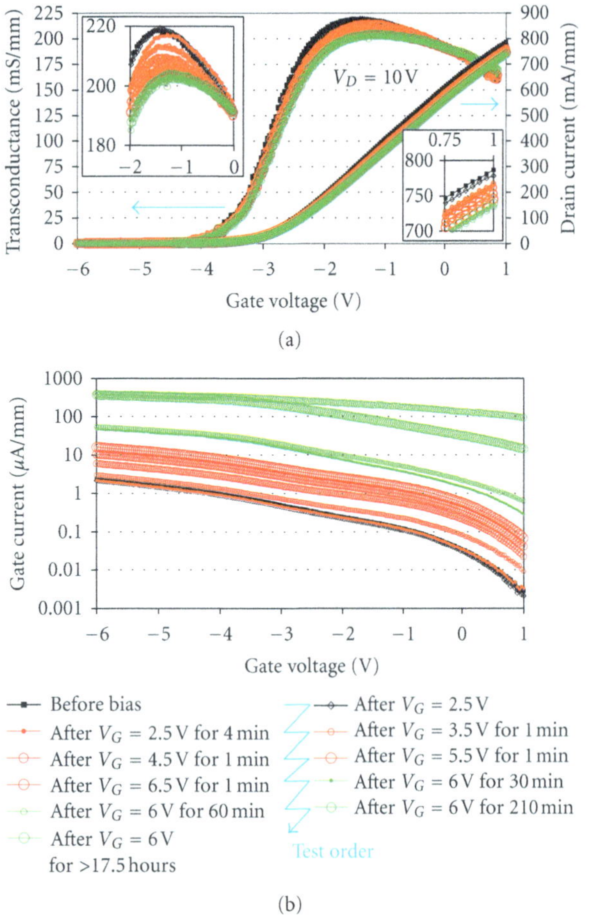


Figure 1: Transfer and transconductance curves (a) and associated gate current in absolute value (b) of the device as measured during characterizations between gate stressing events. Insets show detail at regions of interest in the same data sets. Extra gate current is seen in (b) above $V_G \approx -3.5$ V after 210-minute stress (top curve) that is not seen after longer stress time (second-to-top curve). It is not known if there was a temporary test issue or if that is indeed real.

demonstrated a high level of allowable forward gate stress. The drain current failure criterion in [9] was −10% $I_{DSS}$. The nominal industry gate leakage failure criterion is 1 mA/mm. Neither failure criterion was reached in the characterizations of the present testing (see Figure 1). Although the gate leakage increased more than two orders of magnitude (see Figure 1(b)), $I_{D\max}$, $V_{\mathrm{Gon}}$, and the breakdown voltage were not significantly affected. Of course, the maximum allowable degradation in any particular parameter depends on the specific application.

We should clarify that we do not anticipate continuous long-term bias in real-world operation at the high $V_G$ conditions we have tested and are not stating that this has been shown to be practical. Instead, we showed that brief excursions, or short-term operation, up to $V_G = +6.5$ V may be feasible. Based on the stress time endured for this testing, such excursions would not be catastrophic.

To estimate device lifetime due to high forward gate stress, an appropriate accelerant and acceleration model

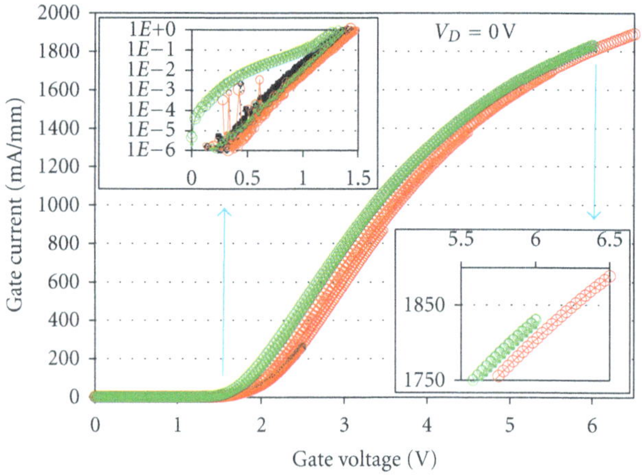


Figure 2: Gate diode curves during the stressing. Insets show additional detail for regions of interest of the same curves as the main plot and share the same units (i.e., mA/mm and V) as the main figure. The data were collected at stress times represented in Figure 1. Black curves represent the initial $V_G = +2.5$ V gate stresses. Red curves represent the gate voltage stress ramps of increasing magnitude collected just prior to the red curves of Figure 1. Green curves are gate voltage stress ramps collected just after the total stress times represented by the green curves of Figure 1.

would need to be determined (since testing at use conditions is impractical) through additional testing and analysis. In [3], the authors state that gate degradation due to high reverse gate bias is weakly dependent on temperature and strongly dependent on gate voltage. Gate voltage may also be an accelerant for forward gate stress. Physical failure analysis would also be required to understand the degradation due to high forward gate stress.

## 4. Conclusion

The mere survival of the device tested at $I_G \geq +1.8$ A/mm and $V_G \geq +6.0$ V for >17.5 hours is remarkable, in addition to the modest degradation in drain current that appears to saturate over stress time. The results reported herein are reproducible as evidenced by the similar responses of three devices, in addition to the fact that these parts are of standard commercial design from a baseline fabrication process. The results observed indicate that the GaN HEMTs tested are extremely robust to high forward gate bias and current. Devices based on the tested structure show the potential to withstand the rigors of forward gate bias and current during RF operation, and the high $I_G$ tolerance seen may allow extra latitude to circuit designers. Further investigations are required to understand the time-temperature-$V_G$ trade space and the full extent of degradation due to high forward bias and current under RF operation and over very long time periods (thousands of hours).

## Disclosure

The views expressed in this paper are those of the authors and do not reflect the official policy or position of the United

States Air Force, Department of Defense, or the US Government.

## Acknowledgments

The authors thank the Air Force Research Laboratory High-Reliability Electronics Virtual Center team for their support and Kelson Chabak, Antonio Crespo, Brian Poling, and Steve Tetlak for helpful suggestions. This research was funded by the Aerospace Components and Subsystems Division, Sensors Directorate, Air Force Research Laboratory (AFRL/RYD).

## References

[1] J. Joh and J. A. del Alamo, "Critical voltage for electrical degradation of GaN high-electron mobility transistors," *IEEE Electron Device Letters*, vol. 29, no. 4, pp. 287–289, 2008.

[2] J. A. del Alamo and J. Joh, "GaN HEMT reliability," *Microelectronics Reliability*, vol. 49, no. 9–11, pp. 1200–1206, 2009.

[3] D. Marcon, T. Kauerauf, F. Medjdoub et al., "A comprehensive reliability investigation of the voltage-, temperature- and device geometry-dependence of the gate degradation on state-of-the-art GaN-on-Si HEMTs," in *Proceedings of IEEE International Electron Devices Meeting (IEDM '10)*, pp. 20.3.1–20.3.4, San Francisco, Calif, USA, December 2010.

[4] K. Joshin and T. Kikkawa, "Recent progress of high power GaN-HEMT for wireless application," in *Proceedings of the Asia-Pacific Microwave Conference (APMC '06)*, pp. 1027–1032, Yokohama, Japan, December 2006.

[5] A. Bettidi, F. Corsaro, A. Cetronio, A. Nanni, M. Peroni, and P. Romanini, "X-band GaN-HEMT LNA performance versus robustness trade-off," in *Proceedings of the European Microwave Conference (EuMC '09)*, pp. 1792–1795, Rome, Italy, October 2009.

[6] H. Xu, C. Sanabria, A. Chini, S. Keller, U. K. Mishra, and R. A. York, "A C-band high-dynamic range GaN HEMT low-noise amplifier," *IEEE Microwave and Wireless Components Letters*, vol. 14, no. 6, pp. 262–264, 2004.

[7] J. Joh, L. Xia, and J. A. Del Alamo, "Gate current degradation mechanisms of GaN high electron mobility transistors," in *Proceedings of IEEE International Electron Devices Meeting (IEDM '07)*, pp. 385–388, Washington, DC, USA, December 2007.

[8] J. D. Brown, D. S. Green, S. R. Gibb et al., "Performance, reliability, and manufacturability of AlGaN/GaN high electron mobility transistors on silicon carbide substrates," *ECS Transactions*, vol. 3, no. 5, pp. 161–179, 2006.

[9] S. Lee, R. Vetury, J. D. Brown et al., "Reliability assessment of AlGaN/GaN HEMT technology on SIC for 48V applications," in *Proceedings of IEEE International Reliability Physics Symposium (IRPS '08)*, pp. 446–449, Phoenix, Ariz, USA, May 2008.

[10] G. Meneghesso, G. Verzellesi, F. Danesin et al., "Reliability of GaN high-electron-mobility transistors: state of the art and perspectives," *IEEE Transactions on Device and Materials Reliability*, vol. 8, no. 2, pp. 332–343, 2008.

# 12

# Tunable Lowpass Filter with RF MEMS Capacitance and Transmission Line

**Shimul C. Saha,[1,2] Ulrik Hanke,[3] Håkon Sagberg,[4] Tor A. Fjeldly,[1] and Trond Sæther[1]**

[1] *Department of Electronics and Telecommunications, Norwegian University of Science and Technology, 7491 Trondheim, Norway*
[2] *Electronics and Nanoscale Engineering Research Division, University of Glasgow, Glasgow G12 8QQ, UK*
[3] *Institute for Micro and Nano Systems Technology, Vestfold University College, 3103 Tønsberg, Norway*
[4] *Department of Microsystems and Nanotechnology, SINTEF ICT, P.O. Box 124, Blindern, 0314 Oslo, Norway*

Correspondence should be addressed to Shimul C. Saha, shimul.saha@iet.ntnu.no

Academic Editor: Jiun Wei Horng

We have presented an RF MEMS tuneable lowpass filter. Both distributed transmission lines and RF MEMS capacitances were used to replace the lumped elements. The use of RF MEMS capacitances gives the flexibility of tuning the cutoff frequency of the lowpass filter. We have designed a low-pass filter at 9–12 GHz cutoff frequency using the theory of stepped impedance transmission lines. A prototype of the filter has been fabricated using parallel plate capacitances. The variable shunt capacitances are formed by a combination of a number of parallel plate RF MEMS capacitances. The cutoff frequency is tuned from C to X band by actuating different combinations of parallel capacitive bridges. The measurement results agree well with the simulation result.

## 1. Introduction

Radio frequency microelectromechanical systems (RF MEMSs) technology offers an attractive capability for RF systems, particularly in support of switching and tuning functions. One such component is a microelectromechanical voltage tuneable capacitor, which can enable a wide tuning range and high-quality (Q) factor. In this paper, we present a design of a tuneable low-pass filter combining RF MEMS capacitors and high impedance transmission lines. The low-pass filter has been designed by means of the theory of stepped-impedance transmission line filters. The filter can be developed with surface micromachining without the need for any external inductors. This will give flexibility in integration and easy fabrication processing.

Compared with solid-state varactors, MEMS tuneable capacitors have the advantage of lower loss and a potentially larger tuning range [1]. The interconnection loss and noise can also be reduced compared with the use of off-chip solid-state RF components. Among all the MEMS capacitors developed to date, the parallel plate configuration is the most commonly used. The actuation mechanism is mostly electrostatic as it consumes very little power during actuation. In standard filter design at low frequency, the MEMS varactor is used as a lumped element with off-chip inductors. At high frequency, some low pass filter designs were reported with integrated transmission lines and MEMS switches (see Cai et al. [2] and Fang et al. [3]). In [2] the tuning of the filter is obtained by changing the transmission line length using MEMS switches. In [3] planar spiral inductor and shunt capacitor are used but without any tuning capability. Also, low-pass and band-pass filters integrating capacitive MEMS switches and short inductive transmission lines have been realised above 10 GHz [4–6]. In [4], the tuning of the cut-off frequency was obtained by switching between different combination of inductance and capacitance. In [5, 6], the cut-off frequency was tuned by changing the inductance using MEMS switching.

In this paper, we have shown a design of a low-pass filter with a high-impedance transmission line and RF MEMS capacitors, where the tuning is obtained by capacitance tuning. The MEMS switches were designed in shunt configuration on a coplanar wave guide (CPW). The MEMS capacitive shunt switch was used in a stepped-impedance low-pass filter instead of a low-impedance transmission line. The series inductor in the low-pass filter was implemented using a short section of a high impedance transmission line [4]. A preliminary theory of low-pass filters using transmission lines and MEMS capacitances is presented in [7, 8]. This filter is much more elegant and compact than filters based on traditional $\lambda/8$ long stub and transmission line filters [9]. Capacitance tuning is much more efficient than the inductance and transmission line tuning. To change the inductance and transmission line, additional MEMS switch is required, which increases the complexity and loss in the systems [2–6]. The designed filter has a large tuning range, over 60% of its nominal cut-off frequency. Although step impedance filters do not have a very sharp rolloff, they can be used for certain applications where compact size and a wide tuning range are required and less rolloff can be accepted. The low-pass filter, for example, can be used in harmonic suppression above the fundamental frequency in frequency synthesizer. The high tuning range will be very useful in the systems with multiple frequency band operation.

The theory of MEMS capacitor and third-order 3 dB ripple chebyshev low-pass filter is described in Section 2. The filter was designed and optimised with advanced design system (ADS) from Agilent [10]. A prototype of the filter was fabricated by our own facility at SINTEF MiNaLab in Oslo, Norway. Three parallel plate bridges were used as tuneable capacitances. The tuning of the capacitance was obtained by actuating different combinations of bridges. The fabrication of the filter is presented in Section 3. The measurement of the filter is presented in Section 4. A comparison of the measurement result with simulation result of the filter, taking into account the effects of process constraints, has been also discussed in Section 4. Finally, the paper is concluded.

## 2. Theory of Capacitive Switch and Third-Order Filter

*2.1. Theory of Parallel Plate Capacitive Switch.* An MEMS capacitive switch consisting of a suspended top plate and a fixed bottom electrode plate with an overlap area $A_e = L_e \times W_e$ is shown in Figure 1. $L_e$ and $W_e$ are the length and the width of the electrode, respectively. A DC bias voltage, $V_{DC}$, is applied between the bridge and the two separate actuation electrodes on the two sides of the centre conductor as shown in Figure 1. All electrodes are covered with a dielectric with thickness $t_d$ and relative permittivity $\varepsilon_r$. The initial gap is $g_0$ at the centre of middle electrode when no DC voltage is applied. When a DC voltage is applied between the bridge and the actuation electrodes, the initial gap reduces according to the actuation voltage [11]. If the fringing field effect is neglected, the value of the capacitance between the suspended bridge and centre conductor for a deflection $x$ of the bridge can be determined by

$$C(x) = \varepsilon_0 \cdot \frac{A_e}{(g_0 - x) + t_d/\varepsilon_r}. \quad (1)$$

When $V_{DC}$ is applied between actuation electrodes and bridge, an attractive electrostatic force $F_e$ is generated with the value:

$$F_e = \frac{1}{2} \frac{C_a(x) V_{DC}{}^2}{(g_{a0} - x + t_d/\varepsilon_r)}. \quad (2)$$

Here, $C_a(x)$ is the total capacitance between the bridge and the two actuation electrodes, $x$ is the distance the top plate has moved from its initial position, $V_{DC}$ is the actuation voltage, and $g_{ao}$ is the initial gap between actuation electrodes and suspended bridge. This electrostatic force tries to pull the top plate down. The beam has a mechanical spring constant (stiffness) which tries to pull the beam back to the upstate position. For small deflections, the electrostatic and elastic force are equal in magnitude. When the beam is deflected one-third of the initial gap, the electrostatic force becomes stronger than the mechanical restoring force. At this point the beam collapses fully to the down state and this voltage is called the pull-down voltage. The down state capacitance is given by

$$C_d = \varepsilon_0 \varepsilon_r \frac{A_e}{t_d}. \quad (3)$$

*2.2. Theory of the Stepped-Impedance Filter.* It is well known [9] that if the length of the transmission line is short ($\beta l < \pi/4$) and the characteristic impedance is high, the transmission line can be represented by the equivalent circuit shown in Figure 2(a) where the reactance is given by

$$X_L = Z_0 \beta \ell. \quad (4)$$

Here, the characteristic impedance of the line is $Z_0$, the length is $l$, and the propagation constant is $\beta$. So, a series inductor can be represented by a short length of high-characteristic-impedance transmission line. In the lumped elements filter model, the shunt capacitors have been replaced by an RF MEMS capacitor which will provide the capability of tuning cut-off frequency.

*2.3. A Third-Order Lowpass Filter Topology Design.* A third-order low pass filter topology is shown in Figure 2(b). The basic filter elements are extracted from the Chebyshev equal ripple low pass filter prototype [7–9]. The normalized filter elements for a 3 dB ripple Chebyshev are given by $g_1 = 3.3487$, $g_2 = 0.7117$, $g_3 = 3.3487$, and $g_4 = 1.00$ which is the load impedance [9]. A ripple of 3 dB is used to give a sharper roll-off above the cut-off frequency. We have chosen a nominal cut-off frequency at 7.0 GHz. With a maximum transmission line impedance of 80 Ω, the length of the transmission line becomes 25. 5° and the nominal capacitance value becomes 1.52 pF when the cut-off frequency is 7 GHz. The proposed filter is fabricated on a 280 μm think high-resistivity silicon substrate. The relative dielectric constant of the substrate is 11.9 and the resistivity is 8000 Ω-cm. After optimization of the filter cut-off frequency in ADS, the nominal shunt capacitor value becomes 1.1 pF and the length of the 80 Ω line becomes 1467 μm (30.7°) for the desired responses [10]. For a 50 Ω CPW transmission line, the dimensions are 80 μm/140 μm/80 μm ($G/W_c/G$).

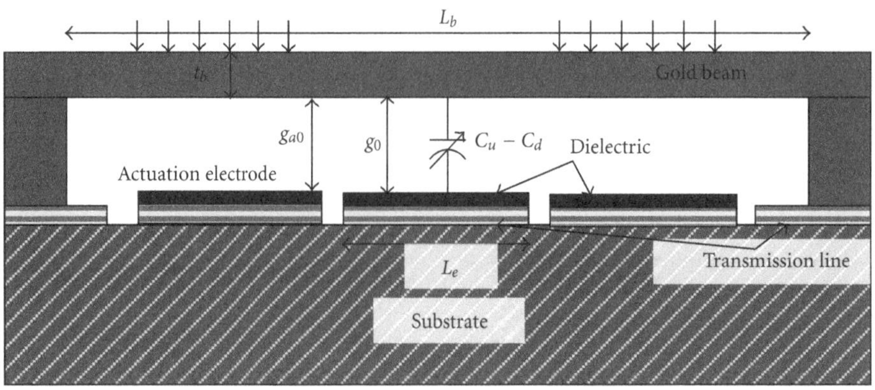


FIGURE 1: Side view of a traditional parallel plate capacitive switch in CPW configuration with separate actuation electrodes.

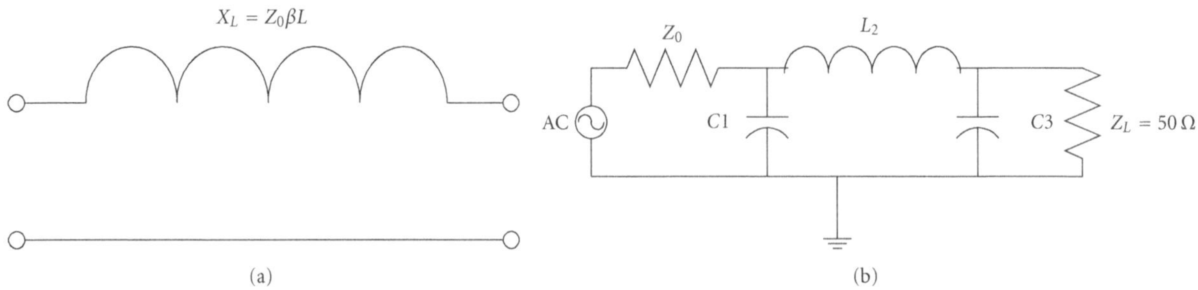


FIGURE 2: (a) Equivalent circuit for short length and high characteristic impedance transmission line and (b) third order low-pass filter topology.

For an 80 Ω CPW transmission line, the dimensions are 160 μm/53 μm/160 μm (*G*/*W*/*G*). A higher value, such as 100 Ω or more, can be chosen for the impedance of the series inductance. This will make the transmission line very narrow and increase the loss in pass band.

## 3. Fabrication of the Filter

A schematic view of the fabricated filter is shown in Figure 3. The fabrication was performed at SINTEF MiNaLab in Oslo. Each shunt capacitance was a combination of two 12 μm wide (on the two sides) and one 15 μm wide (middle) shunt bridge. The variation of the capacitance can be obtained by actuating individual bridges, or combining one or more bridges. The thickness of the centre electrode and the actuation electrode dielectric was ($Si_3N_4$) 220 nm. Aluminium was used as a transmission line.

A four-mask process was used to fabricate the proposed filter. A 280 μm thick silicon wafer with 4–8 kΩ-cm (specified by supplier) resistivity was used for the fabrication. A 500 nm thick oxide was grown on the silicon wafer to obtain lower loss. Then a 500 nm thick Al was deposited by sputtering and patterned by wet etching. A 220 nm thick $Si_3N_4$ was deposited as dielectric by plasma-enhanced chemical vapour deposition (PECVD) and patterned by dry etch. HiPR 6517 photoresist was used as a sacrificial layer. The resist thickness was 2.55 μm and patterned by standard lithography and special baking methods as described in detail in [12]. Then, 1.1 μm thick gold was sputtered and patterned by wet etch. The gold was sputtered at 500 watt and a pressure of 20 mTorr to obtain a low tensile stress [13]. When the gold was sputtered, the transmission line was also sputtered with 1.1 μm additional gold on aluminum. Low tensile stress ensures low-pull-down voltage and gives better reliability. Finally, the shunt capacitive bridges were released by dissolving the resist in acetone and dried in a critical point dryer [12]. Scanning electron microscope (SEM) images of the fabricated filter are shown in Figure 4. As shown in Figure 4, the actuation electrodes are separated from the RF signal electrode, and each bridge has its own set of actuation electrodes. The centre narrow line is the 80 Ohm transmission line, which is equivalent to a series inductor. The capacitors are at both ends of the 80 Ohm transmission line, shunting a 50 Ohm transmission line. In the figure, the three parallel bridges of shunt capacitors can be seen. The extended 50 Ohm transmission lines are used for measurement purpose with a CPW probe.

## 4. Measured Results and Simulation in the Light of Fabrication Constraint

The fabricated filters were measured in a vector network analyser. The filter characteristic was measured up to 30 GHz. Standard line-reflect-reflect-match (LRRM) methods were used for calibration of the network analyser. The measured filter *S* parameters are shown in Figure 5. We measured

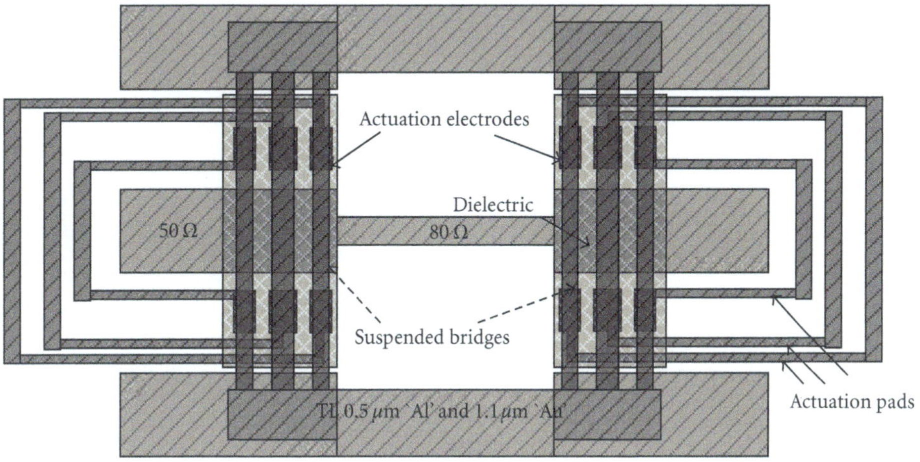


FIGURE 3: The top view of the proposed low pass filter with actuation mechanism.

(a)

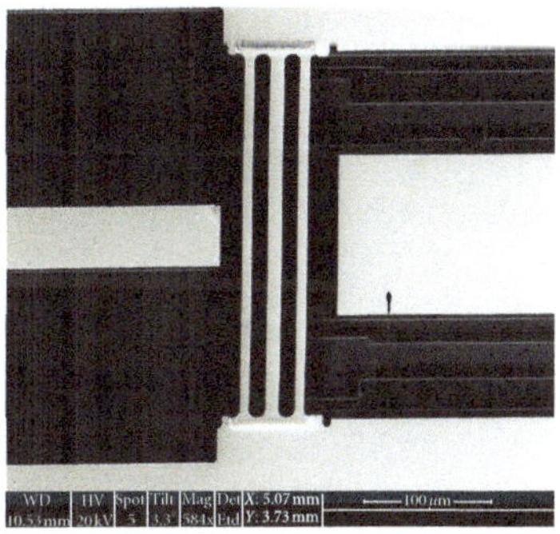

(b)

FIGURE 4: (a) A SEM image of the fabricated low-pass filter (top view). Three narrow bridges at both sides represent the tunable capacitance. Chip size (2.3 mm × 1 mm) and (b) an enlarged image of the low-pass filter at the capacitance area.

the filter characteristics with one bridge down (15 μm wide centre bridge), and 2 bridges down (15 μm and 12 μm wide bridges). Due to lack of actuation facility, we could not actuate all three parallel bridges together. The cut-off frequency for one bridge down is 12 GHz, and two bridges down 9 GHz. The filter is very compact compared to the traditional half-wave-length transmission line filter. In addition, the loss in the pass band is reasonably small.

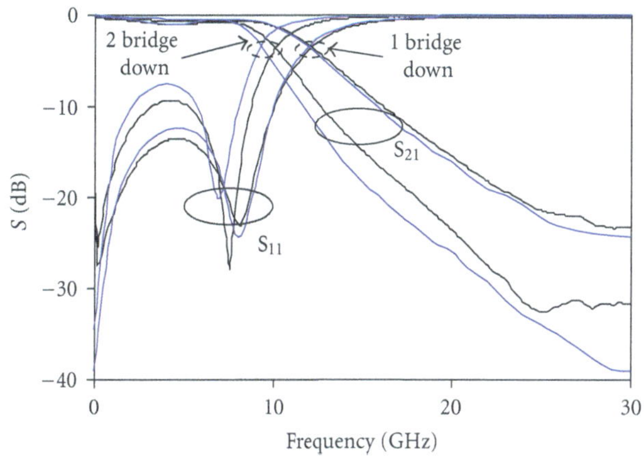


FIGURE 5: Comparison of the filter performances between simulated (blue line) and measured results (black line).

The cut-off frequency was slightly higher than designed frequency. After fabrication, the filters were inspected in an optical interferometer to measure the actual width and height of the bridge. The actual initial heights of the bridges were 2.25 μm and the width was reduced on average 1 μm from each side owing to overetching. Gold is usually smooth and it was found that the average roughness of the gold was 7 nm. The $Si_3N_4$ average roughness was 9 nm, but the main contribution to the roughness was from aluminium. This is because of the high-temperature (300°C) $Si_3N_4$ deposition. From an atomic force microscope (AFM) measurement, it was found that the height of the aluminium bumps was about 80 nm. As the nitride was deposited on aluminium, the total roughness for the downstate capacitance was 96 nm. Considering the overetch and roughness, the downstate capacitance value is 0.38 pF for the 15 μm bridge (13 μm real width). The filter was resimulated with the above parameters. The simulation was compared with measurement results of the fabricated filter as shown in Figure 5. A higher resolution of the measured $S$ parameters is also shown in Figure 6. The measured insertion loss is less than 1.25 dB in the

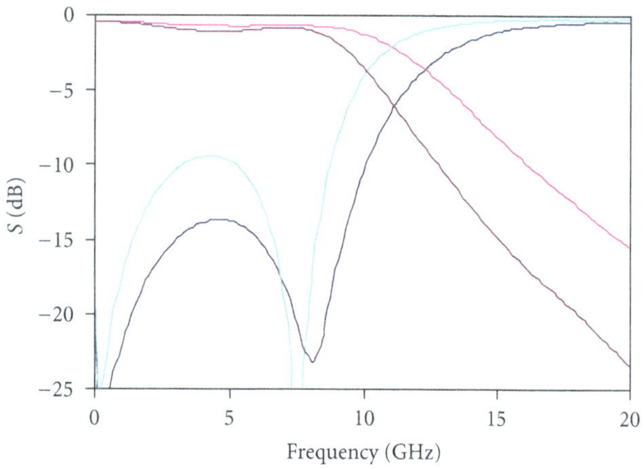


FIGURE 6: The measured filter performances at higher resolution in scale.

pass band for both configurations. The filter has a small amount of ripple in the pass band. The measured results agree well with the simulated results, given the effect of overetching and roughness. There is a slightly higher cut-off frequency in the measured filter. This can happen because of nonplanarity of the bridge at downstate, which will further reduce the downstate capacitance. By using the same roughness as mentioned earlier, the estimated downstate capacitance becomes 0.30 pF for the 12 $\mu$m bridge (10 $\mu$m real width). The total capacitance at downstate becomes 0.68 pF when two bridges are down. The simulated and measured results for this state were compared and are shown in Figure 5. From this figure, it can be seen that the measured filter again has a higher cut-off frequency for the reason discussed above. Also, the narrower bridge will have relatively more overetching and less planarity.

## 5. Conclusions

We have presented a tuneable low-pass filter design with transmission line and MEMS capacitance. The tuning of the cut-off frequency was obtained by actuation of several parallel bridges. The measured results agree well with the simulated result considering the fabrication effects. The filter rolloff is not very sharp but can be improved with the use of high-impedance transmission line and a higher-order of the filter. However, this may have some drawbacks. As only the capacitance was used for tuning and the inductance was fixed, higher order filters would have reduce the tuning range. Also, it would be complicated to tune all the capacitor simultaneously. The fabricated filter is very compact and has very good control of the tuning frequency.

## Acknowledgments

The authors are grateful to the Norwegian Research Council for sponsoring the work through the SMiDA project (no. 159559/130) and the IRRFT Project (no. 159259/I40). The authors also thank MiNa-Lab personnel for their support in the clean room.

## References

[1] J. Zou, C. Liu, J. Schutt-Aine, J. Chen, and S. M. Kang, "Development of a wide tuning range MEMS tunable capacitor for wireless communication systems," in *Proceedings of the International Electron Devices Meeting, Technical Digest (IEDM '00)*, pp. 403–406, December 2000.

[2] M. Cai, X. L. Guo, Y. Li, L. Liu, and L. Zong-sheng, "Design and modeling of reconfigurable MEMS low pass filter," in *Proceedings of the 8th International Conference on Solid-State and Integrated Circuit Technology (ICSICT '06)*, pp. 563–565, October 2006.

[3] J. Fang, Z. W. Liu, Z. M. Chen, L. T. Liu, and Z. J. Li, "Realization of an Integrated Planar LC low-pass filter with Modified Surface Micromachining technology," in *Proceedings of the IEEE Conference on Electron Devices and Solid-State Circuits (EDSSC '05)*, pp. 729–731, December 2005.

[4] D. Peroulis, S. Pacheco, K. Sarabandi, and L. P. B. Katehi, "Tunable lumped components with applications to reconfigurable MEMS filters," in *International Microwave Symposium Digest (IEEE-MTT-S '01)*, vol. 1, pp. 341–344, May 2001.

[5] S. Lee, J. M. Kim, J. M. Kim, Y. K. Kim, and Y. W. Kwon, "Millimeter-wave MEMS tunable low pass filter with reconfigurable series inductors and capacitive shunt switches," *IEEE Microwave and Wireless Components Letters*, vol. 15, no. 10, pp. 691–693, 2005.

[6] J. H. Park, S. Lee, J. M. Kim, H. T. Kim, Y. Kwon, and Y. K. Kim, "Reconfigurable millimeter-wave filters using CPW-based periodic structures with novel multiple-contact MEMS switches," *Journal of Microelectromechanical Systems*, vol. 14, no. 3, pp. 456–463, 2005.

[7] S. C. Saha and T. Sæther, "Modeling and Simulation of Low Pass Filter using RF MEMS Capacitance and Transmission line," in *Proceedings of the International Microelectronics And Packaging Society (IMPS '05)*, pp. 155–159, 2005.

[8] S. C. Saha, U. Hanke, and T. Sæther, "Modeling, design and simulation of tunable band pass filter using RF MEMS capacitance and transmission line," in *Microelectronics: Design, Technology, and Packaging II*, vol. 6035 of *Proceedings of SPIE*, Brisbane, Australia, December 2005.

[9] D. M. Pozar, *Microwave Engineering*, chapter 8, John Wiley & Sons, New York, NY, USA, 2nd edition, 1998.

[10] Advance Design System, from Agilent.

[11] G. M. Rebeiz, *RF MEMS Theory, Design and Technology*, chapter 2, John Wiley & Sons, New York, NY, USA, 2nd edition, 2003.

[12] S. C. Saha, H. Sagberg, E. Poppe, G. U. Jensen, T. A. Fjeldly, and T. Sæther, "Tuning of resist slope with hard-baking parameters and release methods of extra hard photoresist for RF MEMS switches," *Sensors and Actuators, A*, vol. 143, no. 2, pp. 452–461, 2008.

[13] S. C. Saha, H. Sagberg, E. Poppe, G. U. Jensen, and T. Sæther, "Metallization scheme and release methods for fabrication of RF MEMS switches," in *Proceedings of 33rd Micro- and Nano-Engineering conference (MNE '07)*, 2007.

# 13

# Time-Domain Predistortion Method Based on Raised Cosine Signaling in Real Transmission Channels

**Bretislav Sevcik**

*Faculty of Electrical Engineering and Communication, Brno University of Technology, 61600 Brno, Czech Republic*

Correspondence should be addressed to Bretislav Sevcik, xsevci34@stud.feec.vutbr.cz

Academic Editor: G. Ghibaudo

The concept of time-domain predistortion method based on raised cosine signaling is applied in real transmission channels. The proposed PWM-RC method uses raised-cosine pulse shaping instead of conventional rectangular digital signals and pulse-width modulation (PWM) scheme to achieve better output channel data response in harsh channel environment. The conventional predistortion methods based on pulse amplitude adjusting are not compatible with modern low-power CMOS design. Currently existing time-domain predistortion methods which are only based on a PWM scheme show many highfrequency signal harmonic components for both fast and slow signal transitions. It can cause more system crosstalk susceptibility if the crosstalk is dominant factor in transmission channel. In this case, the additional preemphasis boosted undesirable high-frequency components. Finally, the real channel transfer functions in connection with ADS Agilent development studio are used to compare the performance of proposed method with other predistortion methods.

## 1. Introduction

Today's high-speed multicore microprocessors and memory interfaces require ever increasing interconnect bandwidth. Stringent requirements for a reduced board space, miniature connectors, efficient printed circuit board (PCB) design, and low susceptibility to noise are associated with the serial communications links architecture. Many existing serial I/O standards operate at multi-Gbps speeds such as well-known PCI Express (PCIe). Next generation of the standard PCIe Gen 3 is able to operate at 8 Gbps in a single lane. As data rates are increasing, their susceptibility to damage is more critical. It is caused by the nonideal aspects of transmission lines, such as crosstalks and losses as well as energy dissipation caused by reflections and radiation [1]. All these issues can cause significant problems in signal integrity and timing. In such severe environments, sophisticated pulse-shaping techniques such as the equalization or preemphasis (equalization at the transmitter is often called transmitter preemphasis to reflect the effect of the filter operation) need to be employed to increase the data rates [2]. In this paper, for the first time the PWM-RC (raised cosine) predistortion method is applied in real transmission channels; see Figure 1.

The concept of proposed signaling method is briefly described in [3]. It is shown that PWM-RC method significantly reduces the high-frequency content in pre-distorted high-speed signals. The concept of PWM signaling scheme is proposed in [4] and is applied to the coaxial cable loss equalization and simple single PCB trace equalization. It is shown that the PWM preemphasis provides higher maximum loss compensation (24 dB) than the commonly used 2-tap FIR filter (18 dB). The transfer function of PWM method is able to provide a better adaptation to the copper channel; see results in Agilent ADS [5]. However, the implementation of the predistortion methods based on pulse amplitude shaping can be significantly restricted by noise margin because modern low-voltage CMOS devices use supply voltage, which will decrease according to CMOS scaling [6]. However, the crosstalk susceptibility of conventional PWM scheme based on rectangular pulse shaping can cause additional high-frequency content in equalized data

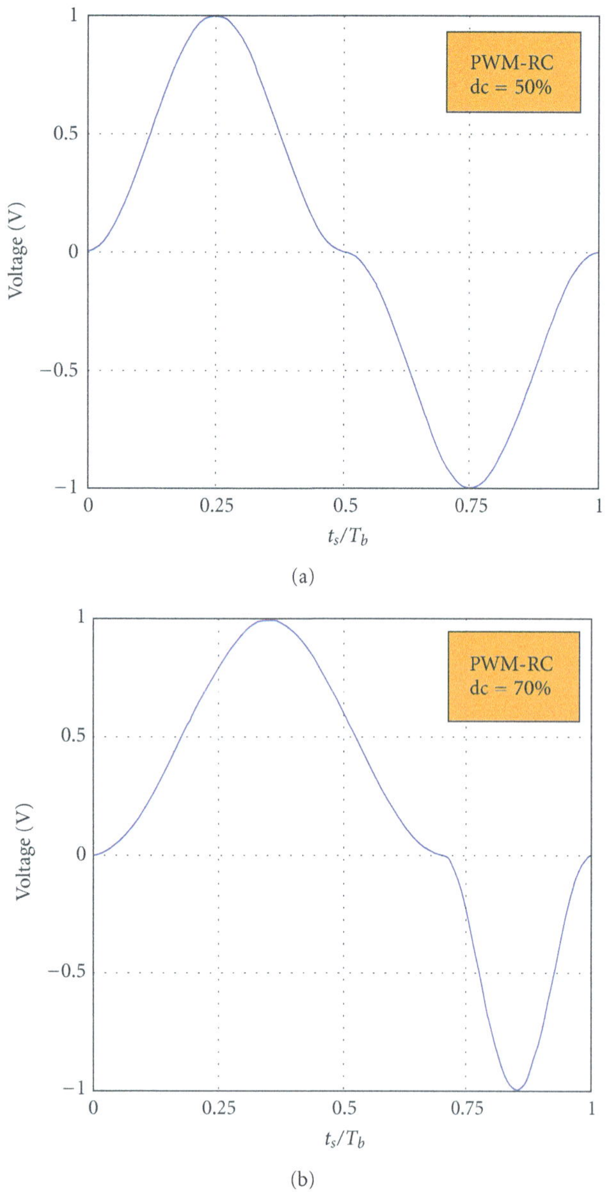


FIGURE 1: PWM-RC signal shaping.

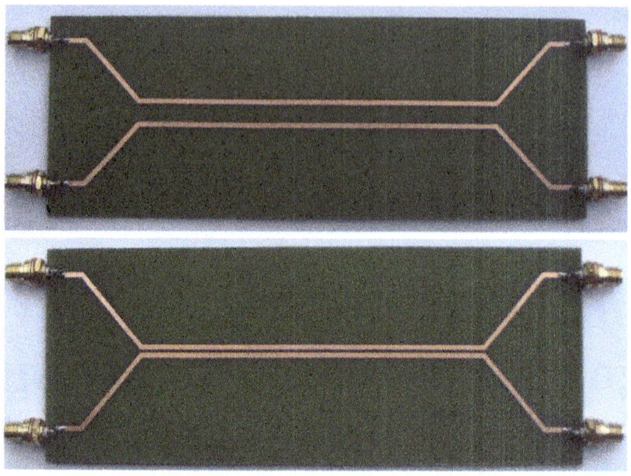

FIGURE 2: Two experimental analyzed 30 cm long backplanes.

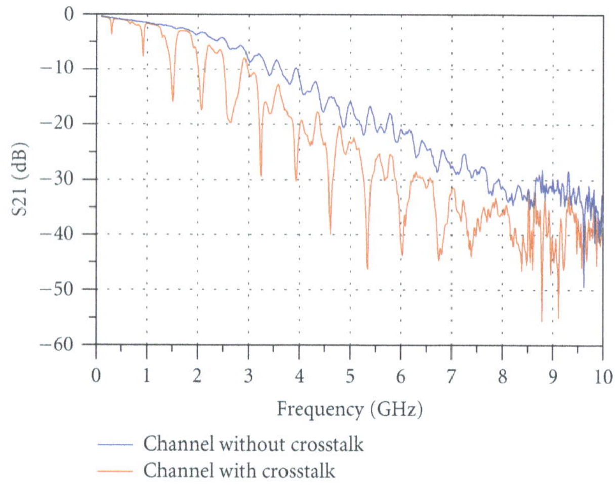


FIGURE 3: Transfer functions of both analyzed experimental channels.

stream. Thus, the loss compensation can be significantly reduced. The combination of PWM predistortion technique and appropriate pulse shaping method (PWM-RC method) can provide an effective reduction of high-frequency components of the pulses. It is shown in [3] that the equalized channel transfer functions, when the PWM-RC method is used, have the better performance at higher frequencies due to the additional variability to adjust high frequency loss compensation. It can play a critical role for optimal preemphasis design. The experimental results presented in this paper were obtained by using the interaction between the Agilent E5071C Network analyzer and Agilent ADS Studio. Thus, the real channel transfer functions can be used together with proposed PWM-RC signaling scheme.

## 2. Channel Properties

In this case two experimental 30 cm long backplanes for transfer function analysis are used. These backplanes were created within diploma thesis [7]; see Figure 2.

The first backplane prototype was designed with relatively large distance between transmission lines, $d = 6$ mm. In this case the microstrip conductor and dielectric losses are dominant factors that must be taken into account. The second backplane prototype is designed with significant distance reduction between transmission lines, $d = 0.5$ mm. In this case a significant crosstalk effect which should be taken into account in real communication systems occurs. This situation is clearly seen in Figure 3 where transfer functions (S21 parameters) for both channels are shown. It is obvious that significant high slopes can occur when crosstalk channel is analyzed. Moreover, the second backplane has intentional disturbance in GND layer. Thus, the effect of vias impact can be imitated.

Some of the final channel output eye diagrams for both measured channels are shown in Figure 4. It is obvious that the data signal passing through the crosstalk channel has almost the same performance as data signal with double data

Table 1: Eye diagram performance for both channels.

| Data rate (Gbps) | No crosstalk | | Crosstalk | |
|---|---|---|---|---|
| | Eye height (mV) | Eye width (ps) | Eye height (mV) | Eye width (ps) |
| 1 | 163.9 | 979.9 | 103.1 | 689.2 |
| 2 | 154.1 | 475.6 | 95.2 | 328.2 |
| 3 | 142.2 | 362.2 | 88.7 | 278.2 |
| 5 | 78.3 | 198.1 | 6.2 | 110.2 |
| 6 | 57.6 | 140.9 | Eye closed | |
| 8 | Eye closed | | Eye closed | |
| 10 | Eye closed | | Eye closed | |

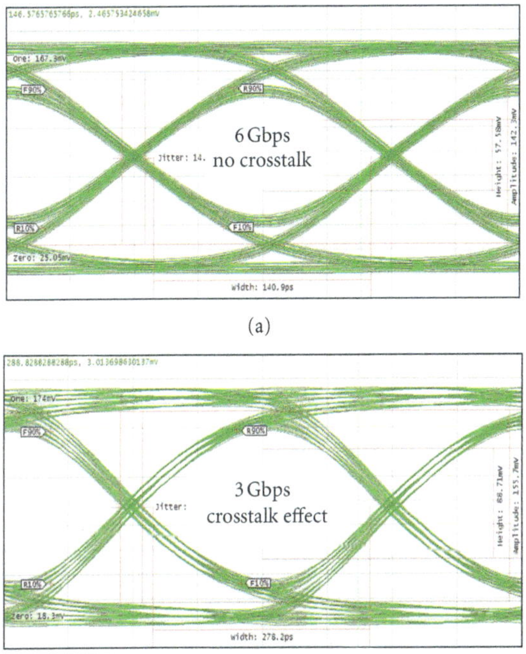


(a)

(b)

Figure 4: Example of eye diagram results for both analyzed channels.

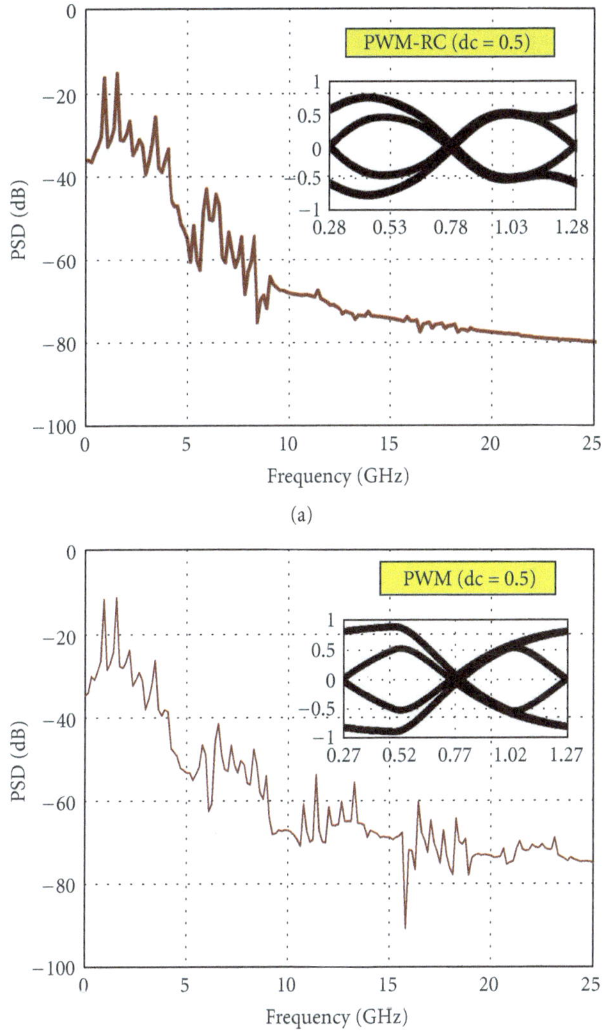


(a)

(b)

Figure 5: Signal spectrum analysis at the transmitter output, 5 Gbps data rate.

rate for channel without crosstalk and GND disturbance. The experimental measurements are summarized in Table 1 where the significant differences between both analyzed channels are noticeable. Voltage amplitude is set to 200 mV.

In this case it is necessary to use a sophisticated method for signal conditioning. The simulations results realized in MathCAD show significant increase in crosstalk for relatively long coupling lengths (this corresponds with realized backplane prototypes measurements) and decreasing in aggressor distance; see [8]. The aggressor represents separated single-ended circuit with the random bit sequences and the same data rate as victim circuit.

## 3. Preemphasis Performance

In [3] it is shown significant high-frequency component reduction for the proposed PWM-RC predistortion method. The power spectral density (PSD) analysis results are shown in Figures 5 and 6. In both cases, the conventional PWM method with proposed PWM-RC method is compared for the same dc (duty-cycle) settings. In the first case the signal spectrum is analyzed at the transmitter output without additional transmission lines effects. Especially the near end crosstalk (NEXT) at the connectors can be amplified by additional high-frequency content. In Figure 5 the maximum preemphasis effect is set. The proposed method significantly reduces high-frequency harmonic components of preemphasized data signal. In this case at frequency 10 GHz the second signal harmonic is considered. Compare both eye diagrams shaping where a random bit sequence is analyzed.

The main disadvantage of conventional preemphasis methods is described in [3]. The strong susceptibility to crosstalk is evident [4, 6]. In [6] by using a simple Fourier series calculation it is shown that the output spectrum of

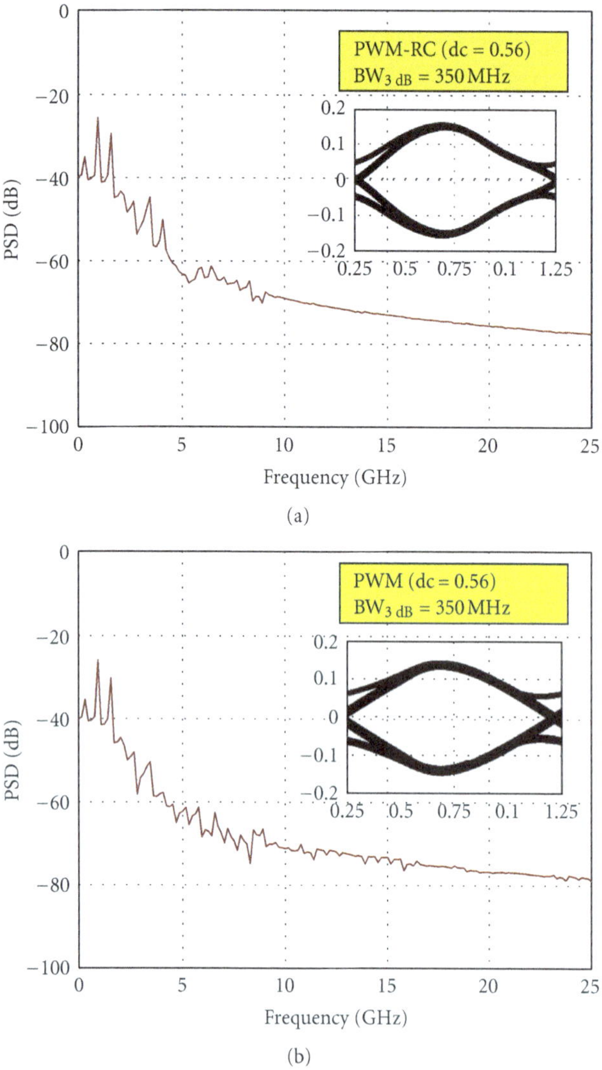


FIGURE 6: Signal spectrum analysis at the channel output, 5 Gbps data rate.

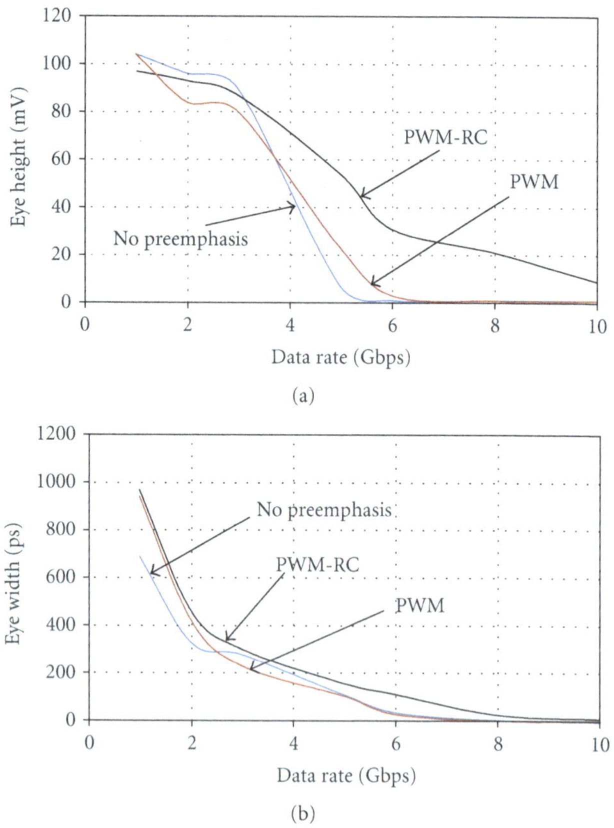


FIGURE 7: Eye diagram opening analysis for various data rates if crosstalk channel is used.

conventional filter exhibits more high-frequency components if low-frequency (LF) pattern (transmitting many 1's in succession) is transmitted. It can cause complications if channel which has been equalized is predisposed to crosstalk formation. In Figure 6 the frequency spectrum analysis at the band-limited channel output for both PWM methods is shown. In this case the optimal preemphasis setting is selected. The real random high-speed 5 Gbps data stream is formed. The proposed PWM-RC method still reduces high-frequency harmonic components at channel output but many high-frequency components are absorbed by the channel itself. The simulated LF pattern was transmitted over the band-limited channel too. The significant reduction in high-frequency components is seens; each of the figures show one bit period where on $x$ axis the bit period is shown. It can be concluded that the proposed PWM-RC scheme is able to sufficiently compensate the additional high-frequency components also for lower switching frequency. This behavior is close to conventional FIR filter where the filter exhibits less power for lower switching frequency. For the HF pattern the difference between FIR and PWM scheme is only in phase (time shift) [4].

Finally, the PWM-RC signaling scheme was implemented in Agilent ADS simulation environment. The measured channel transfer function was used as channel environment for both PWM and PWM-RC method testing. Both main eye diagram parameters are analyzed in Figure 7. It is obvious that PWM-RC method improves eye diagram opening. Significant high slopes occur approximately at Nyquist frequencies for 3, 4, 5, and so forth, Gbps data rate. The data rate is analyzed from 1 Gbps to 10 Gbps with step 1 Gbps. However the main decisive factor is the average value of losses around the observed Nyquist frequencies. In this case the additional 6 dB loss is sufficient to completely close the eye. Note that the 6 dB rule is independent of voltage swing and frequency. This can be seen at Nyquist frequency 2.5 GHz (5 Gbps data rate) where approximately 6 dB channel loss occurs. The significant eye closing is evident; see Figure 7. What does depend on frequency is where the 6 dB loss point actually occurs. One of the decisive parameters for system reliability is a receiver sensitivity that can be used to calculate a transmitted voltage swing. The signaling power can be expressed as [4]. The expression shows that the minimum transmitted signal swing depends on the receiver sensitivity and the channel's frequency response [9]. In our case, the signal swing requirement assumes intersymbol interferences (ISI) compensation by equalization techniques such as preemphasis at the transmitter, ideally for minimum bit error rate (BER) which is typically $10^{-9} - 10^{-14}$.

## 4. Conclusion

In this paper the perspective digital time-domain predistortion technique based on the pulse-width modulation scheme and using the raised cosine signaling (PWM-RC) is analyzed in real transmission channels. In the main part of the paper, the proposed PWM-RC method and conventional PWM method are compared. It is shown that the proposed PWM-RC scheme is capable to open eye more than conventional PWM signaling scheme if channel crosstalk occurs. The signal spectrum analysis results show high-frequency content for PWM scheme especially if channel effect is not considered. The frequency-domain analysis [3] shows better performance of proposed method for the equalization of transmission channels where the attenuation is not monotonically increasing. Additional variability to adjust high-frequency loss compensation can play a critical role for optimal preemphasis setting; see Figure 7 where eye height is significantly improved if PWM-RC scheme is enabled.

The main disadvantage of proposed method lies in a problematic circuit realization. In [10, 11] it is possible to find the solutions for a raised-cosine signaling where only the conventional NRZ pulses are used. Future work will be focused on practical testing of proposed pulse shaping predistortion technique over the real band-limited transmission channel to verify the better method performance. To create the proposed PWM-RC signaling scheme, a software-implemented signal realization will be used.

## References

[1] S. H. Hall and H. Heck, *Advanced Signal Integrity for High-Speed Digital Design*, John Wiley & Sons, Hoboken, NJ, USA, 2009.

[2] Y. Yuminaka and K. Yamamura, "Equalization techniques for multiple-valued data transmission and their application," in *Proceedings of the 37th International Symposium on Multiple-Valued Logic (ISMVL '07)*, pp. 26–29, Oslo, Norway, May 2007.

[3] B. Ševčík and L. Brančík, "Time-domain pre-distortion technique using raised cosine shaping for high-speed serial signaling," *ElectroScope*, no. 4, pp. 1–6, 2011.

[4] J. H. R. Schrader, *Wireline equalization using pulse-width modulation [Ph.D. thesis]*, 2007.

[5] B. Ševčík, L. Brančík, and M. Kubíček, "Analysis of pre-emphasis techniques for channels with higher-order transfer function," *International Journal of Mathematical Models and Methods in Applied Sciences*, vol. 5, no. 3, pp. 433–444, 2011.

[6] Y. Yuminaka, Y. Takahashi, and K. Henmi, "Multiple-valued data transmission based on time-domain pre-emphasis techniques in consideration of higher-order channel effects," in *Proceedings of the 39th International Symposium on Multiple-Valued Logic (ISMVL '09)*, pp. 250–255, Okinawa, Japan, May 2009.

[7] J. Ulrych, "Problematika integrity signálů ve smíšených elektronických systémech," Vysoké učení technické v Brně, Fakulta elektrotechniky a komunikačních technologií, 71 s, Vedoucí diplomové práce doc. Ing. Lubomír Brančík, CSc, Brno, Czech Republic, 2007.

[8] B. Ševčík and L. Brančík, "Analysis and modeling of signal integrity problems in the serial communication devices," in *Proceedings of the International Conference on Mathematical Models for Engineering Science (MMES '10)*, pp. 172–178, IEEE.AM, Canary Islands, Spain, November 2010.

[9] T. B. Wendemagegnehu and M. Aleksic, "A Study of Optimal Data Rates of High-Speed Channels," *DesignCon 2011*, Santa Clara, Calif, USA, 2011.

[10] R. Bashirullah, L. Wentai, and R. Cavin, "Cross-talk reduction for interconnect-limited bus based on raised cosine signaling," in *Proceedings of the Interconnect Technology Conference*, Burlingame, Calif, USA, 2001.

[11] Y. Yuminaka, "High-speed data transmission techniques using raised cosine signaling," *Key Engineering Materials*, vol. 459, pp. 252–259, 2011.

# 14

# A Novel Nanoscale FDSOI MOSFET with Block-Oxide

**Jyi-Tsong Lin, Yi-Chuen Eng, and Po-Hsieh Lin**

*Department of Electrical Engineering, National Sun Yat-Sen University, No. 70 Lien-hai Road, Kaohsiung 80424, Taiwan*

Correspondence should be addressed to Jyi-Tsong Lin; jtlin@ee.nsysu.edu.tw

Academic Editor: Kuan-Wei Lee

We demonstrate improved device performance by applying oxide sidewall spacer technology to a block-oxide-enclosed Si body to create a fully depleted silicon-on-insulator (FDSOI) nMOSFET, which overcomes the need for a uniform ultrathin silicon film. The presence of block-oxide along the sidewalls of the Si body significantly reduces the influence of drain bias over the channel. The proposed FDSOI structure therefore outperforms conventional FDSOI with regard to its drain-induced barrier lowering (DIBL), on/off current ratio, subthreshold swing, and threshold voltage rolloff. The new FDSOI structure is in fact shown to behave similarly to an ultrathin body (UTB) SOI but without the associated disadvantages and technological challenges of the ultrathin film, because a thick Si body allows for reduced sensitivity to self-heating, thereby improving thermal stability.

## 1. Introduction

Semiconductor science, triggered by the impetus of a growing market for faster, more reliable, and less costly chips, has been undergoing a rapid technological development [1]. Many of these new technologies, however, suffer from undesirable side effects. For example, as the gate length of CMOS—the bulk complementary metal-oxide semiconductor—is decreased, short-channel effects (SCEs), such as drain-induced barrier lowering (DIBL) and threshold voltage ($V_{\mathrm{TH}}$) rolloff, become a significant problem because S/D encroachment begins to limit the gate's ability to control the channel. Also, due to the existence of the PN junction between the Si substrate and the S/D regions, a large junction leakage current prevents the use of scaled-down transistors in low standby power (LSTP) applications. Moreover, the parasitic capacitance of the transistor may strongly affect the characteristics of CMOS devices [2–4]. Therefore, the use of planar technology for ultralarge-scale integrated (ULSI) circuits becomes more challenging.

Recently, silicon-on-insulator (SOI) technology has demonstrated promise for nano-CMOS scaling. Compared to its bulk Si counterparts, SOI offers reduced capacitance and lower OFF-state leakage current ($I_{\mathrm{OFF}}$), mainly due to the presence of a buried oxide (BOX) layer under the Si active layer [5]. This can be attributed to the fact that the BOX can be seen as a "blocking layer" to reduce the drain electric field. Also, because the active region is fully isolated, it avoids the latch-up problem of classical CMOS devices.

The benefits of SOI technology, however, are not without associated problems. A partially depleted (PD) SOI transistor cannot achieve an improved performance in future sub-45 nm semiconductor devices, due to the electrical properties of PDSOI MOSFET relative to the thickness of the Si layer. It has been proved that a thick Si film makes reducing DIBL very difficult [6]. The fully depleted (FD) ultrathin body (UTB) SOI MOSFET is able to improve the short-channel characteristics but suffers from high S/D series resistance, because the semiconducting Si layer is, by definition, ultrathin. Also, when the gate-to-source overdrive voltage ($V_{\mathrm{GT}}$) is high enough, the self-heating effect causes a negative incremental conductance in the UTBSOI MOSFET associated with thermal instability. Consequently, the reliability of SOI devices will be degraded by these issues [7–11].

Our previous studies revealed that some of the unique features of the FDSOI MOSFET with block-oxide (bFDSOI) could allow it to replace the UTBSOI structure for CMOS scaling [12–14]. The device parameters of bFDSOI and UTBSOI were not optimized; however, neither bFDSOI nor UTBSOI showed high drain ON-state current ($I_{\mathrm{ON}}$). Now, equipped with a block-oxide-enclosed Si body using oxide sidewall spacer technology, this paper presents a fully depleted silicon-on-insulator (FDSOI) nMOSFET to improve device performance, without using ultrathin silicon film. Also, the authors have optimized the device parameters to enhance current drive without losing the desired electrical characteristics.

This paper is organized as follows: the device structures and their corresponding simulations are described in Section 2. In Section 3, the electrical characteristics are compared between the bFDSOI-FETs and their counterparts in the FDSOI-FET, UTBSOI-FET, and elevated S/D (E-S/D) UTBSOI-FET before a brief summary is carried out in Section 4.

## 2. Device Structures and Simulations

ISE-TCAD was used to design and simulate the thin S/D and the recessed S/D bFDSOI-FETs. The key steps for processing the bFDSOI are shown in Figure 1. First, the Si body was patterned with e-beam (see Figure 1(a)), and for thin S/D bFDSOI-FET, 60 nm thick oxide was deposited using chemical vapor deposition (CVD) as a blocking layer. The deposited oxide layer was then etched back to form the sidewall spacers of the Si body (see Figure 1(b-1)). A layer of poly-Si (5 nm thick) was then deposited to act as an active layer (see Figure 1(c-1)). For the recessed S/D bFDSOI-FET, a 15 nm thick oxide was deposited and etched back to form the sidewall spacers of the Si body (see Figure 1(b-2)). Next, poly-Si was deposited and planarized with chemical mechanical polishing (CMP). Poly-Si (5 nm thick) was then redeposited to form the active region (see Figure 1(c-2)). For both bFDSOI-FETs, a 2.3 KeV boron difluoride ($BF_2$) channel implantation at a dosage of $1.15 \times 10^{12}\,cm^{-2}$ was performed, followed by rapid thermal annealing. Hence, the values of $V_{TH}$ for both devices were determined. Then, 1.4 nm thick oxide was grown and a 50 nm thick poly-Si layer was deposited, followed by gate patterning. After gate patterning, a layer of nitride was deposited and etched back by dry etching. Next, 10 nm oxide was deposited as an implantation screen layer. In order to form S/D regions for bFDSOI-FETs, arsenic ions were implanted at a dosage of $2.1 \times 10^{14}\,cm^{-2}$ and an ion implantation energy of 14 KeV, followed by thermal annealing. As a result, the source/drain resistance for both devices was determined playing an important role in drain currents. The second sidewall spacers were then formed by dry etching. After contact formation, bFDSOI-FETs were achieved.

In this paper, some technical "tricks", such as S/D extension (SDE) implants [15], asymmetric halo [16], and retrograde channel profiles [17], were not used for our bFDSOI-FETs because the purpose of this work is to emphasize the importance of the block-oxide in reducing the SCEs. The device parameters used are listed as follows. For thin S/D and recessed S/D bFDSOI-FETs, typical values of Si body thickness ($T_{Si}$) and poly-Si channel thickness ($T_{CH}$) were 30 nm and 5 nm, respectively. Additionally, three types of SOI MOSFETs (FDSOI, UTBSOI, and E-S/D UTBSOI) were designed; the parameters were based on the same conditions for the simulation. For the FDSOI-FET, the typical value of $T_{Si}$ is 30 nm. For the UTBSOI-FET, the typical value of $T_{Si}$ (= $T_{CH}$) is 5 nm. For the E-S/D UTBSOI-FET, the typical value of $T_{Si}$ (= $T_{CH}$) is 5 nm, and the raised S/D thickness is 35 nm. The other parameters, BOX thickness ($T_{BOX}$) and front-gate oxide thickness ($T_{GOX}$) are 50 nm and 1.4 nm, respectively.

## 3. Results and Discussion

The physics models in this paper, including the generation and recombination model, the effective intrinsic density model, and the basic mobility models, were specified for the $I_{DS}$-$V_{GS}$ characteristics of bFDSOI-FETs and its counterparts. Among them, the mobility models include a doping-dependence model and a high-field-saturation model (velocity saturation for electrons), based on the hydrodynamic Canali model. Besides the models used in $I_{DS}$-$V_{GS}$ simulation, a local-carrier temperature-dependent impact-ionization model is attached to the generation and recombination models, a transverse-field dependence model is attached to the basic mobility models, and a hydrodynamic model is used to simulate numerically the $I_{DS}$-$V_{DS}$ characteristics of bFDSOI-FETs and its counterparts [18].

For bFDSOI-FETs, owing to the presence of single crystal Si body, the poly-Si active layer can be recrystallized after annealing. In the simulation study, the crystalline Si values used for bFDSOI-FETs were the same as those used for FDSOI and UTBSOI devices. The major parameters for the comparison are listed as follows. $V_{DS}$ is the voltage bias applied between drain and source, $V_{TH,lin}$ is the linear threshold voltage at $V_{DS}$ = 0.05 V, and $V_{TH,sat}$ is the saturation threshold voltage at $V_{DS}$ = 1.0 V. The threshold voltage is extracted using the constant current method at $I_{DS}$ – 1 $\mu A/\mu m$. The saturation current ($I_{ON}$) is the drain current at $V_{GS} = V_{DS}$ = 1.0 V. The leakage current ($I_{OFF}$) is the drain current at $V_{GS}$ = 0.0 V and $V_{DS}$ = 1.0 V. DIBL is the difference between the $V_{TH,lin}$ and the $V_{TH,sat}$. In order to adjust the $V_{TH}$ and optimize the $I_{ON}/I_{OFF}$, the barrier of doped poly-Si (gate electrode) is chosen as 0.45 eV because this is the difference between the poly-Si extrinsic Fermi level and the Si intrinsic Fermi level [18].

Figure 2 shows the surface potential plots for both the bFDSOI and SOI MOSFETs with gate length of $L_G$ = 30 nm. As shown in Figure 2, there is no significant change in the potential for bFDSOI-FETs as $V_{DS}$ increases, although a slight change in the potential is observed. However, the FDSOI MOSFET (green line) still shows an increase due to the SCEs. This means that the thick FD scheme cannot effectively handle the issue of DIBL. These results confirm that the influence of drain bias upon the channel current has been reduced. As a consequence, the block-oxide-enclosed Si body helps suppress the SCEs, leading to an improved subthreshold swing and a decrease in $I_{OFF}$. Unfortunately, the $V_{DS}$ appears to drop across the S/D regions rather than the channel regions, which leads to smaller effective $V_{DS}$ which then drops across the channel region. This is not the ideal condition for a comparison of the properties of the intrinsic devices designed. The reason for this is that we have

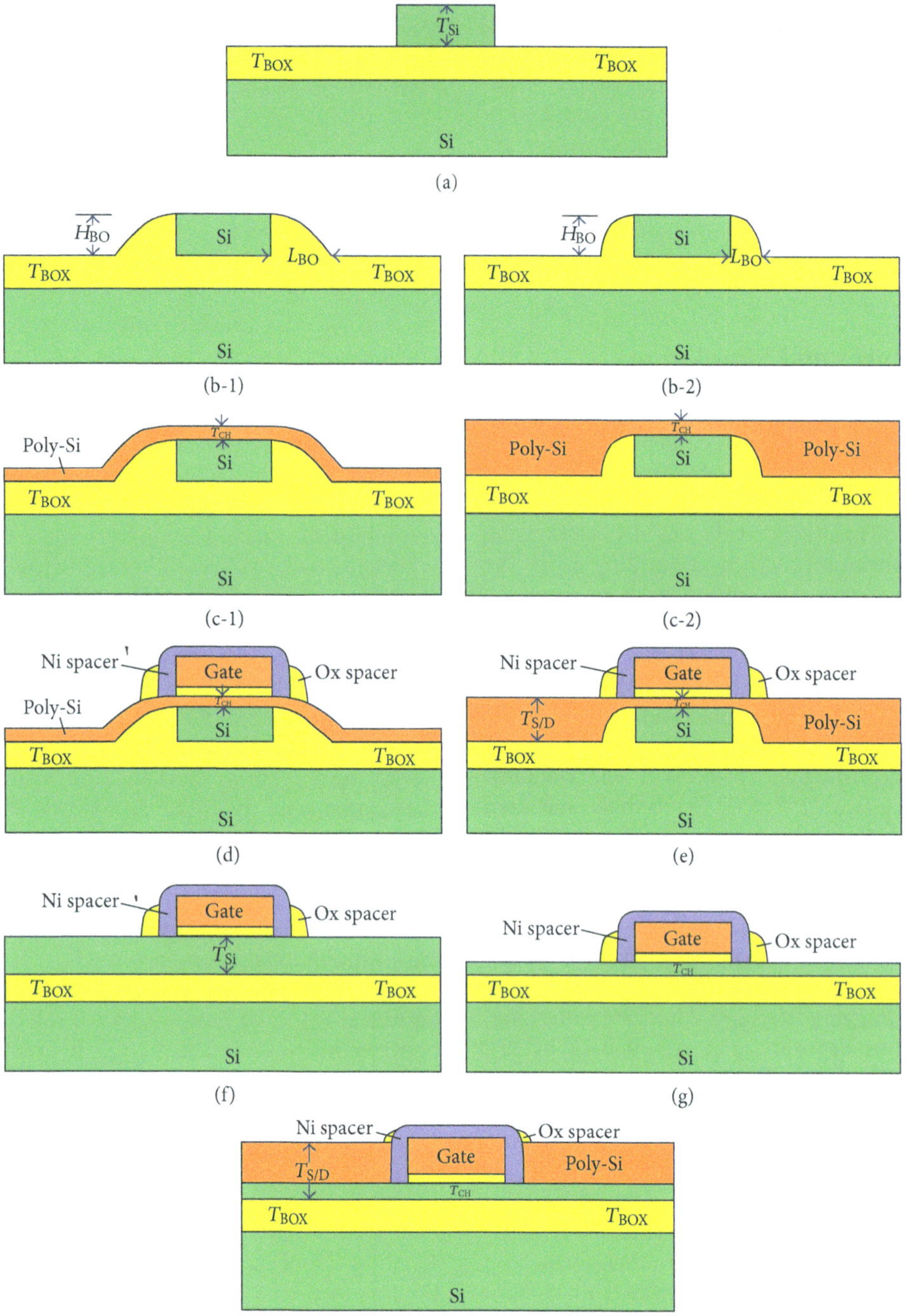


FIGURE 1: Schematic of the process flow of bFDSOI-FETs. (a) Si body patterning. (b-1) Block-oxide formation, where the height of the block-oxide ($H_{BO}$) is equal to the Si body thickness ($T_{Si}$) and the length of the block-oxide ($L_{BO}$) is 60 nm for thin S/D bFDSOI-FET. (b-2) Block-oxide formation, where the $H_{BO}$ is equal to the $T_{Si}$ and the $L_{BO}$ is 15 nm for recessed S/D bFDSOI-FET. (c-1) Poly-Si deposition. (c-2) Poly-Si deposition, planarization, and deposition again. (d) A thin S/D bFDSOI-FET. (e) A recessed bFDSOI-FET. (f) An FDSOI-FET. (g) A UTBSOI-FET. (h) An E-S/D UTBSOI-FET. For thin S/D and recessed S/D bFDSOI-FETs, typical values of $T_{Si}$ and poly-Si channel thickness ($T_{CH}$) are 30 nm and 5 nm, respectively. For FDSOI-FET, the typical value of $T_{Si}$ is 30 nm. For UTBSOI-FET, the typical value of $T_{Si}$ (= $T_{CH}$) is 5 nm. For E-S/D UTBSOI-FET, typical values of $T_{Si}$ (= $T_{CH}$) and poly-Si raised S/D thickness are 5 nm and 30 nm, respectively. For other parameters, BOX thickness ($T_{BOX}$) and front-gate oxide thickness ($T_{GOX}$) are 50 nm and 1.4 nm, respectively.

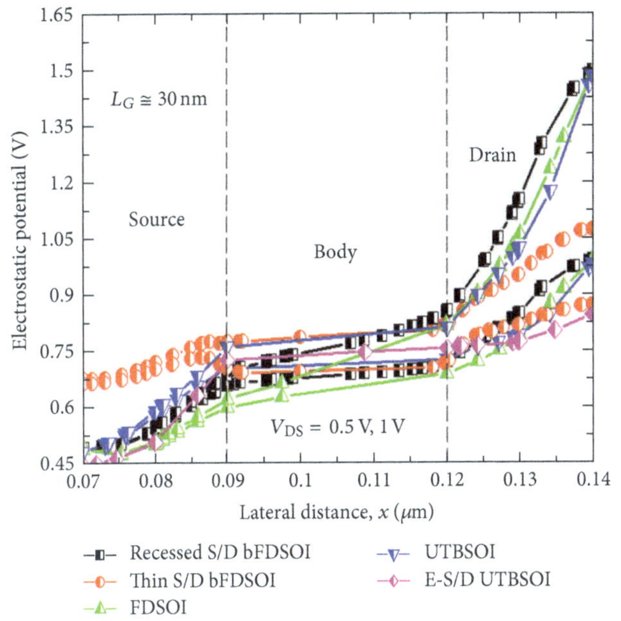


FIGURE 2: Simulated electrostatic potential (V) as a function of lateral distance along the FETs interface with gate length $L_G = 30$ nm.

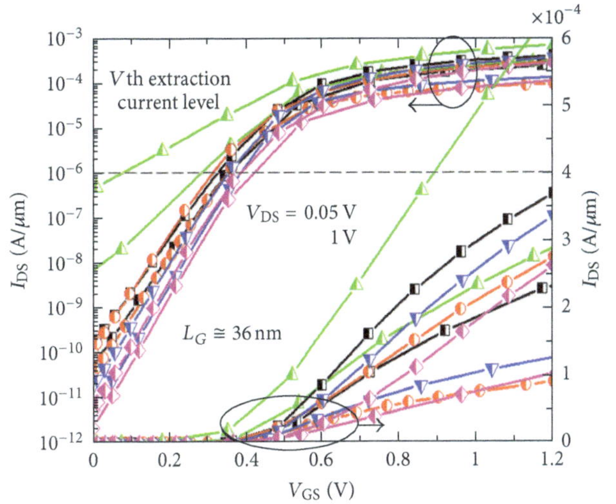


FIGURE 3: Comparison of bFDSOI and SOI MOSFETs $I_{DS}$-$V_{GS}$ characteristics with gate length $L_G$ = 36 nm. The symbols are the same as in Figure 2.

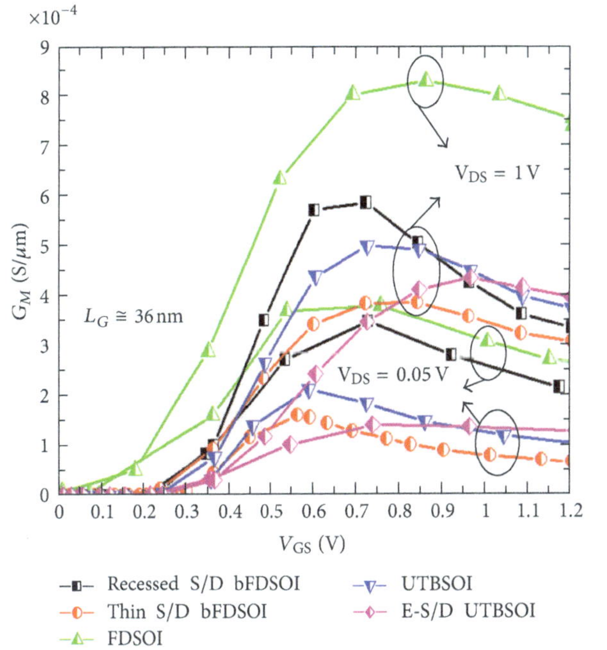


FIGURE 4: Comparison of bFDSOI and SOI MOSFETs $G_M$-$V_{GS}$ characteristics with gate length $L_G$ = 36 nm.

not optimized the S/D engineering in this study so that the characteristics are strongly dependent on the source/drain parasitic resistances. However, this study does predict the general trends for device scaling. Figure 3 shows the transfer characteristics of FETs with gate length $L_G$ = 36 nm. Because of the block-oxide-enclosed Si body, bFDSOI-FETs show that DIBL is suppressed and subthreshold swing is improved. Furthermore, both thin S/D and recessed S/D bFDSOI-FETs show similar results compared to the UTBSOI-FET as well as the E-S/D UTBSOI-FET—additionally, those results are all better than those for the FDSOI-FET. In the UTBSOI-FET or E-S/D UTBSOI-FET, the subthreshold leakage current is the lowest among the transistors. Although FDSOI-FET shows the highest $I_{ON}$ and transconductance ($G_{M,\text{ MAX}}$ in Figure 4), it is very difficult to alleviate the SCEs in a thick body structure. For a given MOS device, transconductance is proportional to the square root of the drain-to-source current $I_{DS}$ [19]. This suggests that thick S/D regions need to be introduced for an SOI to reduce series resistance. Yet a thick S/D structure alone cannot effectively reduce SCEs. Note that in Figure 5, the DIBL of bFDSOI-FETs is much smaller than that of the FDSOI-FET because the influence of drain bias on the channel current is reduced, as discussed earlier. Also, Figure 5 shows that as the gate length is reduced, the DIBL characteristics of the thin S/D bFDSOI-FET become slightly better than the recessed S/D bFDSOI-FET because the thin S/D structure possesses an effective method of suppressing SCEs. Nevertheless, either of the bFDSOI-FETs can alleviate the requirement of using an ultrathin channel to control SCEs in future nanodevices.

Figure 6 shows the $I_{OFF}$ versus $I_{ON}$ for the MOSFETs at different gate lengths. Thanks to its improved subthreshold swing, the bFDSOI-FETs show a lower $I_{OFF}$ when compared with the FDSOI-FET. It should be noted that the FDSOI-FET shows the highest $I_{ON}$ among transistors. As discussed earlier, thick S/D structure alone was unable to effectively reduce source and drain punch-through leakage, leading to the increase in $I_{OFF}$. On the one hand, UTBSOI-FET achieves the lowest $I_{OFF}$, mainly owing to the ultrathin S/D structure that suppresses punch-through and reduces the leakage current. This ultrathin S/D structure also allows the UTBSOI-FET to better control SCEs (see Figure 3), by reducing the effects of charge sharing. But the UTBSOI-FET has a poor $I_{ON}$ because of the high series resistance caused by its ultrathin

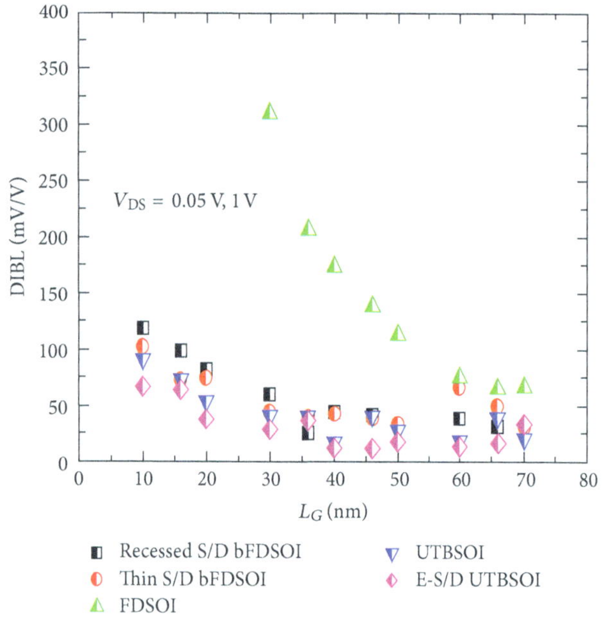


FIGURE 5: Simulated DIBL as a function of the gate length $L_G$ for a low drain bias ($V_{DS}$ = 0.05 V) and high drain bias ($V_{DS}$ = 1.0 V). It is shown that the DIBL characteristics of the bFDSOI-FETs are better than those of the FDSOI-FET and similar to those of the UTBSOI-FET.

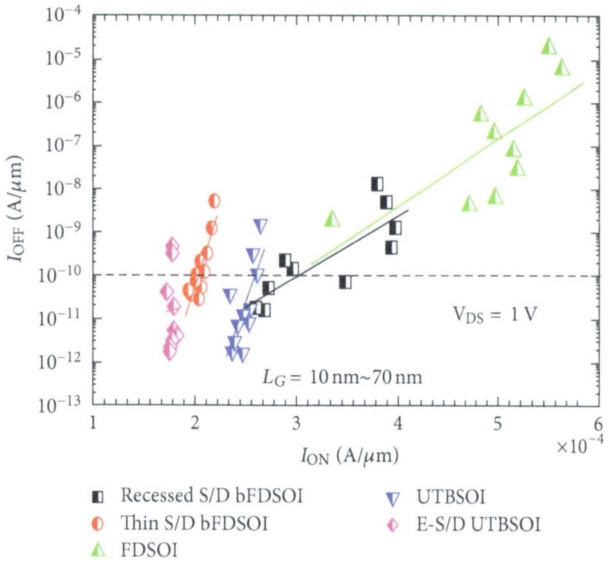


FIGURE 6: Simulated $I_{OFF}$ versus $I_{ON}$ for the MOSFETs at $V_{DS}$ = 1.0 V. Mainly due to improved subthreshold swing, bFDSOI-FETs produce higher on/off current ratio than FDSOI-FET.

S/D regions. In the case of the E-S/D UTBSOI-FET, due to the nonoptimized S/D doping (because we use the same process conditions and parameters to fabricate all devices except those unique respective structures), this results in the longest channel length among all five transistors, and a poor $I_{ON}$ is obtained. That is why the E-S/D UTBSOI-FET shows a better subthreshold swing and lower leakage current compared to

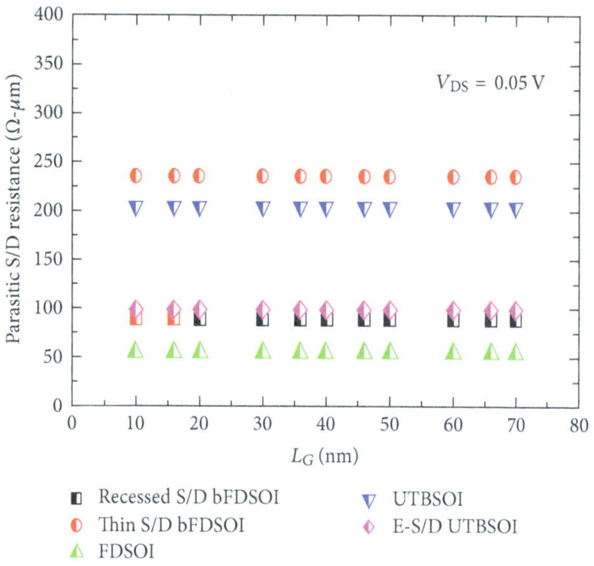


FIGURE 7: Simulated $R_{S/D}$ in different transistors for a low drain bias ($V_{DS}$ = 0.05 V) and high gate bias ($V_{GS}$ = 5.0 V). As shown in the figure, other than the thin S/D bFDSOI and UTSOI devices, devices show lower $R_{S/D}$ because of thick S/D regions.

the UTBSOI-FET as also shown in Figure 3 and smaller DIBL with decreasing gate length as shown in Figure 5. Therefore, the long channel leads to the small channel current and transconductance as shown in Figure 4. Both of the bFDSOI-FETs have similar results to the UTBSOI-FET. This is because the combined applications of a thick Si body and a block-oxide in a FDSOI MOSFET are used to minimize the effects of charge sharing.

Figure 7 shows the S/D series resistance ($R_{S/D}$) for the different transistors, which was extracted at $V_{DS}$ = 0.05 V and $V_{GS}$ = 5.0 V [20]. Compared with the thin S/D bFDSOI-FET and the UTBSOI-FET, both of the recessed S/D bFDSOI-FET and the FDSOI-FET have a very low $R_{S/D}$ because of their thick S/D regions. Furthermore, owing to the raised S/D scheme, the E-S/D UTBSOI-FET reveals a similar $R_{S/D}$ when compared to the recessed S/D bFDSOI-FET. We also consider that due to the recessed S/D scheme, the bFDSOI-FET can get a relatively lower $R_{S/D}$ than that of the E-S/D UTBSOI-FET. In other words, the junction depth that is different for both devices may be a reason why the $R_{S/D}$ of the recessed S/D bFDSOI-FET is relatively smaller. The low $R_{S/D}$ is desirable for SOI devices in high-performance applications, but ultrathin S/D regions have difficulty achieving a low $R_{S/D}$. Although the FDSOI-FET shows the lowest $R_{S/D}$ among the transistors, it is difficult to reduce DIBL and other SCEs. For the recessed S/D bFDSOI-FET, the combined applications of a thick S/D and a block-oxide are an effective way of reducing $R_{S/D}$ as well as DIBL. For the thin S/D bFDSOI-FET, it is difficult to reduce $R_{S/D}$ because of its thin S/D regions, but the thinness of these regions makes it easy to reduce DIBL. On the other hand, the E-S/D UTBSOI-FET also exhibits a smaller $R_{S/D}$ compared to the UTBSOI-FET; however, the Miller capacitance is one of the most important issues for high-frequency applications.

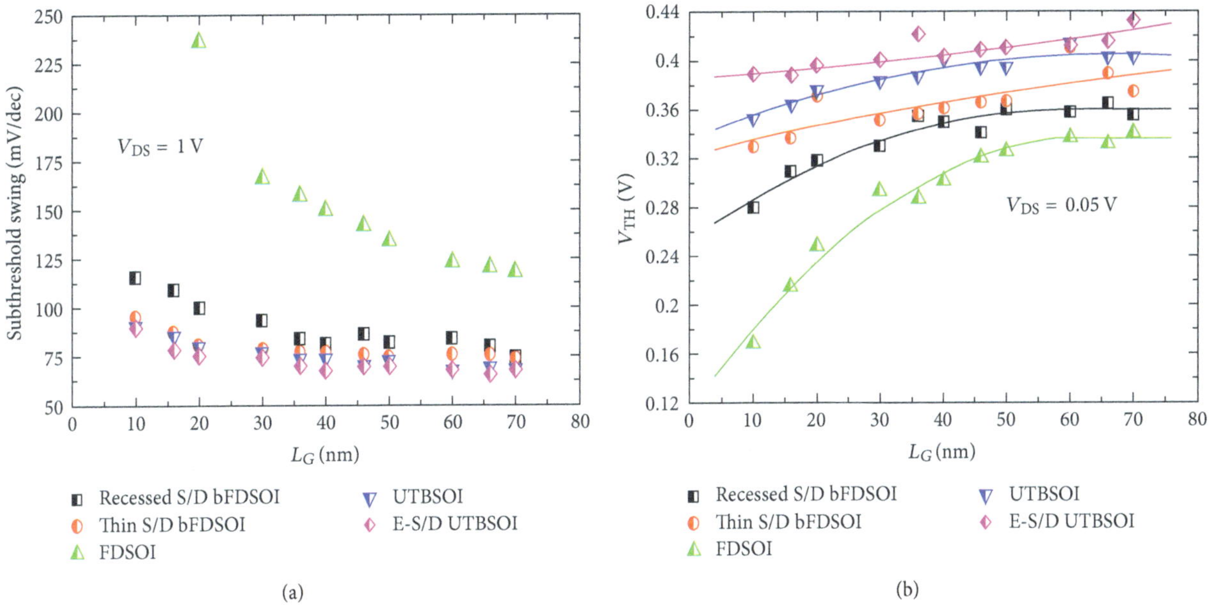


FIGURE 8: Simulated (a) subthreshold swing and (b) $V_{TH}$ dependence on the gate length $L_G$. The bFDSOI-FETs show improved subthreshold swing and better $V_{TH}$ rolloff than FDSOI-FET, similar to the UTBSOI-FET.

The subthreshold swing and $V_{TH}$ of MOSFETs, as a function of the gate length, are presented in Figure 8. Because of the good gate controllability over the active channel, the bFDSOI-FET has an improved subthreshold swing and a better $V_{TH}$ rolloff when compared with the FDSOI-FET. According to recent trends, SOI FETs limit the device performance itself since a uniform film thickness below 10 nm is needed for improving the subthreshold swing [21, 22]. In brief, SOI devices need to conform to the rule that $T_{Si} \leq L_G/4$ [23]. The use of a block-oxide in FDSOI design can alleviate the requirement for a uniform ultrathin Si film, thereby making the characteristics of the bFDSOI-FETs like those of the UTBSOI-FET with or without an E-S/D scheme, even though the total body thickness (including the single crystal Si body and the poly-Si film deposited afterward) is 35 nm in the bFDSOI-FETs.

One of the key issues associated with SOI-based transistors is self-heating effects (SHEs), because the reliability of SOI devices is severely affected by thermal instability. To investigate the influence of SHEs on the device structures, various gate-to-source overdrive voltages ($V_{GT}$) are applied to the devices. With $V_{GT}$ = 1.0 V, thin S/D bFDSOI-FET still shows good behavior in suppressing SHEs. However, FDSOI, UTBSOI, and E-S/D UTBSOI devices, in contrast, suffer from serious thermal effects because the self-heating-induced negative differential conductance (NDC) is observed in the output curves, as shown in Figure 9. The authors also found that the 35 nm thick body bFDSOI structures can help transistors to endure more heat generated in the channel. Although the body of the FDSOI is thicker than that of the UTBSOI, a higher drain current also results in more heat energy and eventually leads to a reduction of the drain current $I_{DS}$ when the FDSOI enters into the saturation region. For the recessed bFDSOI-FET, due to the higher drain current compared to that of the thin S/D bFDSOI-FET, the self-heating-induced NDC is also observed in the output curves. However, the recessed S/D bFDSOI-FET still provides better self-heating immunity than SOI devices owing to its thick body scheme. Additionally, due to the lower current in the bFDSOI device than in the UTBSOI, self-heating is not dominant. If the drain current is high enough, self-heating is still a significant problem for the bFDSOI-FETs, as it is in UTBSOI devices. Nevertheless, owing to the thick body and the block-oxide schemes, requiring a uniform UTB

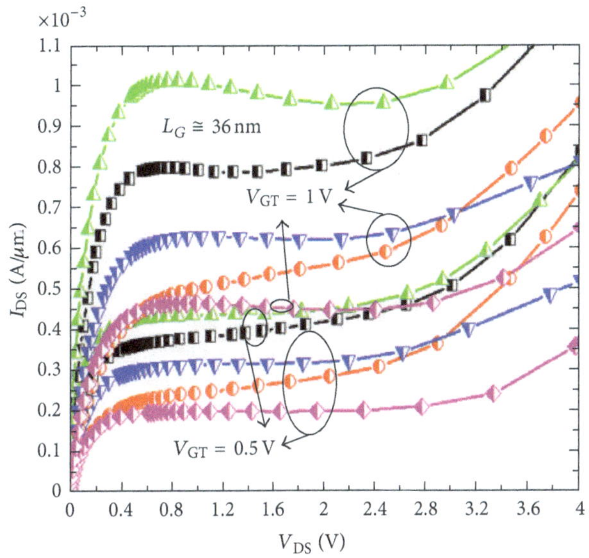


FIGURE 9: Comparison of bFDSOI and SOI MOSFETs $I_{DS}$-$V_{DS}$ characteristics with gate length $L_G$ = 36 nm. The symbols are the same as in Figure 2.

structure for suppression of SCEs can be excluded. This relaxes the technology requirement in a UTB application. In addition, the power supply voltage will eventually decrease in order to reduce the power consumption in the IC circuits. Therefore, the thermal instability in the bFDSOI-FETs can be diminished via a choice of low power supply (i.e., $V_{cc} < 1.0\,V$).

In the case of low drain bias, self-heating is not significant; hence the NDC is not obviously observed in the output characteristics. The better control of SCEs the transistor has, the higher output resistance ($r_o$) the transistor possesses. That is why the UTBSOI with or without an E-S/D structure can produce a higher $r_o$ compared to the FDSOI, with or without a block-oxide scheme. In addition, because large drain current results in large $G_M$, both FDSOI and recessed bFDSOI devices show a larger $G_M$ than those of other transistors. In order to compare the incremental voltage gain ($A_V = G_M/G_D = G_M r_o$), it is believed that the UTSOI with or without E-S/D devices can produce a higher voltage gain as compared to the FDSOI with or without block-oxide scheme, despite the fact that their $G_M$ is lower. It is because the poor short-channel behavior leads to a small $r_o$ observed in output curves for the FDSOI with or without the block-oxide scheme. If the self-heating is dominant, it is not a fair comparison chiefly owing to the NDC phenomenon that results in negative $r_o$.

Figure 10 shows the electron temperature along the MOSFETs' channel surface with gate length $L_G = 36$ nm. For bFDSOI-FETs, the thermal stability can be improved with the addition of a thick body, which results in a lower electron temperature in the channel. Although an ultrathin S/D structure alone shows excellent subthreshold characteristics, SHEs will become a severe problem for SOI-based transistors in particular, as SHEs would likely block the use of SOI MOSFETs in high-performance CMOS applications. It is worthwhile noting that a high-temperature electron possessing high transportation, rotation, and vibration energy will result in the increased probability of phonon scattering and surface scattering, thereby increasing thermal resistance and aggravating self-heating. In contrast to both of these SOI-based transistors, bFDSOI-FETs can alleviate the requirement of uniform ultrathin films to control SCEs, thereby leading to high device reliability. As a result, the thermally induced vibrations of the atoms in the bFDSOI-FETs are ameliorated by its thick body. The body thickness of the FDSOI is thicker than that of the UTBSOI, but a higher $I_{DS}$ also results in more heat energy and eventually leads to a serious reduction of the carrier mobility.

The simulation results in this paper suggest that a metal-gate material and a high-k dielectric should be introduced into the bFDSOI-FET devices. The use of a metal gate would allow the $V_{TH}$ to be optimized even for a lightly doped ultrathin body. In addition, the metal gate, when used with a high-k dielectric, would allow the $I_{ON}/I_{OFF}$ ratio to be improved [24, 25]. This is because the work function of the metal gate can help transistors to adjust the $V_{TH}$ while the high-k dielectric can also relax the requirement of ultrathin gate oxide to reduce the gate-tunneling leakage current.

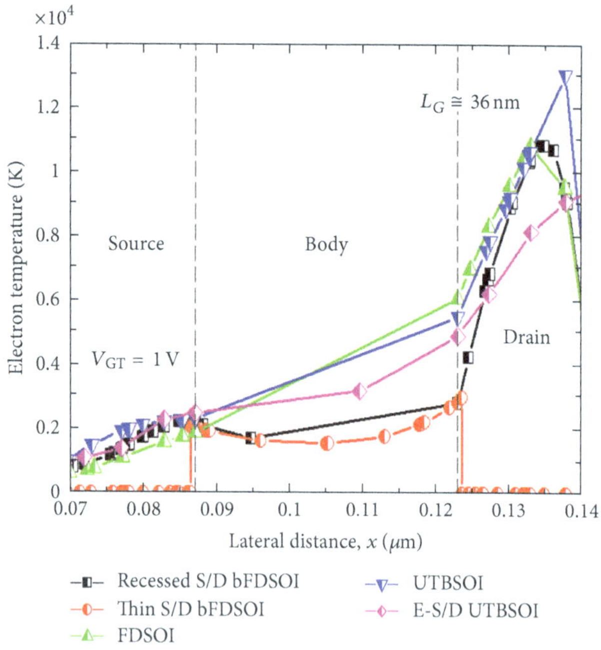


FIGURE 10: Simulated lateral electron temperature (K) as a function of lateral distance along the MOSFETs channel surface with gate length $L_G = 36$ nm.

Two important issues caused by the bFDSOI-FETs are addressed here: (1) the quality of poly-Si and (2) the non-self-aligned process. In the bFDSOI-FETs, the Si film used in this study for comparison is the same as that used in the FD and UTB SOI devices. The mobility of bFDSOI-FETs is actually affected by the poly-Si film. Although the lattice scattering caused by phonons in the bFDSOI-FETs can be reduced by its thick body, the poor quality of poly-Si also results in reduced mobility. Some methods of improving the quality of poly-Si are now described. In general, after making S/D regions by ion implantation, it is believed that poly-Si can be recrystallized because the poly-Si is directly connected to the single crystal Si body. Moreover, advanced recrystallization techniques can also be applied to the bFDSOI process to improve the quality of the poly-Si. Another key issue is the non-self-aligned process used in bFDSOI-FETs. In fact, self-aligned technology is not applicable in the bFDSOI-FETs presented. The misalignment problem will limit the bFDSOI-FETs' performance. All of these issues, including the quality of poly-Si channel and the self-alignment of bFDSOI-FETs, will be addressed in our future research work.

## 4. Conclusion

In this paper a new planar FDSOI MOSFET with block-oxide has been presented and analyzed. The bFDSOI-FET is equipped with a block-oxide on the sidewall of Si body that helps improve the control of SCEs without requiring a uniform UTBSOI structure. As indicated by the two-dimensional (2D) simulation results, the authors found that the characteristics of the bFDSOI-FETs (a reduced DIBL, a higher on/off current ratio, an improved subthreshold swing, and a better $V_{TH}$ rolloff behavior as compared with the FDSOI-FET) are similar to the UTBSOI-FET, because the block-oxide-enclosed Si body helps to diminish the influence of $V_{DS}$ upon the channel current, resulting in desirable device characteristics. Although the short-channel properties of bFDSOI-FETs are somewhat worse than those of UTBSOI-FETs, the results are acceptable. Moreover, both types of bFDSOI-FETs also exhibit significantly lower channel temperature due to their thick bodies. It should be noted that this thick body, including the single crystal Si body and the poly-Si film deposited afterwards, are used to tolerate much of the heat being generated in the channel when compared with the UTBSOI-FET. As a result, the thermal stability of the bFDSOI-FETs can be improved by reducing the scattering of lattice atoms. Compared with the UTBSOI-FET, a thickness requirement of below 10 nm and uniform ultrathin film are excluded for the bFDSOI-FETs to diminish the charge-sharing effect without increasing self-heating. The bFDSOI-FETs are therefore found to improve the reliability of SOI CMOS devices and somewhat relax the critical technology requirement for potential applications.

## References


[1] M. Quirk and J. Serda, *Semiconductor Manufacturing Technology*, Prentice-Hall, Englewood Cliffs, NJ, USA, 2001.

[2] M. J. Kumar and A. Chaudhry, "Two-dimensional analytical modeling of fully depleted DMG SOI MOSFET and evidence for diminished SCEs," *IEEE Transactions on Electron Devices*, vol. 51, no. 4, pp. 569–574, 2004.

[3] O. Thomas, M. Belleville, F. Jacquet, and P. Flatresse, "Impact of CMOS technology scaling on SRAM standby leakage reduction Techniques," in *Proceedings of the IEEE International Conference on Integrated Circuit Design and Technology (ICICDT '06)*, pp. 2–6, May 2006.

[4] M. J. Kumar, V. Venkataraman, and S. K. Gupta, "On the parasitic gate capacitance of small-geometry MOSFETs," *IEEE Transactions on Electron Devices*, vol. 52, no. 7, pp. 1676–1677, 2005.

[5] O. Faynot, T. Poiroux, F. Andrieu et al., "Advanced SOI technologies: advantages and drawbacks," in *Proceedings of the Extended Abstracts of the 6th International Workshop on Junction Technology (IWJT '06)*, pp. 200–203, May 2006.

[6] J. T. Lin, K. C. Lin, T. Y. Lee, and Y. C. Eng, "Investigation of the novel attributes of a vertical MOSFET with internal block layer (bVMOS): 2-D simulation study," in *Proceedings of the 25th International Conference on Microelectronics (MIEL '06)*, pp. 488–491, May 2006.

[7] Y. Omura, H. Konishi, and S. Sato, "Quantum-mechanical suppression and enhancement of SCEs in ultrathin SOI MOSFETs," *IEEE Transactions on Electron Devices*, vol. 53, no. 4, pp. 677–684, 2006.

[8] S. Nuttinck, "Ultrathin-body SOI devices as a CMOS technology downscaling option: RF perspective," *IEEE Transactions on Electron Devices*, vol. 53, no. 5, pp. 1193–1199, 2006.

[9] C. G. Ahn, W. J. Cho, K. J. Im et al., "Recesseed source-drain (S/D) SOI MOSFETs with low S/D extension (SDE) external resistance," in *Proceedings of the IEEE International SOI Conference*, pp. 207–208, October 2004.

[10] K. Komiya, T. Kawamoto, S. Sato, and Y. Omura, "Impact of high-k plug on self-heating effects of SOI MOSFETs," *IEEE Transactions on Electron Devices*, vol. 51, no. 12, pp. 2249–2251, 2004.

[11] Z. Sun, L. Liu, and Z. Li, "Self-heating effect in SOI MOSFET's," in *Proceedings of the 5th International Conference on Solid-State and Integrated Circuit Technology*, pp. 572–574, October 1998.

[12] J. T. Lin, Y. C. Eng, K. D. Huang, T. Y. Lee, and K. C. Lin, "Ultra-short-channel characteristics of planar MOSFETs with block oxide," in *Proceedings of the 13th International Symposium on the Physical and Failure Analysis of Integrated Circuits (IPFA '06)*, pp. 146–149, July 2006.

[13] J. T. Lin, Y. C. Eng, K. D. Huang, T. Y. Lee, and K. C. Lin, "A novel FDSOI MOSFET with block oxide enclosed body," in *Proceedings of the IEEE International Conference on Integrated Circuit Design and Technology (ICICDT '06)*, pp. 145–148, May 2006.

[14] Y. C. Eng, J. T. Lin, K. D. Huang, T. Y. Lee, and K. C. Lin, "An investigation of the effects of Si thickness-induced variation of the electrical characteristics in FDSOI with block oxide," in *Proceedings of the 8th International Conference on Solid-State and Integrated Circuit Technology (ICSICT '06)*, pp. 61–64, October 2006.

[15] E. Yuri and L. Jinning, "Precision implant requirements for SDE Junction Formation in sub-65 nm CMOS devices," in *Proceedings of the Extended Abstracts of the 6th International Workshop on Junction Technology (IWJT '06)*, pp. 21–24, May 2006.

[16] A. Bansal and K. Roy, "Asymmetric halo CMOSFET to reduce static power dissipation with improved performance," *IEEE Transactions on Electron Devices*, vol. 52, no. 3, pp. 397–405, 2005.

[17] S. Venkatesan, J. W. Lutze, C. Lage, and W. J. Taylor, "Device drive current degradation observed with retrograde channel profiles," in *Proceedings of the International Electron Devices Meeting (IEDM '95)*, pp. 419–422, December 1995.

[18] User's Manual, ISE-TCAD, 2004.

[19] A. S. Sedra and K. C. Smith, *Microelectronic Circuits*, Oxford University Press, Oxford, UK, 4th edition, 1988.

[20] A. Dixit, A. Kottantharayil, N. Collaert, M. Goodwin, M. Jurczak, and K. De Meyer, "Analysis of the parasitic S/D resistance in multiple-gate FETs," *IEEE Transactions on Electron Devices*, vol. 52, no. 6, pp. 1132–1140, 2005.

[21] H. Shang, J. Rubino, B. Doris et al., "Mobility and CMOS devices/circuits on sub-10 nm (110) ultra thin body SOI," in *Proceedings of the Symposium on VLSI Technology*, pp. 78–79, June 2005.

[22] K. Samsudin, B. Cheng, A. R. Brown, S. Roy, and A. Asenov, "UTB SOI SRAM cell stability under the influence of intrinsic parameter fluctuation," in *Proceedings of the 35th European Solid-State Device Research Conference (ESSDERC '05)*, pp. 553–556, September 2005.

[23] S. Deleonibus, B. De Salvo, L. Clavelier et al., "CMOS devices architectures for the end of the roadmap and beyond," in *Proceedings of the 8th International Conference on Solid-State and Integrated Circuit Technology (ICSICT '06)*, pp. 51–54, October 2006.

[24] J. B. Kuo and C. H. Lin, "Capacitance behavior of nanometer FD SOI CMOS devices with $HfO_2$ high-k gate dielectric

considering gate tunneling leakage current," in *Proceedings of the 25th International Conference on Microelectronics (MIEL '06)*, pp. 59–61, May 2006.

[25] X. Yu, M. Yu, and C. Zhu, "Advanced HfTaON/$SiO_2$ gate stack with high mobility and low leakage current for low-standby-power application," *IEEE Electron Device Letters*, vol. 27, no. 6, pp. 498–501, 2006.

# Voltage-Mode Four-Phase Sinusoidal Generator and Its Useful Extensions

**Sudhanshu Maheshwari**

*Department of Electronics Engineering, Z. H. College of Engineering and Technology, A.M.U., Aligarh 202 002, India*

Correspondence should be addressed to Sudhanshu Maheshwari; sudhanshu_maheshwari@rediffmail.com

Academic Editor: Ali Umit Keskin

This paper introduces a new voltage-mode second-order sinusoidal generator circuit with four active elements and six passive elements, including grounded capacitors. The frequency and condition of oscillation can be independently controlled. The effect of active element's nonidealities and parasitic effects is also studied; the proposed topology is good in absorbing several parasitic elements involved with the active elements. The circuit is advantageous for generating high frequency signals which is demonstrated for 25 MHz outputs. Several circuit extensions are also given which makes the new proposal useful for real circuit adoption. The proposed theory is validated through simulation results.

## 1. Introduction

Four-phase sine-wave generators with voltage outputs progressively separated by 90° apart find useful applications in communication and instrumentation systems and hence have been well covered in open literature [1–4]. Some of the earlier works based on transconductance-$C$ approach [1] offered compact realization with low transistor counts, but limited in frequency when compared to current conveyor based works [2]. Very novel addition to the literature witnessed compact bread-boarding solutions with scope of future integration of new active elements, like DO-CIBA [3]. Meanwhile, traditional approach of employing band-pass filter for quadrature oscillator realization, capable of generating four-phase outputs, continues to attract recent attention [4]. The four-phase voltage-mode circuit of [5] uses five opamps and five passive components, most of which are in floating form. The works as mentioned so far [1–5] falls in the category of second-order networks. Another variety of circuits employed third-order networks to generate quadrature voltage outputs, with a possibility of extension to four phases by additional active elements [6]. The active-$C$ network of [6] used current controlled conveyors and three capacitors. Though the circuit in [6] was intended to generate four-phase current outputs, two quadrature voltage outputs were simultaneously available, a fact not mentioned therein. The discussion on quadrature oscillators' review is worth a voluminous work and can be restricted here, in view of a recent work [3, and cited therein]. In spite of this restriction, there are many works which find mention for their value, without belittling hundreds of other works, not mentioned herein [7–14]. For instance the work in [10] was a first attempt to use DVCC for oscillator application. Another work later [11] presented voltage-mode quadrature oscillator based on two DDCC and five grounded passive components, which can be extended for four phases by employing additional active building blocks.

This work is based on the realization of four-phase voltage-mode oscillator employing band-pass filter as reported in a recent work [15]. Although, the available work proposed a four-phase oscillator with voltage outputs, which was based on a third-order network [15], the new proposed circuit is based on a second-order network, thus providing a new unreported solution with effective results. The proposed circuit is based on four differential voltage current conveyors (DVCCs), four resistors, and two capacitors, unlike the available work, where three resistors and capacitors each were used [15]. The similarity of the new proposed circuit with available one [15] may only be deceptive, since the two works differ in order by "one," as also mentioned in the following. Circuit parasitic considerations, nonideal issues,

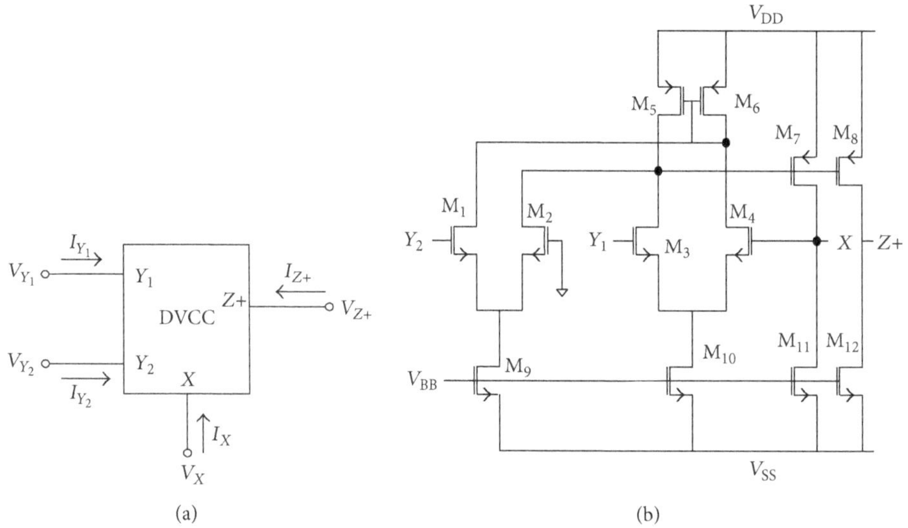


FIGURE 1: Differential voltage current conveyor with $Z+$ output. (a) Symbol and (b) CMOS implementation.

and circuit variations along with the workability are given to support the technical novelty of the proposal.

## 2. Circuit Description

The proposed voltage-mode four-phase sinusoidal generator circuit is based on DVCCs, whose symbol and circuitry are shown in Figure 1, followed by the actual oscillator circuit in Figure 2. The DVCC of Figure 1 is described by the relationship

$$I_{Y1} = I_{Y2} = 0, \qquad V_X = V_{Y1} - V_{Y2}, \qquad I_{Z+} = I_X. \qquad (1)$$

It may be noted that the circuit of DVCC with only $Z+$ stage requires 12 transistors only and can be used as DDCC, by ungrounding the gate of $M_2$, naming it as $Y_3$, with the modified $V_X = V_{Y1} - V_{Y2} + V_{Y3}$, along with $I_{Y3} = 0$. The DVCC/DDCC based analog circuit design continues to be a favorable research topic till recently [4, 16–18]. The basic scheme along with the proposed circuit is shown in Figures 2 and 3, respectively. The circuit requires four active elements and six passive elements, with the advantage of both grounded capacitors. Inverting band-pass transfer function of the basic scheme of Figure 2 is realized from node $Y_3$ of DDCC-1 to node $V_B$ in Figure 3. The inverting gain factor ($-k$) of Figure 2 is realized using DVCC-4 along with $R_3$ and $R_4$ in Figure 3. It may be noted that the inverting amplifier used is one of the several analog blocks as available in [15]. It may be noted that the DDCC-1 has unused $Z$ terminal, which need not to be implemented, thus reducing the implementation of DDCC-1 by two transistors. Routine analysis of the circuit of Figure 3 (along with the block diagram of Figure 2) yields the following characteristic equation:

$$s^2 + sb(1-k) + a = 0,$$
$$\text{where, } a = \frac{4}{R_1R_2C_1C_2}, \qquad (2)$$
$$b = \frac{2}{R_1C_1}, \quad k = \frac{2R_4}{R_3}.$$

The frequency of oscillation (FO) and the condition of oscillation (CO) are found from (3) as follow:

$$\text{FO}: f_o = \frac{1}{\pi\sqrt{R_1R_2C_1C_2}}, \qquad \text{CO}: 2R_4 \geq R_3. \qquad (3)$$

Equation (3) can be easily interpreted for the variation of FO independent of CO, by the frequency determining passive components; whereas, the CO requires setting of separate resistors. The four-phase voltage outputs are related as follows:

$$V_A = -\frac{sR_1C_1}{2}V_B, \qquad V_B = -\frac{sR_2C_2}{2}V_C, \qquad V_D = -V_B. \qquad (4)$$

At the frequency of oscillation, the previous equation can be expressed as follows:

$$V_A = -j\frac{\omega_o R_1C_1}{2}V_B, \qquad V_B = -j\frac{\omega_o R_2C_2}{2}V_C, \qquad (5)$$
$$V_D = -V_B.$$

For a design with equal resistors ($R_1 = R_2 = R$) and equal capacitors ($C_1 = C_2 = C$), the angular frequency of oscillation is obtained from (3) as $\omega_o = 2/\text{RC}$; then (5) reduces to

$$V_A = -jV_B, \qquad V_B = -jV_C, \qquad V_D = -V_B. \qquad (6)$$

Equation (6) shows equal amplitudes for the four outputs, which are separated in phase progressively by 90°.

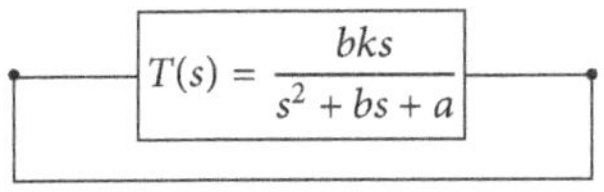


FIGURE 2: Basic scheme of oscillator realization.

## 3. Parasitic Considerations

The circuit of Figure 3 is next analyzed for the involvement of DVCC parasitic elements. It is a well-known fact that the parasitic effects are circuit-topology dependent. A current conveyor circuit with resistive $X$ terminations is suited for absorbing the $X$ terminal intrinsic resistance ($R_X$), although the net value does increase by $R_X$. Similarly, a current conveyor-based network with capacitive terminations at $Y$ and/or $Z$ terminals is good from the point of absorption of parasitic capacitance at these terminals ($C_Y$ and/or $C_Z$, resp.), though the net value at those nodes does increase. Resistive terminations at $Z$ terminal with moderate design values (few KΩ's) also does not hinder the circuit behavior much from the ideal, as the parasitic resistance at $Z$ ($R_Z$) is relatively much higher than the externally connected resistor, appearing in shunt. Coming to the circuit of Figure 3, it may be noted that the various external passive elements are modified due to the parasitics as follows:

$$C_1' = C_1 + C_Z + 3C_Y, \qquad C_2' = C_2 + C_Z + 2C_Y,$$
$$R_I' = R_I + R_X, \quad I = 1, 2, 3, \tag{7}$$
$$R_4' = R_4 // R_P, \qquad R_P = \left(\frac{R_Y}{2}\right) // R_Z.$$

The nonideal FO and CO are thus found as follows:

$$\text{FO} : f_o' = \frac{1}{\pi\sqrt{R_1' R_2' C_1' C_2'}}, \qquad \text{CO} : 2R_4' \geq R_3'. \tag{8}$$

Equations (7)-(8) suggest that the actual FO would be slightly smaller than the desired FO, but the circuit topology does provide the designer scope of nullifying the parasitic effects, by predistorting the passive components. The actual errors as a result of parasitic elements would be given in the results' section.

## 4. Nonideal Transfer Gain Effects

A nonideal DVCC is characterized by the following relationship:

$$V_X = \beta_1 V_{Y1} - \beta_2 V_{Y2}, \qquad I_{Y1} = I_{Y2} = 0, \qquad I_{z+} = \alpha I_x. \tag{9}$$

Here, the voltage transfer gains from $Y_i$, $i = 1, 2$, to $X$ terminal is denoted by $\beta_i$, and current transfer gain form $X$ terminal to $Z+$ terminal is $\alpha$. The proposed circuit's topology (Figure 3) is such it nullifies the transfer gain effect from $Y_1$ to $X$, thus easing the nonideal expression rid of $\beta_1$. An analysis of the circuit using the nonideal description yields the oscillator's characteristic equation as follows:

$$s^2 + s\frac{(1+\alpha)\beta^2}{R_1 C_1}\left[1 - (1+\alpha)\beta\frac{R_4}{R_3}\right] + \frac{(1+\alpha)^2\beta^3}{R_1 R_2 C_1 C_2} = 0. \tag{10}$$

It may be noted that the $\beta_2$ for different DVCCs is taken identical to simplicity of the expression, thus dropping the suffix. The nonideal FO and CO in light of the previous equation thus become

$$\text{FO} : f_o = \frac{(1+\alpha)\sqrt{\beta^3}}{2\pi\sqrt{R_1 R_2 C_1 C_2}}, \qquad \text{CO} : (1+\alpha)\beta R_4 \geq R_3. \tag{11}$$

Equation (11) needs to be first interpreted in terms of frequency errors, as the numerator would be slightly smaller than the ideal expression (3), thus causing a decrement in the actual frequency of oscillation in comparison to the designed value. A similar effect was found in previous section after incorporation of parasitic elements. Thus it is to be appreciated that the FO would be slightly smaller than the desired value, to be further commented upon in results' section. Secondly, (11) has been derived with the assumption that the transfer gains for various DVCCs would be identical, which is quite practical, especially for integrated realizations. Now, the appearance of $\beta$ with power 3/2 may be quite annoying from sensitivity perspective, but actually it is equal to three different $\beta_2$s for different conveyors. With this clarification, the sensitivity of FO with respect to all active and passive elements is found to be within unity in magnitude, ensuring good performance.

## 5. Verification Results

The four-phase oscillator of Figure 3 was simulated using the DVCC of Figure 1 using PSPICE tool. The parameters' details are listed in many recent works, hence not repeated here [14, 15]. The circuit was designed with capacitive elements of 10 pF and frequency determining resistors ($R_1 = R_2$) as 3 KΩ, thereby yielding a theoretical frequency of oscillation as 10.6 MHz from (3). The CO was adjusted through setting $R_3$ in the vicinity of 4 KΩ, retaining $R_4 = 2$ KΩ. The output waveforms are shown in Figure 4, where the FO is found as 10.25 MHz, resulting in an error in FO of approximately 3%. The spectrum of the four outputs is shown in Figure 5; the THD is found to be around 1.1%. The proposed circuit provides interesting options for tunability through either $R_1$ and/or $R_2$ without restriction on their ratios or involvement in CO, a preferred method exhibited in most of the related literatures on the subject. However, the circuit topology used [15] has a scope of being used as a potential analog cell for field programmable analog arrays, where programmable capacitor arrays are employed. In such applications, an array of equal capacitors is available, which need to be switched so as to vary a circuit parameter. With this view, the tunability of the proposed oscillator is studied by using capacitive elements of values 8 pF, 6 pF, and 4 pF so as to result in FO as 12.5 MHz,

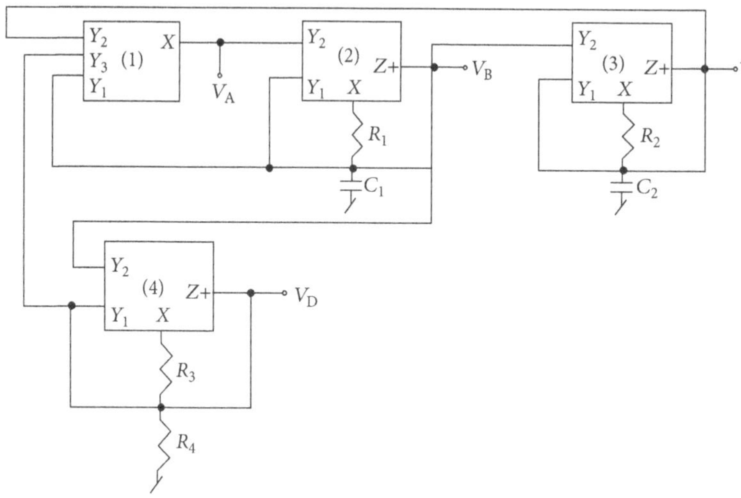


FIGURE 3: Proposed four-phase oscillator circuit.

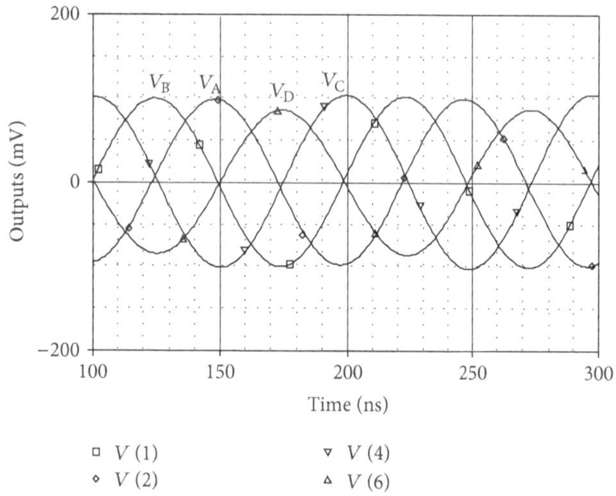


FIGURE 4: Voltage outputs of the proposed circuit.

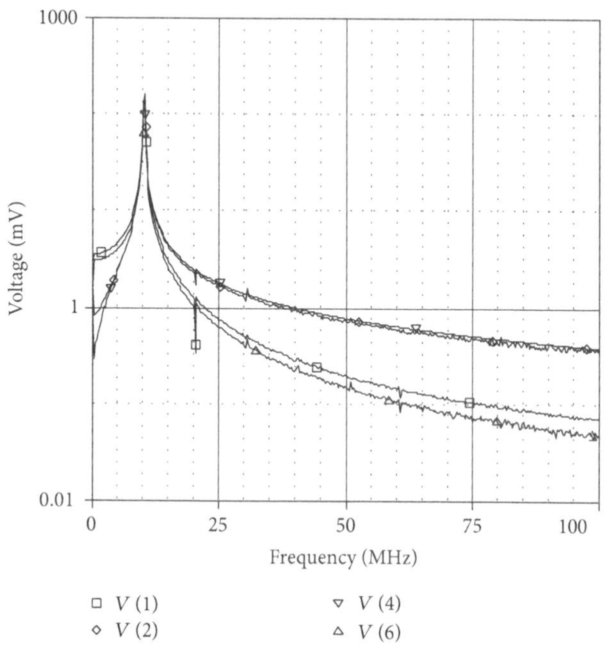


FIGURE 5: Fourier spectrum of outputs with FO = 10.25 MHz.

17.5 MHz, and 25 MHz, respectively. The spectrum of the output waveforms is shown in Figure 6, which confirms the practical utility of the proposed circuit.

## 6. Circuit Extensions

This section presents some useful circuit extensions by simple modifications to the proposed circuit of Figure 3. Several variations with distinct features are given as useful future adoptions.

*6.1. Eight Outputs' Circuit.* The voltage-mode four-phase oscillator of Figure 3 can be extended for explicit four-phase current outputs by employing $Z$-stages. Additional $Z+$ and $Z-$ stages incorporation in DVCC-2 and DVCC-3 each result in four current outputs spaced progressively 90 degrees apart. With this modification (addition), the circuit can be used as a complete four-phase oscillator with both voltage as well as current outputs. The resulting circuit is unique in light of the available literature, by simultaneously generating eight outputs in all. The previously referred circuit (with eight outputs capability) is also simulated to support the theory. Additional $Z+$ and $Z-$ stages each in DVCC-2 and DVCC-3 are implemented so as to result in four-phase current outputs. For the presentation of results, the $Z-$ stages are incorporated in the above-mentioned conveyors so as to generate the quadrature current outputs as shown in Figure 7, with the same design as given earlier for voltage outputs. The Fourier spectrum is next shown in Figure 8, with the FO as 10.25 MHz. It may again be noted that two additional outputs (phase inverted with respect to the ones shown in Figure 7) would be available at $Z+$ terminals of the conveyors.

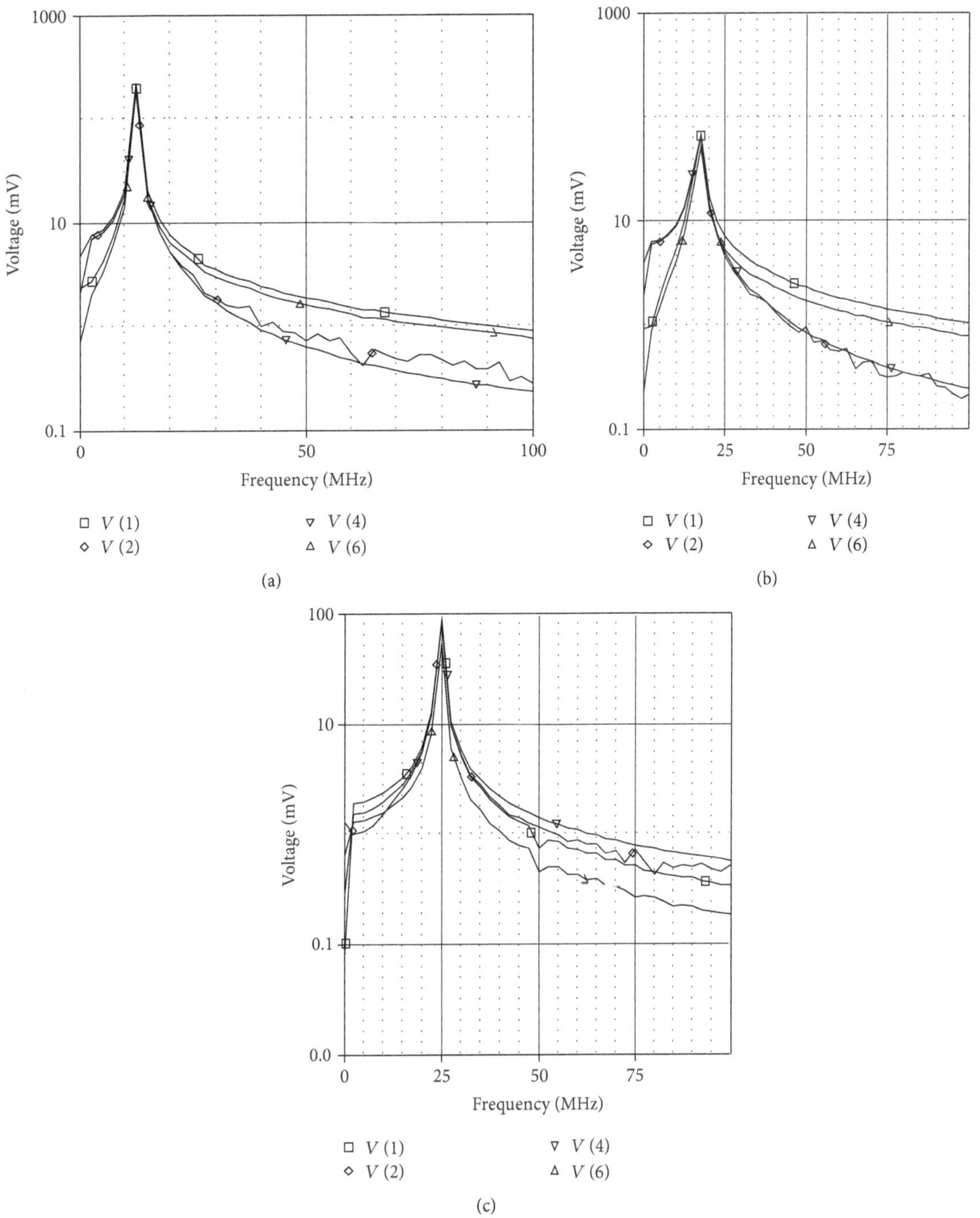


FIGURE 6: Fourier spectrum for different "C." (a) $C = 8$ pF, FO = 12.5 MHz; (b) $C = 6$ pF, FO = 17.5 MHz; (c) $C = 4$ pF, FO = 25 MHz.

*6.2. Circuit with Simpler Configuration.* Circuit with simpler configuration can be obtained by leaving the $Z$ terminal unused while retaining the circuit's functionality. This modification results in an FO half of what is obtained for Figure 3, and the CO reduces to $R_4 \geq R_3$. The resulting circuit is simpler in the sense that $Z$ stages need not to be implemented (please note that current outputs as mentioned in following section will not be available here). More importantly, the parasitic capacitance $C_z$ in (7) disappears as a result of this modification, enabling reduction in errors in FO, for a given capacitor ($C_1$, $C_2$) based design as compared to the circuit of Figure 3. This also provides a design guideline to use smaller capacitance-based design for a permissible FO error. On the other hand, in order to obtain the same FO range as in circuit of Figure 3, the modified circuit would necessitate smaller resistance values, which makes "$R_x$" induced error more influential, when compared to the circuit of Figure 3. Such discussions often lead to unending arguments, best

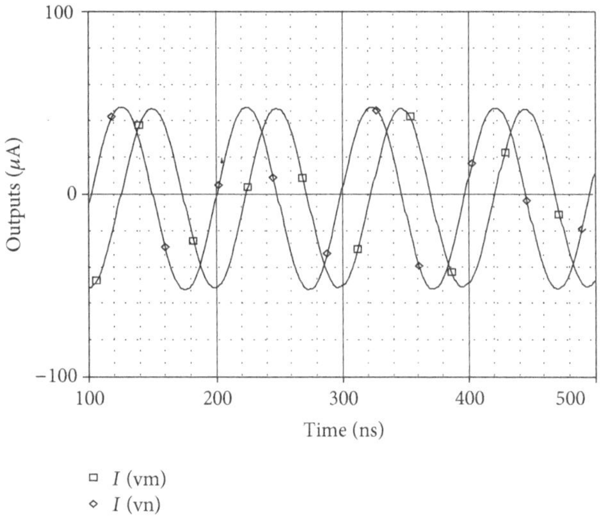


FIGURE 7: Quadrature current outputs for extended circuit.

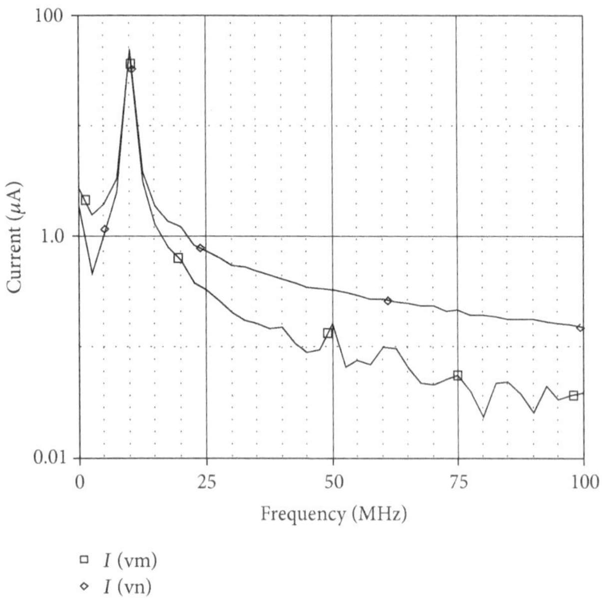


FIGURE 8: Fourier spectrum of the current outputs with FO = 10.25 MHz.

attributed to the artistic mysteries of analog circuit design. Notwithstanding, newer topologies are actually synthesized, as a result of such discussions.

*6.3. Circuits with Tunable DVCCs.* Another circuit enhancement is the use of voltage controlled DVCC as was introduced in [19]. In such case, the $X$-terminated resistors are eliminated, and instead the intrinsic $X$ terminal resistance becomes effective, which is tunable through the bias voltage as has been shown in [19]. Though limited tenability is achieved, the circuit does become a voltage controlled oscillator. Alternatively, current controlled DVCCs can be employed in place of conventional DVCCs. The $X$ terminal resistors are again replaced by intrinsic resistance. The circuit's frequency of oscillation can then be tuned through the bias currents. However, the relative complexity involved will almost double, as the CC-DVCC is a combination of CCCII and DVCC, as has been very well demonstrated in some useful novel works [20, 21].

*6.4. Use of Current Gain Variable DVCC.* Very interesting circuit advancement can be obtained by employing current gain variable DVCCs. It may be noted that conventional DVCC has unity current transfer gain from $X$ to $Z$ terminal. This may be varied to obtain a current-gain variable DVCC. The concept of current-gain variable current conveyors has its origin from some earlier works [22–25]. The proposed circuit of Figure 3 is a promising candidate for the purpose, as is evident from the nonideal expressions (10)-(11). These suggest that the oscillator's frequency and condition are a function of current transfer gain "$\alpha$." If $\alpha$ is made variable, then the FO can be tuned through it, rather than the external passive elements. A simple way to achieve this is to vary the aspect ratio of transistors comprising the $Z$ stage as was also given in [14], but for amplitude control in that particular work. A more practical way is to tune the current gain and hence the frequency of the oscillator by digital means as given in the following.

*6.5. Digitally Controlled Circuit.* A last future enhancement is the use of a digitally controlled current transfer gain stage, which has been recently employed in some work [26, and cited therein]. The current transfer gain "$\alpha$" can be digitally controlled through an $n$-bit control work so as to vary the frequency of the oscillator. In such an oscillator, none of the external passive elements need to be varied for frequency tuning. The complete circuit digital control and analog can thus be made compatible to standard IC technologies. Further details of the above-mentioned circuit extensions are not to be further elaborated for brevity reasons.

## 7. Some Interesting Remarks

The topic of this paper assumes significance in light of some interesting works which continue to appear till recently in the open literature [27, 28]. As far as the quadrature oscillator's (with two quadrature outputs) extension to four outputs is concerned, additional inverters can be employed in such cases. A DVCC with input voltage at $Y_2$ and the output at $X$ (with rest unused terminals) can perform the function of an inverter. The $X$ terminal resistance ($R_x$) along with output capacitance ($C_x$) by way of forming a low-pass filter (lossy integrator) results in finite (although) high pole frequency, thus in a way limiting high frequency operation with poorer THD, as compared to the proposed circuit, where lossless integrators provide infinite pole frequency. Thus the distortions at high frequency would not be as much in the proposed circuit as in a quadrature oscillator with inverters to generate four signals. This justifies the use of the proposed structure, as compared to the one with additional inverters. Coming to the possible application of quadrature oscillators, their utility in quadrature mixers, single sideband generation vector voltmeters, and so forth have often been highlighted in literature. The circuits with quadrature outputs along with their inverted versions (four phases) are of special interest as two floating output sources of quadrature signals. In such an application, the circuit can also be referred to as a differential output quadrature oscillator. Other possible applications of four-phase quadrature signals can be in quadrature phase shift keying, as well. Besides, when four quadrature signals are passed through comparators, they become a source of four-phase clock generators. In nutshell, the proposed circuit and their extensions presented in the preceding section cover useful application areas.

## 8. Conclusion

The paper deals with a new proposal for voltage-mode four-phase oscillator circuit with independent control of FO as well as CO and use of grounded capacitors and a circuit topology suited to minimizing the parasitic effects and applicability to high frequency. Nonideal and parasitic study is also given. Some useful circuit extensions for a simpler topology, for simultaneous four-phase current outputs, voltage/current tunable oscillator, and digitally controlled circuit are also given. The new proposals are enrichment to the available knowledge on the subject.

## Acknowledgment

The author is thankful to the anonymous reviewers for positive feedback on the paper. The author also thanks Professor Ali Umit Keskin, Academic Editor, for recommending the paper.

## References

[1] I. A. Khan and S. Khwaja, "An integrable gm-C quadrature oscillator," *International Journal of Electronics*, vol. 87, no. 11, pp. 1353–1357, 2000.

[2] P. Beg, I. A. Khan, and M. T. Ahmed, "Tunable four phase voltage mode quadrature oscillator using two CMOS MOCCIIs," in *Proceedings of the International Multimedia, Signal Processing and Communication Technologies (IMPACT '09)*, pp. 155–157, Aligarh, India, March 2009.

[3] V. Biolková, J. Bajer, and D. Biolek, "Four-phase oscillators employing two active elements," *Radioengineering*, vol. 20, no. 1, pp. 334–339, 2011.

[4] S. Maheshwari, J. Mohan, and D. S. Chauhan, "High input impedance voltage-mode universal filter and quadrature oscillator," *Journal of Circuits, Systems and Computers*, vol. 19, no. 7, pp. 1597–1607, 2010.

[5] T. Parveen, I. A. Khan, and M. T. Ahmad, "Realization of four phase sinusoidal quadrature oscillator using operational amplifier," *Journal of Active and Passive Electronic Devices*, vol. 2, pp. 171–176, 2007.

[6] S. Maheshwari and I. A. Khan, "Current controlled third order quadrature oscillator," *IEE Proceedings—Circuits, Devices and Systems*, vol. 152, no. 6, pp. 605–607, 2005.

[7] A. M. Soliman, "Simple sinusoidal RC oscillators using current conveyors," *International Journal of Electronics*, vol. 42, pp. 61–77, 1975.

[8] A. M. Soliman, "Two integrator loop quadrature oscillators: a review," *Journal of Advanced Research*, vol. 4, no. 1, pp. 1–11, 2013.

[9] R. Senani, "New canonic single-resistance-controlled sinusoidal oscillator using a single current conveyor," *Electronics Letters*, vol. 15, no. 18, pp. 568–569, 1979.

[10] S. S. Gupta and R. Senani, "Grounded-capacitor current-mode SRCO: novel application of DVCCC," *Electronics Letters*, vol. 36, no. 3, pp. 195–196, 2000.

[11] M. Kumngern and K. Dejhan, "DDCC-based quadrature oscillator with grounded capacitors and resistors," *Active and Passive Electronic Components*, vol. 2009, Article ID 987304, 4 pages, 2009.

[12] D. Biolek, A. Ü. Keskin, and V. Biolkova, "Quadrature oscillator using CDTA-based integrators," *WSEAS Transactions on Electronics*, vol. 3, no. 9, pp. 463–469, 2006.

[13] J. W. Horng, C. L. Hou, C. M. Chang, Y. T. Lin, I. C. Shiu, and W. Y. Chiu, "First-order allpass filter and sinusoidal oscillators using DDCCs," *International Journal of Electronics*, vol. 93, no. 7, pp. 457–466, 2006.

[14] S. Maheshwari and B. Chaturvedi, "High output impedance CMQOs using DVCCs and grounded components," *International Journal of Circuit Theory and Applications*, vol. 39, no. 4, pp. 427–435, 2011.

[15] S. Maheshwari, "Analogue signal processing applications using a new circuit topology," *IET Circuits, Devices and Systems*, vol. 3, no. 3, pp. 106–115, 2009.

[16] H. O. Elwan and A. M. Soliman, "Novel CMOS differential voltage current conveyor and tits applications," *IEE Circuits Devices Systems*, vol. 144, pp. 856–860, 1997.

[17] W. Chiu, S. I. Liu, H. W. Tsao, and J. J. Chen, "CMOS differential current conveyors and their applications," *IEE Circuits Devices Systems*, vol. 143, pp. 91–96, 1996.

[18] K. Pal, "Modified current conveyors and their applications," *Microelectronics Journal*, vol. 20, no. 4, pp. 37–40, 1989.

[19] S. Maheshwari, "A canonical voltage-controlled VM-APS with grounded capacitor," *Circuits, Systems, and Signal Processing*, vol. 27, no. 1, pp. 123–132, 2008.

[20] P. Prommee, M. Somdunyakanok, and S. Toomsawasdi, "CMOS-based current-controlled DDCC and its applications," in *Proceedings of the IEEE International Symposium on Circuits and Systems: Nano-Bio Circuit Fabrics and Systems (ISCAS '10)*, pp. 1045–1048, June 2010.

[21] P. Prommee and M. Somdunyakanok, "CMOS-based current-controlled DDCC and its applications to capacitance multiplier and universal filter," *AEU—International Journal of Electronics and Communications*, vol. 65, no. 1, pp. 1–8, 2011.

[22] C. A. Papazoglou and C. A. Karybakas, "Non interactive electronically tunable CCII-based current mode biquadratic filters," *IEE Circuits Devices Systems*, vol. 144, pp. 178–184, 1997.

[23] C. A. Karybakas and C. A. Papazoglou, "Low-sensitive CCII-based biquadratic filters offering electronic frequency shifting," *IEEE Transactions on Circuits and Systems II*, vol. 46, no. 5, pp. 527–539, 1999.

[24] W. Surakampontorn and P. Thitimajshima, "Integrable electronically tunable current conveyors," *IEE Proceedings, Part G*, vol. 135, no. 2, pp. 71–77, 1988.

[25] W. Surakampontorn and K. Kumwachara, "CMOS-based electronically tunable current conveyor," *Electronics Letters*, vol. 28, no. 14, pp. 1316–1317, 1992.

[26] I. A. Khan and M. T. Simsim, "A novel impedance multiplier using low voltage digitally controlled CCII," in *Proceedings of*

*the IEEE GCC Conference and Exhibition (GCC '11)*, pp. 331–334, Dubai, UAE, February 2011.

[27] A. M. Soliman, "Transformation of oscillators using Op Amps, unity gain cells and CFOA," *Analog Integrated Circuits and Signal Processing*, vol. 65, no. 1, pp. 105–114, 2010.

[28] S. Maheshwari and R. Verma, "Electronically tunable sinusoidal oscillator circuit," *Active and Passive Electronic Components*, vol. 2012, Article ID 719376, 6 pages, 2012.

# 16

# Electronically Tunable Sinusoidal Oscillator Circuit

**Sudhanshu Maheshwari and Rishabh Verma**

*Department of Electronics Engineering, Zakir Hussain College of Engineering and Technology, Aligarh Muslim University, Aligarh 202002, India*

Correspondence should be addressed to Sudhanshu Maheshwari, sudhanshu_maheshwari@rediffmail.com

Academic Editor: Winai Jaikla

This paper presents a novel electronically tunable third-order sinusoidal oscillator synthesized from a simple topology, employing current-mode blocks. The circuit is realized using the active element: Current Controlled Conveyors (CCCIIs) and grounded passive components. The new circuit enjoys the advantages of noninteractive electronically tunable frequency of oscillation, use of grounded passive components, and the simultaneous availability of three sinusoidal voltage outputs. Bias current generation scheme is given for the active elements used. The circuit exhibits good high frequency performance. Nonideal and parasitic study has also been carried out. Wide range frequency tuning is shown with the bias current. The proposed theory is verified through extensive PSPICE simulations using 0.25 $\mu$m CMOS process parameters.

## 1. Introduction

The sinusoidal oscillators constitute an important building block which finds numerous applications in analogue signal processing. This electronic function provides standard test and carrier signals for communication and instrumentation circuits and also acts as the starting signal for generation of several other types of test signals. Similarly, the multiphase oscillators also find interesting applications in communication and instrumentation systems. For these highly precise processes, third-order sinusoidal oscillators are best suited because they enjoy lower harmonic distortions as compared to the second-order sinusoidal oscillators [1].

When the repertoire of technical literature is surveyed, many oscillator circuits are available based on one or the other type of active devices [2–4], but the third-order oscillators are not available in profusion. The third-order oscillators were first proposed in [5] using operational transconductance amplifiers (OTAs), but the dynamic range of the circuits based on OTAs is very limited [6]. Also, the frequency of oscillation cannot be controlled without affecting the condition of oscillation. The reported third-order oscillator in [7] provided four quadrature current outputs, all at high impedance, which makes it suitable for current-mode analogue signal processing. Two current-mode third-order sinusoidal oscillators are proposed in [8] which require three OTAs and three grounded capacitances. In the second oscillators circuit, the frequency of oscillation cannot be controlled without affecting the condition of oscillation. Also, for a large variation in the bias current, it gives a relatively less variation in the frequency of oscillation. The circuit proposed in [9] has limited upper frequency because of external resistor termination at X-terminals, which when reduced would increase the frequency errors due to the parasitic X-resistance becoming appreciable in comparison with the external ones. The circuit proposed in [10] provided two quadrature current outputs and required fewer components. The third-order oscillator proposed in [11] provided four quadrature voltage outputs but it required three floating resistors. The reported circuit in [12] uses CMOS CCCIIs and grounded capacitors, provides both quadrature current as well as voltage outputs, and also enjoys electronic tunability. Lastly, the quadrature oscillator reported in [13] requires three CDTAs and three grounded capacitors. This quadrature oscillator, despite being third-order, imparts very high THD values. One work based on two multioutput DVCCs and two grounded capacitors provide high impedance current outputs, but falls in second

TABLE 1: Transistor aspect ratios of CMOS CCCII.

| Transistors | $W$ ($\mu$m) | $L$ ($\mu$m) |
|---|---|---|
| $M_5$, $M_9$ | 10 | 0.5 |
| $M_6$, $M_{10}$ | 16 | 0.5 |
| All other NMOS | 6 | 0.5 |
| All other PMOS | 10 | 0.5 |

order category [14]. Recently, novel oscillator circuit for voltage outputs using new active element (DO-CIBA) and grounded capacitors has been reported, which also requires a number of floating resistors [15]. Another very novel work reported multiphase oscillator circuit operating in current mode and was based on log-domain approach [16].

In this paper, a third-order sinusoidal oscillator is synthesized from the simple topology employing cascaded first-order low-pass filters coupled with gained feedback. The active element used is CMOS Current Controlled Conveyors (CCCIIs). The frequency of oscillation can be electronically tuned independently of the condition of oscillation. This circuit offers the advantages of the simultaneous availability of three sinusoidal voltage outputs and the use of grounded passive components which makes it suitable for IC implementation.

## 2. Proposed Circuit

The newly proposed third-order sinusoidal oscillator is based on the use of cascaded current-mode first order low pass filters coupled with gain block in feedback. The basic block diagram of this topology is shown in Figure 1. For the realization of the sinusoidal oscillator, the current-mode low pass filter is used and the value of "K" is taken to be negative; which can be realized using a current-mode inverting amplifier. The basic block diagram is shown in Figure 1. The basic blocks of Figure 1 are realized using CCCIIs and grounded components. The CCCII was proposed in [17] and ever since its introduction, it has been used for the realization of numerous useful and high performance analogue circuits [17–21]. This device possesses the advantages of the electronic tunability through the parasitic resistance appearing at the terminal X, good dynamic range, and greater linearity. As far as the basic scheme of employing low-pass filters for multiphase oscillators is concerned, the same was first introduced and used for OTA-based realizations [22, 23]. The CMOS implementation of CCCII is shown in Figure 2 and the aspect ratios of all the MOS transistors used are listed in Table 1.

The CCCII is described by the following current and voltage relationships:

$$V_X = V_Y + I_X R_X, \quad I_Y = 0, \quad I_{Z+} = I_X, \quad I_{Z-} = -I_X. \tag{1}$$

Here,

$$R_X \approx \frac{1}{\sqrt{8\mu C_{ox}(W/L)I_B}}, \tag{2}$$

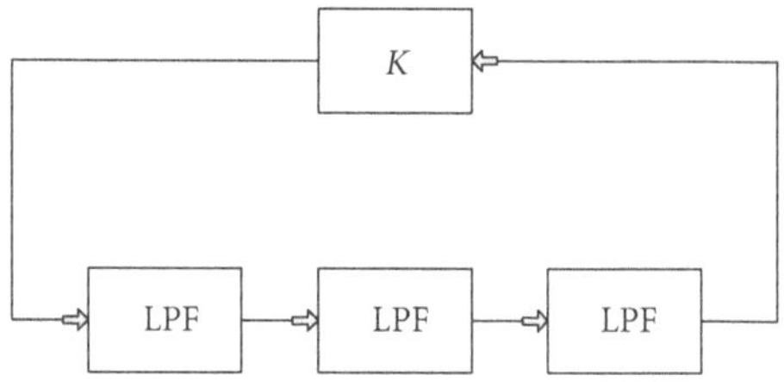


FIGURE 1: Block Diagram of the topology for third-order oscillator.

where $\mu$, $C_{ox}$, $W$, and $L$ are, respectively, the surface mobility, oxide capacitance, channel width, and length of the transistors $M_9$ and $M_{10}$, assuming that both transistors are matched.

The circuit diagram of the newly proposed third-order sinusoidal oscillator, based on the topology shown in Figure 1, is given in Figure 3. The circuit requires four CCCIIs, three capacitors, and a resistor. On performing the routine analysis using the properties of CCCII defined in (1), the characteristic equation of the proposed third-order sinusoidal oscillator shown in Figure 3 is given by

$$as^3 + bs^2 + cs + d = 0, \tag{3}$$

where $a = R_{X1}R_{X2}R_{X3}C_1C_2C_3$, $b = R_{X1}R_{X2}C_1C_2 + R_{X1}R_{X3}C_1C_3 + R_{X2}R_{X3}C_2C_3$, $c = R_{X1}C_1 + R_{X2}C_2 + R_{X3}C_3$, $d = (R_1 + R_{X4})/R_{X4}$.

One simple design is possible by assuming $C_1 = C_2 = C_3 = C$ and $R_{X1} = R_{X2} = R_{X1} = R_X$, that is, $I_{B1} = I_{B2} = I_{B3} = I_B$. Substituting $s = j\omega$, and equating real and imaginary parts on the two sides of (3), we get the frequency and condition of oscillation a

$$\text{F.O.}: \omega_0 = \frac{\sqrt{3}}{R_X C}, \quad \text{C.O.}: R_1 = 8R_{X4}. \tag{4}$$

It is evident from (4) that $\omega_0$ can be varied independent of the condition of oscillation by varying $I_B$. It may be noted that three current outputs can further be obtained by using additional Z+ stages in CCCII (1–3). Incorporation of Z− stages in each of these conveyors will further add to the three more outputs as their phase inverted version, so as to make six distinct current outputs. It is worth mentioning that the new proposed circuit differs from one CCCII-based work [21] which also presents a multiphase oscillator, by way of using grounded X terminals, contrary to the available work, where each X-terminal has capacitive terminations. It is a well-known fact that current conveyors-based circuits with capacitive terminations at X-port have limited frequency performance, which is also evident from [21], where maximum oscillation frequency is well below 1 MHz. The proposed work, in contrast, can provide signal frequencies in tens of MHz, as would be evident from the later sections.

## 3. Bias Current Generation

As evident from (4), the circuit requires frequency control through $R_x$, which in turn needs $I_B$ control. In practice,

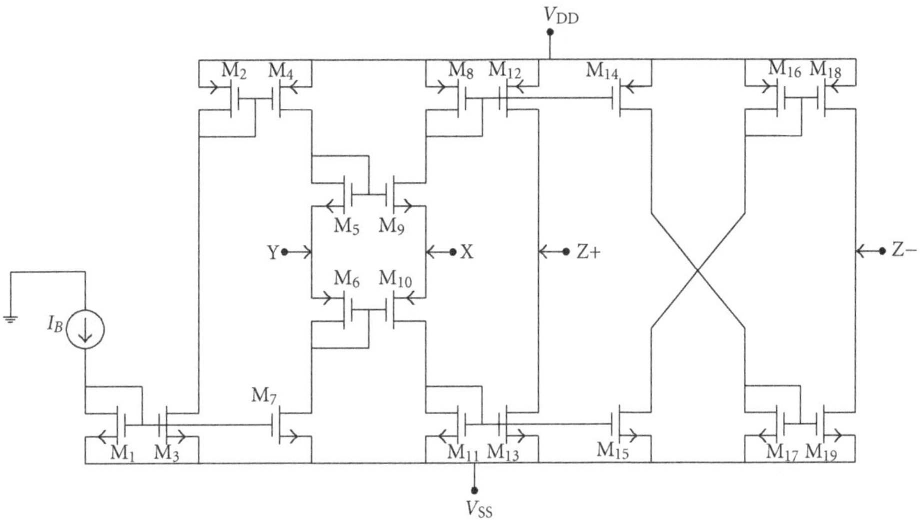


FIGURE 2: CMOS implementation of CCCII.

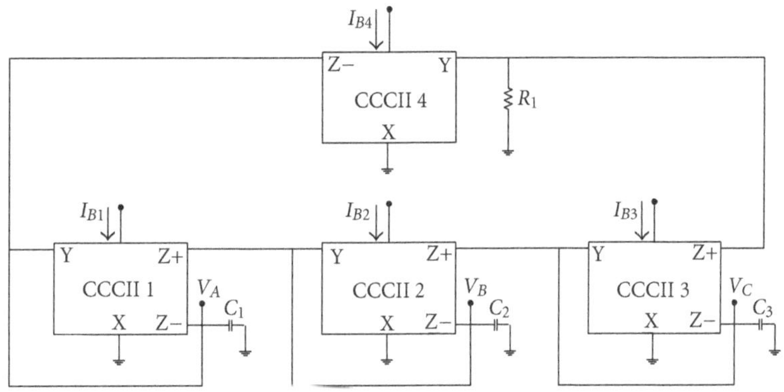


FIGURE 3: Third-order sinusoidal oscillator.

such an equal bias current for the three conveyors can be generated using the transistorized circuitry. The scheme for generation of same bias current $I_B$ for CCCII (1–3) is shown in Figure 4. The bias current is mirrored through an NMOS current mirror and then three current sources are realized using the PMOS current mirror and steering transistors. It may be noted that the current mirror transistors use matched aspect ratio. The circuitry of Figure 4 allows the variation of frequency of oscillation through a single bias current.

## 4. Nonideal Analysis

The performance of any current conveyor-based circuit may deviate from the ideal one due to nonidealities of the active elements. The non-ideal CCCII is described by the following current & voltage relationships:

$$V_X = \beta_k V_Y + I_X R_X,$$
$$I_Y = 0, \qquad I_{Z+} = \alpha_k I_X, \qquad I_{Z-} = -\alpha_k I_X, \qquad (5)$$

where $\beta_k = 1 - \varepsilon_{Vk}$, $\alpha_k = 1 - \varepsilon_{Ik}$.

Here, $\varepsilon_{Vk}$ ($|\varepsilon_{Vk}| \ll 1$) and $\varepsilon_{Ik}$ ($|\varepsilon_{Ik}| \ll 1$) represent the voltage tracking error from Y-terminal to X-terminal and current tracking error from X-terminal to Z-terminal of the kth CCCII, respectively. Generally, these tracking factors remain constant and frequency independent in low- to-medium frequency range. But, at higher frequencies, these tracking factors become frequency dependent.

On reanalyzing the circuit shown in Figure 3 using the nonideal equations of CCCII given in (5), the simplified expressions for frequency of oscillation (FO) and condition of oscillation (CO), assuming matched CCCIIs exhibiting same current and voltage transfer gains (denoted as $\alpha$ and $\beta$, resp.) are given by

$$\text{F.O.}: f_o = \frac{\alpha\beta\sqrt{3}}{2\pi R_x C}, \qquad \text{C.O.}: R_1 = \frac{8R_{x4}}{\alpha\beta}. \qquad (6)$$

Clearly, the frequency of oscillation is modified due to the nonidealities of CCCII, but as the values of $\alpha$ and $\beta$ remain close to unity till tens of MHz, the performance of the circuit is not severely affected. It is also worth mentioning that some frequency deviations would also be expected due to the current mirror transfer errors used in the bias current generation circuitry.

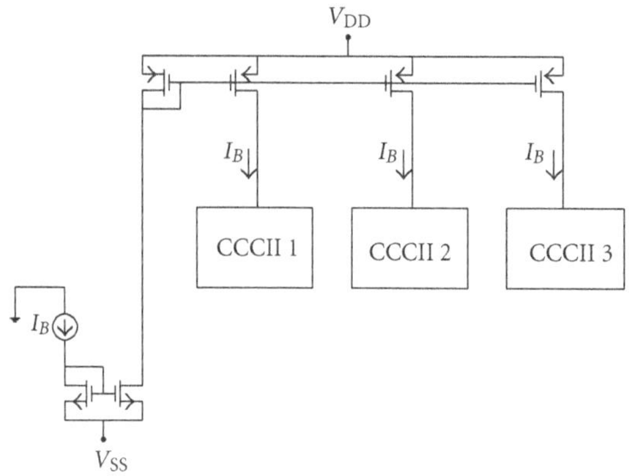


FIGURE 4: Bias current generation scheme for CCCII (1–3).

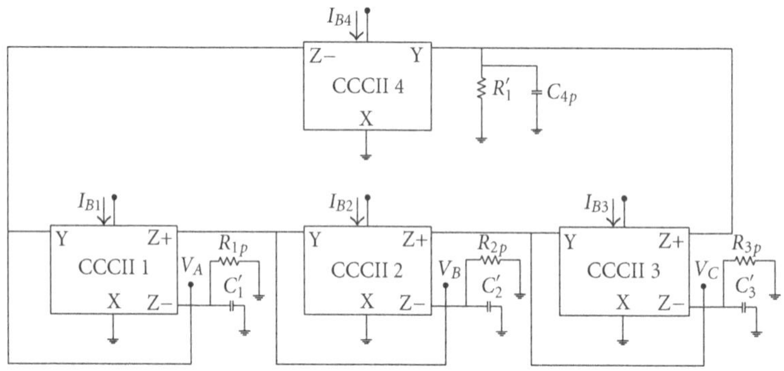


FIGURE 5: Third-order sinusoidal oscillator including parasitics.

## 5. Parasitic Study

The performance of any circuit is also affected due to the presence of various parasites at the terminals of the device. In the presence of the parasites of CCCII, the circuit given in Figure 3 modifies to Figure 5 where $C_1' = C_1 \| C_{y1} \| C_{z1} \| C_{z4}$, $C_2' = C_2 \| C_{y2} \| C_{z2} \| C_{z1}$, $C_3' = C_3 \| C_{y3} \| C_{z3} \| C_{z2}$, $C_{4p} = C_{y4} \| C_{z3}$, $R_{1p} = R_{y1} \| R_{z1} \| R_{z4}$, $R_{2p} = R_{y2} \| R_{z2} \| R_{z1}$, $R_{3p} = R_{y3} \| R_{z3} \| R_{z2}$, $R_1' = R_1 \| R_{y4} \| R_{z4}$.

On analyzing the circuit shown in Figure 5 considering the parasites outlined above, the expression for the frequency of oscillation is as given below:

$$\text{F.O.}: \omega_0 = \sqrt{\frac{a+b+c}{C_1'C_2'C_3'\left(R_{1p}R_{2p}R_{3p}{}^2R_1{}'R_{X4}/R_{X3}\right)}}, \quad (7)$$

$$\begin{aligned}
\text{C.O.}: (a+b+c)&\left[C_1'C_2'\left(\frac{R_{1p}R_{2p}R_{3p}R_1{}'R_{X4}}{R_{X3}}\right)\right.\\
&+C_1'C_3'\left\{\left(\frac{R_{1p}R_{3p}{}^2R_1{}'R_{X4}}{R_{X3}}\right)\left(1+\frac{R_{1p}}{R_{X1}}\right)\right\}\\
&\left.+C_2'C_3'\left\{\left(\frac{R_{2p}R_{3p}{}^2R_1{}'R_{X4}}{R_{X3}}\right)\left(1+\frac{R_{2p}}{R_{X2}}\right)\right\}\right]\\
&=\left\{C_1'C_2'C_3'\left(\frac{R_{1p}R_{2p}R_{3p}{}^2R_1'R_{X4}}{R_{X3}}\right)\right\}\\
&\times\left[\left(\frac{R_1{}'R_{3p}R_{X4}}{R_{X3}}\right)\left\{1+\left(\frac{R_{2p}}{R_{X2}}\right)+\left(\frac{R_{1p}R_{2p}}{R_{X1}R_{X2}}\right)\right\},\right.\\
&\left.+\left(\frac{R_{1p}R_{2p}R_{3p}}{R_{X1}R_{X2}R_{X3}}\right)\right],
\end{aligned} \quad (8)$$

where $a = C_1'\{(R_{1p}R_{3p}R_1'R_{X4}/R_{X3})(1+R_{1p}/R_{X1})\}$, $b = C_1'\{(R_{2p}R_{3p}R_1'R_{X4}/R_{X3})(1+R_{2p}/R_{X2})\}$, $c = C_3'(R_{3p}^2R_1'R_{X4}/R_{X3})\{1+(R_{1p}/R_{X1})+(R_{1p}R_{2p}/R_{X1}R_{X2})\}$

It may be clarified here, that the above defined variables are not the same as used in Section 2.

It is apparent from this analysis that despite including the effect of parasitics, the order of the characteristic equation of the oscillator has not changed. This is due to the fact that the X-terminal of all the CCCIIs is grounded. This leads to the interpretation that the higher order distortions are not introduced in the circuit within the desired frequencies of operation. This is an additional advantage of this circuit. A careful investigation of (7) reveals that the effects of various parasites can be compensated by selecting little deviated values of the input bias currents from their calculated values.

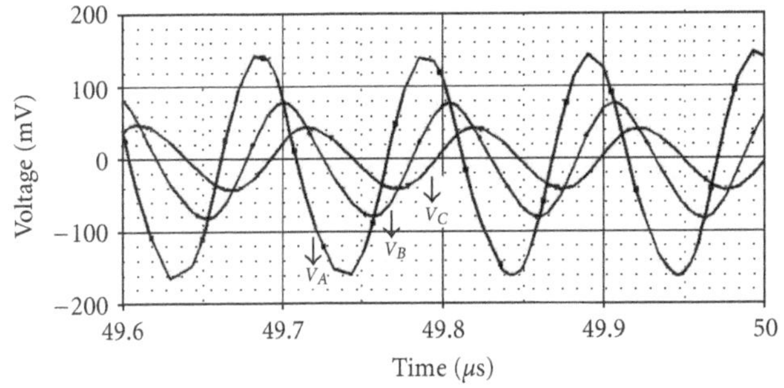


FIGURE 6: Output waveforms of sinusoidal oscillator.

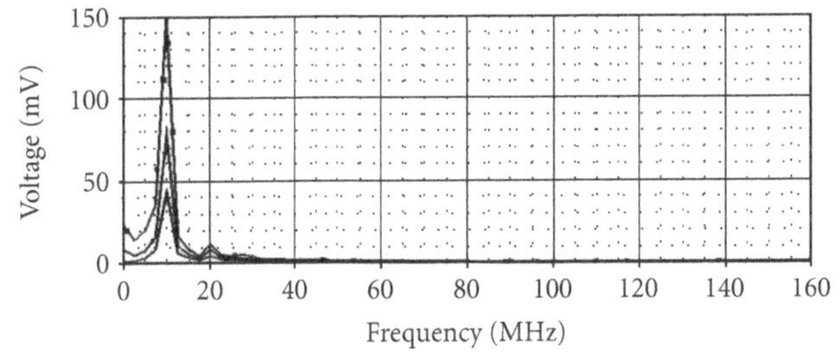


FIGURE 7: Fourier spectrum of output voltages.

## 6. Simulation Results

In order to prove the practical utility of the proposed circuit, the PSPICE simulations using the Level 3 model of TSMC MOSIS 0.25 μm CMOS process parameters [24] are performed. The supply voltages used are $V_{DD} = -V_{SS} = 1.25$ V. The third-order sinusoidal oscillator based on the circuit in Figure 3 is designed for frequency of 10 MHz by selecting $C_1 = C_2 = C_3 = C = 20$ pF, $R_1 = 16$ K, $I_{B1} = I_{B2} = I_{B3} = I_B = 40\,\mu$A, and $I_{B4} = 13\,\mu$A. The three output sinusoidal waveforms of the oscillator circuit are shown in Figure 6. The simulated frequency of oscillation is around 9.78 MHz, thereby giving 0.2% error, which is quite low and acceptable. It may be noted that the three outputs have different amplitudes because of the basic structure being based on low-pass filters. Additionally, it is well known that the oscillators can be augmented for automatic amplitude control using diodes and resistor network, which are also popularly called the limiter circuits [25]. The simulated THD value is obtained to be 2.9%, which is moderately low. The Fourier spectrum of the voltage outputs is shown in Figure 7, where the second harmonic seems to be the main contributor to distortions, while higher harmonics are almost negligible. The frequency of oscillation can be varied by changing the input bias current $I_B$, independent of the condition of oscillation. The graph showing the variation in frequency of oscillation with the variation in bias current $I_B$ for different values of capacitance "$C$" is given in Figure 8. The frequency of oscillation shows almost two and a half-fold variations for the variation in bias current from 10 μA to 80 μA.

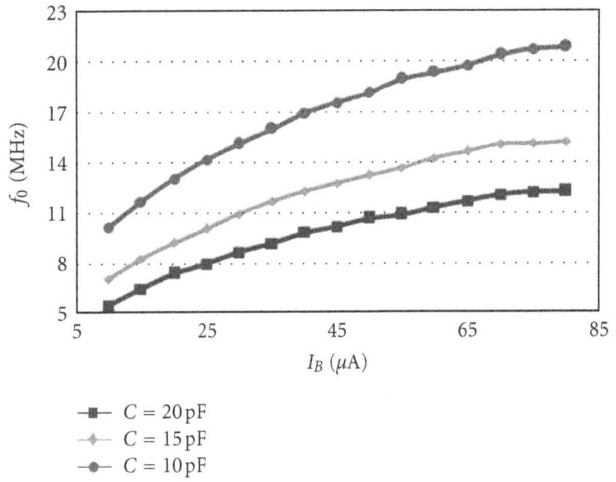


FIGURE 8: Variation in $f_0$ with $I_B$ for different values of capacitors.

## 7. Conclusions

In this paper, a novel electronically tunable third-order sinusoidal oscillator using CCCIIs and grounded components is proposed. This new circuit is synthesized from a simple topology employing cascaded first-order low-pass filters coupled with gained feedback. This circuit offers the advantages of the simultaneous availability of three sinusoidal voltage outputs and the use of grounded passive components which makes it suitable for IC implementation. The frequency of oscillation can be electronically tuned independently of the condition of oscillation through the bias currents of the active element. The generation circuitry of appropriate bias currents for the oscillator is given. Nonideal and parasitic considerations are included. The proposed circuit shows good high frequency performance and a wide-range electronic tuning of frequency. The proposed circuit is the novel addition to the already vast repertoire of literature on the subject and its IC implementation is a natural future problem.

## Acknowledgment

The authors thank reviewers for recommending the paper with some interesting suggestions, which made the paper more elegant. The authors are also thankful to the Associate Editor, Dr. Winai Jaikla for recommending this paper.

## References

[1] M. E. Van Valkenburg, *Analog Filter Design*, Holt Sounders International, 1987.

[2] A. M. Soliman, "Novel oscillators using current and voltage followers," *Journal of the Franklin Institute*, vol. 335, no. 6, pp. 997–1007, 1998.

[3] A. M. Soliman, "Current mode CCII oscillators using grounded capacitors and resistors," *International Journal of Circuit Theory and Applications*, vol. 26, no. 5, pp. 431–438, 1998.

[4] I. A. Khan, P. Beg, and M. T. Ahmed, "First order current mode filters and multiphase sinusoidal oscillators using CMOS

MOCCIIs," *Arabian Journal for Science and Engineering*, vol. 32, no. 2, pp. 119–126, 2007.

[5] P. Prommee and K. Dejhan, "An integrable electronic-controlled quadrature sinusoidal oscillator using CMOS operational transconductance amplifier," *International Journal of Electronics*, vol. 89, no. 5, pp. 365–379, 2002.

[6] E. Sanchez-Sinencio, J. Ramirez-Angulo, B. Linares-Barranco, and A. Rodriguez-Vazquez, "Operational transconductance amplifier-based nonlinear function syntheses," *IEEE Journal of Solid-State Circuits*, vol. 24, no. 6, pp. 1576–1586, 1989.

[7] S. Maheshwari and I. A. Khan, "Current controlled third order quadrature oscillator," *IEE Proceedings*, vol. 152, no. 6, pp. 605–607, 2005.

[8] T. Tsukutani, Y. Sumi, and Y. Fukui, "Electronically controlled current-mode oscillators using MO-OTAs and grounded capacitors," *Frequenz*, vol. 60, no. 11-12, pp. 220–223, 2006.

[9] S. Maheshwari, "Quadrature oscillator using grounded components with current and voltage outputs," *IET Circuits, Devices and Systems*, vol. 3, no. 4, pp. 153–160, 2009.

[10] S. Lawanwisut and M. Siripruchyanun, "High output-impedance current-mode third-order quadrature oscillator based on CCCCTAs," in *Proceedings of the IEEE Region 10 International Conference (TENCON '09)*, November 2009.

[11] S. Maheshwari, "Analogue signal processing applications using a new circuit topology," *IET Circuits, Devices and Systems*, vol. 3, no. 3, pp. 106–115, 2009.

[12] S. Maheshwari, "Current-mode third-order quadrature oscillator," *IET Circuits, Devices and Systems*, vol. 4, no. 3, pp. 188–195, 2010.

[13] J. W. Horng, H. Lee, and J. Y. Wu, "Electronically tunable third-order quadrature oscillator using CDTAs," *Radioengineering*, vol. 19, no. 2, pp. 326–330, 2010.

[14] S. Maheshwari, "High output impedance current-mode all-pass sections with two grounded passive components," *IET Circuits Devices and Systems*, vol. 2, no. 2, pp. 234–242, 2008.

[15] V. Biolková, J. Bajer, and D. Biolek, "Four-phase oscillators employing two active elements," *Radioengineering*, vol. 20, no. 1, pp. 334–339, 2011.

[16] P. Prommee, N. Sra-ium, and K. Dejhan, "High-frequency log-domain current-mode multiphase sinusoidal oscillator," *IET Circuits, Devices and Systems*, vol. 4, no. 5, pp. 440–448, 2010.

[17] A. Fabre, O. Saaid, F. Wiest, and C. Boucheron, "High frequency applications based on a new current controlled conveyor," *IEEE Transactions on Circuits and Systems I*, vol. 43, no. 2, pp. 82–91, 1996.

[18] M. T. Abuelma'atti and N. A. Tasadduq, "A novel single-input multiple-output current-mode current-controlled universal filter," *Microelectronics Journal*, vol. 29, no. 11, pp. 901–905, 1998.

[19] M. T. Abuelma'atti and M. A. Al-Qahtani, "A current-mode current-controlled current-conveyor-based analogue multiplier/divider," *International Journal of Electronics*, vol. 85, no. 1, pp. 71–77, 1998.

[20] T. Tsukutani, Y. Sumi, and N. Yabuki, "Versatile current-mode biquadratic circuit using only plus type CCCIIs and grounded capacitors," *International Journal of Electronics*, vol. 94, no. 12, pp. 1147–1156, 2007.

[21] C. Loescharatatamdee, W. Kiranon, W. Sangpisit, and W. Yadum, "Multiphase sinusoidal oscillator using translinear current conveyors and grounded passive components," in *Proceedings of the 33rd Southeastern Symposium on System Theory*, pp. 59–563, 2001.

[22] M. T. Abuelma'atti and R. H. Almaskati, "A new OTA-based active-C oscillator," *International Journal of Electronics*, vol. 63, no. 3, pp. 331–334, 1987.

[23] I. A. Khan, M. T. Ahmed, and N. Minhaj, "Tunable OTA-based multiphase sinusoidal oscillators," *International Journal of Electronics*, vol. 72, no. 3, pp. 443–450, 1992.

[24] P. Prommee, K. Angkeaw, M. Somdunyakanok, and K. Dejhan, "CMOS-based near zero-offset multiple inputs max-min circuits and its applications," *Analog Integrated Circuits and Signal Processing*, vol. 61, no. 1, pp. 93–105, 2009.

[25] A. S. Sedra and K. C. Smith, *Microelectronics Circuits*, Oxford University Press, 2011.

# 17

# Novel Power Reduction Technique for ReRAM with Automatic Avoidance Circuit for Wasteful Overwrite

**Takaya Handa,[1] Yuhei Yoshimoto,[1] Kazuya Nakayama,[2] and Akio Kitagawa[1]**

[1] *School of Electrical and Computer Engineering, College of Science and Engineering, Kanazawa University, Kakuma, Kanazawa 920-1192, Japan*
[2] *School of Health Sciences, College of Medical, Pharmaceutical and Health Sciences, Kanazawa University, 5-11-80 Kodatsuno, Kanazawa 920–0942, Japan*

Correspondence should be addressed to Kazuya Nakayama, knaka@kenroku.kanazawa-u.ac.jp

Academic Editor: Daisaburo Takashima

Low-power operations can be great advantageous for ReRAM devices. However, wasteful overwriting such as the SET operation to low-resistance state (LRS) device and the RESET operation to high-resistance state (HRS) device causes not only an increase in power but also the degradation of the write cycles due to repeatedly rewriting. Thus, in this paper, we proposed a novel automatic avoidance circuit for dealing with wasteful overwriting that uses a sense amplifier and estimated the energy consumption reduction rate by conducting a circuit simulation. As a result, this circuit helped to reliably avoid the wasteful overwriting operation to reduce about 99% and 97% of wasteful energy using VSRC and CSRC, respectively.

## 1. Introduction

Resistance random access memory (ReRAM) has been widely expected for use as the next generation nonvolatile memory because of its superior characteristics, such as low voltage operation, high-speed performance, and low power. ReRAM has a Metal-Insulator-Metal (MIM) structure in which the transition metal oxide is sandwiched between two electrodes as a storage element. This device has two switching modes: (i) unipolar switching in which the device depends on the pulse width and amplitude of the applied voltage and (ii) bipolar switching in which the device depends on the polar character [1]. In this paper, we used a bipolar switching device that made the best use of this proposed circuit. Figure 1 shows (a) the structure of the ReRAM device and (b) the definition of the SET and RESET operations.

In this research, SET is defined as an operation conducted to change from HRS to LRS, and RESET is an operation to change from LRS to HRS. ReRAM has the great advantage of being a low power device. However, in order to have this advantage, not only the necessary device characteristics but also peripheral circuits, such as read/write circuits, are needed. On one hand, wasteful overwrite operations such as SET for LRS devices or RESET for HRS not only increase the power consumption but also degrade the write cycles [2]. One solution is for the circuit to judge whether or not the device needs to be written after the circuit initially reads the resistance state like verifying operation in conventional NAND flash memory [3]. However, in this way, both the power consumption and processing time increase because an additional full read operation is required for the write operation, as shown in Figure 2(a). Therefore, in this paper, we propose a new circuit that can automatically judge whether or not the circuit needs to write by using the avoidance function of the wasteful overwrite to reduce the wasteful power consumption by improving the conventional sense amplifier [4], as shown in Figure 2(b). In this proposed scheme, write operation can be conducted during read operation, so full read operation is not necessary, and this circuit has some advantage in aspects of speed and power. And also this scheme has merits of using the result of sense amplifier to write operation and adding the "write function" to conventional "read circuit," and so read and write circuit needs not to be separated and achieving wasteful overwrite avoidance with added only 4 MOSFET against conventional sense amplifier.

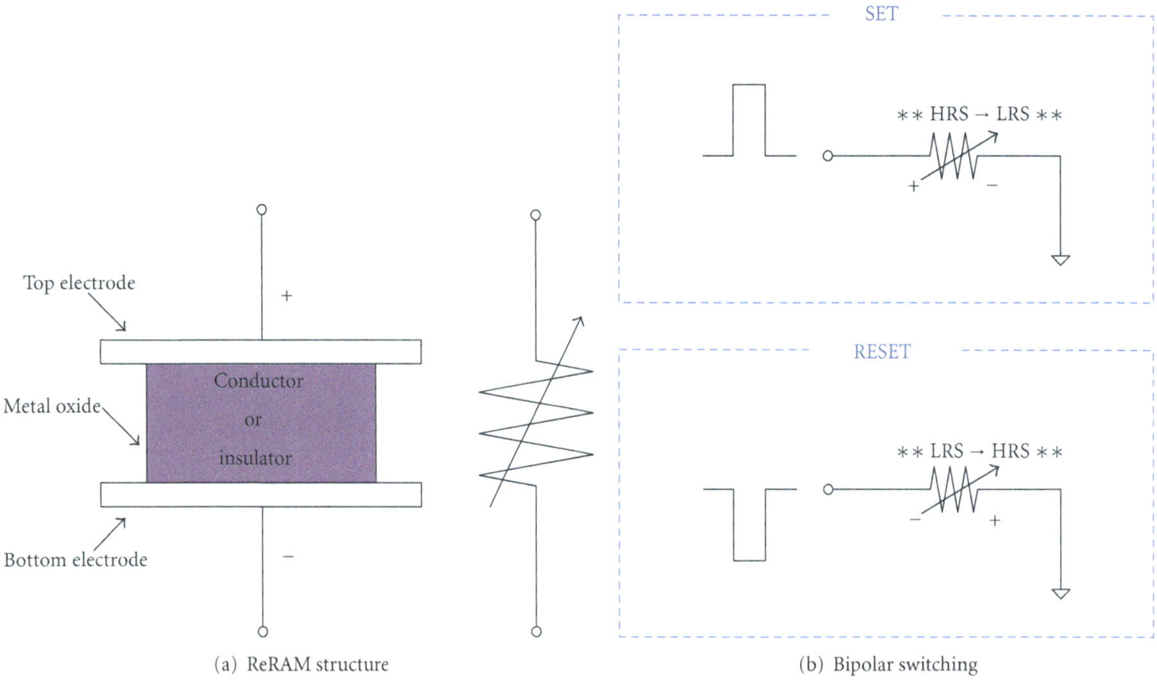


FIGURE 1: (a) Structure of ReRAM device and (b) definition of SET and RESET operations.

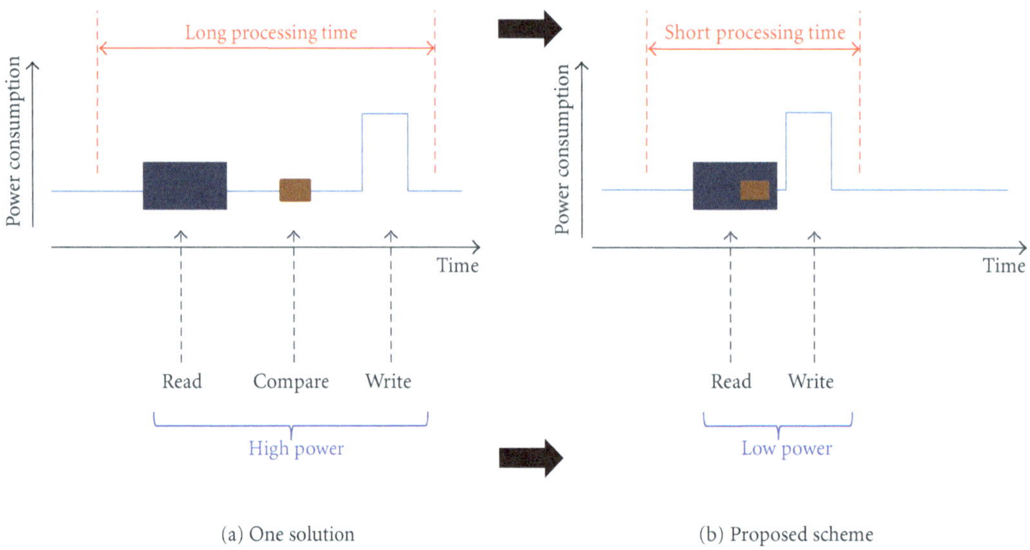


FIGURE 2: Sequences of (a) one solution and (b) proposed scheme.

This paper is organized as follows. An explanation about the circuit configuration and the principle of the operations are presented in Section 2. A Verilog-A empirical model of a ReRAM device and our simulation results are presented in Section 3. The conclusion is summarized in Section 4.

## 2. Circuit Structures

We propose two write circuits to avoid the wasteful overwrite that are composed of a voltage sense amplifier (VSA) and a current sense amplifier (CSA) because we compare the reduction power between these two circuits. These circuits have both functions, as a write decision circuit and a sense amplifier [4]. Here, we call these circuits a voltage sense rewriting circuit (VSRC) and a current sense rewriting circuit (CSRC).

### *2.1. Circuit Structure and Operations*

*(1) VSRC.* Figure 3(a) shows the schematic of the VSRC, and 3(b) shows the symbol. The read and write current paths are drawn as solid and dashed lines, respectively.

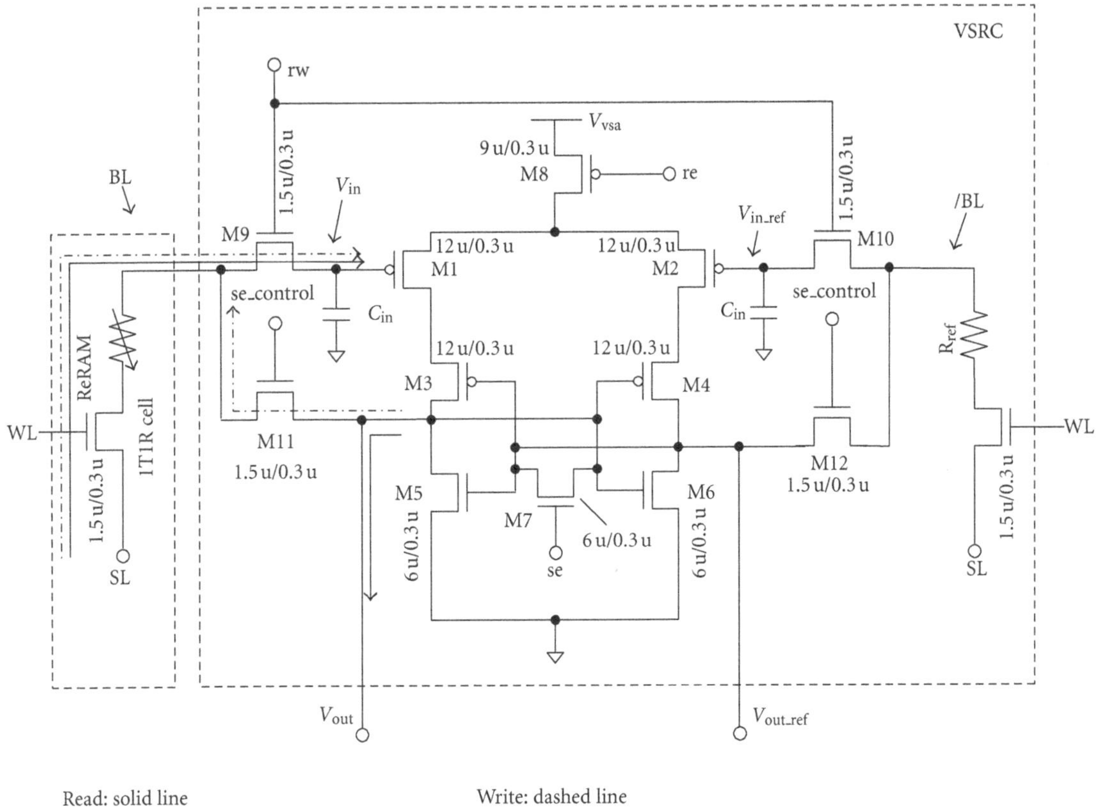


(a) Schematic

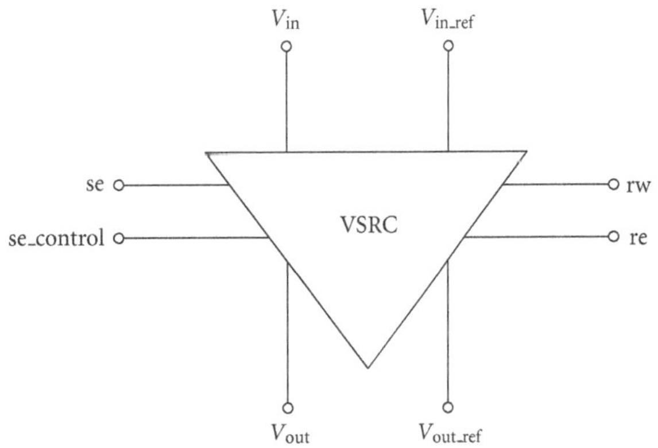


(b) Symbol

FIGURE 3: (a) Schematic and (b) symbol of VSRC.

Figure 4 shows a timing diagram for the VSRC. Operation of the VSRC is divided into three phases: (i) cell reading, (ii) amplification, and (iii) precharge.

(i) In the cell-reading phase, the Vout and Vout_ref of the VSRC are equalized. The input voltage dependent on the state of the ReRAM device is temporarily applied to gates of M1 and M2.

(ii) In the amplification phase, both read and write can be operated by charging Cin. The read process is operated through the solid line shown in Figure 2. In this phase, SL is set to the read voltage (Vread), rw to "H" (vdd), and se_control to "L" (gnd), and the potential difference between Vin and Vin_ref in the equalization phase is amplified and outputted in the latching part. Meanwhile, the write process is operated through the dashed line, and rw is set to "L", and se_control is set to "H". Here, SL is asserted to "L" or "H" in accordance with the SET or RESET operations, respectively.

(iii) In the precharge phase, the power supply of the VSRC is stopped because M8 is OFF in synchronization with the rising edge of the re signal (power saving mode).

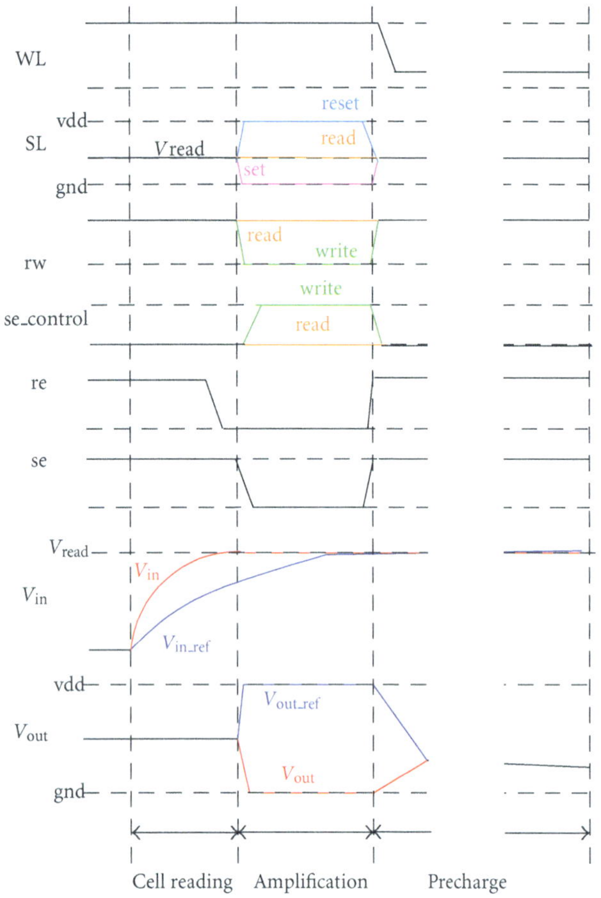


FIGURE 4: Timing diagram for VSRC operation.

Here, as shown in Figure 4, the re signal is asserted to "L" just before the end of cell-reading phase to suppress wasteful DC current in the sense amplifier. Therefore, DC current that flows through sense amplifier is minimized even at long-time cell reading due to long bit-line lengths when large array size is adopted.

*(2) CSRC.* The schematic of the CSRC is shown in Figure 5. In this circuit, the rw signal is used to control the switching between read and write operations. The CSRC operations are also divided into the (i) cell reading and (ii) amplification phase in both the read and write operations. Figure 6 shows a timing diagram of the CSRC. In the initial condition, control signals such as se and rw are set to "H", and re is set to "L".

(i) In the cell reading phase, the control signals are the same for both the read and write operations. By setting re to "H", the precharge of the current path for the read operation is started. Both the se and rw signals are set to "H" and SL is set to "L". Output nodes Vout and Vout_ref are equalized to vdd/2, where the equalized voltage can be calibrated by optimizing the size of M3 and M4. Here, the precharge current flows through the memory cells depending on the ON resistance of M5, M6, and M8, so the size of MOSFET must be carefully decided on to prevent any disturbance.

(ii) In the amplification phase, the rw signal is set to "L" in the read operation and "H" in the write operation. By asserting se to "L", M7 is OFF, and the current difference between the accessed cell and the reference cell is amplified and latched. If a ReRAM device is in HRS, the Vgs of M3 is smaller than that of M4. Thus, Vout is asserted to "H", and Vout_ref is asserted to "L" because the discharge capacity of the Vout node is lower than that of Vout_ref. On the other hand, SL is fixed to "L" in the SET operation and asserted to "H" in the RESET operation.

Here, as shown in Figure 6, DC current that flows through sense amplifier is minimized even at long-time cell reading due to long bit-line lengths by controlling the re signal like VSRC.

*2.2. Principle of Automatic Avoidance of Wasteful Overwrite Using VSRC and CSRC Architecture.* The relationship between the resistance state and BL voltage is summarized as follows:

(i) ReRAM < Rref ⇒ BL → "L", /BL → "H"

(ii) ReRAM > Rref ⇒ BL → "H", /BL → "L".

Thus, using the output of a sense amplifier depending on the current resistance state, a multiple overwrite of the same logic value can be avoided, as shown in Figure 7. In the SET operation, SL is set to "L" in the amplification phase. When the device is in HRS, BL becomes "H", and so, the SET voltage is applied to the device. When the device is in LRS, BL becomes "L" and the SET voltage is not applied to the device. On the other hand, in the RESET operation, SL is set to "H". The reset voltage is not applied to the device because BL becomes "H" when the device is in HRS and is applied because BL becomes "L" when the device is in LRS. We summarize the relationship between SL and BL, and principle of automatic avoidance of wasteful overwrite as follows:

(i) SET Operation:

ReRAM (LRS) < Rref ⇒ SL → "L", BL → "L"
(SET current does not flow : avoidance)
ReRAM (HRS) > Rref ⇒ SL → "L", BL → "H"
(SET current flows through ReRAM : write).

(i) RESET Operation:

ReRAM (LRS) < Rref ⇒ SL → "H", BL → "L"
(RESET current flows through ReRAM : write)
ReRAM (HRS) > Rref ⇒ SL → "H", BL → "H"
(RESET current does not flow : avoidance).

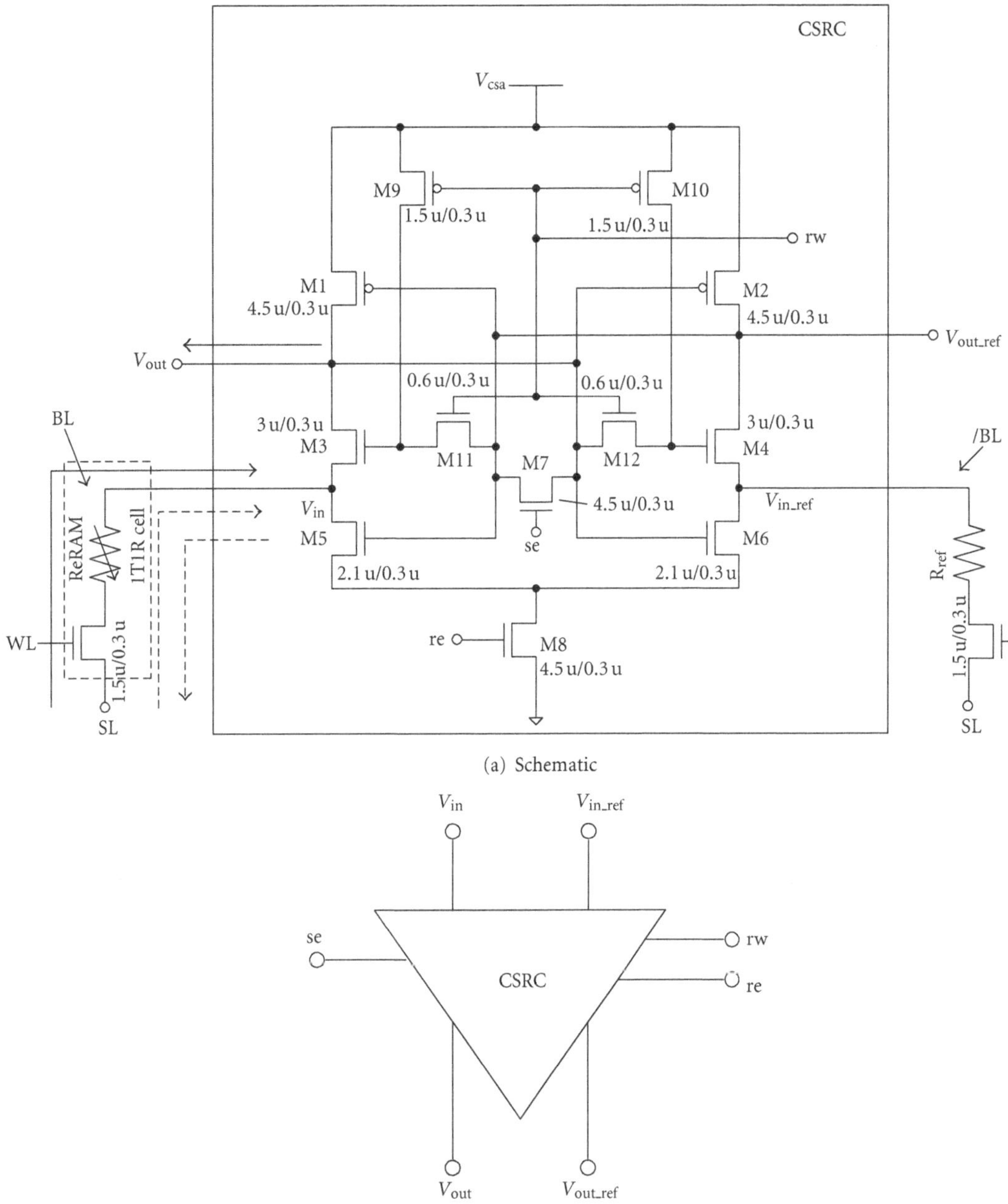


FIGURE 5: (a) Schematic and (b) symbol of CSRC.

## 3. Simulation

*3.1. Simulation Condition.* The device characteristics of the ReRAM used in this paper and the simulation conditions are listed in Tables 1 and 2, respectively [4].

*3.2. Verilog-A Modeling.* The behavioral device model of ReRAM was developed using a Verilog-A based on the characteristics in Table 1. In this paper, we described the transient response of the bipolar operation of a ReRAM device. Figure 8 shows the simulation results of the transient response in the write and read operations using only the Verilog-A model.

Using this Verilog-A model makes it possible to simulate more a realistic circuit behavior while taking the phenomenon typically found in ReRAM devices such as sharp resistance switching in the write operation into consideration.

*3.3. Simulation Circuit Structure.* The circuit structure for our simulation is shown in Figure 9. Here, we omitted the description of the other signals, such as the se and re signals. Here, peripheral control circuits, such as address decoding and timing generation aren't included in our simulation condition.

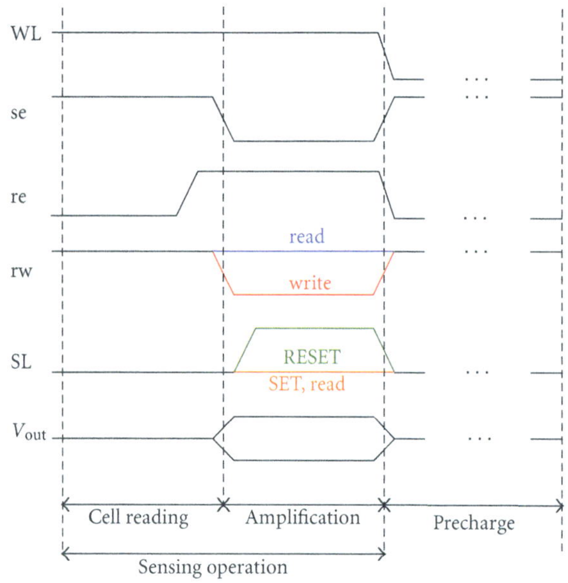


FIGURE 6: Timing diagram for CSRC operation.

TABLE 1: ReRAM device characteristics.

| | |
|---|---|
| High-resistance value | 80 kΩ |
| Low-resistance value | 20 kΩ |
| SET transition time [1] | ~50 ns |
| RESET transition time | ~50 ns |
| SET voltage | More than 2.2 V |
| RESET voltage | Less than −1.4 V |
| Read voltage | Less than 0.5 V |

TABLE 2: Simulation condition.

| | |
|---|---|
| Simulator | HSPICE |
| Fabrication process | 180-nm CMOS |
| Power supply | 3.3 V |
| Temperature | 27°C |
| Amplification time | 70 ns |
| Bit-line capacitance (Cb) | 200 fF |

The read/write sequence was given as (a) READ → SET → READ → SET → READ for an HRS device and (b) READ → RESET → READ → RESET → READ for an LRS device. We conducted a simulation with conventional circuit and proposed circuit as shown in Figure 10. Here, we used the reversed logic-VSRC and CSRC as conventional circuit. Reversed logic-VSRC and CSRC was designed to overwrite intentionally as RESET operation to HRS device and SET operation to LRS device by replacing Vout with Vout_ref in VSRC and CSRC as shown in Figure 3 and Figure 5, respectively. Therefore, we can compare the energy consumption using conventional and proposed circuit.

TABLE 3: Results of power reduction ratio per cell.

| | Operation | SET | RESET |
|---|---|---|---|
| | w/o Avoid [fJ] | 11865 | 3405 |
| VSRC | w/o Avoid [fJ] | 48 | 135 |
| | Reduction ratio [%] | 99.5 | 96.0 |
| CSRC | w/Avoid [fJ] | 288.4 | 351.2 |
| | Reduction ratio [%] | 97.5 | 89.7 |

*3.4. Simulation Results.* The results of the simulation are shown in Figure 11.

For example, if the state of the ReRAM device is initially in the LRS condition, the resistance state of this device transits in the first RESET operation. However, in the second RESET operation, this operation is unnecessary because the resistance state of this device has already transited to HRS in the first RESET operation. In the first RESET operation as shown in Figure 11(a), we found that the RESET pulse (negative pulse) is applied to both ends of the ReRAM device, and the device transits to the HRS. On the other hand, in the second RESET operation, this operation is wasteful, and, thus, the RESET pulse is not applied, and we can confirm that the useless write operation is automatically avoided.

*3.5. Considerations for Power Consumption.* From the simulation results shown in Figure 11, the amount of power consumption per one ReRAM device that the VSRC and CSRC can reduce is shown in Figure 12.

In Figure 12(a) with no avoidance, RESET voltage is applied to ReRAM and consumes wasteful power (Area 1). On the other hand, in Figure 12(b) with avoidance, RESET voltage is not applied to ReRAM because BL voltage is changed dynamically depending on the current resistance state, and, so, the voltage difference between ReRAM becomes zero. Here, we defined the power reduction ratio as

$$\text{reduction\_ratio} = \frac{\text{Energy(Area1)} - \text{Energy(Area2)}}{\text{Energy(Area1)}}, \quad (1)$$

where Energy(Area1) is an energy caused by wasteful overwrite with no avoidance, and Energy(Area2) is an overhead energy caused by using proposed circuit.

We evaluated the reduction ratio of the consumed energy using the proposed circuit (w/avoidance) against the conventional wasteful overwrite energy (w/o avoidance).

We calculated the energy reduction ratio for the SET → SET and RESET → RESET operation sequences, as shown in Figure 13. Table 3 shows each energy consumption and reduction ratio in RESET and SET operation. Here, Energy(Area1) and Energy(Area2) in Figure 12 correspond to 3405 [fJ] and 135 [fJ] in Table 3, respectively. We achieved almost a 90% power reduction using the proposed circuit as shown in Table 3. Optimizing the timing of the se_control and SL makes it possible to reduce the power consumption of the CSRC although the power reduction ratio of the CSRC was smaller than that of the VSRC. Here, the speed overhead using proposed circuit can be controlled to only 1.1

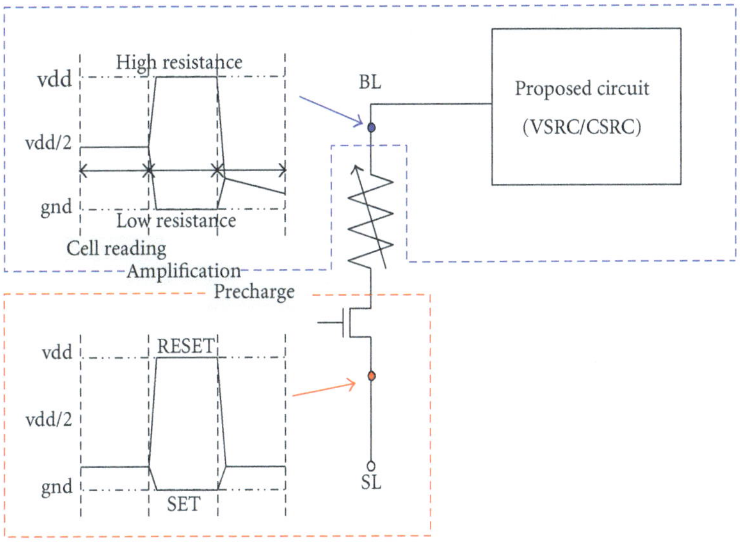


Figure 7: Principle of write operation with automatic avoidance of wasteful overwrite using VSRC and CSRC architecture.

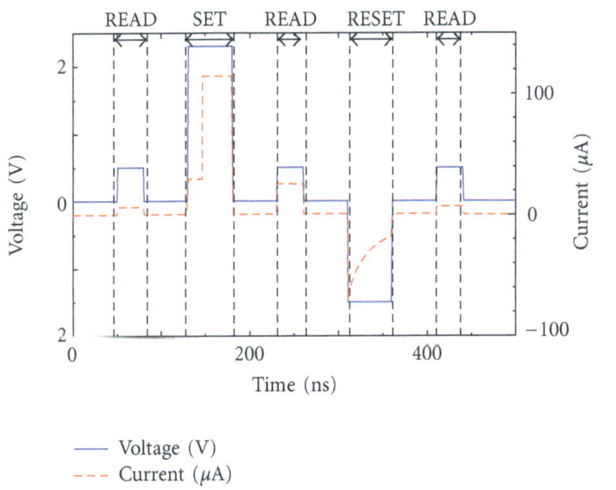


Figure 8: Transient response characteristics of ReRAM device model with Verilog-A.

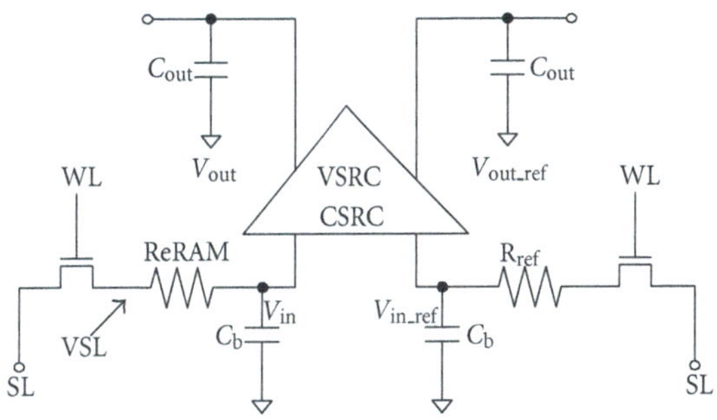


Figure 9: Circuit configuration for simulation.

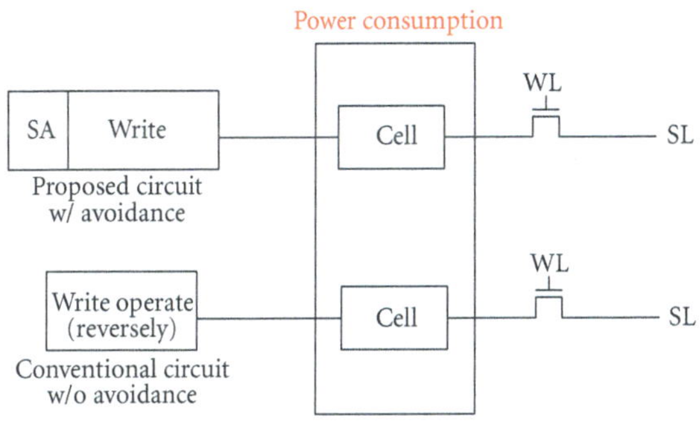


Figure 10: Comparison of energy consumption between conventional and proposed circuit.

times increase of time compared to conventional one because equalization time (10 ns) is 7 times smaller than amplification time (70 ns).

Generally, the time constant of the BL load depends on the resistances of ReRAM devices, the parasitic capacitances of the cell transistors, the distributed wire resistance, capacitance of BL, and the array size. These variations are solved by carefully controlling the timing of sensing (cell reading time). In practice, the sensing behaviors of these circuits depend on the data patterns stored in the memory cell array because the different data patterns on a column cause various BL loads. The waveform of the BL voltage is fluctuated within 2% for any data pattern in the condition that assumes 1024 cell per bitline.

The power reduction ratio in Table 3 was estimated for the consumed energy only in ReRAM array including activated memory cells and word-line/bit-line voltage swing. The energy consumption in the sense amplifier is mainly due to the DC current through the transistor M8 in VSRC and CSRC, and the DC current path is formed when both re and se signals are asserted during the cell-reading phase. Assuming that the period of the DC current flow is 10 ns, the approximate energy reduction ratio considering DC current consumed in sense amplifier is about 84% for the VSRC and CSRC though we haven't done complete optimization

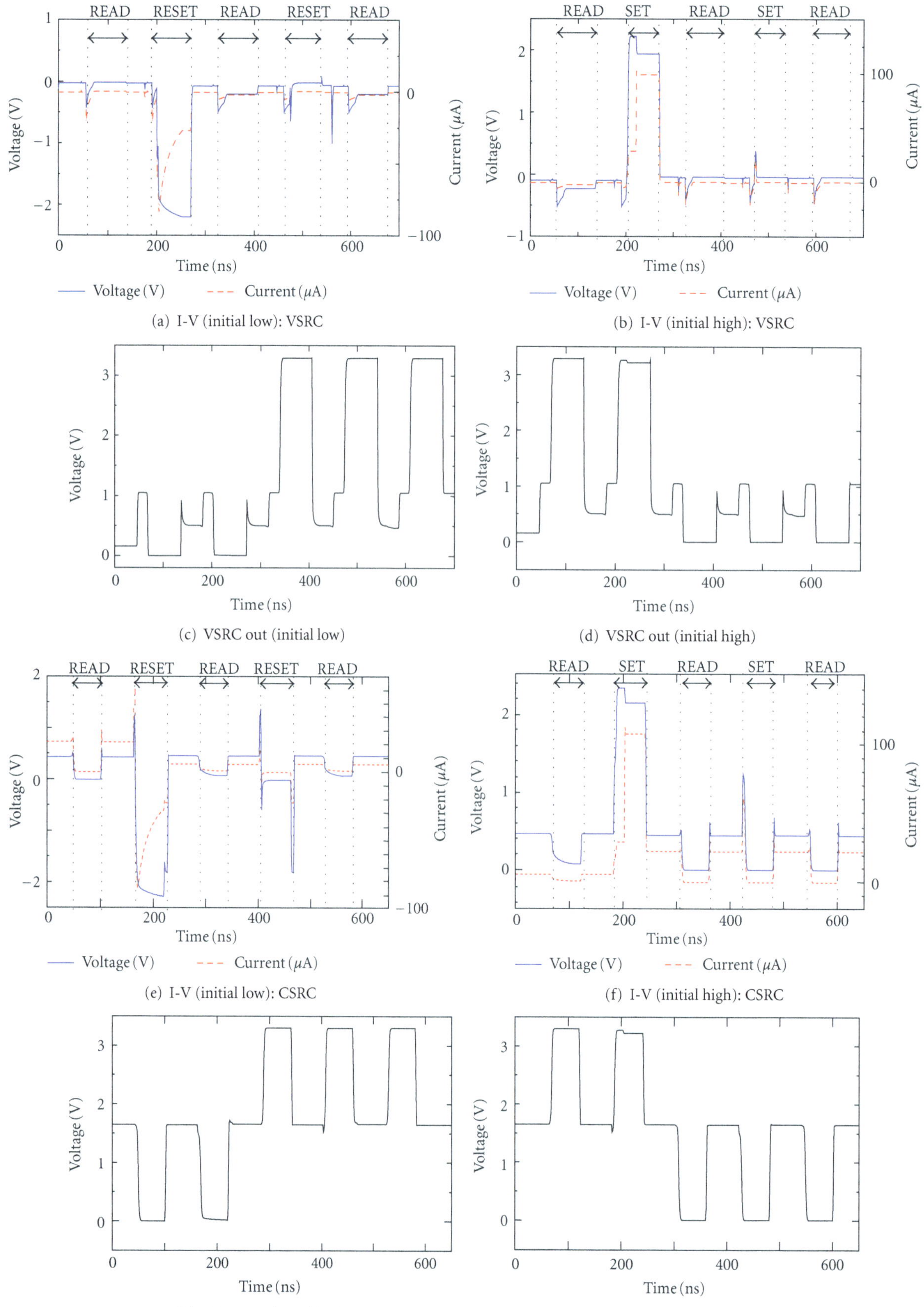


(a) I-V (initial low): VSRC

(b) I-V (initial high): VSRC

(c) VSRC out (initial low)

(d) VSRC out (initial high)

(e) I-V (initial low): CSRC

(f) I-V (initial high): CSRC

(g) CSRC out (initial low)

(h) CSRC out (initial high)

FIGURE 11: Simulation results.

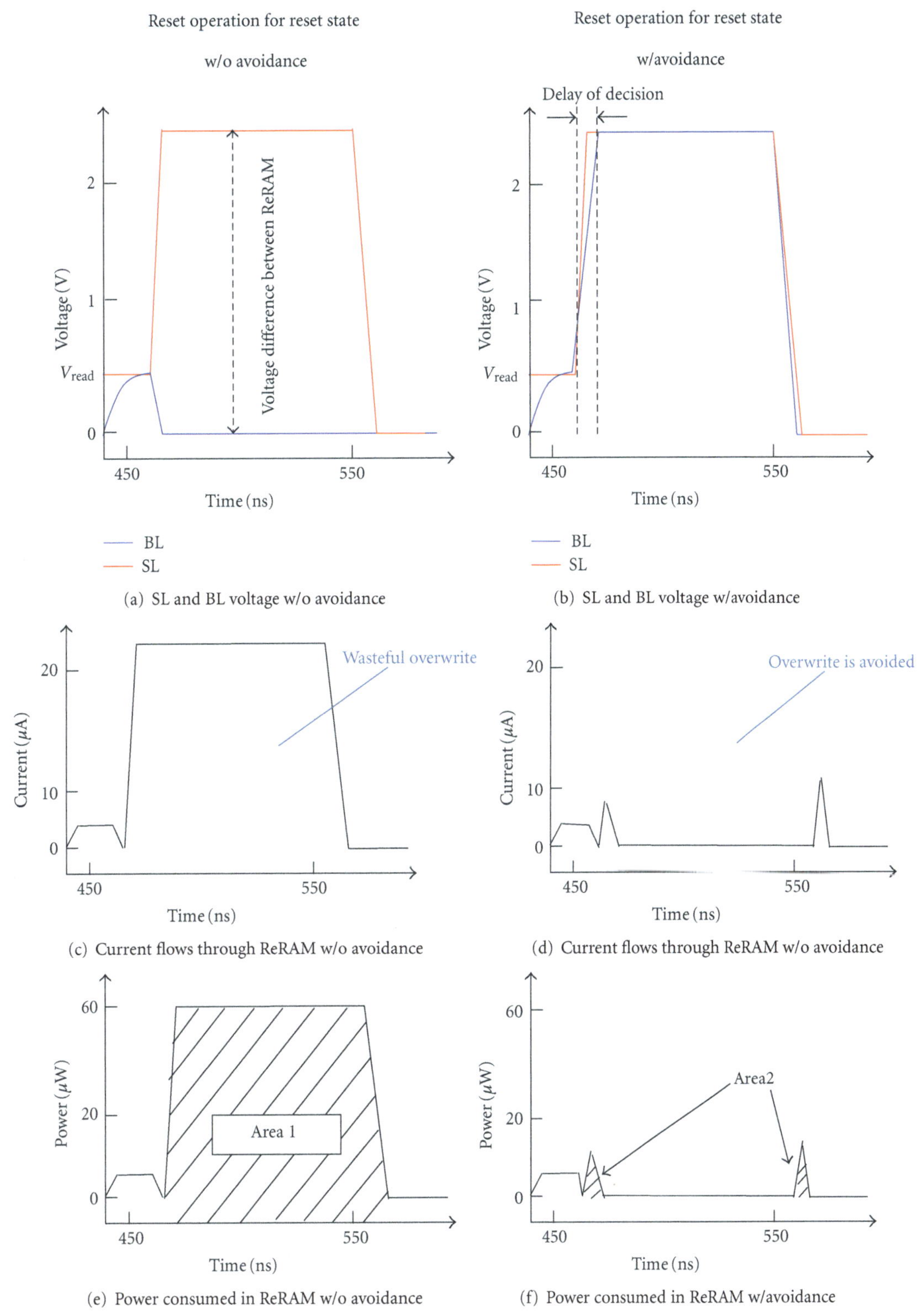


FIGURE 12: Concept of power reduction using automatic avoidance circuit against conventional nonavoidance scheme.

of this circuit in consideration of the minimum DC current in the sense amplifier. The DC current can be suppressed by further optimization of the circuit and the timing of re signal.

## 4. Conclusion

We proposed an automatic avoidance circuit of the wasteful overwrite and evaluated the proposed circuit. As a result,

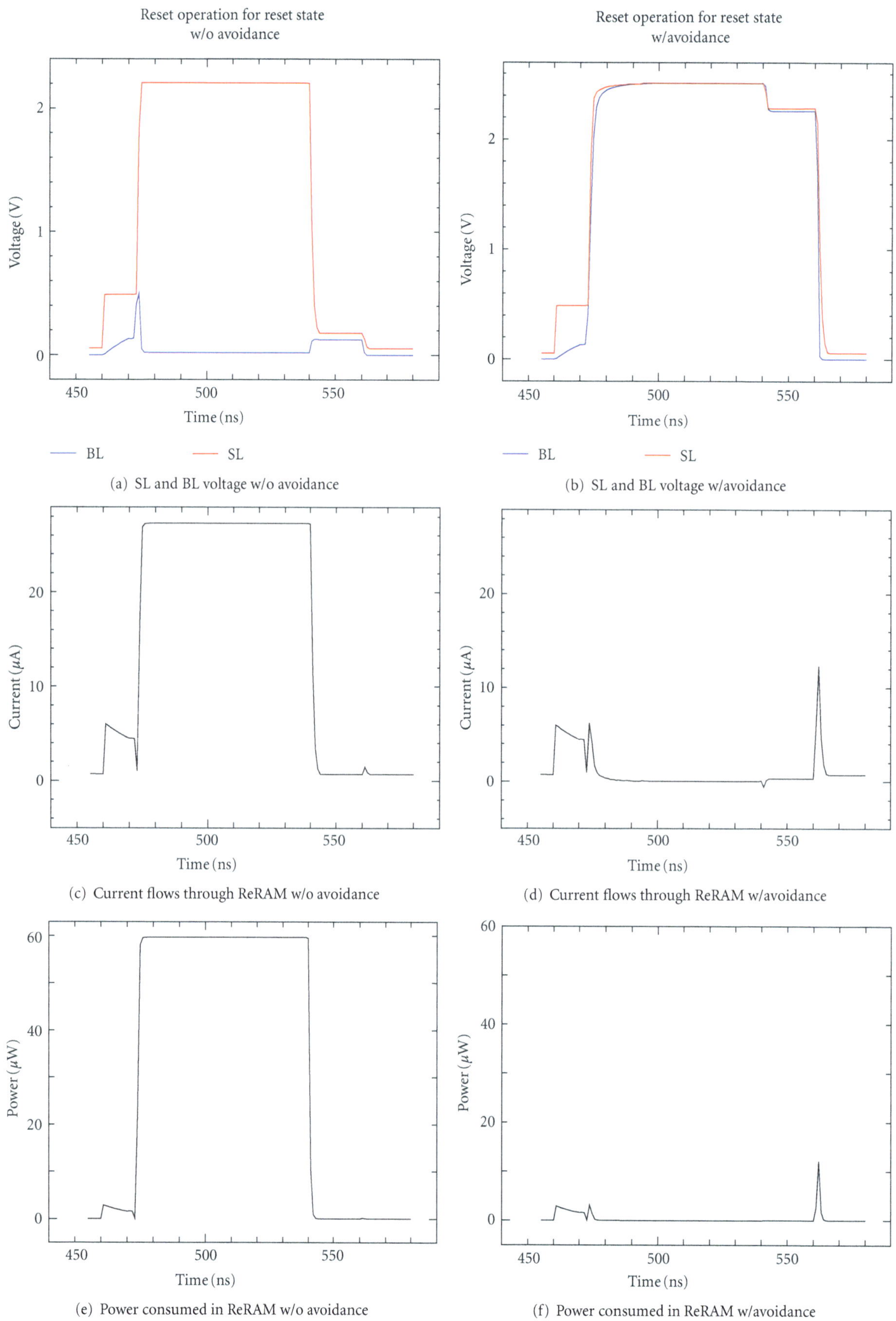


(a) SL and BL voltage w/o avoidance

(b) SL and BL voltage w/avoidance

(c) Current flows through ReRAM w/o avoidance

(d) Current flows through ReRAM w/avoidance

(e) Power consumed in ReRAM w/o avoidance

(f) Power consumed in ReRAM w/avoidance

Figure 13: Simulation results.

we achieved about a 99% and a 97% energy reduction ratio using the VSRC and CSRC, respectively.

## Acknowledgments

This work was supported by VLSI Design and Educational Center (VDEC), the University of Tokyo in collaboration with Cadence Corporation and Mentor Graphics, Inc. The VLSI chip in this study has been fabricated in the chip fabrication program of VDEC, the University of Tokyo in collaboration with Rohm Corporation and Toppan Printing Corporation. This work was also supported by Grant-in-Aid for Scientific Research (C) (20510116, 23560391), Grant-in-Aid for Exploratory Research (23651136) of Japan Society for the Promotion of Science (JSPS), and Adaptable & Seamless Technology Transfer Program through Target-driven R&D (AS2121327) of Japan Science and Technology Agency (JST).

## References

[1] S. Kawabata, M. Nakura, S. Yamazaki et al., "CoOx-RRAM memory cell technology using recess structure for 128 Kbits memory array," in *Proceedings of the IEEE International Memory Workshop (IMW'10)*, pp. 1–2, Seoul, Korea, May 2010.

[2] Y. S. Chen, H. Y. Lee, P. S. Chen et al., "Overview and future challenges of hafnium oxide ReRAM," in *Proceedings of the International Conference on Solid State Devices and Materials (SSDM'10)*, pp. 1106–1107, 2010.

[3] C. C. Chung, H. Lin, Y. M. Shen, and Y. T. Lin, "A multilevel sensing and program verifying scheme for Bi-NAND flash memories," in *Proceedings of the IEEE VLSI-TSA International Symposium on VLSI Design, Automation and Test (VLSI-TSA-DAT'05)*, pp. 267–270, April 2005.

[4] Y. Ueda, Y. Iwata, and T. Inaba, "Design of low read bias voltage and high speed sense amplifier for STT-MRAM," Tech. Rep., pp. 7–12, IEICE, April 2007.

# 18

# Composite Right- and Left-Handed Traveling-Wave Field-Effect Transistors

**Koichi Narahara**

*Graduate School of Science and Engineering, Yamagata University, 4-3-16 Jonan, Yonezawa, Yamagata 992-8510, Japan*

Correspondence should be addressed to Koichi Narahara, narahara@yz.yamagata-u.ac.jp

Academic Editor: Egidio Ragonese

We introduce a composite right- and left-handed travelling-wave field-effect transistor (CRLH TWFET) for developing large-scale platform to support left-handed waves. The device represents two electromagnetically coupled CRLH transmission lines by capacitance and FET transconductance. Owing to the couplings, two different modes can support waves in CRLH TWFETs. It was experimentally established that waves supported by one of the modes were amplified, while those supported by the other mode were significantly attenuated. To quantify the wave propagation in CRLH TWFETs, we developed a numerical model based on the transmission line theory that well simulated measured results. This paper discusses the results of numerical calculations that validate the design criteria of CRLH TWFETs.

## 1. Introduction

Composite right- and left-handed (CRLH) transmission lines (TLs) have been extensively studied and several important breakthroughs have been achieved in the management of electromagnetic continuous waves [1]. CRLH TLs have a single series and shunt inductance-capacitance (LC) pair in each unit cell. They exhibit lefthandedness; that is, at certain frequencies, their phase velocity has a sign opposite to that of their group velocity. Figure 1 shows the unit cell of a CRLH TL, where $L_R$, $C_L$, $L_L$, and $C_R$ are the series inductance, series capacitance, shunt inductance, and shunt capacitance, respectively. The series LC pair consists of $L_R$ and $C_L$, and the shunt pair consists of $L_L$ and $C_R$. These parameters define the upper and lower zero-wavelength frequencies of the CRLH TL, which are given by $\omega_u = \max(1/(C_L L_R)^{1/2}, 1/(C_R L_L)^{1/2})$ and $\omega_l = \min(1/(C_L L_R)^{1/2}, 1/(C_R L_L)^{1/2})$, respectively. The line exhibits righthandedness at frequencies greater than $\omega_u$ and left-handedness at frequencies lower than $\omega_l$. A frequency band gap exists between $\omega_l$ and $\omega_u$ in which all supporting modes are evanescent.

The use of CRLH TLs may sometimes be impractical because of their finite electrode resistance and substrate current leakage, and may require loss compensation schemes. Recently, several authors have investigated methods for amplifying LH waves in 1D using transistors [2–4].

In particular, we analyse the use of two CRLH TLs that interact continuously via transistor transconductance together with both capacitive and inductive couplings, which we call a CRLH travelling-wave field-effect transistor (TWFET), or CRLH TWFET for brevity.

There are two different schemes to achieve loss compensation or amplification of left-handed waves. One of the methods is to interconnect passive CRLH sections with amplifier section. Based on this methodology, Casares-Miranda et al. [2] fabricated a leaky-wave antenna, whose gain was much enhanced compared to its passive counterpart. Multiple cascades of the interconnecting sections enable extension of the effective propagation length of a LH wave. However, a phase jump occurs at the amplifier section. Another method is continuous amplification. Si et al. [3] proposed a microwave CRLH TL incorporating ideal gain devices in each unit cell. They demonstrated a simple circuit cell consisting of two FETs operated as such gain devices. On the other hand, Maezawa et al. [4] proposed a CRLH TL periodically loaded with resonant tunnel diode (RTD) pairs. The negative differential resistance (NDR) of RTDs can compensate the loss in passive CRLH TLs. It is suggested that the line can amplify signals traveling along the line by the NDR of the RTD. In either case, the unit-cell structure is much simpler than CRLH TWFETs.

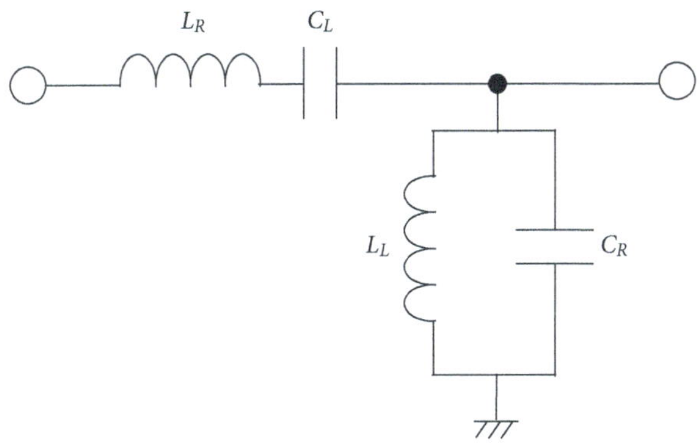


FIGURE 1: Unit cell of CRLH TL.

The CRLH TL has inductance-capacitance pairs in each unit cell. With the simple introduction of an NDR device, these pairs can exhibit resonant behaviour. We carried out the stability analysis with respect to CRLH TWFETs to establish the fact that they succeed in suppressing instability and amplifying the LH waves [5]. On the other hand, as far as we know, such stability analysis has not been carried out for the schemes that employ NDR device in each unit cell. Table 1 summarizes the merits and demerits of the above-mentioned three platforms for compensating attenuation of LH waves.

In the following, we first clarify our design concept of CRLH TWFETs. We then discuss several fundamental properties of a CRLH TWFET including the device configuration, dispersion relationship, and design criteria for wave amplification and termination schemes. Next, we describe the numerical model to simulate the measured results. Then, we numerically confirm the validity of the design criteria of a CRLH TWFET.

## 2. Design Concept of CRLH TWFETs

Figure 2 illustrates our design concept of CRLH TWFETs. Owing to couplings, two different propagation modes, called the c- and $\pi$-modes, can support waves in CRLH TWFETs. Each mode has its own speed, characteristic impedance, and the voltage fraction between the gate and drain lines. In general, the c-mode waves have the even parity between the gate and drain lines. On the other hand, the $\pi$-mode waves have the odd parity between the gate and drain lines.

Our design concept stems from the fact that a CRLH TWFET can amplify the waves carried by one of two modes, and attenuate ones carried by the other mode significantly. We can exclude the influences caused by waves carried by the attenuating mode to suppress modal dispersions. Moreover, wave reflection at the ends can be minimized by the matched terminations with the characteristic impedances corresponding to the amplifying mode.

At present, we consider the case where the $\pi$-mode is the unique one to be amplified. Then, the terminal resistances $R_{\mathrm{Tg}}$ and $R_{\mathrm{Td}}$ are set equal to the $\pi$-mode characteristic impedances for the gate and drain lines, respectively. In order to develop the waves carried by the $\pi$-mode, the input ports

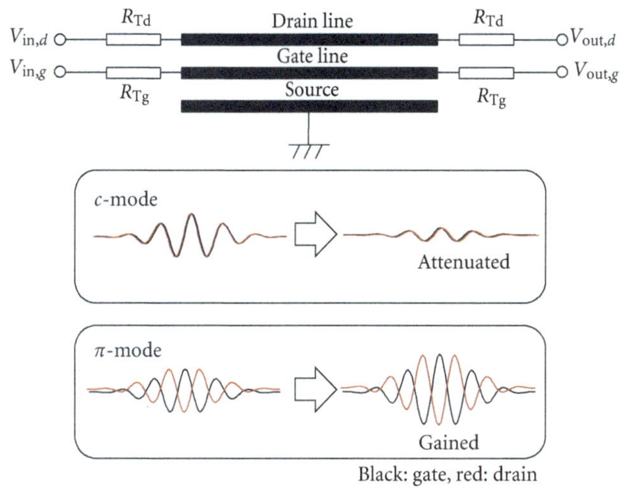


FIGURE 2: Design concept of CRLH TWFETs.

$V_{\mathrm{in},g}$ and $V_{\mathrm{in},d}$ are excited by the waves satisfying the $\pi$-mode voltage fraction. In this study, we describe the results of numerical calculations using a model of a CRLH TWFET based on the elemental transmission line theory. The model well simulates the measured properties of the device and can evaluate the contributions of the finite series inductors to both LH and RH waves travelling in a line. The operation bandwidth is limited by the FET cutoff frequency. Moreover, the influences of parasitic inductance and capacitance, which cannot be assumed a priori, increase at high frequencies. However, as far as the line is based on the print-circuit-board (PCB) technologies, we can well suppress the influences of parasitic elements, whose operation bandwidth presently extends the microwave frequencies.

## 3. CRLH TWFETS

Figure 3 shows the circuit configuration of the unit cell of a CRLH TWFET. Each unit cell includes an FET. For the gate (drain) line, $L_{Rg(d)}, C_{Lg(d)}$, $L_{Lg(d)}$, and $C_{Rg(d)}$ represent the series inductor, series capacitor, shunt inductor, and shunt capacitor, respectively. The gate line is connected with the gate electrode of the FET, whose drain electrode is connected with the drain line. The source electrode is grounded. Note that the biasing voltages to the gate ($V_G$) and drain ($V_D$) of the FET can be rendered via $L_{Lg}$ and $L_{Ld}$, respectively. Every inductor accompanies finite parasitic resistance, which is denoted as $R_G$, $R_D$, $R_{\mathrm{in},g}$, and $R_{\mathrm{in},d}$ for $L_{Rg}$, $L_{Rd}$, $L_{Lg}$, and $L_{Ld}$, respectively. The drain-source current of an FET is modelled as $I_{ds}(V_G) = \beta(V_G - V_{\mathrm{th}})^2$, where $\beta$, $V_{th}$, and $V$ are the transconductance coefficient, threshold voltage, and the gate line voltage, respectively. For analytical evaluations, we consider the small-signal counterpart of $I_{ds}$, that is, $I_{ds}(V_G) = G_m(V_G - V_{\mathrm{th}})$, where $G_m$ is the transconductance given by $2\,\beta(V_G - V_{\mathrm{th}})$. Moreover, an FET includes the gate-source capacitance $C_{gs}$, gate-drain capacitance $C_{gd}$, and drain-source capacitance $C_{ds}$. These FET parasitic capacitances can be embedded in elements in Figure 2. Concretely, $C_{gs}$, $C_{ds}$, and $C_{gd}$ can be included in $C_{Rg}$, $C_{Rd}$, and $C_m$, respectively.

TABLE 1: Figure of merits of amplifying LH waves.

| | Cascades of CRLH TL and amplifier sections | Unit cell with NDR devices/circuits | CRLH TWFET |
|---|---|---|---|
| Design flexibility | + | + | − |
| Phase continuity | − | + | + |
| Verification | + | − | + |
| Stability | + | ± | + |

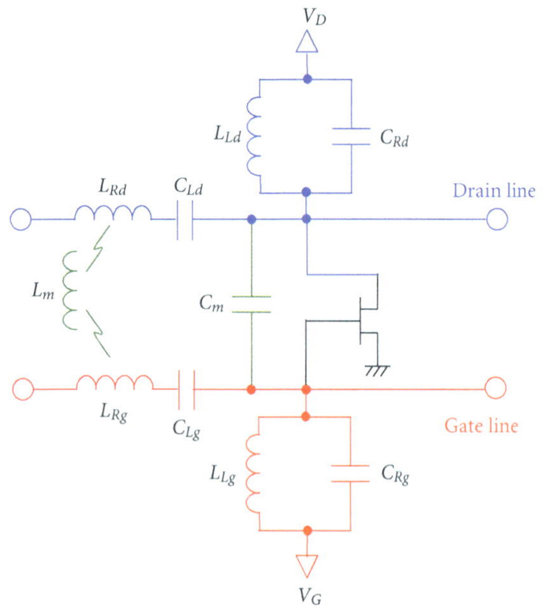


FIGURE 3: Unit cell of CRLH TWFET. Both the gate and drain lines have the CRLH TL structures. These two lines are coupled via the mutual inductance $L_m$, mutual capacitance $C_m$, and FET contributions.

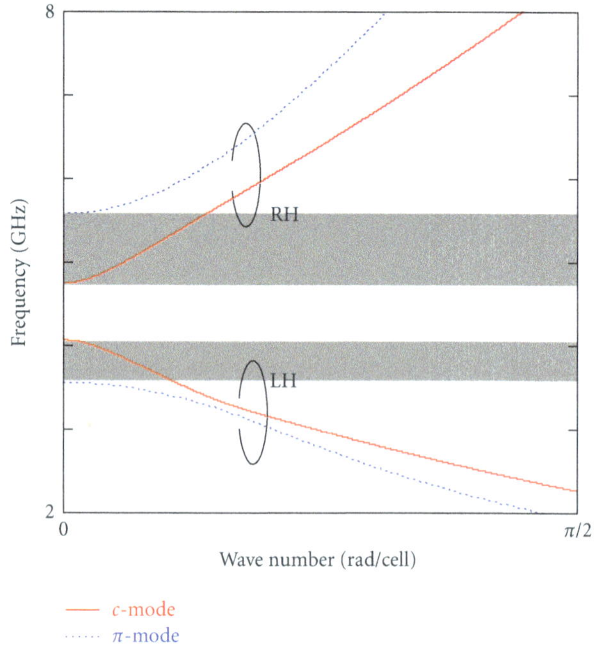


FIGURE 4: Dispersive property of CRLH TWFET. The filled frequency regions are sandwiched by zero-wavelength frequencies, which are 3.6, 4.1, 4.7, and 5.6 GHz in ascending order.

Owing to the presence of the passive coupling between the gate and drain lines, two different propagation modes can carry the waves [6].

Each mode has its own dispersion, characteristic impedance, and the voltage fraction between the lines. Figure 4 shows a typical dispersion relationship of each mode. To obtain this, we set $L_{Rg}$, $C_{Lg}$, $L_{Lg}$, and $C_{Rg}$ to 0.75 nH, 1.5 pF, 1.25 nH, and 1.0 pF, respectively. Moreover, $L_{Rd}$, $C_{Ld}$, $L_{Ld}$, $C_{Rd}$, and $C_m$ are set to 1.0 nH, 2.0 pF, 0.5 nH, 1.5 pF, and 0.2 pF, respectively. There are mostly two different modes for a given frequency, although only the unique mode is allowed in the filled frequencies in Figure 4. For definiteness, we call the fast and slow modes the $\pi$ and $c$, respectively. Accordingly, the solid and dotted curves correspond to the $c$- and $\pi$-modes, respectively. The upper (lower) two exhibit RH (LH) properties.

By inputting the voltage signals having the $c$- ($\pi$-) mode voltage fraction between the lines, they are carried by the $c$- ($\pi$-) mode. In addition, the matched termination with the $c$- ($\pi$-) mode characteristic impedance suppresses the multiple reflections that terribly distort the waves in the line. When the voltage fraction is positive (negative), the corresponding mode becomes in phase (out of phase).

We consider the cases when a small $G_m$ is introduced. We can then obtain the $c$- ($\pi$-) mode per-unit-cell gain $\alpha_{c(\pi)}$ by series-expanding the complex dispersion with respect to $G_m$, which is given by

$$\begin{aligned}\alpha_c &= -\frac{v_c L_m G_m}{4(1/v_c^2 - 1/v_\pi^2)}\left(\frac{1}{v_c^2} - \frac{1}{v_\pi^2} + \frac{1}{v_m^2}\right),\\ \alpha_\pi &= -\frac{v_\pi L_m G_m}{4(1/v_c^2 - 1/v_\pi^2)}\left(\frac{1}{v_c^2} - \frac{1}{v_\pi^2} - \frac{1}{v_m^2}\right),\end{aligned} \tag{1}$$

where $v_{c(\pi)}$ represents the phase velocity of the $c$- ($\pi$-) mode. Moreover, $v_m$ is defined by

$$\begin{aligned}\frac{1}{v_m^2} = &-\frac{2C_m}{C_{Lg}C_{Ld}L_m\omega^4}\left(\frac{\omega^2}{\omega_{\mathrm{seg}}^2} - 1\right)\left(\frac{\omega^2}{\omega_{\mathrm{sed}}^2} - 1\right)\\ &+ C_m\left(\frac{1}{C_{Lg}}\left(\frac{1}{\omega_{\mathrm{seg}}^2} - \frac{1}{\omega^2}\right) + \frac{1}{C_{Ld}}\left(\frac{1}{\omega_{\mathrm{sed}}^2} - \frac{1}{\omega^2}\right)\right)\\ &+ \frac{\omega_{Lg}^2}{\omega^4}\left(1 - \frac{\omega^2}{\omega_{\mathrm{sed}}^2} - \frac{\omega^2}{\omega_{\mathrm{shg}}^2}\right) + \frac{\omega_{Ld}^2}{\omega^4}\left(1 - \frac{\omega^2}{\omega_{\mathrm{sed}}^2} - \frac{\omega^2}{\omega_{\mathrm{shd}}^2}\right)\\ &+ \frac{1}{\omega_{Rg}^2} + \frac{1}{\omega_{Rd}^2},\end{aligned} \tag{2}$$

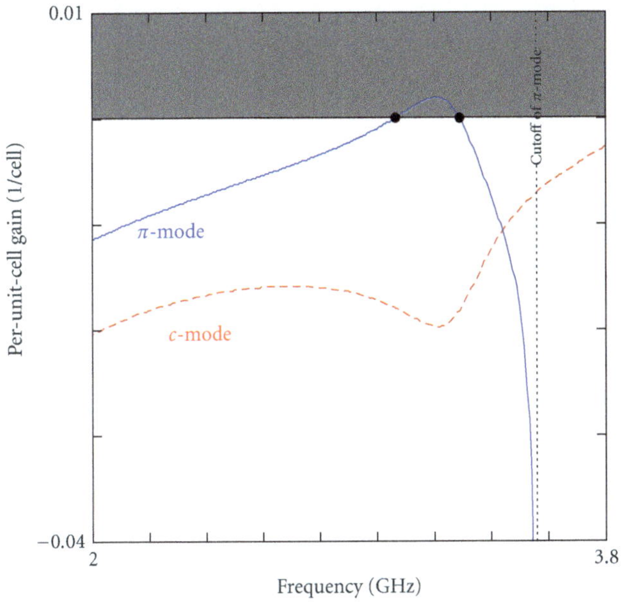


Figure 5: Per-unit-cell gain for LH frequencies. The solid and dashed curves represent the $\pi$- and $c$-modes, respectively. The filled region corresponds to amplification.

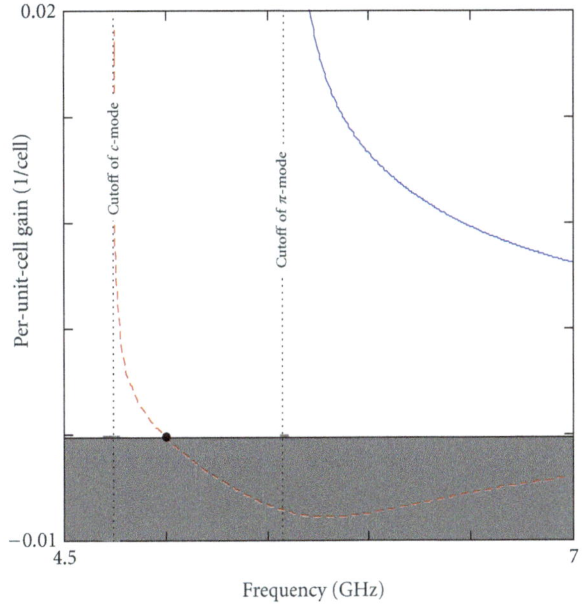


Figure 6: Per-unit-cell gain for RH frequencies. The solid and dashed curves represent the $\pi$- and $c$-modes, respectively. The filled region corresponds to amplification.

where $\omega_{Rg} = 1/(L_{Rg}C_{Rg})^{1/2}$, $\omega_{Rd} = 1/(L_{Rd}C_{Rd})^{1/2}$, $\omega_{Lg} = 1/(L_{Lg}C_{Lg})^{1/2}$, $\omega_{Ld} = 1/(L_{Ld}C_{Ld})^{1/2}$, $\omega_{\text{seg}} = 1/(L_{Rg}C_{Lg})^{1/2}$, $\omega_{\text{sed}} = 1/(L_{Rd}C_{Ld})^{1/2}$, $\omega_{\text{shg}} = 1/(L_{Lg}C_{Rg})^{1/2}$, and $\omega_{\text{shd}} = 1/(L_{Ld}C_{Rd})^{1/2}$. Note that $v_m$ can become imaginary for frequencies for which the RHS of (3) becomes negative. Moreover, (3) becomes singular for $L_m = 0$. In such a situation, $\alpha_{c(\pi)}$ should be written as

$$\alpha_c = -\frac{v_c C_m G_m \left(1 - \omega^2/\omega_{\text{sed}}^2\right)\left(1 - \omega^2/\omega_{\text{seg}}^2\right)}{2\omega^4 C_{Lg} C_{Ld} \left(1/v_c^2 - 1/v_\pi^2\right)}, \tag{3}$$

$$\alpha_\pi = -\frac{v_\pi C_m G_m \left(1 - \omega^2/\omega_{\text{sed}}^2\right)\left(1 - \omega^2/\omega_{\text{seg}}^2\right)}{2\omega^4 C_{Lg} C_{Ld} \left(1/v_c^2 - 1/v_\pi^2\right)}. \tag{4}$$

In the above, $\alpha_{c(\pi)}$ is defined to be negative for amplifying RH waves. Correspondingly, the amplitude of the wave increases if $\alpha_{c(\pi)}$ becomes positive for LH waves, because the phase velocity has the opposite sign from the group velocity. By definitions, $v_\pi$ is always greater than $v_c$; therefore, $\alpha_c\alpha_\pi < 0$ for any frequencies if $L_m$ is negligible. It is concluded that one of the two modes can be uniquely amplified. More specifically, the $c$- ($\pi$-) mode can be uniquely amplified for LH (RH) waves if $\omega < \omega_{\text{sel}}$ or $\omega > \omega_{\text{seu}}$, where $\omega_{\text{seu}} = \max(\omega_{\text{seg}}, \omega_{\text{sed}})$ and $\omega_{\text{sel}} = \min(\omega_{\text{seg}}, \omega_{\text{sed}})$. On the other hand, the $c$- ($\pi$-) mode can be uniquely amplified for RH (LH) waves if $\omega_{\text{sel}} < \omega < \omega_{\text{seu}}$.

By the introduction of any small but finite resistance $R$, we obtain the form $\alpha_{c,\pi} = p_{c,\pi}G_m + q_{c,\pi}R$ with some coefficients $p_{c,\pi}$ and $q_{c,\pi}$ for complex dispersion. Irrespective of $q_{c,\pi}$, only one of the two modes is possibly amplified, if $p_c p_\pi < 0$. Thus, the key criteria that a CRLH TWFET can be designed to satisfy the condition for the unique mode to be amplified are equally valuable even with any loss or leakage.

In general, the resistance becomes larger for higher frequencies. It tends to reach a saturated value. Instead of modelling this frequency dependence, we employ the saturated resistance for all the frequencies for overestimating. We assigned the values of parasitic resistance to be proportional to the inductance, such that the resistance for 1.0-nH inductor becomes 0.15 Ω. By adding the linearized contribution of resistances in per-unit-cell gain, we obtain Figures 5 and 6, setting $G_m$ to 2.0 mS. Figure 5 shows the per-unit-cell gain for low frequencies corresponding to the LH branches. The amplified waves are expected only for the $\pi$-mode at the frequencies between 3.0 and 3.3 GHz. The gain becomes maximal at 3.2 GHz, where the discrepancy between $v_c$ and $v_\pi$ becomes minimal. Figure 6 is for high frequencies corresponding to the RH branches. The amplified waves are expected only for the $c$-mode at the frequencies greater than 4.95 GHz.

## 4. Numerical Simulation of Test Line

Recently, we experimentally characterized a CRLH TWFET using discrete components on a breadboard [7]. We built a 90-section device. For convenience, all the series inductances were eliminated. As a result, the test line exhibited LH property for all the frequencies allowed to travel. The transistors used were Toshiba 2SK30A, whose typical threshold voltage was −1.8 V. The inductors and capacitors were implemented using the TDK SPT0406 series and the TDK FK series, respectively. The values of $C_{Lg}$, $C_{Rg}$, $C_{Ld}$, $C_{Rd}$, and $C_m$ were set to 470, 47, 470, 470, and 47 pF, respectively. Moreover, the values of $L_{Lg}$ and $L_{Ld}$ were set to 10.0 and 4.7 μH, respectively. The gate and drain lines were fed by the envelope pulses generated by an NF WF1974 two-channel function generator. The waveforms were detected by Agilent

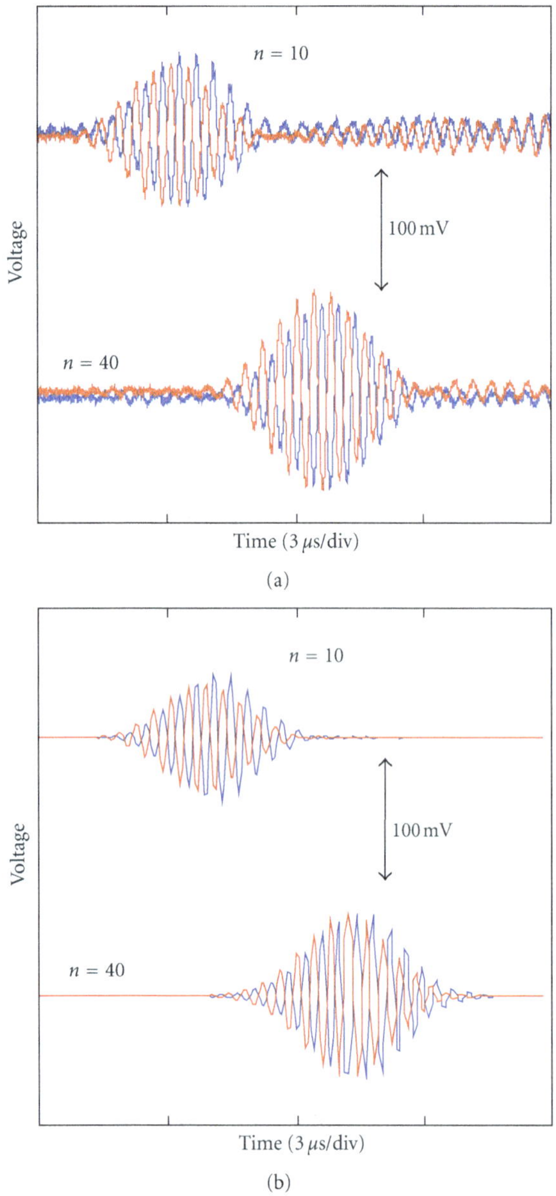


FIGURE 7: Time-domain waveforms in test CRLH TWFET. (a) Measured and (b) calculated ones for the $\pi$-mode envelope pulses.

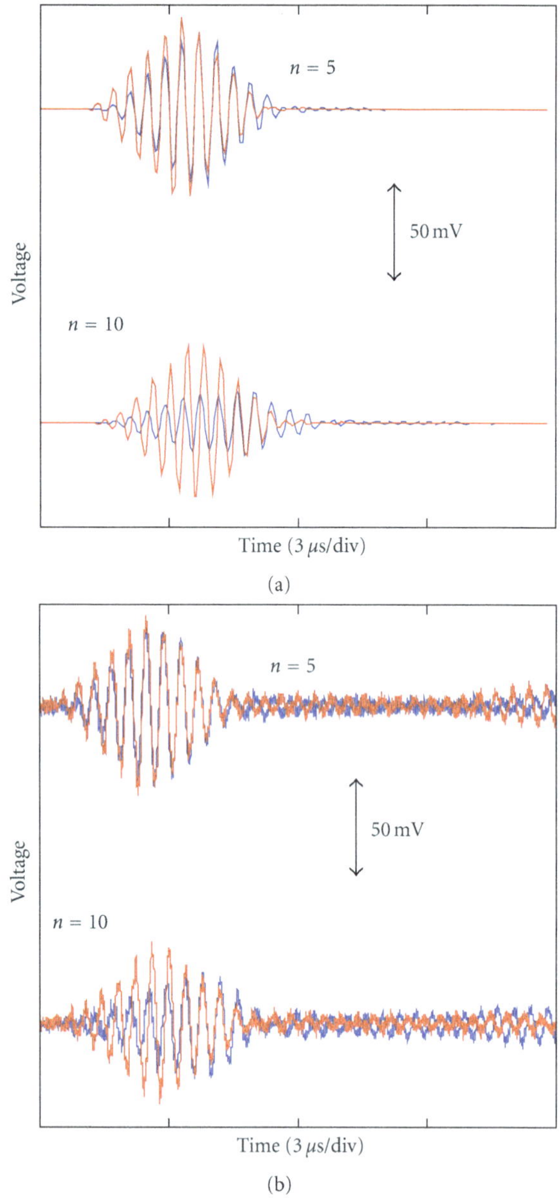


FIGURE 8: Time-domain waveforms in test CRLH TWFET. (a) Measured and (b) calculated ones for the $c$-mode envelope pulses.

10073C passive probes and monitored in the time domain by using an Agilent DSO90254A oscilloscope. Through the measurements, we found that the $\pi$-mode gains amplitudes, while the $c$-mode loses them. Moreover, when the test line is terminated with the $\pi$-mode characteristic impedance, the multiple reflections of waves could be effectively suppressed.

To reinforce these observations, we simulate the measured results by solving the transmission equations of the test line using finite-difference time-domain method. The equations are given by

$$
\begin{aligned}
& L_{Rg}\frac{d^2 I_n}{dt^2} + L_m \frac{d^2 J_n}{dt^2} \\
& \quad = -\frac{I_n}{C_{Lg}} - R_g \frac{dI_n}{dt} - \frac{d}{dt}\left(V_n - V_{n-1}\right), \\
& L_{Rd}\frac{d^2 J_n}{dt^2} + L_m \frac{d^2 I_n}{dt^2} \\
& \quad = -\frac{J_n}{C_{Ld}} - R_d \frac{dJ_n}{dt} - \frac{d}{dt}\left(W_n - W_{n-1}\right), \\
& \frac{W_n - V_D}{L_{Ld}} \\
& \quad = \left(\frac{d}{dt} + \frac{R_{\text{in},d}}{L_{Ld}}\right)\left(-\left(C_{Rd} + C_m\right)\frac{dW_n}{dt}\right. \\
& \qquad \left. + C_m \frac{dV_n}{dt} + J_n - J_{n+1} - I_{ds}(V_n)\right), \\
& \frac{V_n - V_G}{L_{Lg}} = \left(\frac{d}{dt} + \frac{R_{\text{in},g}}{L_{Lg}}\right) \\
& \qquad \times \left(-\left(C_{Rg} + C_m\right)\frac{dV_n}{dt} + I_n - I_{n+1} C_m \frac{dW_n}{dt}\right),
\end{aligned} \tag{5}
$$

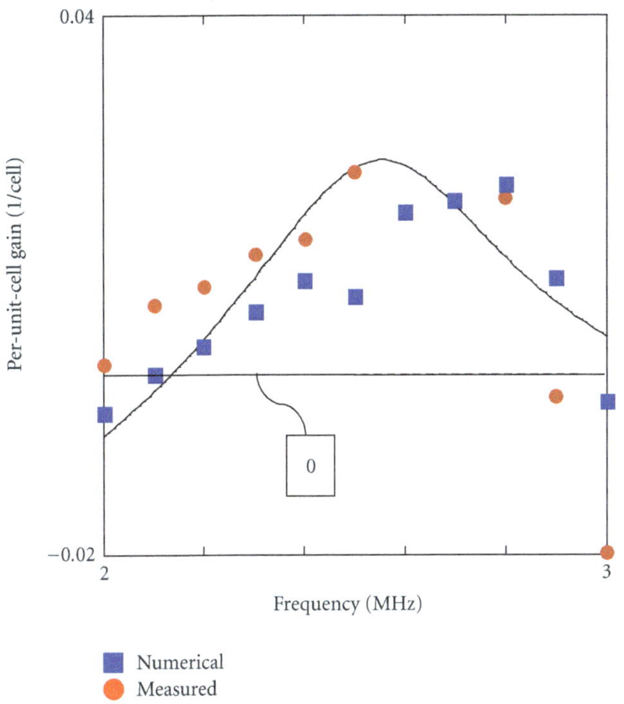


FIGURE 9: Frequency dependence of per-unit-cell gain. Circles and squares represent the measured and numerical dependences, respectively. Positive gains correspond to wave amplification. The solid curve plots (5) for reference.

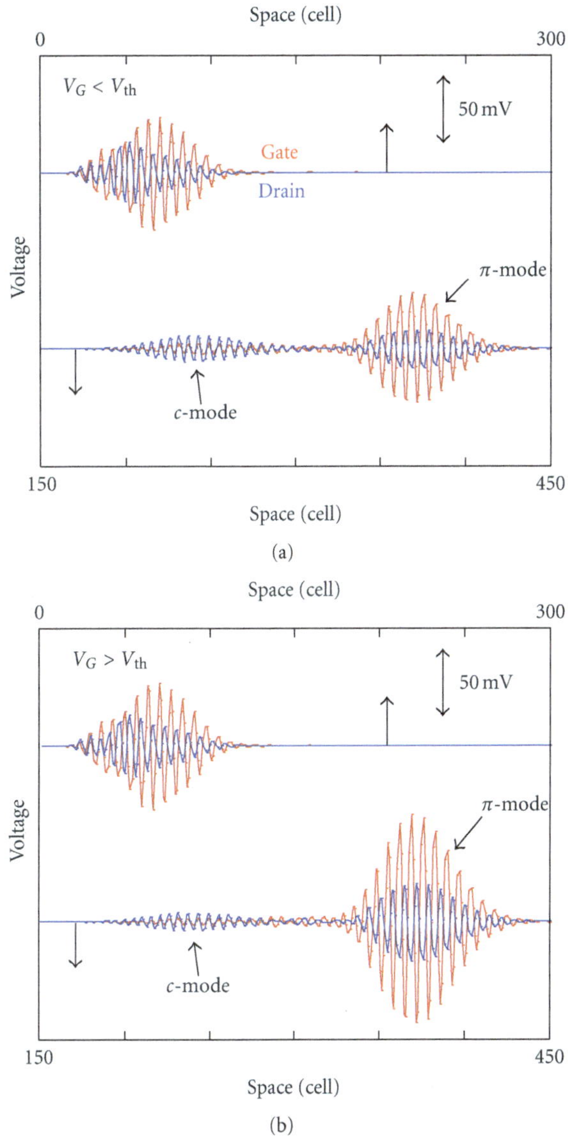


FIGURE 10: Numerically obtained envelope pulses in large CRLH TWFET. The total cell number is set to 500 without altering line parameters that were used to obtain Figures 6(b) and 7(b).

where $V_n$ and $W_n$ represent the gate and drain line voltages at the $n$th cell, respectively. Moreover, $I_n$ and $j_n$ represent the gate and drain line current flowing into the $n$th cell, respectively. Figures 7(a) and 7(b) show the measured and calculated waveforms, respectively, when gaussian envelope pulses with a carrier frequency of 2.5 MHz were input at one of the ends with the $\pi$-mode voltage fraction between the gate and drain lines. The upper and lower waveforms were monitored at the 10th and 40th cells, respectively. We can clearly see that the waves gain amplitude during propagation in Figure 7(a). Similar comparison between the measured and calculated waveforms is carried out for the $c$-mode envelope pulses in Figure 8. The upper and lower waveforms were monitored for the carrier frequency of 2.5 MHz at the 5th and 10th cells, respectively. In contrast to the $\pi$-mode waves, the $c$-mode ones lose amplitude during propagation. In both cases, the calculated waveforms well simulate the measured ones, including the amplitude gain and dispersive waveform distortions. By varying the carrier frequency of the input envelope pulses and voltage fractions, we obtain the frequency dependence of the per-unit-cell gain in Figure 9. Circles and squares represent the measured and calculated results for the $\pi$-mode inputs, respectively. The similarity between the measured and calculated results is established, so that we conclude that the numerical evaluations predict sufficiently the quantitative performance of CRLH TWFETs.

To discriminate the $c$- and $\pi$-modes clearly, we carried out calculations of a large-size line. We set the total cell number to 500 and an envelope pulse with the carrier frequency of 2.3 MHz was input to the 1st cell of the gate line. The pulse was split into the $c$- and $\pi$-mode components, which were well separated in space after the long-length propagation. Figures 10(a) and 10(b) show the calculated results for $V_G < V_{th}$ and $V_G > V_{th}$, respectively. Upper and lower waveforms correspond to the pulses before and after the long-length propagation, respectively. The $\pi$-mode components preceded the $c$-mode ones and was amplified when $V_G > V_{th}$. On the other hand, it is clearly seen that the $c$-mode components was more attenuated for $V_G > V_{th}$ than for $V_G < V_{th}$.

## 5. Numerical Performance of CRLH TWFETs

We evaluate numerically the performance of a CRLH TWFET including series inductors. We employed the same reactance values as those used to obtain Figures 5 and 6.

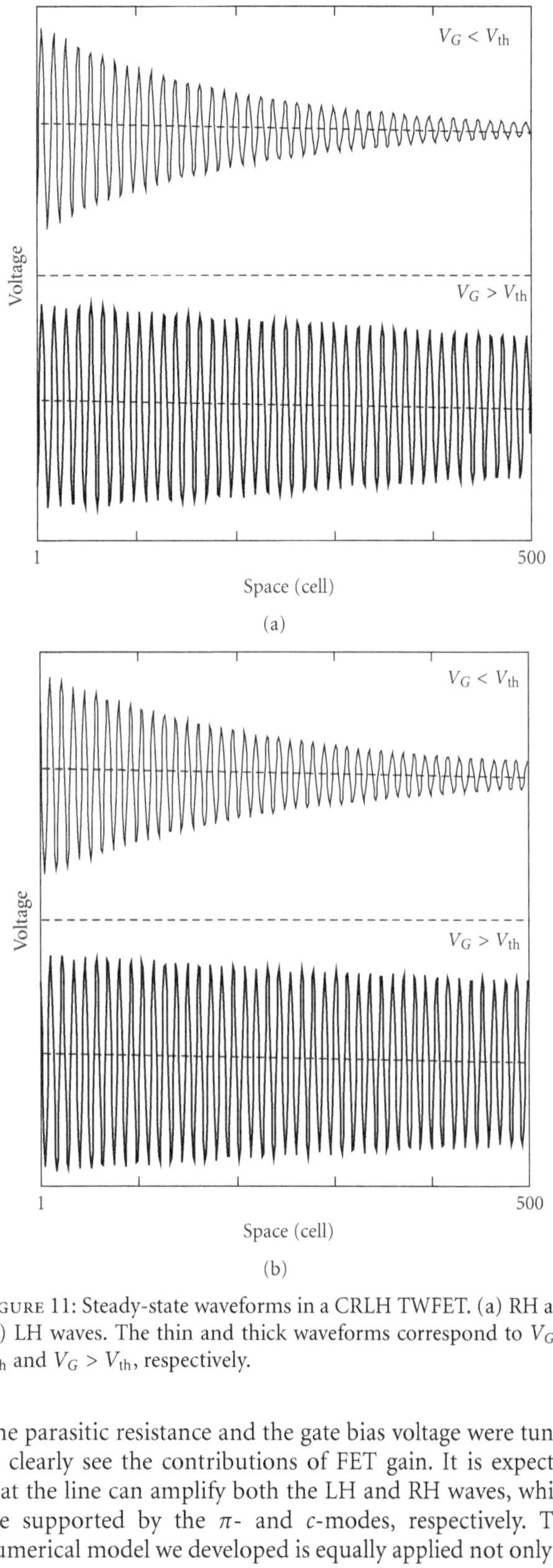


FIGURE 11: Steady-state waveforms in a CRLH TWFET. (a) RH and (b) LH waves. The thin and thick waveforms correspond to $V_G < V_{th}$ and $V_G > V_{th}$, respectively.

The parasitic resistance and the gate bias voltage were tuned to clearly see the contributions of FET gain. It is expected that the line can amplify both the LH and RH waves, which are supported by the $\pi$- and $c$-modes, respectively. The numerical model we developed is equally applied not only to the MHz circuits but also to the microwave circuits, when the print-circuit-board techniques are effectively utilized; therefore, the results obtained by our model can be considered to be sufficiently reliable for these parameter values. Responses to the sinusoidal inputs were evaluated, considering practical situations where CRLH TWFETs would be applied. The thick and thin waveforms in Figure 11(a) correspond to the steady-state LH sinusoidal waves for $V_G > V_{th}$ and for $V_G < V_{th}$, respectively. The frequency was set to 3.2 GHz, where the $\pi$-mode was supposed to be uniquely amplified. We thus designed terminal resistances to be matched with the $\pi$-mode characteristic impedances. Wave attenuation was well compensated by the FET contribution.

Finally, the properties of steady-state RH sinusoidal waves are shown in Figure 11(b). The frequency was set to 5.8 GHz. According to Figure 6, the $c$-mode is uniquely amplified, so that we set the terminations to be matched with the c-mode impedance. Compensation of attenuation was also established for the RH waves.

In both cases, it was confirmed that development of standing waves was well suppressed by the matched terminations.

## 6. Conclusions

We characterize waves in CRLH TWFETs. We found that one of the possible two propagation modes is uniquely amplified for wide range of frequency. By the terminations with the characteristic impedance corresponding to the amplifying mode, the multiple reflections of waves can be effectively suppressed. These design criteria may be practically useful to produce large-scale platform to support both the LH and RH waves.

## References

[1] C. Caloz and T. Itoh, *Electromagnetic Metamaterials: Transmission Line Theory and Microwave Applications*, Wiley-Interscience, New York, NY, USA, 2006.

[2] F. P. Casares-Miranda, C. Camacho-Peñalosa, and C. Caloz, "High-gain active composite right/left-handed leaky-wave antenna," *IEEE Transactions on Antennas and Propagation*, vol. 54, no. 8, pp. 2292–2300, 2006.

[3] L.-M. Si, T. Jiang, K. Chang, X. Lv, L. Ran, and H. Xin, "Active microwave metamaterials incorporating ideal gain devices," *Materials*, vol. 4, no. 1, pp. 73–83, 2010.

[4] K. Maezawa, K. Kasahara, and M. Mori, "A traveling wave amplifier based on composite right/left handed (CRLH) transmission lines periodically loaded with resonant tunneling diode pairs," in *Proceedings of the 22nd International Conference on Indium Phosphide and Related Materials (IPRM '10)*, pp. 362–365, June 2010.

[5] S. Nakagawa and K. Narahara, "Characterization of left-handed traveling-wave transistors," *IEICE Transactions on Electronics*, vol. E92-C, no. 11, pp. 1396–1400, 2009.

[6] K. C. Gupta, R. Garg, and I. J. Bahl, *Microstrip Lines and Slotlines*, Artech, London, UK, 1979.

[7] S. Nakagawa and K. Narahara, "Compensation of wave attenuation in left- handed traveling-wave field-effect transistors," *Progress in Electromagnetics Research Letters*, vol. 28, pp. 195–205, 2012.

# 19

# Adaptive Gain and Analog Wavelet Transform for Low-Power Infrared Image Sensors

**P. Villard,[1] T. Thabuis,[1] M. Belleville,[1] G. Sicard,[2] G. Decaens,[3] and M. Zecri[3]**

[1] *CEA LETI, MINATEC Campus, 38054 Grenoble, France*
[2] *TIMA Laboratory, 38031 Grenoble, France*
[3] *SOFRADIR, 38113 Veurey-Voroize, France*

Correspondence should be addressed to P. Villard, patrick.villard@cea.fr

Academic Editor: Gianluca Traversi

A decorrelation and analog-to-digital conversion scheme aiming to reduce the power consumption of infrared image sensors is presented in this paper. To exploit both intraframe redundancy and inherent photon shot noise characteristics, a column based 1D Haar analog wavelet transform combined with variable gain amplification prior to A/D conversion is used. This allows to use only an 11-bit ADC, instead of a 13-bit one, and to save 15% of data transfer. An $8 \times 16$ pixels test circuit demonstrates this functionality.

## 1. Introduction

Modern high-performance infrared sensors, like CdHgTe-based ones, require low-power consumption and digital output to reduce their cost and increase their ease of use, by avoiding the need for analog components on proximity board. However, when developing large format sensors (e.g., $1280 \times 1024$ pixels) the bottlenecks of analog-to-digital conversion and data transfer for low-power compliance worsen. Thus, the first two main contributors to power consumption, to consider for minimizing it, are the analog-to-digital converters (ADC-) and the drivers for data transfer off the chip.

Several digital read-out circuits have been demonstrated, relying on pixel-level [1], column-level [2], or array-level A/D conversion. In such sensors, power optimization is focused on the ADC itself, and each pixel signal is treated as completely independent, in time and space, from the others. Thus no specific transfer rate optimization is implemented. Moreover, the ADC noise figure is defined with regards to the lowest-input signal noise, without considering the signal and noise dependency in the case of photons; this leads to over-conservative conversion for large input fluxes.

To target the data transfer power, compression is a well-known technique used in image processing to reduce the bit rate. Compression algorithms are composed of two steps: firstly, data are decorrelated using either a predictor or a transformation, then entropy coding is applied to reduce the bit rate. Implementations of compression are mostly digital; however, decorrelation schemes can also be implemented in the analog domain [3–5].

This paper, by exploiting the input signal characteristics as well as the inherent spatial redundancy, targets a decrease of both the ADC resolution and the amount of data transfer. It presents a decorrelation scheme based on a modified first-level Haar decorrelation combined with a variable gain applied to its coefficients accordingly to their probability density function (PDF).

This paper is organised as follows. Section 2 discusses the main noise contributions in an infrared image sensor. In Section 3, considering the pixels spatial correlations, the Haar wavelet decorrelation is introduced and its output on a real image discussed. Section 4 presents the proposed mixed wavelet transform and adaptive gain principle. Section 5 details its integrated electronic implementation, while the associated test results are discussed in Section 6.

## 2. Signal-to-Noise Ratio in Image Sensors

The classical read-out chain of a digital image sensor is composed of a photodiode, a current integrator, a pixel follower, and an out-of-focal-plane ADC. Photons whose energy falls

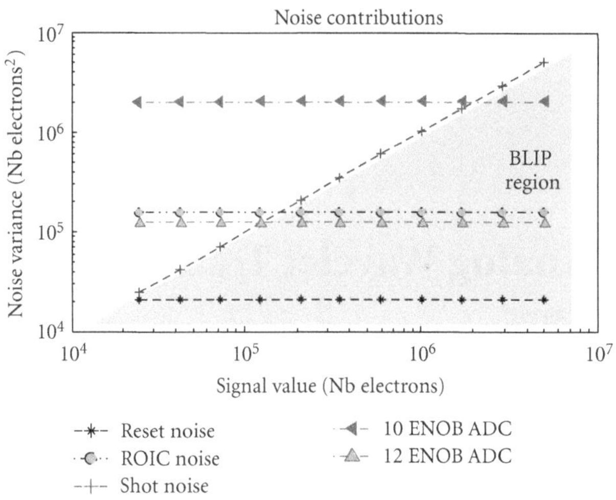


FIGURE 1: Noise contributions in an infrared imager.

into the photodiode material bandgap are converted to electrons. These electrons generate a photocurrent in the reverse biased photodiode. After integration, the resulting charge creates a voltage difference which is then transferred and A-to-D converted. The main noise sources [6] all along the chain are as follows: the shot noise which comes from photodiode current and has a noise power equal to the number of electrons generated during the photo conversion; the kTC or reset noise which appears at the reset of the integration capacitor; the read-out circuit noise which depends on the circuit architecture and technology used; the quantization noise which is usually considered to be a white noise inversely proportional to the ADC effective number of codes.

The calculation results on Figure 1 show how these noise contributions change with the input signal, in the case of a typical infrared imager characterized by a 500 fF integrating capacitor, a 1.6 V full scale at the ADC input and a 130 $\mu V_{rms}$ read-out noise (ROIC). The latter is a realistic value for a commercial product with resolution and frame rate higher than 640 × 480 pixels at 50 Hz. The highest part of the dynamic range is dominated by the photocurrent shot noise whereas the quantization noise limits the SNR for low photon flux (Figure 1). The goal of infrared detection is to keep the overall circuit noise contribution below photon shot noise over the dynamic range of interest, to ensure background limited infrared photodetection (BLIP).

## 3. Pixel-to-Pixel Correlation

Real life images exhibit significant spatial redundancies. This is already widely used in image standards like JPEG to minimize data storage. To efficiently take benefit from those redundancies, image transformation is needed. A transformation changes the domain of analysis from spatial domain to frequency domain in the case of discrete cosine transform DCT [5] or to wavelet domain in the case of discrete wavelet transform DWT [4, 7]. Most of the information is contained in the low-frequency coefficients for DCT or the high-level ones for DWT.

The case of the two-dimensional integer Haar wavelet decorrelation is considered in this study for its low hardware requirement: it is only based on adders and substractors. A 1st order Haar wavelet transform converts the raw output data of a 2 × 2 group of pixels $(a, b, c, d)$ in one binning coefficient $A$ and 3 detail coefficients $B, C,$ and $D$, according to the following equations [4]:

$$A = a + b + c + d, \qquad C = a - b + c - d,$$
$$B = a + b - c - d, \qquad D = a - b - c + d. \qquad (1)$$

$A$ corresponds to the mean value of the four pixels, while $B$, $C$, and $D$ depend on the gradients within the 2 × 2 group.

Figure 2 presents a raw infrared image and its associated pixel value PDF (probability density function). The corresponding $A, B, C,$ and $D$ coefficients PDFs are shown Figure 3. It clearly appears that the detail coefficients exhibit a small deviation around 0, as neighbouring pixels have correlated values.

## 4. Mixed Wavelet Transform and Adaptive Gain Principle

In this work, pixel-to-pixel correlation is exploited thanks to a decorrelation scheme based on a 1st order Haar wavelet transform. Contrary to the case of visible images, in the infrared domain reference video streams are not available for benchmarking of image processing solutions. Thus we built our own set of infrared video data by acquiring images in various conditions including indoor and outdoor, night and day, static and moving scenes. A statistical analysis of these infrared images shows that more than 99% of the detail coefficients lie in the [−FS/8, FS/8] interval, FS being the full scale, which confirms their small deviation around 0. Hence they can be amplified by 4 prior to digitization without inducing saturation.

In addition, compared to the signal power, the shot noise power follows a square-root law. Consequently the ADC quantization noise has more impact at low flux. This

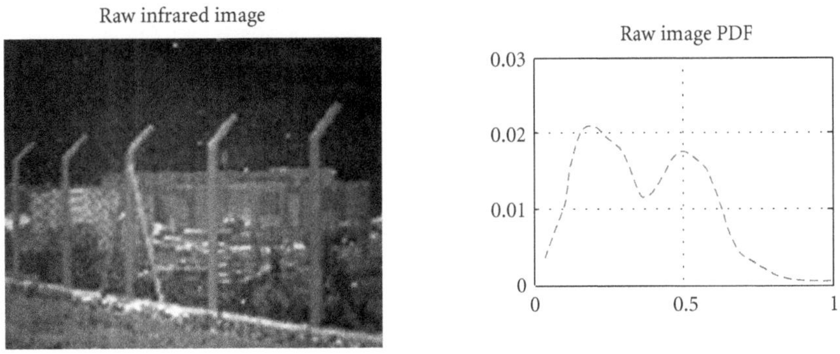


Figure 2: Raw infrared image and its associated PDF.

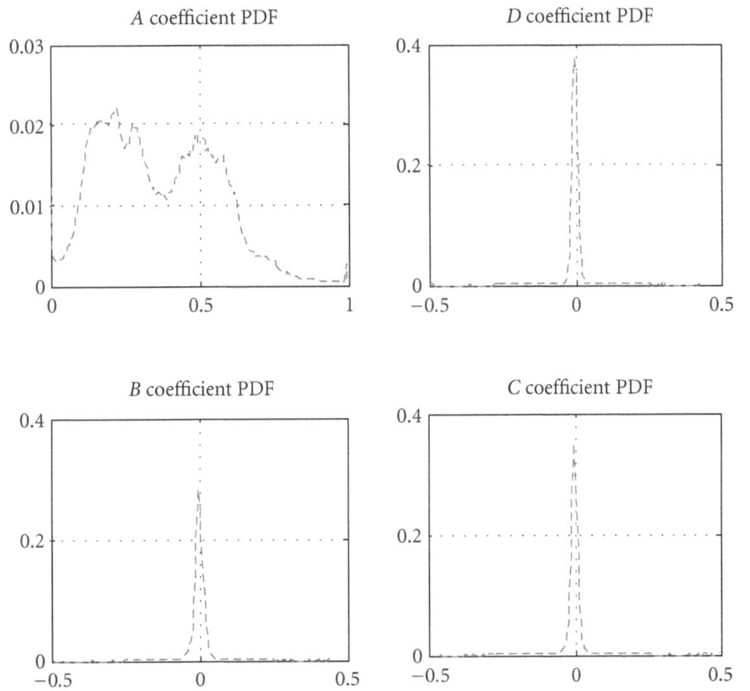


Figure 3: PDFs of the binning and detail coefficients of the raw infrared image.

impact can be reduced by preamplifying the signal or the binning coefficient which is its spatial mean, with high gain at low flux and low gain at high flux. As binning coefficients are spatially correlated, a predictive scheme can be used to anticipate their values. The predictor can be a combination of spatiotemporal neighboring binning coefficients previously determined. However, taking the last binning coefficient calculated as a prediction presents the advantage of an easy implementation and offers good approximation performance. In this proposed solution, we chose to implement such an adaptive gain prior to the AD Conversion. We selected gain values of 4, 2, and 1 for the ranges defined by the intervals [0, FS/4], [FS/4, FS/2], and [FS/2, FS], respectively. This allows to decrease the ADC effective number of bits (ENOB) by 2, provided the pre-amplifier's noise is low enough.

Combining these two ideas led us to the following new concept depicted on Figure 4: one binning and three detail coefficients are generated per $2 \times 2$ group of pixels by the discrete wavelet transform (DWT). For the detail coefficients the gain is set to 4 and for the binning coefficient it is set to 1, 2, or 4 depending on an estimation from the previous group of pixels in the column, thus further exploiting the spatial correlation within the image.

## 5. Silicon Demonstrator

This concept has been implemented in a test chip, on a 3.3 V, 0.35 $\mu$m CMOS process (Figure 5). The matrix format is 16 rows by 8 columns made of typical infrared 25 $\mu$m by 25 $\mu$m pixels. For testability purpose each pixel voltage can be set to an arbitrary value within the dynamic range ("electrical image"). The variable gain algorithm is simple (Figure 6): within the fixed set $\{1, 2, 4\}$ the gain used for the next $2 \times 2$ block is twice, equal to or half the one used for the current $2 \times 2$ block when the current value falls, respectively

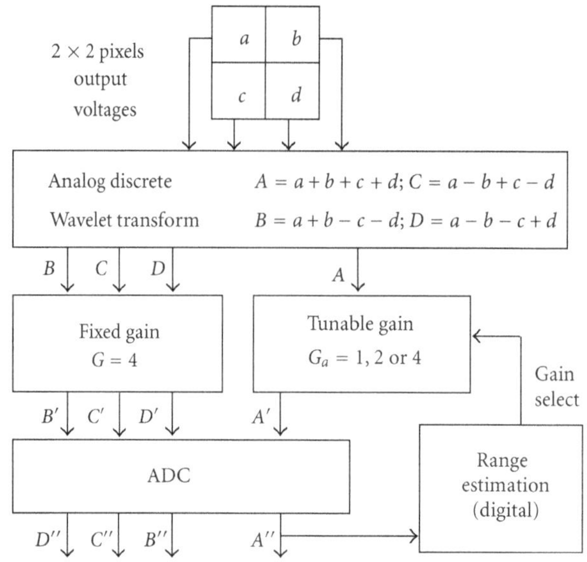


Figure 4: Proposed read-out principle.

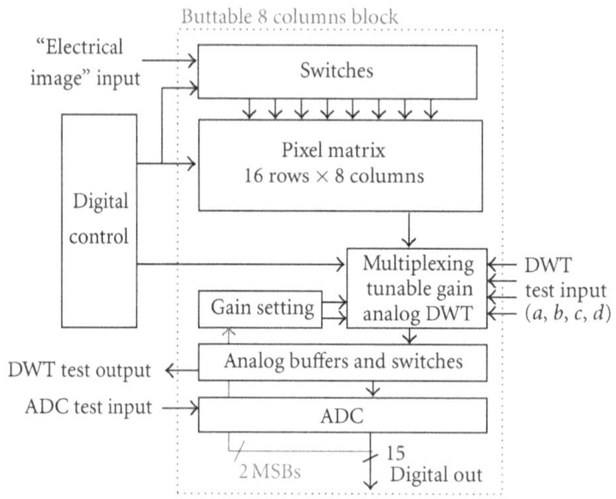


Figure 5: Test chip block diagram.

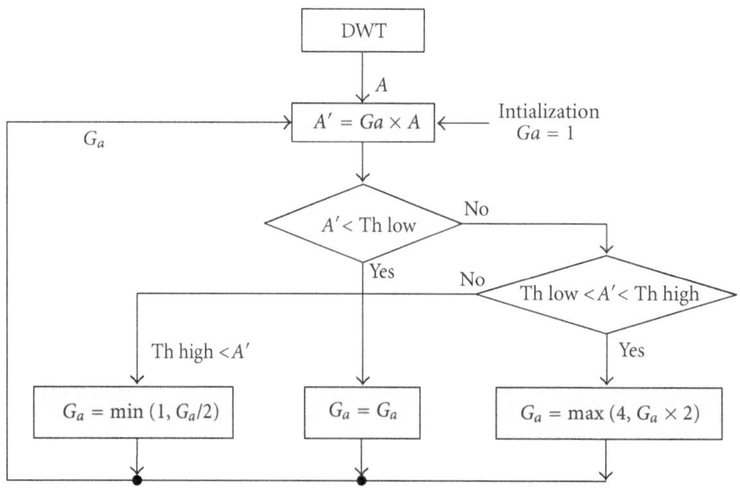


Figure 6: Binning coefficient adaptive gain algorithm.

TABLE 1: Theoretical and measured characteristics of binning and detail coefficients. Conditions: input $a = b = c = d$ swept along full dynamic range.

| | Conditions | Theoretical | Measured |
|---|---|---|---|
| Binning coefficient slope | Gain setting = 1 | 1 | 0.98 |
| | Gain setting = 2 | 2 | 1.94 |
| | Gain setting = 4 | 4 | 3.78 |
| Binning coefficient offset | Gain setting = 1 | 0 V | 16 mV |
| | Gain setting = 2 | 0 V | 8 mV |
| | Gain setting = 4 | 0 V | −4 mV |
| Detail coefficient slope | Uniform "electrical image" | 0 | 0.007 |
| Detail coefficient offset | Uniform "electrical image" | 0 V | 9 mV |

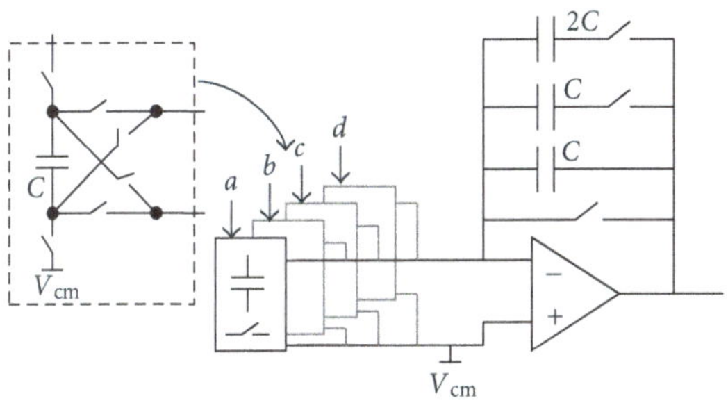


FIGURE 7: Analog DWT and gain stage implementation.

FIGURE 8: Test chip photograph. The chip size is 4.7 mm × 1.3 mm.

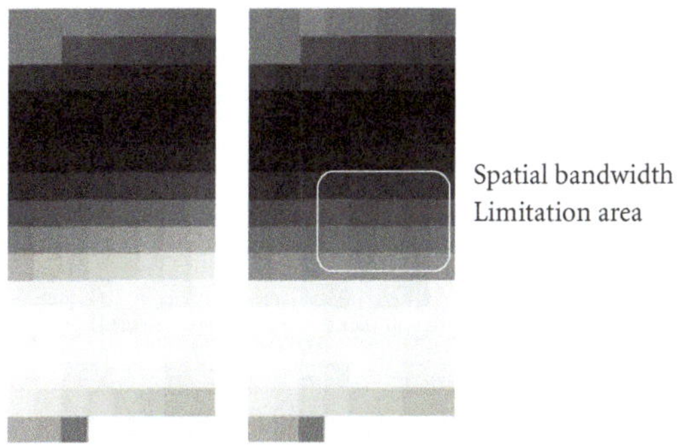


(a) Test pattern (b) Reconstructed image

FIGURE 9: Measured input test image and reconstructed image.

in the low, mid, or high part of the dynamic range. These parts are defined by two tunable digital thresholds. At each new frame, the gain is reset to its minimum value. The analog DWT/gain block is based on a standard switched capacitor schematic (Figure 7). Sample capacitors are first precharged to the voltage ($V_x - V_{cm}$, $x = a$ then $b, c,$ and $d$). Then the charges are transferred to the amplifier's feedback capacitors with appropriate switch conditions to ensure the desired polarity and gain. The resulting voltage is then sampled hold, and converted by a 11-bit pipeline ADC which is shared by 8 columns.

The building blocks of this small format image sensor are designed to be compatible with a standard-size component (e.g., 640 × 512): the number of lines can be increased by simply adding 2 × 2 pixel blocks and the number of columns can be increased by abutting blocks composed of 1 ADC and 8 columns.

Thanks to a configuration register the chip can be set up in several test modes allowing to independently characterize the "electric image" writing process, the image analog read-out, the analog wavelet transform, and the AD conversion.

## 6. Test Results

The chip (Figure 8) has been tested at typical infrared imager operating temperature. Binning and detail coefficients measured results are presented Table 1. Slopes and offsets are derived by linear regression on the measured "coefficient versus input voltage" curves. The measurement was performed by applying the same value to the four inputs ($a = b = c = d$) and sweeping this value over the whole dynamic range. The theoretical results for such conditions are $A = G_a \times a$, with $G_a = 1, 2,$ or 4, and $B = C = D = 0$. The measured values exhibit some deviations which are due to nonsymmetrical parasitic capacitance effects (e.g., capacitance between substrate and bottom plate of a poly-poly capacitor) in the switched capacitor block. Like in standard image sensors, off chip offset and gain corrections can partly compensate for this; this has not been used in this study.

An example of measurements on an "electrical image" with a specific pattern is given Figure 9. The reconstructed image is obtained from the measured DWT coefficients by applying (off-chip) the inverse-DWT without any digital calibration.

Figure 9 shows that the reconstructed image matches the original one, but with an attenuated contrast in the transition region between the darkest and the brightest zones. This effect is due partly to the above-mentioned deviations and mostly to the read-out concept. Indeed, in high-contrast regions, the high-gain-to-low-gain transition might be spatially too slow. Also, some detail coefficients might lie outside the [−FS/8, FS/8] interval. Both cases will lead to saturation. Figure 9 input image was built to illustrate those effects. On real images, this quite improbable case will mostly occur around the "dead" pixels and will not excessively affect the image quality.

Also due to the parasitic capacitance effects, each $2 \times 2$ block in the contrasted regions of the reconstructed image exhibits a column-oriented fixed pattern. This effect can be significantly reduced by both design optimization and off-chip calibration.

The analog blocks were designed to be compatible with a $640 \times 512$ image sensor working at 60 frames per s. The measured power consumption of the DWT/gain block is 100 $\mu$W. The ADC exhibits an 11-bit ENOB at 246 kHz for 600 $\mu$W. This 700 $\mu$W proposed solution is intended to replace a 13-bit ENOB ADC which consumes 2.4 mW. Thus, we can expect a significant (>70%) power saving on the ADC contribution. In addition, switching from 13 bits to 11 bits, the dynamic power consumption due to data transfer will decrease by 15%; since the 2-bit gain information does not need to be transferred, it can be computed offchip by running the same gain setting algorithm on the digital data.

In the worst case, that is, gain of 4 (low flux), the output referred noise due to the DWT/gain block circuitry (amplifiers, charge injection, kT/C noise...) was measured to be 200 $\mu V_{rms}$. Quadratically added to the 310 $\mu V_{rms}$ quantization noise of the 11-bit ADC (on a 2.2 V full scale), this leads to a total 370 $\mu V_{rms}$ noise voltage added to the signal previously amplified by 4. So, when scaled to the original signal, the equivalent noise voltage is 92 $\mu V_{rms}$, which corresponds to the quantization noise of a 12.8-bit ENOB on a 2.2 V full scale. In other words, compared to a 13-bit solution (77.5 $\mu V_{rms}$ quantization noise), the SNR loss, that is, the increase of the low end of the BLIP flux range, is 19%.

## 7. Conclusion

This paper presents a column-wise decorrelation scheme for high-resolution IR imagers and its silicon implementation and tests. This architecture exploits both intraframe redundancy and inherent photon shot noise characteristics to achieve a reduced power consumption while preserving the background limited infrared photodetection objective. It is based on a first-level Haar wavelet transform with predictive estimation of binning coefficients dynamic range. Combining into an analog implementation, an adaptive gain for the binning coefficients and a fixed gain for the detail coefficients allow to use a lower resolution ADC thus reducing the power consumption. Further power reduction is achieved thanks to the reduced amount of data to be transmitted. Overall, an image quality equivalent to a 12.8 bit direct quantization scheme is achievable. Those results prove the worthiness of the adaptive gain and analog wavelet transform approach to decrease the power consumption of infrared image sensors by allowing to relax the ADC resolution and, in addition, by, accordingly, decreasing the output switching activity.

## References


[1] X. Wang, W. Wong, and R. Hornsey, "A high dynamic range CMOS image sensor with inpixel light-to-frequency conversion," *IEEE Transactions on Electron Devices*, vol. 53, no. 12, pp. 2988–2992, 2006.

[2] M. F. Snoeij, A. J.P. Theuwissen, K. A.A. Makinwa, and J. H. Huijsing, "Multiple-ramp column-parallel ADC architectures for CMOS image sensors," *IEEE Journal of Solid-State Circuits*, vol. 42, no. 12, pp. 2968–2977, 2007.

[3] W. D. León-Salas, S. Balkir, K. Sayood, N. Schemm, and M. W. Hoffman, "A CMOS imager with focal plane compression using predictive coding," *IEEE Journal of Solid-State Circuits*, vol. 42, no. 11, Article ID 4362102, pp. 2555–2572, 2007.

[4] Q. Luo and J. G. Harris, "A novel integration of on-sensor wavelet compression for a CMOS imager," in *Proceedings of the IEEE International Symposium on Circuits and Systems (ISCAS'02)*, pp. III/325–III/328, May 2002.

[5] S. Kawahito, M. Yoshida, M. Sasaki et al., "A CMOS image sensor with analog two-dimensional DCT-based compression circuits for one-chip cameras," *IEEE Journal of Solid-State Circuits*, vol. 32, no. 12, pp. 2030–2041, 1997.

[6] H. Tian, B. Fowler, and A. El Gamal, "Analysis of temporal noise in CMOS photodiode active pixel sensor," *IEEE Journal of Solid-State Circuits*, vol. 36, no. 1, pp. 92–101, 2001.

[7] A. Olyaei and R. Genov, "CMOS wavelet compression imager architecture," in *Proceedings of the 7th IEEE Emerging Technologies Workshop: Circuits and Systems for 4G Mobile Communications (ETW'05)*, pp. 104–107, June 2005.

# 20

# Fabrication and Characterization of New Ti-$TiO_2$-Al and Ti-$TiO_2$-Pt Tunnel Diodes

**Yaksh Rawal, Swaroop Ganguly, and Maryam Shojaei Baghini**

*Department of Electrical Engineering, IIT-Bombay, National Center for Photovoltaic Research and Education (NCPRE), Mumbai 400076, India*

Correspondence should be addressed to Yaksh Rawal, yakshr@ee.iitb.ac.in

Academic Editor: Junseok Chae

Remotely empowered wireless sensor networks use different energy resources including photovoltaic solar cells, wireless power transmission, and batteries. As another option the electromagnetic energy available in the ambient can be harvested to power these remote sensors. This is particularly valuable if it is desirable to harvest the ambient energy available in the wide range of electromagnetic spectrum. This has motivated the research for developing energy harvesting devices which can absorb this energy and produce a DC voltage. Rectenna, an antenna coupled with a rectifier, is the main component used for absorbing electromagnetic radiation at GHz and THz frequencies. Rectifying MIM tunnel diodes are able to operate at tens and hundreds of GHz frequency. As the preliminary steps towards development of high-frequency rectifiers, this paper presents fabrication and DC characterization of two new MIM diodes, Ti-$TiO_2$-Al and Ti-$TiO_2$-Pt. *G-V* analysis of the fabricated diodes verifies tunneling. Brinkman-Dynes-Rowell model is used to extract oxide thickness of which the derived value is around 9 nm. Ti-$TiO_2$-Pt diode exhibits rectification ratio of 15 at 0.495 V, which is more than rectification ratio reported in earlier works.

## 1. Introduction

Recently electromagnetic collectors are being explored as one of possible means of harvesting ambient energy, available in the wide range of solar spectrum [1]. RF energy harvesting is a special case of this broad approach [2–4]. This kind of energy can be used for low-power applications such as remote powering of sensors; charging mobile phones and laptops anywhere. Electromagnetic waves, received by an antenna, are converted into DC voltage with the help of high-frequency rectifiers, filters, and voltage multipliers. The heart of the rectifier is the diode. This diode can be a Schottky diode or a metal-insulator-metal (MIM) tunnel diode. The Schottky diode has limitations such as long reverse recovery time, and hence it cannot provide faithful rectification at high frequencies. MIM tunnel diode works on the principle of quantum mechanical tunneling. More often antenna and rectifier element are integrated as one component, called rectenna. Brown from Raytheon Company, for the first time used the term "rectenna" for rectifying-antenna, which was invented to absorb the microwave radiation and simultaneously convert it to DC power [5]. In 1964, he demonstrated a microwave powered helicopter which lasted for 10 hours with the helicopter hovering at an altitude of 60 ft. In 1972 Bailey proposed a new concept, called electromagnetic wave energy converter (EWEC), using unique pyramidal solar radiation absorber-converter elements [6]. Hoofring et al. have reported the fabrication and characterization of thin film metal-oxide-metal tunnel diodes for the use in far-infrared rectenna [7]. Recently, the Clean Energy Research Center, University of South Florida, has reported harvesting energy of infrared radiations from solar energy [8, 9]. They have also reported application of MIM tunnel diodes in millimeter wave detection [10]. Millimeter wave imaging systems are the next generation of imaging systems being developed for security and surveillance purposes.

Importance of developing MIM diodes for future highly efficient rectenna is the motivation behind the research presented in this paper. The paper presents fabrication flow and DC characterization of two new MIM diodes, Ti-$TiO_2$-Al and Ti-$TiO_2$-Pt. Brinkman-Dynes-Rowell $G$-$V$ model is used to extract the oxide thickness, which is calculated around 9 nm. Behavior of native and plasma oxide devices is compared. Ti-$TiO_2$-Pt diode exhibits rectification ratio of 15 at 0.495, which is more than rectification ratios reported in earlier works [11].

## 2. Fabrication of MIM Tunnel Diodes

Figure 1 shows the process flow for native oxidized sample (left) and plasma oxidized sample (right). The top electrode was deposited by the shadow mask technique. The squares of sizes 200 microns, 100 microns, and 50 microns were deposited. Two different fabrication techniques are used to form $TiO_2$: native oxide growth and plasma oxidation. This section explains details of the process flow for both techniques. Diagrams of both process flows shown in Figures 1(a) and 1(b) show the structure of MIM tunnel diode and its microscopic top view image.

*2.1. Fabrication Flow for Native $TiO_2$ Growth.* A silicon wafer is used as a substrate for the MIM stack. The bottom layer of the metal is titanium, which is sputtered by RF sputtering at 150 W forwarding power up to a thickness of 70 nm. The sample is kept in class 1000 clean room for not more than a week to form a native titanium dioxide on it. Then the top electrode is formed by the thermally evaporating aluminum up to a thickness of 150 nm at base pressure of $3.5 \times 10^{-6}$ mbar. Another sample is made in which the top electrode is evaporated with platinum using electron beam evaporation at base pressure of $3 \times 10^{-6}$ mbar. The deposition of the top electrode is done with the help of shadow mask which has the squares of the sizes 200 microns, 100 microns, and 50 microns. Titanium can readily be oxidized, forming a native oxide when exposed to the oxygen in the air for some time. This fact is used to get the native oxide on the titanium and the selection of counter electrode to get the work function difference.

*2.2. Fabrication Flow for Plasma Oxidized Ti.* The alternate method for the titanium dioxide formation which can be highly controlled is plasma oxidation. Titanium is deposited by electron beam evaporation up to a thickness of 80 nm and then exposed to the oxygen plasma for 30 seconds. The forwarding RF power to form the plasma is 1500 W, and the standard oxygen flow in the chamber is 3000 sccm. The Applied Materials Etch Centura is used for the plasma oxidation. Other recipes were also tried by exposing the titanium for 15 seconds and also 15 and 30 seconds variant with RF power of 1000 W. The only sample that was working was the plasma oxidation, performed at 1500 W for 30 seconds. Oxygen flow is kept constant at 3000 sccm, which is minimum value to sustain the oxygen plasma. The kinetics of the native oxide formation has been thoroughly explained by Mott [12–14] and Cabrera and Mott [15]. The kinetics of plasma oxidation has been explained by Adrianus with respect to magnetic tunnel junctions [16].

## 3. Characterization of MIM

The MIM tunnel diode is characterized using semiconductor characterization system from Keithley probe station 4200SCS. Both the contacts are made from the top. The Al or Pt mesas are contacted from the top and the bottom electrode of Ti is contacted by little scratching through the thin oxide formed on the surface. The current density versus voltage plot for the MIM diode with native oxide is shown in Figure 2. Figure 3 shows the current density versus voltage plot for the MIM with plasma oxidized titanium dioxide. The higher current density of Figure 3 can be explained by the higher work function difference between titanium and platinum electrodes and the small area of the devices. The better quality of plasma oxidized titanium dioxide over native oxide of titanium is also a reason for higher current density.

The $dI/dV = I'$ is the conductivity of the MIM diode. The rate of change of conductivity is defined as $d^2I/dV^2 = I''$. The ratio of the second derivative of current and the first derivative of current is called the sensitivity of the MIM diode, $S = I''/I'$. This ratio defines the nonlinearity of the diode and hence the diode performance. The first derivative of current defines the rate of change of the conductance and hence the asymmetry. Figure 4 shows the maximum sensitivity of the diode measured at room temperature, which is $18\,V^{-1}$ at dc voltage of 0.09 V. This is good as compared to $15\,V^{-1}$ obtained by Krishnan [11]. The $dI/dV$ plot for the native oxide and plasma oxide of titanium dioxide MIM is shown in Figures 5 and 6, respectively.

According to Brinkman et al., the plot of conductivity $(dI/dV)$ versus voltage will be a parabola [17]. The asymmetry in the parabolic curve is due to the difference between work functions of the metal electrodes used for MIM device. Therefore, similar to forward and reverse biasing of diodes if the applied voltage to the electrode with lower work function is more than the applied voltage to the electrode with higher work function, the MIM diode will be forward biased. As Figures 5 and 6 show, the conductivity is higher in forward-biased region, compared to reverse-biased region of two MIM diodes. Shape of conductivity parabola is not same for two diodes since one of them uses Pt and Ti for electrodes and the other one uses Al and Ti for electrodes. The coefficients of the fitted parabola were used to extract the thickness of the dielectric, which was found to be 9 nm for native oxide and plasma oxide. The current density versus voltage and the conductivity versus voltage plots for both native titanium and plasma oxide were observed. The asymmetry in the conductivity accounted for the work function difference of the dissimilar metal electrodes. Brinkman et al. (BDR fit) [17] gave the equation for the conductivity of the metal-insulator-metal diode as

$$\frac{G(V)}{G(0)} = 1 - \left(\frac{A_0 \Delta\varphi}{16\overline{\varphi}^{3/2}}\right)(qV) + \left(\frac{9A_0^2}{128\overline{\varphi}}\right)(qV)^2 \quad (1)$$

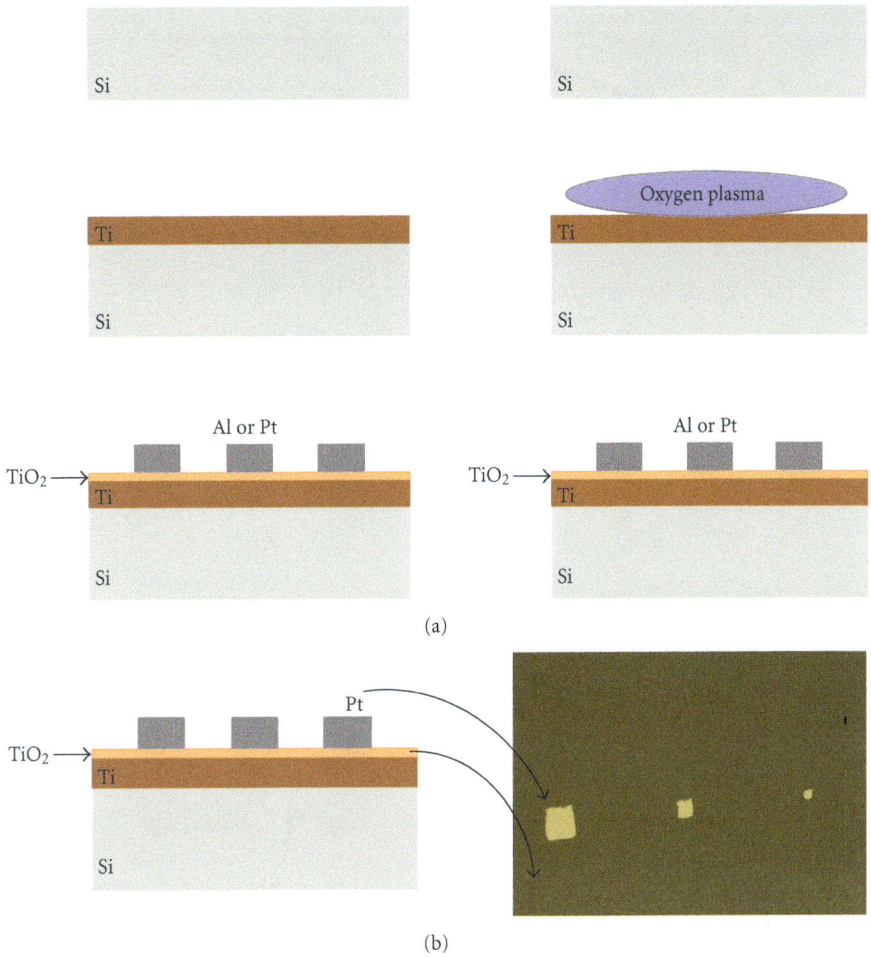


Figure 1: (a) process flow for native oxidized sample (left) and plasma oxidized sample (right). The top electrode was deposited by the shadow mask technique. The squares of sizes 200 microns, 100 microns, and 50 microns were deposited. (b) structure of MIM tunnel diode and its microscopic top view showing the Pt mesas of sizes 200 $\mu$m, 100 $\mu$m, and 50 $\mu$m squares.

or

$$G(V) = \left[G(0)(q^2)\left(\frac{9}{128}\frac{A_0^2}{\overline{\varphi}}\right)\right]V^2 - \left[G(0)(q)\left(\frac{A_0}{16}\frac{\Delta\varphi}{\overline{\varphi}^{3/2}}\right)\right]V + G(0), \quad (2)$$

where $G(0) = (3.16 \times 10^{10})\sqrt{\overline{\varphi}}te^{-1.025t\sqrt{\overline{\varphi}}}$; $A_0 = 4\sqrt{2m}t/3\hbar = (1.2067 \times 10^{19})t$.

In (2), $G(0)$ is conductivity at zero voltage, $m$ is effective mass of an electron in the TiO$_2$, $\Delta\varphi$ is difference between potential barrier heights of the metals, $\overline{\varphi}$ is the average barrier height, $t$ is thickness of insulator, $\hbar$ is reduced plank's constant, $V$ is the applied bias voltage, and $q$ is charge of an electron.

Values of the coefficients in the parabola, given by (2), are calculated by fitting the experimental $dI/dV$ versus voltage plot to a parabola.

Performing some mathematical calculations with the parabolic fit to the experimental data, we get the thickness of the insulator is found to be 9 nm, for both native and plasma oxide. It should be noted here that the absolute mass of an electron in titanium dioxide film is taken to be half of the electron mass. The asymmetry is also defined by the

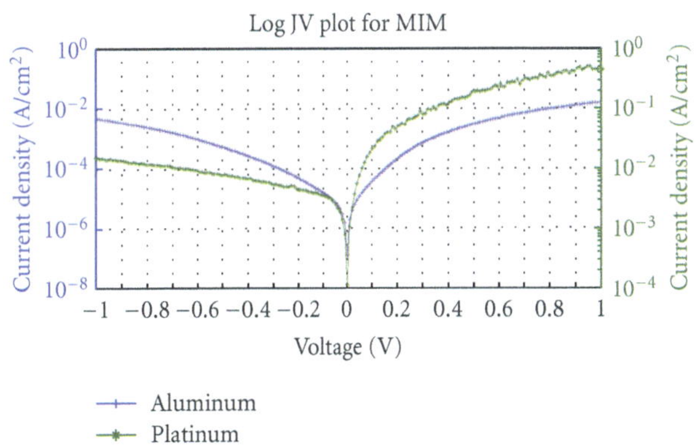


Figure 2: Current density versus voltage characteristics for Ti-TiO$_2$ (native)-Al and Ti-TiO$_2$ (native)-Pt MIM diode. (Logarithmic scale).

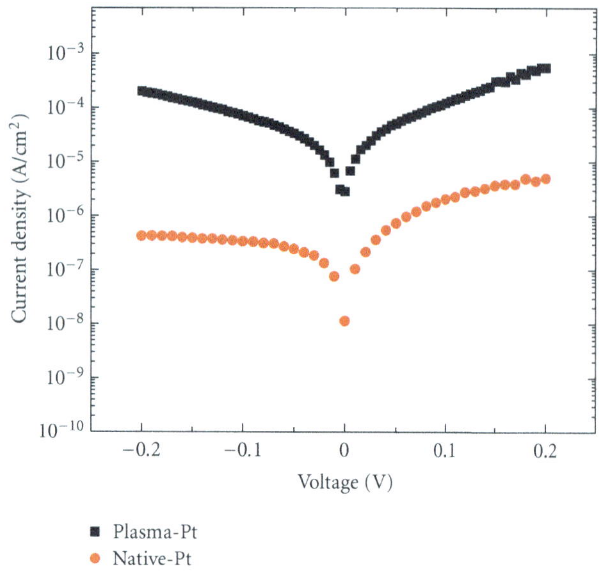


FIGURE 3: Comparison of current density (Logarithmic scale) of MIM Diodes made with native and Plasma $TiO_2$.

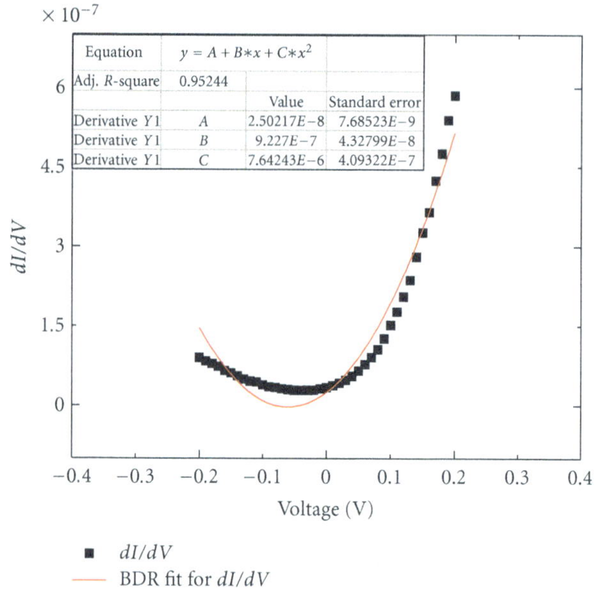


FIGURE 5: $dI/dV$ of Ti-$TiO_2$ (native)-Al MIM diode with fitted parabola.

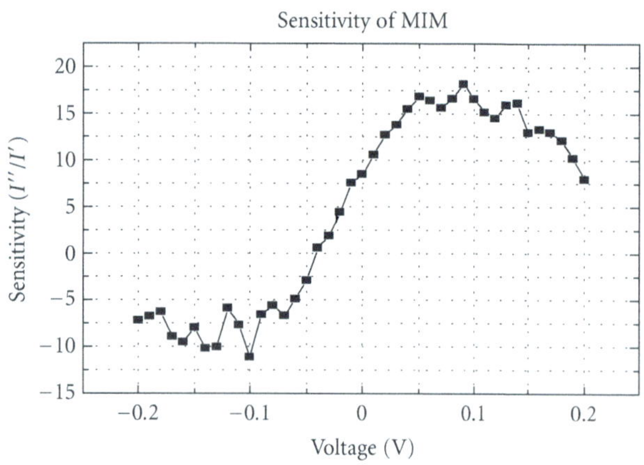


FIGURE 4: Sensitivity of Ti-$TiO_2$ (native)-Al MIM diode.

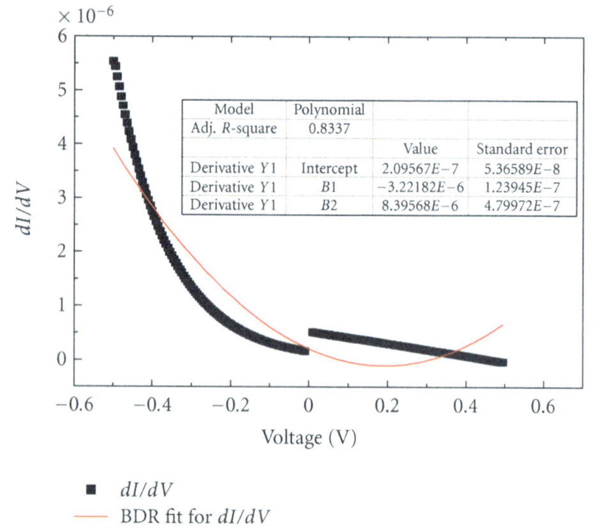


FIGURE 6: $dI/dV$ of Ti-$TiO_2$ (plasma)-Pt MIM diode with fitted parabola.

rectification ratio of the diode which is 15 at 0.495 V for plasma oxide, is more than reported in previous works. The rectification ratio for the native oxide is observed to be 6.5 at 0.33 V. The plot for the rectification ratio is shown in Figures 7(a) and 7(b).

As can be seen from Table 1, the sensitivity and the rectification ratio ($I_f/I_r$) depend on the barrier thickness and the work function of the metals. According to (2) and by defining sensitivity $S = I''/I'$, it is observed that sensitivity depends on parameters $A_0$, $\varphi$, and $\Delta\varphi$. Increasing work function difference will enhance sensitivity as it is also mentioned by Krishnan et al. [8]. Other factors are properties of insulator-metal interface such as trap density and oxide charges. In the presented work, we found that native $TiO_2$ is able to provide better sensitivity. However, more studies and structure examinations are required to precisely explain the mechanisms of sensitivity improvement.

## 4. Conclusion

This paper presents two different new MIM tunnel diodes: Ti-$TiO_2$-Pt using plasma oxidization of Ti of which work function is 1.31 eV less, compared to Pt work function and Ti-$TiO_2$-Al using native oxidization of Ti of which work function is 0.27 eV more, compared to Al. Measurement results show that the diode with native $TiO_2$ shows similar current densities but higher sensitivity as compared to the earlier reported work [11]. The plasma oxidation of Ti leads to higher current densities as compared to the MIM diode formed by native oxide. Parabolic conductivity plot confirmed the tunneling phenomenon. This parabola is used

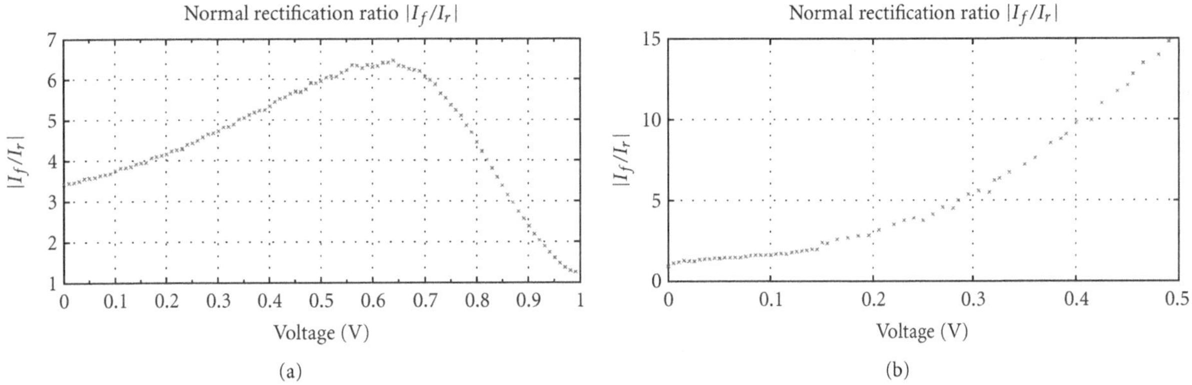


FIGURE 7: Rectification ratios $|I_f/I_r|$ plotted versus absolute value of voltages for (a) native oxide and (b) plasma oxide.

TABLE 1: Compares device characteristics of reported MIM diodes in the literature and this paper.

| Work | Type of mim diode | Thickness of dielectric (method of deposition) | Sensitivity ($V^{-1}$) | $I_f/I_r$ |
|---|---|---|---|---|
| | *Metal-insulator-metal* | | | |
| Hoofring et al. [7] | Ni-NiO-Au (0.64 $\mu m^2$) | 2.2 nm (plasma) | 4.55 | 1.1 |
| Krishnan et al. [8] | Ni-NiO-Cr (1 $\mu m^2$) | 3 nm (reactive sputter) | 5 | ~1.1 |
| Krishnan et al. [9] | Ni-NiO-Cr/Au (100 $\mu m^2$-1 $\mu m^2$) | 3 nm (plasma) | NA | 4.5–6 |
| Krishnan et al. [10] | Ni-NiO-Cr/Au (1 $\mu m^2$ area) | 3 nm (plasma) | 7 | |
| Krishnan et al. [10] | Ni-NiO-Cr/Au (100 $\mu m^2$ area) | 3 nm (plasma) | 15 | NA |
| This work | Ti-$TiO_2$-Al (21287 $\mu m^2$ area) | 9 nm (native) | 18 | 6.5 |

to estimate the barrier thickness, which is around 9 nm for both diodes. Plasma oxidization can be optimized to get ultra thin oxide and hence high current density can be obtained. Ti and Pt, due to their high work function difference prove to be a desirable choice for MIM tunnel diode. In spite of the same thickness, the higher current density for the plasma oxide can be explained by the higher work function of Pt compared to Al.

## Acknowledgment

Financial support of National Center for Photovoltaic Research and Education at IIT-Bombay, funded by MNRE, Govt of India is gratefully acknowledged.

## References

[1] D. K. Kotter, S. D. Novack, W. D. Slafer, and P. J. Pinhero, "Theory and manufacturing processes of solar nanoantenna electromagnetic collectors," *Journal of Solar Energy Engineering*, vol. 132, no. 1, pp. 0110141–0110149, 2010.

[2] M. Arrawatia, M. S. Baghini, and G. Kumar, "RF energy harvesting system from cell towers in 900 MHz band," in *Proceedings of the National Conference on Communications (NCC '11)*, Bangalore, India, January 2011.

[3] M. Arrawatia, M. S. Baghini, and G. Kumar, "RF energy harvesting system at 2.67 and 5.8 GHz," in *Proceedings of the Asia-Pacific Microwave Conference Proceedings (APMC '10)*, pp. 900–903, Yokohama, Japan, 2010.

[4] M. Arrawatia, V. Diddi, H. Kochar, M. S. Baghini, and G. Kumar, "An integrated CMOS RF energy harvester with differential microstrip antenna and on-chip charger," in *Proceedings of the IEEE International Conference on VLSI Design*, Sister Conferences of DAC, Hyderabad, India, 2012.

[5] W. C. Brown, "The history of power transmission by radio waves," *IEEE Transactions on Microwave Theory and Techniques*, vol. 32, no. 9, pp. 1230–1242, 1984.

[6] R. L. Bailey, "A proposed new concept for a solar energy convertor," *Journal of Engineering for Power*, vol. 94, no. 2, pp. 73–77, 1972.

[7] A. B. Hoofring, V. J. Kapoor, and W. Krawczonek, "Submicron nickel-oxide-gold tunnel diode detectors for rectennas," *Journal of Applied Physics*, vol. 66, no. 1, pp. 430–437, 1989.

[8] S. Krishnan, H. La Rosa, E. Stefanakos, S. Bhansali, and K. Buckle, "Design and development of batch fabricatable metal-insulator-metal diode and microstrip slot antenna as rectenna elements," *Sensors and Actuators A*, vol. 142, no. 1, pp. 40–47, 2008.

[9] S. Krishnan, E. Stefanakos, and S. Bhansali, "Effects of dielectric thickness and contact area on current-voltage characteristics of thin film metal-insulator-metal diodes," *Thin Solid Films*, vol. 516, no. 8, pp. 2244–2250, 2008.

[10] S. Krishnan, S. Bhansali, E. Stefanakos, and Y. Goswami, "Thin film metal-insulator-metal junction for millimeter wave detection," *Procedia Chemistry*, vol. 1, no. 1, pp. 409–412, 2009.

[11] S. Krishnan, *Thin film metal-insulator-metal junction for millimeter wave detection*, Ph.D. thesis, University of South Florida, 2008.

[12] N. F. Mott, "A theory of the formation of protective oxide films on metals," *Transactions of the Faraday Society*, vol. 35, pp. 1175–1177, 1939.
[13] N. F. Mott, "The theory of the formation of protective oxide films on metals, II," *Transactions of the Faraday Society*, vol. 35, pp. 472–483, 1940.
[14] N. F. Mott, "The theory of the formation of protective oxide films on metals," *Transactions of the Faraday Society*, vol. 43, pp. 429–434, 1947.
[15] N. Cabrera and N. F. Mott, "Theory of the oxidation of metals," *Reports on Progress in Physics*, vol. 12, p. 163, 1948.
[16] C. Adrianus, *Plasma oxidation of magnetic tunnel junctions*, Ph.D. thesis, Eindhoven University of Technology, 2005.
[17] W. F. Brinkman, R. C. Dynes, and J. M. Rowell, "Tunneling conductance of asymmetrical barriers," *Journal of Applied Physics*, vol. 41, no. 5, pp. 1915–1921, 1970.

# A Smart Infrared Microcontroller-Based Blind Guidance System

**Amjed S. Al-Fahoum, Heba B. Al-Hmoud, and Ausaila A. Al-Fraihat**

*Biomedical Systems and Informatics Engineering Department, Hijjawi Faculty for Engineering Technology, Yarmouk University, Irbid 21163, Jordan*

Correspondence should be addressed to Amjed S. Al-Fahoum; afahoum@yu.edu.jo

Academic Editor: Liang Han

Blindness is a state of lacking the visual perception due to physiological or neurological factors. The partial blindness represents the lack of integration in the growth of the optic nerve or visual centre of the eye, and total blindness is the full absence of the visual light perception. In this work, a simple, cheap, friendly user, smart blind guidance system is designed and implemented to improve the mobility of both blind and visually impaired people in a specific area. The proposed work includes a wearable equipment consists of head hat and mini hand stick to help the blind person to navigate alone safely and to avoid any obstacles that may be encountered, whether fixed or mobile, to prevent any possible accident. The main component of this system is the infrared sensor which is used to scan a predetermined area around blind by emitting-reflecting waves. The reflected signals received from the barrier objects are used as inputs to PIC microcontroller. The microcontroller is then used to determine the direction and distance of the objects around the blind. It also controls the peripheral components that alert the user about obstacle's shape, material, and direction. The implemented system is cheap, fast, and easy to use and an innovative affordable solution to blind and visually impaired people in third world countries.

## 1. Introduction

Many people suffer from serious visual impairments preventing them from travelling independently. Accordingly, they need to use a wide range of tools and techniques to help them in their mobility. One of these techniques is orientation and mobility specialist who helps the visually impaired and blind people and trains them to move on their own independently and safely depending on their other remaining senses. Another method is the guide dogs which are trained specially to help the blind people on their movement by navigating around the obstacles to alert the person to change his/her way. However, this method has some limitations such as difficulty to understand the complex direction by these dogs, and they are only suitable for about five years. The cost of these trained dogs is very expensive, also it is difficult for many of blind and visually impaired persons to provide the necessary care for another living being. There is an international symbol tool of blind and visually impaired people just like the white cane with a red tip which is used to enhance the blind movement. Nowadays, different types of these canes have been used such as the white cane [1], the smart cane [2], and the laser cane [3]. However, this tool has several constraints: long length of the cane, limitations in recognizing obstacles, and also difficulty to keep it in public places. Recently, many techniques have been developed to enhance the mobility of blind people that rely on signal processing and sensor technology. These called electronic travel aid (ETA) devices help the blind to move freely in an environment regardless of its dynamic changes. According to the literature, ETAs are mainly classified into two major aspects: sonar input (laser signal, infrared signals, or ultrasonic signals) [4–8] and camera input systems (consists mainly of a mini CCD camera) [9–13]. Their details are shown in Table 1. The way these devices operate just like the radar system that uses ultrasonic fascicle or laser to identify height, the direction, and speed of fixed and moving objects. The distance between the person and the obstacles is measured by the time of the wave travel. However, all existing systems inform the blind of the presence of an object at a specific distance in front of or near to him. These details permit the user to change his or her way. Information about the object characteristics can create additional knowledge

Table 1: Existing wearable blind guidance systems.

| System type | System name | Reference | Year | Country | Signal input | Signal output | Advantages | Disadvantages |
|---|---|---|---|---|---|---|---|---|
| Sonar systems | NavBelt | [5] | 1998 | USA | Ultrasonic | Auditory display | Can provide range of information of the obstacle | Cannot provide precise information such as shape and motion state of obstacles, and also cannot make judgments to the travel orientation |
| | People sensor | [6] | 1998 | USA | Ultrasonic | Vibrotactile cues | | |
| | 3D space perceptor | [7] | 2003 | Canada | Laser | Auditory display | | |
| | Navigation system based on optical peacans | [8] | 2001 | Japan | Infrared signal | Auditory display | | |
| Camera based systems | vOICe | [9] | 1992 | Netherlands | CCD camera | Auditory display | It provides information not only about obstacles along the traveled path<br>Assists the user in selecting the preferred travel path<br>It can capture whole environment information<br>It can use advanced image processing algorithm to train complex data | larger size, higher cost relatively, and it need two cameras so as distance detection algorithms is more complex than sonar-based ETAs |
| | Tyflos | [10] | 2001 | USA | Camera and laser scanner | Auditory display | | |
| | NAVI | [11] | 2007 | Malaysia | CCD camera | Auditory display | | |
| | AudioMan | [12] | 2007 | China | CCD camera | Auditory display | | |
| | SoundView | [13] | 2009 | China | CCD camera | Auditory display | | |

to enhance space manifestation and memory of the blind. To overcome the above-mentioned limitations, this work offers a simple, efficient, configurable electronic guidance system for the blind and visually impaired persons to help them in their mobility regardless of where they are, outdoor or indoor. The originality of the proposed system is that it utilizes an embedded vision system of three simple IR sensors and brings together all reflective signals in order to codify an obstacle through PIC microcontroller. Hence, in addition to distance the proposed guidance system enables the determination of two main characteristics of the obstacle which are material and shape. Furthermore, the user of the system does not need to carry a cane or other marked tool. He/she can just wear a hat and hand mini stick (size of a pen) just like others. It has high immunity to ambient light and colour of object. It has typical response time about 39 ms, and it is very suitable for real-time applications.

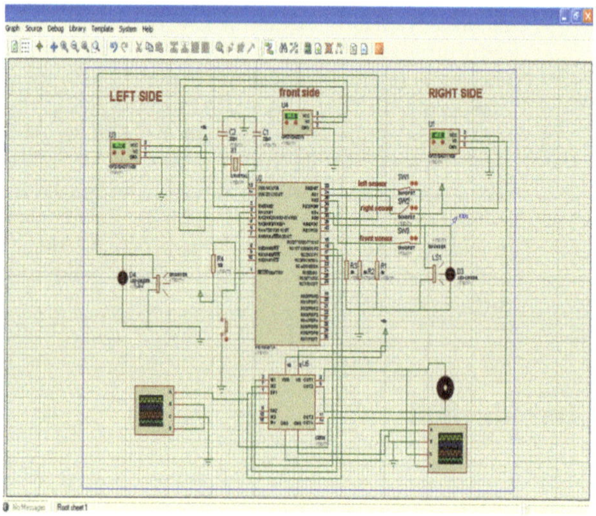


Figure 1: The whole circuit design using Proteus simulation.

## 2. Materials and Methods

*2.1. Block Diagram of the Proposed Smart Microcontroller-Based Blind Guidance System.* Proteus is software for microprocessor simulation, schematic capture, and printed circuit board (PCB) design. Proteus-VSM (Virtual System Modelling) permits the cosimulation of the embedded software for popular microcontrollers alongside with the hardware design [14]. The software Proteus was used to design and analyze both preliminary and final results for each part of the system [14]. Proteus enabled the system to be tested and modified until accurate and expected results were obtained prior to the circuit installation in practice. Figure 1 shows the design and the implementation of the proposed smart blind guidance system. The block diagram of the system is illustrated in Figure 2. It shows three reflective signals that were produced as follow: from front IR sensor, right IR sensor, and from left IR sensor. All signals are inputs for ADC on a PIC microcontroller. These signals are digitized and used as inputs to a specific program implemented in real time within PIC microcontroller. As a result of these signals the microcontroller according to some internal instructions will produce an output which will be transferred from the PIC to the buzzer and motor driver IC to aware the blind about the barriers blocking his/her way.

*2.1.1. PIC Microcontroller 16F877A.* To make wearable obstacle detection system for visually impaired people respond

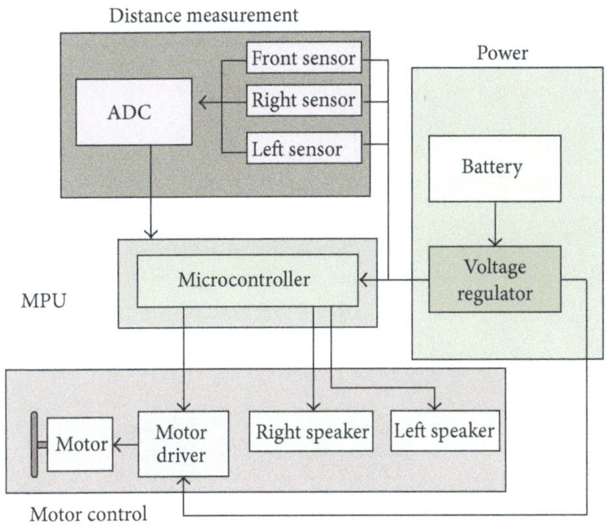


FIGURE 2: Block diagram of the system.

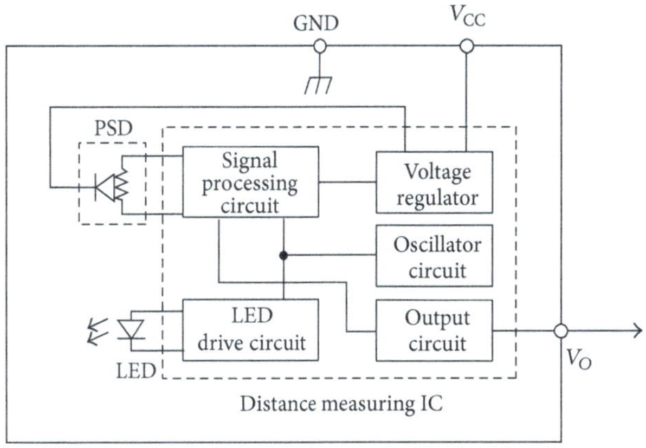


FIGURE 3: The block diagram of IR Sensor.

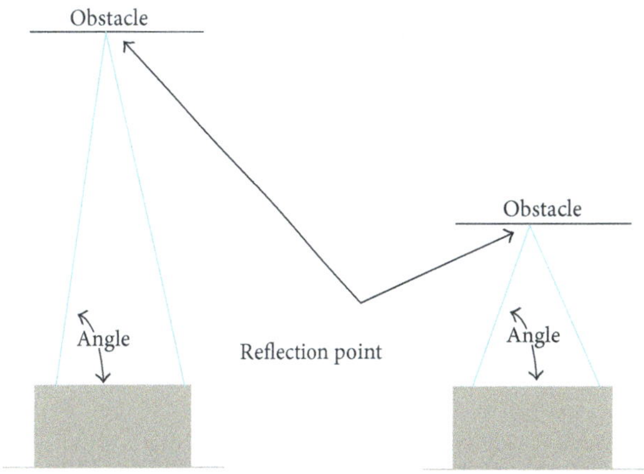


FIGURE 4: Different angles with different distances.

faster, it should be equipped with advanced microcontroller to decrease computational complexity. PIC 16F877A was chosen to detect any switch triggered and generate the audio sounds and vibrations. The PIC does not have an operating system and simply runs the program in its memory when it is turned on. PIC microcontroller is a small computer on a single integrated circuit which stores a set of instructions. It consists of a processor core, memory, and programmable input/output peripherals. PIC is an important component in the proposed system which deals with a MicroC programming code which was installed in it. The system is featured by its small size and low cost when it is compared with other systems that use separate microprocessor, input/output devices, and memory. Mixed signal microcontrollers are common, integrating analog components needed to control nondigital electronic systems. PIC microcontroller operates at +5 V which can be regulated using the voltage regulator (L7805) which conserves voltage at +5 V if the input voltage for it exceeds +5 V. Also PIC cannot run without using its crystal oscillator which is used to execute the programming code. The PIC is used as a real-time processing element; therefore, a high frequency oscillator is used [15].

*2.1.2. IR Sensors.* Three IR sensors are employed to acquire particular details relative to the obstacle categorization. In this paper, the suggested guidance system principle is shown by introducing the main sensors and their ability to detect and identify typical objects. The IR sensors are the main electronic components in the proposed system because it acts as the new eyes for the blind. One IR sensor is located on the hand mini stick to scan the front side, and the other two sensors are located on the hat at both right and left sides. IR sensors will scan all area in there range of IR beam. Any obstacle lies in the scanning range of the IR beam will be reflected and picked back by the receiver unit in the sensor. The distance and angle detection depend on the body that caused beam reflection. Figure 3 shows the block diagram of IR sensor. The range of distance to be scanned is controlled by a switch. Three switches are used to control the range of sensors in the three directions (i.e., when switches are on, the sensors will work at full range; otherwise they will work at their half range). The IR sensors are used to detect an obstacle besides (right and left) and in front of the blind at a distance between 10 and 150 cm.

The basic tools in electronic IR guide are mini hand stick and head hat. These tools have IR sensors on it. These sensors represent the blind eyes so its position is very important to give right decision that depends on their output. Mini hand stick contains IR sensor with a distance range from 10 to 150 cm. This range can be controlled by the PIC microcontroller, so it can be operated at half range (75 cm) via open switch button. What makes the sensor favourable is its small weight which is about 65 gm, also it offers proper protection for the sensor via thin transparent layer of plastic. The head hat is the second tool in the proposed blind electronic guidance system. The hat provides the system the ability to scan areas in the right and left side of the blind via two sensors fixed on it. The two sensors can work either on their full range 150 cm or their half range (75 cm) according to what is preferable by the user who can control it via switches.

*Theory of IR Sensor Operation.* The new IR sensors use a developed approach that not only gives object detection at a longer range, but also offers a range of information. These new domains provide a much better immunity to ambient lighting conditions because of the new method of ranging. These new domains use a small linear CCD array and triangulation to calculate the distance and presence of objects in the field of vision. A pulse of infrared light is released by the emitter. This light travels across the specific area, if there is no obstacles in it, then the light never reflects, and the result shows no object, but if there is an obstacle in their way, then the light hits it, and it will be reflected. This action creates a triangle between the reflection point, the emitter, and the detector as shown in Figure 4. Triangle angles vary based on the distance of the object. The receiving part of these new detectors is in fact a precision lens that transmits the reflected light onto different parts of the enclosed linear CCD array according to the angle of the triangle described previously. The CCD array can determine at what angle the reflected light came back at, and therefore, it can calculate the distance to the object. This new method of ranging almost avoids the interference from ambient light and offers amazing in difference to the colour of object being detected. The infrared light is sent out from a transmitter to the object in front, by passing through a condense lens, so that the light intensity is focused on a certain point. Refraction occurs once the light hits the surface of the object. Part of the refracted light will be sent back to the receiver end, in which another lens will combine these lights and determine the point of impact. The light will then be passed on to an array of phototransistors. The position in which the light falls can be used to calculate the distance ($L$) from the transmitter to the obstacle using the following formula:

$$\frac{L}{A} = \frac{F}{X} \quad \text{Therefore } L = \frac{(F \times A)}{X}. \tag{1}$$

Thus, the distance value from the phototransistors will be sent to the Signal Evaluation Module before it is changed to voltage, resulting in a change of voltage according to the measured distance as shown in Figure 5.

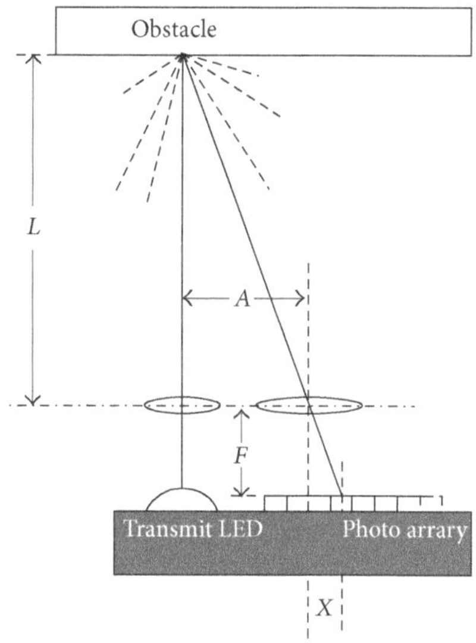

FIGURE 5: Distance measurement.

*2.1.3. Reading IR Sensor Output with A/D Converter.* The IR sensor's output voltages will change according to the detection distance. To calculate the approximate distance, the following formula is used:

$$R = \left(\frac{2914}{[V + 5]}\right) - 1, \tag{2}$$

where $R$ is distance in centimetre unit and $V$ is digital data from $A/D$ converter.

*2.1.4. Speakers.* Speakers are main indicators of the designed system through which the blind can easily determine the shape and material of surrounding bodies around him/her. There are two speakers: left and right. If the tune sound is due to the left speaker then there is an object in the range of left IR sensor, and when tune is due to the right, it means that an object in right side of blind exits. When both speakers give the tune, it means that the obstacle is in front of blind. Another advantage for speakers is their small weight and size, so they do not constitute a burden to blind, and he/she can easily control the sound level via variable resistor to choose the sound level that does not make him/her annoyed.

*2.1.5. Vibration Motor Driver.* L293D is quad push-pull drivers designed to provide bidirectional drive current up to 1 A or 600 mA per channel. All channels are TTL-compatible logic inputs, and each output is a complete totem-pole drive circuit with Darlington transistor sink and pseudo-Darlington source. The main function of L293D in this system is to control the current that is delivered to vibration motor using an enable pin that is connected directly to PIC. Vibration motor indicates how much the detected body in range is closer to blind in both left and right sides, and when the vibrator works with the left or the right speaker, then it indicates that there is a body in front and left or right according to speaker that was working. Figure 6 shows the signal that controls the motor and the incremental change in pulses width that result because of the decreased distance from the barrier.

*2.1.6. Switches.* The function of switches in this system is to control the range of the sensor. When switches are open (off), sensors operate at their half range, and when it is closed (ON), sensors operate at their full range.

## 3. Results and Discussion

The presented system is designed and configured for practical use. The system is able to handle seven states that may face the blind people. The system will respond to each state according to a specific program which is coded and installed in the PIC microcontroller. The first case is when the body is in the right direction of the blind, then just the right speaker will give sound. The second case is when the body is at the left side

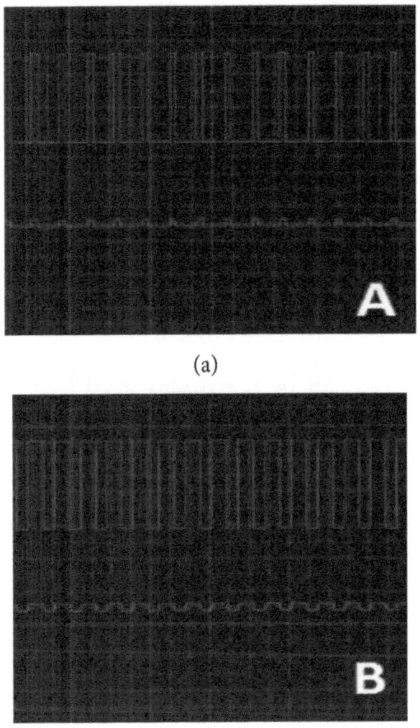

(a)

(b)

FIGURE 6: (a) Narrow ON time pulses (little vibration far distance), (b) wide ON time pulses (strong vibration close distance).

TABLE 2: Mean output voltage of the selected material.

| Material type | Mean output voltage |
|---|---|
| Wood | $0.928 \pm 0.025$ |
| Steel | $0.621 \pm 0.06$ |
| Plastic | $0.27 \pm 0.13$ |
| Glass | $0.75 \pm 0.03$ |

TABLE 3: Results of shape recognition tests.

| Shape of object | Number of trials | Detection ratio |
|---|---|---|
| Triangle pyramid | 10 | 93 |
| Cone | 10 | 93 |
| Cylinder | 10 | 96 |
| Rectangular pyramid | 10 | 94 |

of the blind, then just the left speaker will give sound. The third case is when the body is in front of the blind, then both the left and right speakers will give sound. The forth case is when the body is located in the range of right and front sensors, that is, from front and right, then right speaker and vibration motor will work together. The fifth case is when the body is located in front and left sides of the blind, then left speaker and vibration motor will work. The sixth case is when the bodies are approaching the blind from both right and left directions, then just vibration motor will work. The last case is when the bodies come in all directions (front, right, and left), then both speakers and vibration motor will work (worst case scenario).

*3.1. Trial Results.* To distinguish an obstacle from another, it is crucial to acquire the key features of the objects, such as material and form. The obstacle categorization can sufficiently be differentiated by its characteristics, so that it can be detected. According to the main aim of this work and to make use of simple sensors, the infrared sensors have been examined to identify the obstacle materials and surfaces. All 10 male blind subjects, aged 18–50, participated in the test of the system. Every participant was taught to understand one path (150 m long). The paths were selected to offer multiple obstacles of different materials of interest.

*3.1.1. Materials Recognition.* One of the main characteristics for obstacle different types to be detected is the differentiation by type of material as a first strategy. For four materials steel, glass, wood, and plastic, the emitted signal by the IR sensor was analyzed at a given distances. The microcontroller gathers the reflected information from the obstacles by measuring the width of the echo pulse signal. The pulse width is directly pro portional to the distance of the nearest obstacle. By capturing the value of an internal counter 16 bits at rise and fall time of the reflected pulse, the microcontroller gets a 16 bit value of the distance. Obstacles were located successively from 10 to 150 cm (in the range of 10 cm). The voltage mean values according to the four selected materials are summarized in Table 2.

All the obtained data of the conducted trials were analyzed using SPSS. One way ANOVA followed by Turkey was applied with significance level of 0.05. An important impact on the substance was noticed around the voltage ($F(1.25) = 55.7$; $P < 0,05$). This result indicates that, using the infrared sensor, it can help to make a distinction between the four chosen material types.

However, the output signal of an infrared sensor depends on the material and the distance of the object. The longer the distance of the object, the smaller the output signal. To test the effect on these two factors, the voltage mean value according to the distance and material categories is illustrated in Figure 7. A significant effect of the voltage on the material according to the distance ($F(112) = 77.8$; $P < 0,05$) was noticed. From the figure, it can be clearly seen that wood can be significantly differentiated from other materials at any distance. Regarding the other three materials, plastic, metal, and glass, material differentiation depends on the distance. At a distance less than 57 cm plastic can easily be detected. However, the system can differentiate between steel and glass at distances more than 57 cm from the head of the hand mini stick. Moreover, the system failed to make any discrimination of materials at distances more than 140 cm and less than 20 cm.

*3.1.2. Shape Analysis.* In order to reduce material ambiguity according to the object distance (metal versus glass or plastic versus metal), a second IR sensor can be used to extract other object parameters, such as their shapes. The object form was extracted through voltage variation, while the object is scanned with the IR sensor in the hand mini stick. Table 3 shows that the system was successful in detection of 4 types

TABLE 4: Comparison of different devices.

| System | Power consumption | Range | Portability | Invasiveness | Response time | Ease of usage |
|---|---|---|---|---|---|---|
| The current suggested system | Low | Medium | Yes | Noninvasive | Fast | Friendly (no need for training) |
| NavBelt | Medium | Low | No | Noninvasive | Medium | Not friendly (needs training) |
| The white cane | High | Medium | No | Invasive | Medium | Not friendly (needs training) |
| NAVI | High | High | No | Noninvasive | Slow | Not friendly (needs training) |

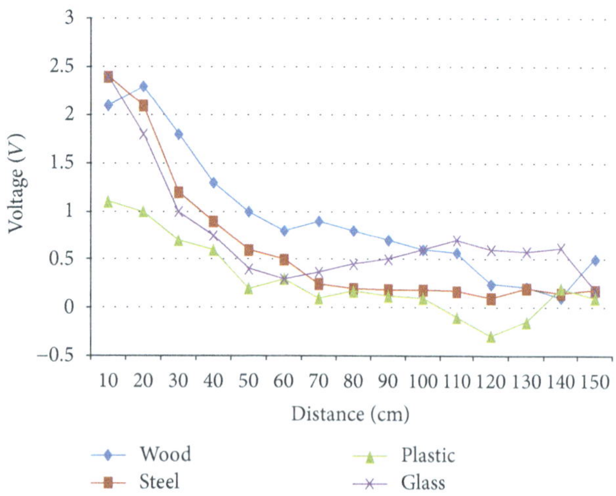


FIGURE 7: Voltage output at the object distance of different material categories.

of 3D different shapes (triangle pyramid, cone, cylinder, and rectangular pyramid). Therefore, the proposed system can be utilized for industrial applications.

*3.1.3. System's Overall Accuracy.* The system was tested on 10 blind subjects using obstacles of different shapes and materials at different distances. The results of the experiment are very encouraging. It revealed an accuracy of 93% which indicates that the system is efficient and unique in its capability in specifying the source and distance of the objects that may encounter the blind.

*3.2. Parameters Specifications.* To evaluate the performance of the proposed guidance system, several parameters should be considered. The first important parameter is the power consumption of the system and how long it will stay working without the need to recharge. The following ratings are considered: consumption of an electrical power of 0–0.5 W is regarded as low power, 0.5–1 W as medium consumption, and higher than 1 W as high consumption. The second parameter is the range of operation. A tool which can find obstacles throughout 0–2 m can be considered as a low range device, 2–4 m as medium range, while higher than 4 m is considered as high range. The third parameter is the response time, and a system sensing and responding 0–100 ms is regarded as fast, 100–200 ms medium and higher than 200 ms as slow. The portability is another important parameter of the system. The system which can be worn and used by the subject for prolonged time is considered as a portable system; otherwise it is regarded as nonportable. The easiness of the system usage is considered as another parameter. An easy to use device is actually easy to get to and an easy to function. Finally the noninvasive utilization of the system is considered as a property of the system. Table 4 compares the aforementioned parametric specifications for different systems.

*3.3. Comparisons with Other Systems.* As mentioned in the introduction, any new technology should be compared with what is already available and up to date technology. The proposed system is compared with the NavBelt system which is worn by the user like a belt around the waistline and through a set of stereo earphones to provide acoustic signals which direct the user around obstacles. However, the NavBelt system suffers from certain limitations such as the difficulty of conveying information to the user to allow rapid walking, on the other hand the NavBelt must be used together with white cane [4, 16]. At the stage of current development, it is clear that the white cane is potential in terms of cost, weight, electrical power, and reliability to reveal obstacles on the floor. However, if the floor is a fully reflective surface, then the cane laser scanning device fails. In addition to the revealed difficulties of the white cane in detecting hidden obstacles, it exhibits difficulty to the user in storing it in a public place [16, 17]. The results show that the suggested system can detect more distant obstacles by using infrared and other sensing devices and can provide more precise range information of the obstacle than the aforementioned sonar systems. Moreover, the system overcomes the limitations of the camera-based system due to its noninvasive nature, low power consumption, cheap cost, simplicity, and ease of customization. However, even at this early stage of the system's implementation, there are many advantages in using the newly proposed system: first of all it is safe, inexpensive and can be worn everywhere; when there are some obstacles, the system alerts the user and allows him/her to make better understanding of the space around him/her due to the detection of the material and shape of the object. Hence, the blind can change his/her way and move more safely and easily. Although the proposed system was able to work successfully, there is some future works which will be focused on replacing the speaker's tune by real human sound to guide the blind exactly. Moreover, shape detection test for objects that move at different rotational speeds across several distances will further be considered.

## 4. Conclusion

A simple, cheap, configurable, easy to handle electronic guidance system is proposed to provide constructive assistant and support for blind and visually impaired persons. The system is designed, implemented, tested, and verified. The real-time results of the system are encouraging; it revealed an accuracy of 93% in detecting different shapes, materials, and distances. The results indicate that the system is efficient and unique in its capability in specifying the source and distance of the objects that may encounter the blind. It is able to scan areas left, right, and in front of the blind person regardless of its height or depth. Therefore, it was favoured by those who participated in the test. The IR sensor has been fully utilized in order to advance the mobility of the blind and visual impaired people in safe and independent way. This system does not require a huge device to be hold for a long distance, and it also does not require any special training. This system also resolves limitations that are related to the most of the movement problems that may influence the blind people in their environment. Future work will be focused on enhancing the performance of the system and reducing the load on the user by replacing the speaker's tune by real human sound to guide the blind exactly. Moreover, shape detection test for objects that move at different rotational speeds across several distances will further be considered.

## References

[1] D. Yuan and R. Manduchi, "Dynamic environment exploration using a Virtual White Cane," in *Proceedings of the IEEE Computer Society Conference on Computer Vision and Pattern Recognition (CVPR '05)*, pp. 243–249, IEEE, San Diego, Calif, USA, June 2005.

[2] A. A. Tahat, "A wireless ranging system for the blind long-cane utilizing a smart-phone," in *Proceedings of the 10th International Conference on Telecommunications (ConTEL '09)*, pp. 111–117, IEEE, Zagreb, Croatia, June 2009.

[3] D. Bolgiano and E. Meeks Jr., "A laser cane for the blind," *IEEE Journal of Quantum Electronics*, vol. 3, no. 6, p. 268, 1967.

[4] S. Shoval, I. Ulrich, and J. Borenstein, "NavBelt and the guide-cane [obstacle-avoidance systems for the blind and visually impaired]," *IEEE Robotics and Automation Magazine*, vol. 10, no. 1, pp. 9–20, 2003.

[5] S. Shoval, J. Borenstein, and Y. Koren, "Auditory guidance with the navbelt-a computerized travel aid for the blind," *IEEE Transactions on Systems, Man and Cybernetics C*, vol. 28, no. 3, pp. 459–467, 1998.

[6] S. Ram and J. Sharf, "The people sensor: a mobility aid for the visually impaired," in *Proceedings of the 1998 Digest of Papers 2nd International Symposium on Wearable Computers*, pp. 166–167, IEEE, October 1998.

[7] E. Milios, B. Kapralos, A. Kopinska, and S. Stergiopoulos, "Sonification of range information for 3-D space perception," *IEEE Transactions on Neural Systems and Rehabilitation Engineering*, vol. 11, no. 4, pp. 416–421, 2003.

[8] K. Magatani, K. Sawa, and K. Yanashima, "Development of the navigation system for the visually impaired by using optical beacons," in *Proceedings of the 23rd Annual International Conference of the IEEE Engineering in Medicine and Biology Society*, pp. 1488–1490, IEEE, October 2001.

[9] P. B. L. Meijer, "An experimental system for auditory image representations," *IEEE Transactions on Biomedical Engineering*, vol. 39, no. 2, pp. 112–121, 1992.

[10] N. G. Bourbakis and D. Kavraki, "An intelligent assistant for navigation of visually impaired people," in *Proceedings of the 2001 IEEE 2nd International Symposium on Bioinformatics and Bioengineering Conference*, pp. 230–235, IEEE, 2001.

[11] G. Sainarayanan, R. Nagarajan, and S. Yaacob, "Fuzzy image processing scheme for autonomous navigation of human blind," *Applied Soft Computing Journal*, vol. 7, no. 1, pp. 257–264, 2007.

[12] Z.-G. Fang, J. Xu, F.-l. Bao, and L.-H. Zhang, "AudioMan: design and implementation of environmental information data mapping," *Chinese Journal of Ergonomics*, vol. 2, article 001, 2007.

[13] M. Nie, J. Ren, Z. Li et al., "SoundView: an auditory guidance system based on environment understanding for the visually impaired people," in *Proceedings of the 31st Annual International Conference of the IEEE Engineering in Medicine and Biology Society: Engineering the Future of Biomedicine (EMBC '09)*, pp. 7240–7243, IEEE, September 2009.

[14] B. Su and L. Wang, "Application of Proteus Virtual System Modelling (VSM) in teaching of microcontroller," in *Proceedings of the International Conference on E-Health Networking, Digital Ecosystems and Technologies (EDT '10)*, pp. 375–378, IEEE, April 2010.

[15] J. Burroughs, "X-10 home automation using the PIC16F877A," *Lamp*, vol. 10, article 10, 2010.

[16] D. Dakopoulos and N. G. Bourbakis, "Wearable obstacle avoidance electronic travel aids for blind: a survey," *IEEE Transactions on Systems, Man and Cybernetics C*, vol. 40, no. 1, pp. 25–35, 2010.

[17] S. Chumkamon, P. Tuvaphanthaphiphat, and P. Keeratiwintakorn, "A blind navigation system using RFID for indoor environments," in *Proceedings of the 5th International Conference on Electrical Engineering/Electronics, Computer, Telecommunications and Information Technology (ECTI-CON '08)*, pp. 765–768, IEEE, May 2008.

# 22

# Gate Stack Engineering and Thermal Treatment on Electrical and Interfacial Properties of Ti/Pt/$HfO_2$/InAs *p*MOS Capacitors

**Chung-Yen Chien, Jei-Wei Hsu, Pei-Chin Chiu, Jen-Inn Chyi, and Pei-Wen Li**

*Department of Electrical Engineering, National Central University, Chungli 32001, Taiwan*

Correspondence should be addressed to Pei-Wen Li, pwli@ee.ncu.edu.tw

Academic Editor: Yeong-Her Wang

Effects of gate stack engineering and thermal treatment on electrical and interfacial properties of Ti/Pt/$HfO_2$/InAs metal insulator semiconductor (MIS) capacitors were systematically evaluated in terms of transmission electron microscopy, energy dispersive X-ray spectroscopy, current-voltage, and capacitance-voltage characterizations. A 10 nm thick Pt metal effectively suppresses the formation of interfacial oxide, $TiO_2$, between the Ti gate and $HfO_2$ gate dielectric layer, enhancing the gate modulation on the surface potential of InAs. An *in situ* $HfO_2$ deposition onto the *n*-InAs channel with an interfacial layer (IL) of one-monolayer InP followed by a 300°C post-metal-anneal produces a high-quality $HfO_2$/InAs interface and thus unravels the annoying Fermi-level pinning, which is evidenced by the distinct capacitance dips in the high-/low-frequency *C-V* characteristics. The interface trap states could be further suppressed by replacing the InP IL by an As-rich InAs, which is substantiated by a gate leakage reduction and a steep voltage-dependent depletion capacitance.

## 1. Introduction

Motivation to study low band-gap InAs and InSb channels for next-generation metal oxide semiconductor (MOS) transistors is strong in light of their superior carrier mobility [1] and established epitaxy techniques [2, 3] among other emerging technologies such as carbon nanotube and graphite. However, the progress of realizing high-performance InAs and InSb MOS transistors has been impeded by the stringent restraint on thermal budget as well as the lack of robust gate dielectrics and suitable surface treatments for unraveling annoying Fermi-level pinning [4] at the oxide/semiconductor interface. Recently encouraging experimental demonstrations of high-*K* gate dielectrics ($Gd_2O_3$, $Al_2O_3$, and $HfO_2$) on GaAs [5–7], InP [8], and InAs [9–11] channels using atomic-layer deposition (ALD) techniques have shed lights and attracted tremendous attentions on this venerable subject. In spite that various chemical and plasma surface treatments [12–14] have been proposed for annihilating surface states of III-V compound semiconductors, it still remains elusive how to produce a clean and well-passivated channel surface for subsequent high-quality gate dielectric growth. The most practical and promising approach for the surface preparation of III-V channels appears to be the *in situ* growth of gate dielectrics on channels.

In this paper, the authors report the interfacial and electrical properties of Ti/Pt/$HfO_2$/InAs *p*MOS capacitors, in which a high-*K* gate dielectric, $HfO_2$ was grown on an *n*-InAs epitaxial layer without breaking vacuum by means of the integration of a molecular beam epitaxy system (Riber-32 MBE) with an atomic layer deposition system (Picoson R-100 ALD). In order to prevent thermal evaporation of InAs, all the device fabrication processes were kept below 300°C. Both Ti and Pt metals were evaluated for the gate materials. The interfacial layer effect arising from one-monolayer InP or As-rich InAs between $HfO_2$ and InAs channel, the deposition conditions of $HfO_2$, and post-metal-anneal (PMA) processes were systematically studied for improving the interfacial and electrical properties of $HfO_2$/InAs *p*MOS capacitors.

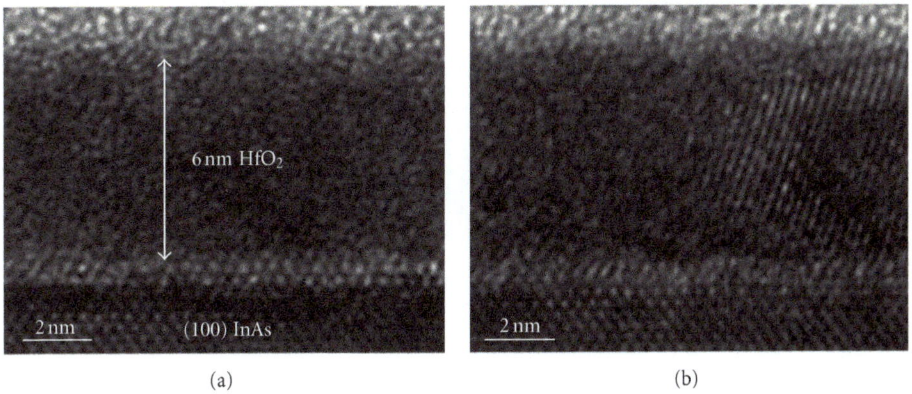


FIGURE 1: CTEM images of $HfO_2$ *in situ* deposited on InP/InAs channel at (a) 200°C and (b) 300°C. Abrupt and native oxide free interface provides a promising interface for MOS applications.

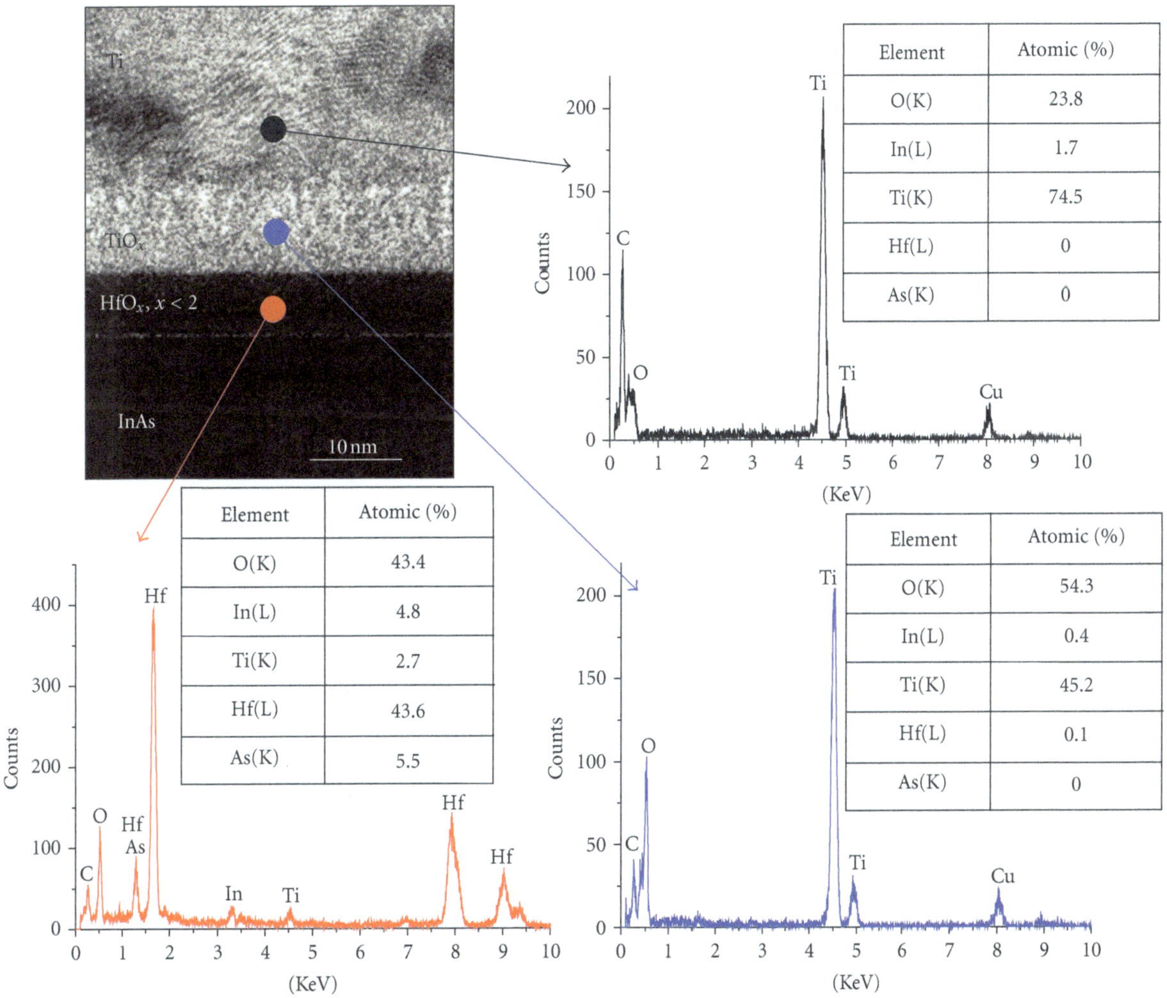


FIGURE 2: Energy dispersive X-ray spectroscopy (EDX) analysis obtained from various locations of Ti/InP/InAs MOS structure. EDX analysis reveals that an interfacial $TiO_x$ ($x \sim 1.2$) and Hf-rich oxide were formed after Ti deposition onto $HfO_2$/InP/InAs structure.

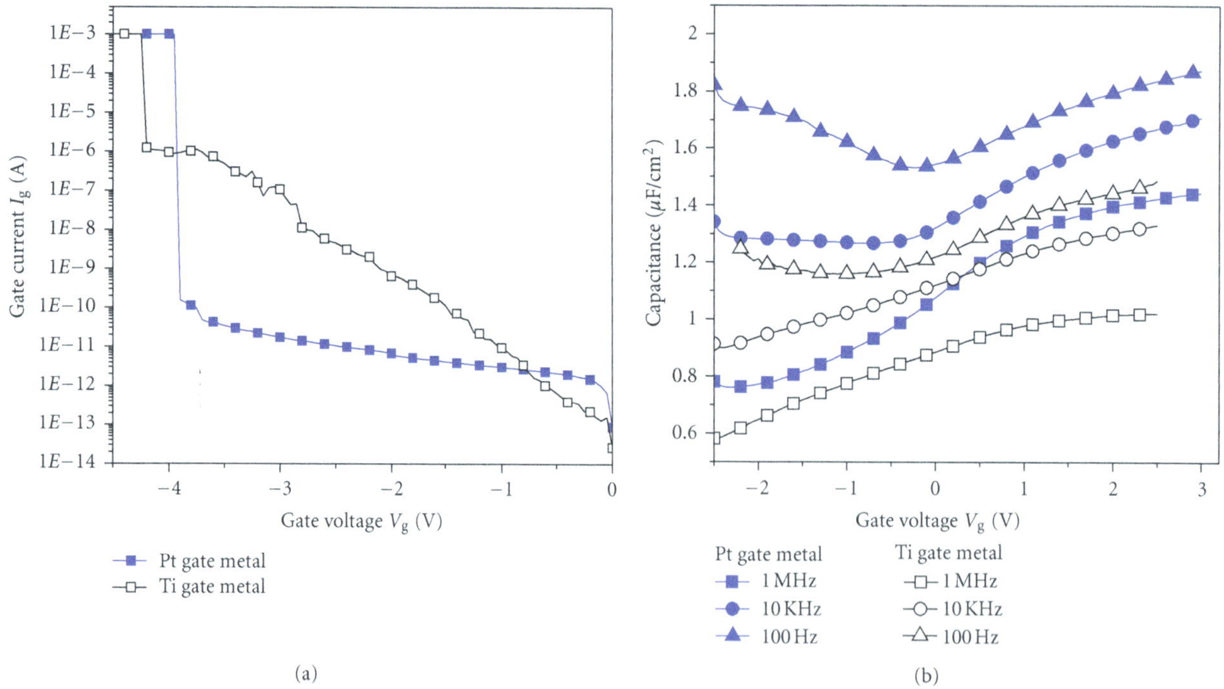


FIGURE 3: (a) $I$-$V$ and (b) $C$-$V$ characteristics of $HfO_2$/InP/InAs MOS capacitors of $100 \times 100\,\mu m^2$ with Ti and Pt gate metal, respectively.

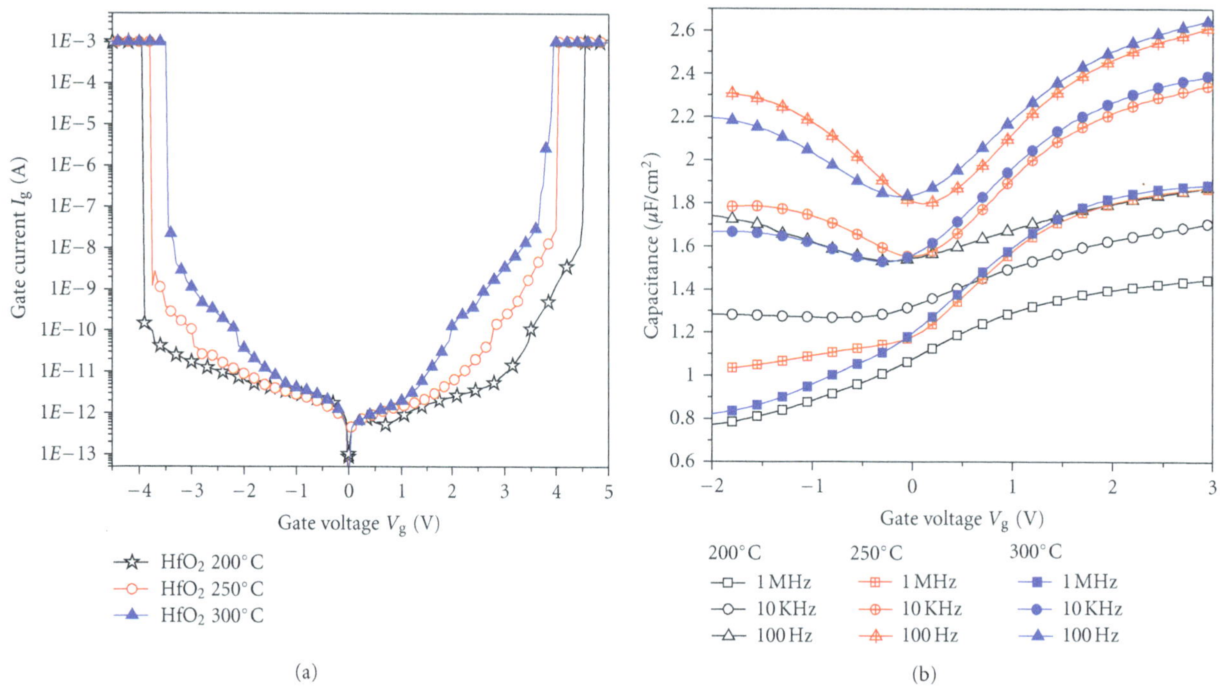


FIGURE 4: (a) $I$-$V$ and (b) frequency-dependent $C$-$V$ characteristics of Pt/$HfO_2$/InP/InAs MOS capacitors in which $HfO_2$ was deposited at 200, 250, and 300°C, respectively.

## 2. Experimental

The fabrication of $HfO_2$/InAs MOS capacitors started with a 2″ (100) $n$-InAs substrate doped with Si [(2–5) $\times$ $10^{17}\,cm^{-3}$], followed by an epitaxial growth of a 200 nm thick $n$-InAs channel layer uniformly doped with Si in a concentration of $2 \times 10^{17}\,cm^{-3}$ and an one-monolayer InP or As-rich InAs IL on the top of InAs channel. After transferring the sample from the MBE growth chamber to ALD system through a high-vacuum

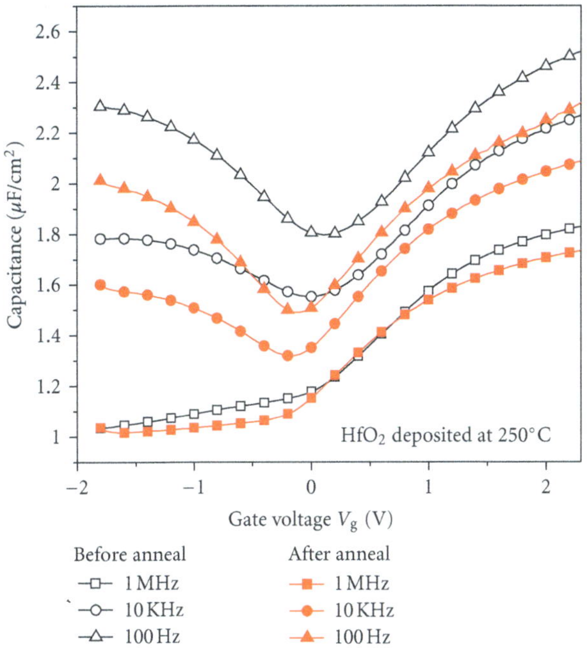


FIGURE 5: Frequency-dependent *C-V* characteristics of Pt/$HfO_2$/InP/InAs MOS capacitors in which $HfO_2$ was deposited at 250°C before and after a subsequent 300°C forming gas PMA.

chamber with a base pressure of $6 \times 10^{-8}$ Torr, a 6 nm thick $HfO_2$ layer was deposited at 200-300°C using water vapor and tetrakis(ethylmethylamino)hafnium [TEMAH, $Hf[N(C_2H_5)(CH_3)]_4$] precursors. A 300 nm thick Ti or a 50 nm thick Pt was evaporated as gate materials followed by the gate definition either using photolithography and $HF/H_2O_2$ chemical etching or through a shadow mask. Finally, the fabrication of MOS capacitors was completed by the evaporation of a 100 nm thick Al onto the substrate as the back-gate electrode and a 300°C PMA for 30 min. The interfacial and structural properties of the Ti/Pt/$HfO_2$/InAs system were examined using high-resolution transmission electron microscopy (HRTEM) and energy dispersive X-ray spectroscopy (EDX). The current-voltage (*I-V*) and capacitance-voltage (*C-V*) characteristics of the $HfO_2$/InAs MOS capacitors were characterized using Keithley-4200 and HP4284 semiconductor analyzers.

## 3. Results and Discussion

Figure 1 shows the cross-sectional TEM (CTEM) images of the gate stack of Ti/$HfO_2$/InP/InAs structure. The *in situ* growth of $HfO_2$ on an InP/InAs channel produces a native-oxide free and abrupt $HfO_2$/InP/InAs interface, indicating dangling bonds or surface states of InP/InAs are suppressed and providing a good interface for charge inversion under gate modulation. Additionally, there appears to be a 10–12 nm thick interfacial layer between a 6 nm thick $HfO_2$ layer and Ti-gate, and EDX analysis reveals that the chemical composition of the interfacial layer is $TiO_x$ ($x \sim 1.2$) and the underlay hafnium oxide becomes a Hf-rich oxide (Figure 2), suggesting Ti catalytically releases oxygen from the stoichiometric $HfO_2$, forming a Hf-rich oxide, and the subsequent reaction of Ti and released oxygen generates $TiO_x$. The undesired interfacial $TiO_x$ and Hf-rich oxide layers not only deteriorate the oxide integrity but also thicken the gate dielectrics thickness, which are evidenced by a high-gate leakage, a poor gate modulation on the surface potential of InAs channel but a slightly higher breakdown voltage as illustrated in the *I-V* and *C-V* characteristics (Figure 3).

An interesting finding that we have made through TEM observation is that a thin Pt barrier metal before Ti deposition effectively suppresses the formation of an interfacial oxide layer and retains the stoichiometry and high quality of $HfO_2$. The dielectric thickness of the Pt-gated MIS capacitor is 6 nm thick $HfO_2$, whereas the Ti-gated MIS capacitors have a gate oxide of 16–18 nm (10–12 nm $TiO_x$ and 6 nm Hf-rich oxide). Although the thinner gate dielectric in the Pt-gated MIS capacitors sustains a smaller breakdown voltage than Ti-gated devices do, the reduction of gate leakage is more than 4 orders in magnitude as compared with the Ti-gated MIS capacitors. As a consequence, there appear a significant increase in the accumulation capacitance for a Pt-gated MIS diode and an enhanced gate modulation from *C-V* characteristics in Figure 3(b).

The gate-oxide integrity appears to be strongly dependent on the $HfO_2$ deposition temperature and the thermal treatment after metal patterning. Increasing the deposition temperature from 200 to 250°C not only enhances the accumulation capacitance because of an increase in the dielectric constant of $HfO_2$ from 11 to 15, but also improves the inversion capacitance as a consequence of a better $HfO_2$/InP/InAs interface. This is evidenced by a steep *C-V* transition in the depletion regime and a distinct capacitance dip when gate voltage modulates the surface potential of the InAs channel from depletion to inversion regimes (Figure 4). However, a higher $HfO_2$ deposition temperature at 300°C degrades the gate leakage and breakdown *E*-field accordingly, possibly originating from the crystallinity of $HfO_2$ transitioning from amorphous to polycrystalline phases from 200 to 300°C (Figure 1(b)). An additional 300°C PMA in $H_2/N_2$ ambient for 30 min further appears to reduce the dangling-bonds and annihilates the surface states at the $HfO_2$/InP/InAs interface, but also improves the frequency-dependence of the accumulation capacitance as a result of the dielectric constant dispersion (Figure 5). The above mentioned experimental results also suggests that the InAs epitaxial layer is of high quality and does not degrade after a long duration (45 min) of 300°C ALD and PMA process.

The interfacial properties and electrical characteristics of InAs MOS capacitors could be further improved by replacing the unrelaxed InP IL with a latticed match As-rich InAs IL, in spite that the bandgap of InAs (0.345 eV) is smaller than InP (1.344 eV). This is evidenced by a reduction of the gate leakage by one order in magnitude as shown in Figure 6(a). This is also substantiated by a steeper slope in the depletion region of the *C-V* characteristics in Figure 6(b). Suppressed surface states by As-rich IL and a high-thermal generation rate in narrow band-gap InAs [15] make the *C-V* curves

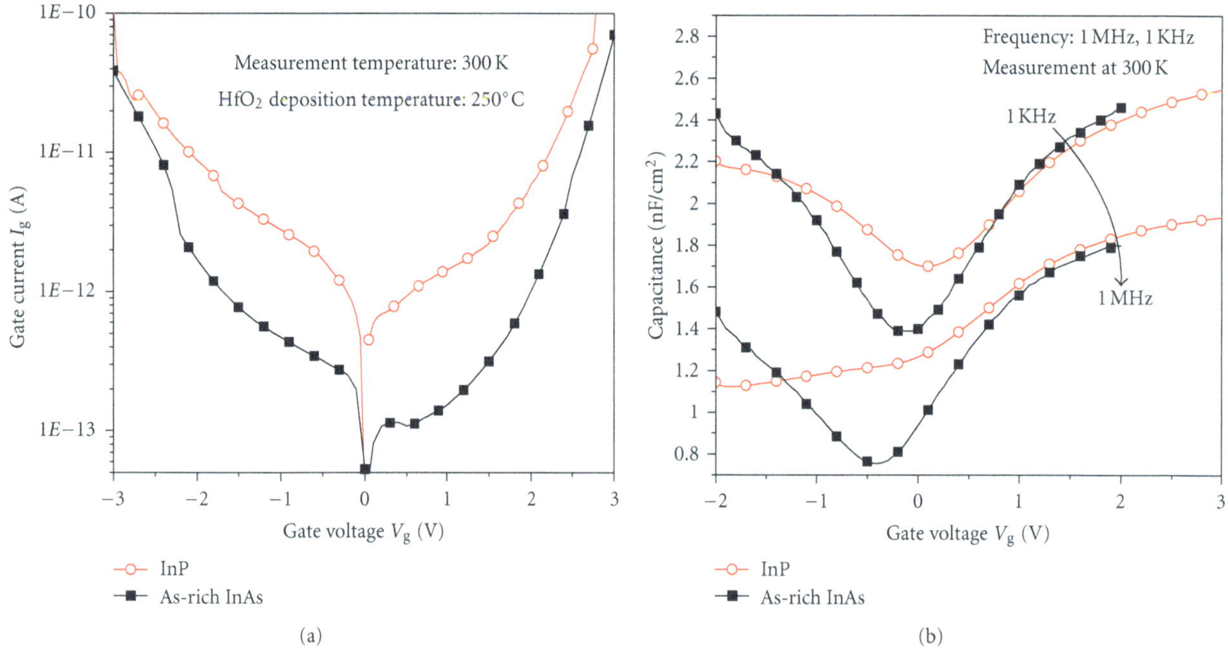


Figure 6: (a) $I$-$V$ and (b) $C$-$V$ characteristics of Pt/$HfO_2$/InAs MOS capacitors with respective InP or As-rich InAs interfacial layers.

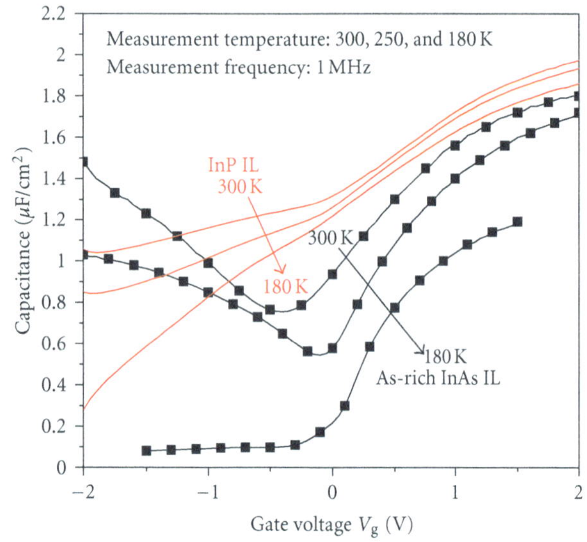


Figure 7: 1 MHz temperature-dependent $C$-$V$ characteristics of InAs MOS capacitor with respective InP or As-rich interfacial layers.

measured at 1 MHz deviate from the conventional high-frequency $C$-$V$ characteristics, in particular, at the inversion regime (Figure 7). For example, the inversion capacitances at 300 and 250 K are voltage dependent and getting close to the value of the gate oxide capacitance, instead of the effective capacitance of gate oxide capacitance in series with the depletion capacitance. Decreasing measurement temperature from 300 to 180 K decreases the thermal-generated carrier population and makes the inversion capacitance value restoring back to the one of typical high-frequency $C$-$V$ characteristics. The extracted interface trap density ($D_{it}$) using a conductance method from the Pt/$HfO_2$/InAs MOS capacitors with InP and As-rich interfacial layer is $1.59 \times 10^{13}$ and $7 \times 10^{12}$ $cm^{-2}$ $eV^{-1}$, respectively, which are comparable with reported data [9–11].

## 4. Conclusion

The *in situ* growth of $HfO_2$ on InAs channel produces an abrupt and oxide-free interface, which is an important prerequisite for high-performance InAs MOS devices. The interfacial and electrical properties of $HfO_2$/InAs MOS capacitors are strongly influenced by the gate metal, $HfO_2$ deposition temperature, the post-metal anneal, and interfacial control layer between $HfO_2$ and InAs channel. A Pt barrier metal effectively suppresses the formation of interfacial $TiO_x$ oxide and thus significantly improves the gate leakage and insulating quality of the gate stack. The gate oxide integrity and the gate dielectric constant of $HfO_2$ could be further improved by a 250°C ALD deposition and a 300°C PMA treatment. An As-rich InAs IL further suppress surface states, evidenced by the reduction of gate leakage, and depletion/inversion capacitances.

## Acknowledgments

This work was supported by the National Science Council, Republic of China under Contract NSC 99-2622-E-008-017-CC1 and Joint Development Project (JDP) with Taiwan Semiconductor Manufacturing Company (TSMC). The authors would like to thank for the valuable discussions with Dr. Y. R. Lin, C. C. Cheng, C. H. Ko, and C. H. Wann from TSMC.

## References

[1] G. Dewey, M. K. Hudait, K. Lee et al., "Carrier transport in high-mobility III-V quantum-well transistors and performance impact for high-speed low-power logic applications," *IEEE Electron Device Letters*, vol. 29, no. 10, pp. 1094–1097, 2008.

[2] L. Goldstein, F. Glas, J. Y. Marzin, M. N. Charasse, and G. Le Roux, "Growth by molecular beam epitaxy and characterization of InAs/GaAs strained-layer superlattices," *Applied Physics Letters*, vol. 47, no. 10, pp. 1099–1101, 1985.

[3] A. J. Noreika, M. H. Francombe, and C. E. C. Wood, "Growth of Sb and InSb by molecular-beam epitaxy," *Journal of Applied Physics*, vol. 52, no. 12, pp. 7416–7420, 1981.

[4] H. S. Kim, I. Ok, M. Zhang et al., "A study of metal-oxide-semiconductor capacitors on GaAs, $In_{0.53}$ $Ga_{0.47}As$, InAs, and InSb substrates using a germanium interfacial passivation layer," *Applied Physics Letters*, vol. 93, no. 6, Article ID 062111, 3 pages, 2008.

[5] G. K. Dalapati, Y. Tong, W. Y. Loh, H. K. Mun, and B. J. Cho, "Impact of interfacial layer control using $Gd_2$ $O_3$ in $HfO_2$ gate dielectric on GaAs," *Applied Physics Letters*, vol. 90, no. 18, Article ID 183510, 3 pages, 2007.

[6] H. C. Lin, T. Yang, H. Sharifi et al., "Enhancement-mode GaAs metal-oxide-semiconductor high-electron-mobility transistors with atomic layer deposited $Al_2O_3$ as gate dielectric," *Applied Physics Letters*, vol. 91, no. 21, Article ID 212101, 3 pages, 2007.

[7] F. S. Aguirre-Tostado, M. Milojevic, K. J. Choi et al., "S passivation of GaAs and band bending reduction upon atomic layer deposition of $HfO_2/Al_2O_3$ nanolaminates," *Applied Physics Letters*, vol. 93, no. 6, Article ID 061907, 3 pages, 2008.

[8] Y. S. Kang, C. Y. Kim, M. H. Cho et al., "Thickness dependence on crystalline structure and interfacial reactions in $HfO_2$ films on InP (001) grown by atomic layer deposition," *Applied Physics Letters*, vol. 97, no. 17, Article ID 172108, 3 pages, 2010.

[9] D. Wheeler, L.-E. Wernersson, L. Fröberg et al., "Deposition of $HfO_2$ on InAs by atomic-layer deposition," *Microelectronic Engineering*, vol. 86, no. 7-9, pp. 1561–1563, 2009.

[10] J. Wu, E. Lind, R. Timm, M. Hjort, A. Mikkelsen, and L.-E. Wernersson, "$Al_2O_3$/InAs metal-oxide-semiconductor capacitors on (100) and (111)B substrates," *Applied Physics Letters*, vol. 100, no. 13, pp. 132905–132907, 2012.

[11] H. D. Trinh, E. Y. Chang, Y. Y. Wong et al., "Effects of wet chemical and trimethyl aluminum treatments on the interface properties in atomic layer deposition of $Al_2O_3$ on InAs," *Japanese Journal of Applied Physics*, vol. 49, no. 11, Article ID 111201, 2010.

[12] H. Oigawa, J. F. Fan, Y. Nannichi, H. Sugahara, and M. Oshima, "Universal passivation effect of $(NH_4)_2S_x$ treatment on the surface of III-V compound semiconductors," *Japanese Journal of Applied Physics*, vol. 30, no. 3, pp. L322–L325, 1991.

[13] H. D. Trinh, E. Y. Chang, P. W. Wu et al., "The influences of surface treatment and gas annealing conditions on the inversion behaviors of the atomic-layer-deposition $Al_2O_3$/n-$In_{0.53}Ga_{0.47}As$ metal-oxide-semiconductor capacitor," *Applied Physics Letters*, vol. 97, no. 4, Article ID 042903, 3 pages, 2010.

[14] C. Marchiori, D. J. Webb, C. Rossel et al., "H plasma cleaning and a-Si passivation of GaAs for surface channel device applications," *Journal of Applied Physics*, vol. 106, no. 11, Article ID 114112, 8 pages, 2009.

[15] P. D. Ye, "Main determinants for III-V metal-oxide-semiconductor field-effect transistors (invited)," *Journal of Vacuum Science and Technology A*, vol. 26, no. 4, pp. 697–704, 2008.

# Noise Performance of Heterojunction DDR MITATT Devices Based on Si ~ $Si_{1-x}Ge_x$ at W-Band

**Suranjana Banerjee,[1] Aritra Acharyya,[2] and J. P. Banerjee[2]**

[1] *Academy of Technology, West Bengal University of Technology, Adisaptagram, Hooghly, West Bengal 712121, India*
[2] *Institute of Radio Physics and Electronics, University of Calcutta, 92 APC Road, Kolkata, West Bengal 700009, India*

Correspondence should be addressed to Aritra Acharyya; ari_besu@yahoo.co.in

Academic Editor: Gerard Ghibaudo

Noise performance of different structures of Si ~ $Si_{1-x}Ge_x$ anisotype heterojunction double-drift region (DDR) mixed tunneling and avalanche transit time (MITATT) devices has been studied. The devices are designed for operation at millimeter-wave W-band frequencies. A simulation model has been developed to study the noise spectral density and noise measure of the device. Two different mole fractions $x = 0.1$ and $x = 0.3$ of Ge and corresponding four types of device structure are considered for the simulation. The results show that the $n - Si_{0.7}Ge_{0.3}$ ~ $P$-Si heterojunction DDR structure of MITATT device excels all other structures as regards noise spectral density ($0.82 \times 10^{-16}$ $V^2$ sec) and noise measure (33.09 dB) as well as millimeter-wave properties such as DC-to-RF conversion efficiency (20.15%) and CW power output (773.29 mW).

## 1. Introduction

Impact avalanche transit time (IMPATT) devices are noisy devices having an average noise level of 30–40 dB [1–4]. The noise in IMPATT devices arises mainly from random nature of carrier generation by impact ionization and avalanche multiplication phenomena. The intrinsic avalanche noise in IMPATT device depends mainly on the carrier ionization rates in the base semiconductor material [5]. The group IV-IV compound semiconductor, $Si_{1-x}Ge_x$ has been used as an important base material for both optoelectronic and microelectronic devices [6–9]. $Si_{1-x}Ge_x$ is a bandgap engineered material whose material properties depend on the Ge mole fraction ($x$) [10].

Mixed tunneling avalanche transit time (MITATT) device is an important member of avalanche transit time (ATT) device family operating at higher millimeter-wave frequencies [11–23]. In 1958 W. T. Read [11] in his very early paper predicted that band-to-band tunneling phenomenon might limit the DC-to-RF conversion efficiency of the IMPATT diodes at high frequencies. Kwok and Haddad [12] reported the effect of tunneling on the negative conductance of the device. They considered that instantaneous carrier generation process through tunneling is equivalent to that of a field-dependent carrier source. This concept gave birth to new modes of IMPATT device, namely, MITATT and tunnel transit time (TUNNETT) modes. Nishizawa et al. [13] described the design and principle of the pulse oscillation characteristics of GaAs TUNNETT diodes. Elta and Haddad [14, 15] analyzed the high frequency performance of IMPATT diode by using a modified Read-type equation and considering dead space correction for impact ionization for the charge carriers. They proposed that the above mentioned three different modes (pure IMPATT, MITATT, and TUNNETT modes) of operation of IMPATT depend on the width of avalanche region. Luy and Kuehnf [16] found that the device efficiency decreases due to tunnel generated carriers. A generalized computer simulation method for MITATT mode based on the model proposed by Roy et al. [17, 18] was later reported [19]. Elta [20] and Kane [21] showed that the performance of IMPATT devices deteriorates due to tunneling-induced phase distortion. Eisele and Haddad [22] also carried out experimental work on TUNNETT mode operation of IMPATT diodes grown on diamond heat sink. They reported highest RF conversion efficiency of TUNNETT devices. According to [23], special designs of

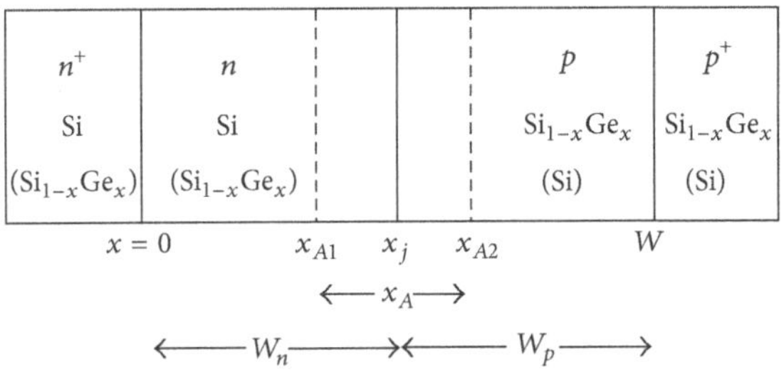


FIGURE 1: One-dimensional schematics of the Si ~ $Si_{1-x}Ge_x$ anisotype heterojunction DDR MITATT devices.

MITATT devices can provide higher efficiency, higher output power, and lower noise as compared to normal IMPATT diodes. The effect of tunneling on the high frequency properties of DDR Si IMPATTs operating at millimeter-wave and terahertz frequencies was studied and reported by Acharyya et al. in their earlier paper [24, 25] which showed that the critical background doping concentration and operating frequency above which tunneling effect becomes predominant are $5.0 \times 10^{23}\ m^{-3}$ and 260 GHz, respectively. The millimeter-wave performance of IMPATTs operating in MITATT mode was studied by Acharyya et al. from the shift of avalanche transit time (ATT) phase delay and reported in [26]. The reported results show that with increasing frequency the shift of ATT phase delay increases which explains the physical reason behind the deterioration of millimeter-wave performance of the device at higher frequencies. Since tunneling is a noiseless phenomenon, it is expected that the noise level in MITATTs will be lower than that in IMPATTs. Furthermore, it is reported [27] that heterojunction IMPATTs have lower noise level than their homojunction counterparts. All these facts have inspired the authors to study the noise performance of four types of Si ~ $Si_{1-x}Ge_x$ anisotype heterojunction DDR structures of MITATT devices at W band. In this study, only two different mole fractions of Ge, $x = 0.1$ and $x = 0.3$, are considered. The performance of different structures of heterojunction DDR MITATTs is compared with their homojunction counterpart based on Si operating at the same frequency band.

## 2. Simulation Method to Study the Noise Properties

One-dimensional model of reverse biased Si ~ $Si_{1-x}Ge_x$ DDR MITATT device shown in Figure 1 is considered for noise analysis. The DC electric field and current density profiles in the depletion layer of the device are obtained from simultaneous numerical solution of fundamental device equations, that is, Poisson's equation, combined carrier continuity equation in the steady state, current density equations, and mobile space charge equation subject to appropriate boundary conditions as discussed in detail in the earlier papers by the authors of [24–26]. A double-iterative simulation method described elsewhere [17] is used to solve these equations and to obtain the electric field and current density profiles. The boundary conditions for the electric field at the depletion layer edges, that is, at $x = 0$ and $x = W$, are given by

$$\xi(0) = 0, \qquad \xi(W) = 0. \tag{1}$$

Similarly, the boundary conditions for normalized current density $P(x) = (J_p(x) - J_n(x))/J_0$ (where the total current density $J_0 = J_n(x) + J_p(x)$, where $J_n(x)$ and $J_p(x)$ are the electron and hole current densities, respectively, at the space point $x$) at the depletion layer edges are given by

$$P(0) = \left(\frac{2}{M_p(0)} - 1\right), \qquad P(W) = \left(1 - \frac{2}{M_n(W)}\right), \tag{2}$$

where $M_n(W)$ and $M_p(0)$ are, respectively, the electron and hole multiplication factors at the depletion layer edges whose values are of the order of $10^6$ under dark or unilluminated condition of the device.

The magnitude of peak field at the junction ($\xi_p$), breakdown voltage ($V_B$), the widths of avalanche and drift zones ($x_A$ and $x_D$, where $x_D = d_n + d_p$), and the voltage drops across these zones ($V_A$, $V_D = V_B - V_A$) are obtained from double-iterative DC simulation program. These values are fed back as input parameters in the small-signal simulation program to obtain the admittance properties of the device such as avalanche resonance frequency ($f_a$), optimum frequency ($f_p$), negative conductance ($G(\omega)$), and corresponding susceptance ($B(\omega)$) as functions of frequency. Two second-order differential equations are framed by resolving the diode impedance $Z(x,\omega)$ into its real part $R(x,\omega)$ and imaginary part $X(x,\omega)$ [18, 19, 24, 25, 28–31]. Here, $Z(x,\omega) = R(x,\omega) + jX(x,\omega)$, where $R(x,\omega)$ and $X(x,\omega)$ are, respectively, the negative specific resistance and specific reactance at the space point $x$, for angular frequency of $\omega$ (frequency $f = 2\pi/\omega$). A double-iterative simulation over the initial choice of the values of $R$ and $X$ described in details in [18] is used to solve simultaneously the two above said second order differential equations subject to appropriate boundary conditions at the depletion layer edges. The spatial profiles of negative specific resistance and specific reactance (i.e., $R(x,\omega)$ versus $x$ and $X(x,\omega)$ versus $x$) for a particular frequency are obtained from the above solution within the depletion layer of the device. The device negative resistance ($Z_R(\omega)$) and reactance ($Z_X(\omega)$) are obtained from the numerical integration of the $R(x,\omega)$ and $X(x,\omega)$ profiles over the space-charge layer width ($W$). Thus,

$$Z_R(\omega) = \int_0^W R(x,\omega)\,dx, \qquad Z_X(\omega) = \int_0^W X(x,\omega)\,dx. \tag{3}$$

The impedance of the device is given by $Z_D(\omega) = Z_R(\omega) + jZ_X(\omega)$ and the device admittance is $Y_D(\omega) = 1/Z_D(\omega) = G(\omega) + jB(\omega)$. The negative conductance and corresponding susceptance at different frequencies are computed from the following expressions:

$$|G(\omega)| = \frac{Z_R(\omega)}{\left(Z_R(\omega)^2 + Z_X(\omega)^2\right)},$$
$$|B(\omega)| = \frac{-Z_X(\omega)}{\left(Z_R(\omega)^2 + Z_X(\omega)^2\right)}. \quad (4)$$

The random nature of the impact ionization process is the main source of noise in avalanche transit time (ATT) devices. This random impact ionization process gives rise to fluctuations in the DC current and DC electric field which appear as small-signal components to their DC values even in the absence of voltage variation across the device. Open circuit condition without any variation of applied voltage is considered for the noise analysis of IMPATT/MITATT device. Starting from the small-signal AC field due to noise $e(x,x') = e_r(x,x') + j\,e_i(x,x')$, two second-order differential equations are framed corresponding to the real $(e_r(x,x'))$ and imaginary $(e_i(x,x'))$ parts of the noise electric field $e(x,x')$. This field is assumed to be due to a noise source $\gamma(x')$ located at space point $x'$ within the depletion region of the device [32–34]. The numerical solution of two simultaneous differential equations involving the real and imaginary parts of noise electric field $e(x,x')$ is carried out by using a double-iterative technique, and Runge-Kutta method subject to the satisfaction of appropriate boundary conditions at the depletion layer edges [32–34]. The noise source $\gamma(x')$ is first considered to be located at one edge of the depletion region. The noise source is then shifted to the next space point and the procedure is repeated until the entire depletion region is covered and the other edge of the depletion layer is reached. Numerical integration of noise electric field $e(x,x')$ over the entire depletion layer provides the terminal voltage $V_T(x')$ produced by noise source, that is,

$$V_T(x') = \int_0^W e(x,x')\,dx. \quad (5)$$

The transfer impedance of the device is defined as

$$Z_T(x') = \frac{V_T(x')}{I_N(x')}, \quad (6)$$

where $I_N(x')$ is the average current generated in the interval $dx'$ due to $\gamma(x')$ located at $x'$. The mean-square noise voltage is obtained from

$$\langle v_n^2 \rangle = 2q^2 \cdot df \cdot A \int \left|Z_T(x')\right|^2 \gamma(x')\,dx'. \quad (7)$$

Mean-square noise voltage per bandwidth is called noise spectral density ($\langle v_n^2\rangle/df$ $V^2$ sec). The noise performance of the device can be known from a parameter called the noise measure (NM) defined as

$$\mathrm{NM} = \frac{\langle v_n^2\rangle/df}{4K_BT(-Z_R - R_S)}, \quad (8)$$

where $K_B$ is the Boltzmann constant ($K_B = 1.38 \times 10^{-23}$ J $K^{-1}$), $T$ is the absolute temperature, $Z_R$ is the device negative resistance, and $R_S$ is the positive parasitic series resistance associated with the device.

## 3. Material Parameters and Design

The realistic field dependence of ionization rates ($\alpha_n$, $\alpha_p$) and drift velocities ($v_n$, $v_p$) of Si and $Si_{1-x}Ge_x$ ($x$ = 0.1 and $x$ = 0.3) at realistic junction temperature of 500 K are taken from the recently published experimental reports [35–40]. Other material parameters of Si and $Si_{1-x}Ge_x$ ($x$ = 0.1 and $x$ = 0.3) such as intrinsic carrier concentration ($n_i$), effective density of states of conduction and valance bands ($N_c$, $N_v$), diffusion coefficients ($D_n$, $D_p$), mobilities ($\mu_n$, $\mu_p$) and diffusion lengths ($L_n$, $L_p$) of charge carriers, and permittivity ($\varepsilon_s$) are taken from the published data given in [10]. The active layer widths ($W_n$, $W_p$) and doping concentrations ($N_D$, $N_A$) of all different structures of MITATTs under consideration are designed following a width modulated design method suggested for MITATT devices in [23] for operation at 94 GHz atmospheric window frequency. The doping concentrations of the $n^+$- and $p^+$-layers ($N_{n^+}$, $N_{p^+}$) are taken to be in the order of $10^{25}$ for W-band operation. The designed doping and structural parameters are listed in Table 1.

## 4. Results and Discussion

The authors have used a double-iterative simulation method [18, 19, 24, 25, 28–31] to study the static and high-frequency properties of homojunction ($N$-Si ~ $P$-Si) and heterojunction ($N$-Si ~ $p$-$Si_{0.9}Ge_{0.1}$, $N$-Si ~ $p$-$Si_{0.7}Ge_{0.3}$, $n$-$Si_{0.9}Ge_{0.1}$ ~ $P$-Si and $n$-$Si_{0.7}Ge_{0.3}$ ~ $P$-Si) DDR IMPATTs operating at 94 GHz atmospheric window. Peak tunneling generation rate ($qG_{\mathrm{Tp}}$), peak avalanche generation rate ($qG_{\mathrm{Ap}}$), and ratio of $G_{\mathrm{Tp}}$ to $G_{\mathrm{Ap}}$ ($G_{\mathrm{Tp}}/G_{\mathrm{Ap}}$ (%)) of all the devices under consideration are listed in Table 2. The ratio $G_{\mathrm{Tp}}/G_{\mathrm{Ap}}$ (%) is very high (31.69%) in Si homojunction DDR IMPATT. Thus, the phase distortion associated with tunneling results in deterioration of RF performance of the Si homojunction device and the device operates in MITATT mode. But the same ($G_{\mathrm{Tp}}/G_{\mathrm{Ap}}$ (%)) is very small around 0.06%–0.29% in the Si ~ $Si_{1-x}Ge_x$ heterojunction DDRs, which indicates that those devices operate in pure IMPATT mode (without considerable amount of band-to-band tunneling) at 94 GHz.

The simulated DC parameters such as peak electric field ($\xi_p$), breakdown voltage ($V_B$), avalanche voltage ($V_A$), avalanche layer width ($x_A$), ratio of avalanche layer width to total depletion layer width ($x_A/W$ (%), where $W = W_n + W_p$) and DC-to-RF conversion efficiency ($\eta$) of all the devices at the respective bias current densities ($J_0$) are given in Table 3. Figure 2 shows the electric field profiles of those devices. It is evident from Figure 2 and Table 3 that the heterojunction DDRs require lower field at breakdown as compared to their Si homojunction counterpart. Breakdown voltage ($V_B$) is obtained by numerical integration of electric field profile over the depletion layer width (i.e., from $x$ = 0 to $x$ = $W$). The simulated values of breakdown voltage of heterojunction

Table 1: Design parameters.

| Structures* | $W_n$ ($\mu$m) | $W_p$ ($\mu$m) | $N_D$ ($\times10^{23}$ m$^{-3}$) | $N_A$ ($\times10^{23}$ m$^{-3}$) | $N_{n^+}$ ($\times10^{25}$ m$^{-3}$) | $N_{p^+}$ ($\times10^{25}$ m$^{-3}$) | $D_j$ ($\mu$m)$^\dagger$ |
|---|---|---|---|---|---|---|---|
| NSPS | 0.40 | 0.38 | 1.20 | 1.25 | 5.0 | 2.7 | 35.0 |
| NSpSG1 | 0.34 | 0.32 | 0.85 | 0.85 | 5.0 | 2.7 | 35.0 |
| NSpSG2 | 0.34 | 0.30 | 0.85 | 1.00 | 5.0 | 2.7 | 35.0 |
| nSGPS1 | 0.32 | 0.32 | 0.78 | 0.90 | 5.0 | 2.7 | 35.0 |
| nSGPS2 | 0.34 | 0.32 | 0.85 | 0.90 | 5.0 | 2.7 | 35.0 |

*NSPS: *N*-Si ~ *P*-Si homojunction DDR MITATT.
*NSpSG1: *N*-Si ~ *p*-$Si_{0.9}Ge_{0.1}$ anisotype heterojunction DDR MITATT.
*NSpSG2: *N*-Si ~ *p*-$Si_{0.7}Ge_{0.3}$ anisotype heterojunction DDR MITATT.
*nSGPS1: *n*-$Si_{0.9}Ge_{0.1}$ ~ *P*-Si anisotype heterojunction DDR MITATT.
*nSGPS2: *n*-$Si_{0.7}Ge_{0.3}$ ~ *P*-Si anisotype heterojunction DDR MITATT.
$^\dagger D_j$: is the diameter of the p-n junction.

Table 2: Values of $qG_{\text{Tp}}$, $qG_{\text{Ap}}$, and $G_{\text{Tp}}/G_{\text{Ap}}$.

| Parameters | NSPS | NSpSG1 | NSpSG2 | nSGPS1 | nSGPS2 |
|---|---|---|---|---|---|
| $qG_{\text{Tp}}$ ($\times10^{12}$ m$^{-3}$ sec$^{-1}$) | 552.87 | 6.3498 | 3.6309 | 3.8271 | 1.4560 |
| $qG_{\text{Ap}}$ ($\times10^{15}$ m$^{-3}$ sec$^{-1}$) | 1.7441 | 4.5615 | 3.5989 | 1.3303 | 2.2810 |
| $G_{\text{Tp}}/G_{\text{Ap}}$ (%) | 31.69 | 0.14 | 0.10 | 0.29 | 0.06 |

Table 3: Millimeter-wave and noise properties.

| Parameters | NSPS | NSpSG1 | NSpSG2 | nSGPS1 | nSGPS2 |
|---|---|---|---|---|---|
| $J_0$ ($\times10^8$ A m$^{-2}$) | 2.8 | 3.0 | 3.2 | 3.3 | 3.6 |
| $\xi_p$ ($\times10^7$ V m$^{-1}$) | 6.0125 | 3.6760 | 3.5666 | 3.5534 | 3.3656 |
| $V_B$ (V) | 23.89 | 12.89 | 11.81 | 11.40 | 11.08 |
| $V_A$ (V) | 16.34 | 6.67 | 6.06 | 4.37 | 4.07 |
| $x_A$ ($\mu$m) | 0.354 | 0.210 | 0.204 | 0.166 | 0.162 |
| $x_A/W$ (%) | 45.39 | 31.82 | 31.86 | 25.94 | 24.55 |
| $\eta$ (%) | 10.06 | 15.36 | 15.49 | 19.64 | 20.15 |
| $f_p$ (GHz) | 106 | 94 | 96 | 95 | 94 |
| $G_p$ ($\times10^7$ S m$^{-2}$) | 4.6593 | 8.3760 | 10.594 | 10.441 | 11.790 |
| $B_p$ ($\times10^7$ S m$^{-2}$) | 17.320 | 12.252 | 9.0710 | 12.102 | 4.2031 |
| $Q_p$ ($=-B_p/G_p$) | 3.72 | 1.46 | 0.86 | 1.16 | 0.36 |
| $Z_R$ ($\times10^{-9}$ $\Omega$ m$^2$) | 1.4484 | 3.8026 | 5.4463 | 4.0869 | 7.5254 |
| $P_{\text{RF}}$ (mW) | 647.44 | 571.47 | 563.22 | 710.86 | 773.29 |
| $\langle v_n^2\rangle/df$ ($\times10^{-16}$ V$^2$ sec) | 39.51 | 3.75 | 1.85 | 1.39 | 0.82 |
| NM (dB) | 40.00 | 37.42 | 36.54 | 34.27 | 33.09 |

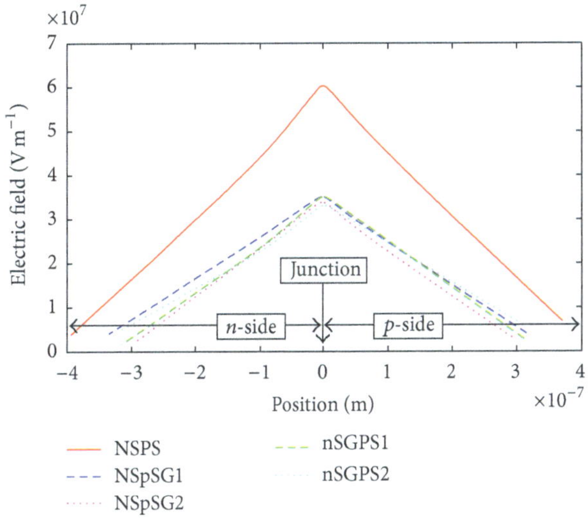


Figure 2: Electric field profiles of the DDR MITATT devices.

devices are lower than that of homojunction device. Again the avalanche voltage drop can be obtained from numerical integration of the electric field profiles over the avalanche region (i.e., from $x = x_{A1}$ to $x = x_{A2}$). The avalanche voltages of heterojunction MITATTs are found to be smaller than that of homojunction device. Narrower avalanche widths of heterojunction MITATTs indicate sharper growth of normalized current density profiles ($P(x)$ versus $x$). The sharper the growth of $P(x)$ profile, the narrower the avalanche zone. This leads to higher DC-to-RF conversion efficiency ($\eta$) of heterojunction devices as compared to their homojunction counterpart based on Si [41]. Table 3 further shows that this efficiency is the highest in *n*-$Si_{0.7}Ge_{0.3}$ ~ *P*-Si heterojunction DDR device (20.15%) of all other types of DDRs.

The admittance characteristics of the devices under consideration are shown in Figure 3. Table 3 shows that magnitude of the peak negative conductance ($G_p$) and negative resistance ($Z_R$) are higher for heterojunction MITATTs as compared to those for Si homojunction counterpart. It may be noted that the above parameters ($G_p$, $Z_R$) are maximum for *n*-$Si_{0.7}Ge_{0.3}$ ~ *P*-Si heterojunction structure (nSGPS2). Also power output ($P_{\text{RF}}$) from that particular structure (nSGPS2) is the highest (773.29 mW) as compared to that from all other structures. The lowest value of *Q*-factor ($Q_p = -B_p/G_p$ =0.36) in that structure (nSGPS2) indicates growth rate and stability of IMPATT oscillation. The spatial variations of negative resistivity of all structures of MITATT devices are shown in Figure 4. All these negative resistivity profiles exhibit two peaks in the two drift regions with a minimum in the avalanche region. The magnitude of negative resistivity peaks in both the drift layers is the highest in that structure (nSGPS2) compared to those of other structures.

Furthermore, it is interesting to observe from Table 3 that, though the breakdown voltage ($V_B$) of *n*-$Si_{0.9}Ge_{0.1}$ ~ *P*-Si and

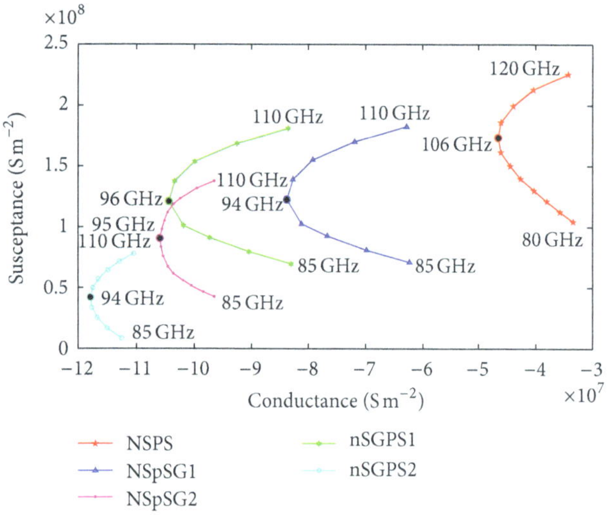


FIGURE 3: Admittance characteristics of the DDR MITATT devices.

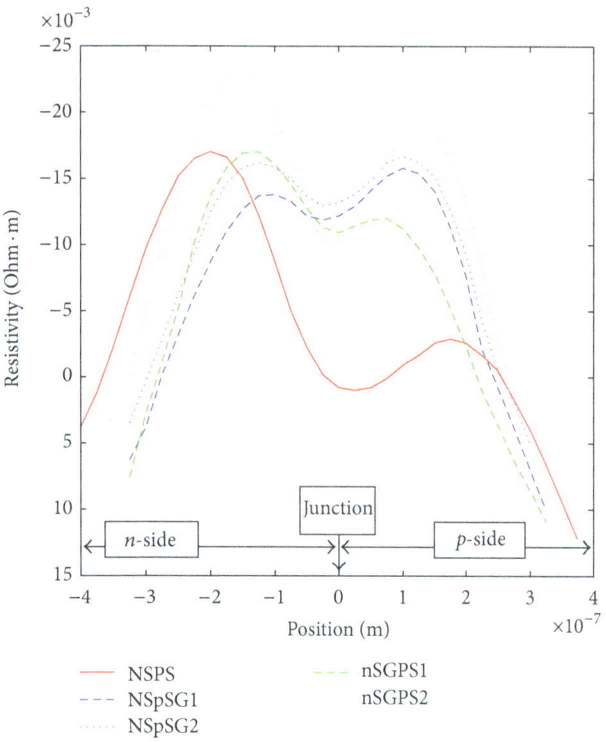


FIGURE 4: Negative resistivity profiles of the DDR MITATT devices.

$n$-$Si_{0.7}Ge_{0.3}$ ~ $P$-Si heterojunction DDRs is almost half of that of Si homojunction DDR due to much smaller breakdown field in $Si_{1-x}Ge_x$ than that of Si [10, 37–39], the RF power output ($P_{RF} = \eta \times V_B \times J_0 \times A_j$) which is proportional to both the breakdown voltage ($V_B$) and DC-to-RF conversion efficiency ($\eta$) of those heterojunction devices is higher as compared to that of their homojunction counterpart due to much larger DC-to-RF conversion efficiency ($\eta$) of those above said heterojunction devices.

Figures 5 and 6 show, respectively, the simulated noise spectral densities ($\langle v_n^2 \rangle / df$) and noise measures (NM) against

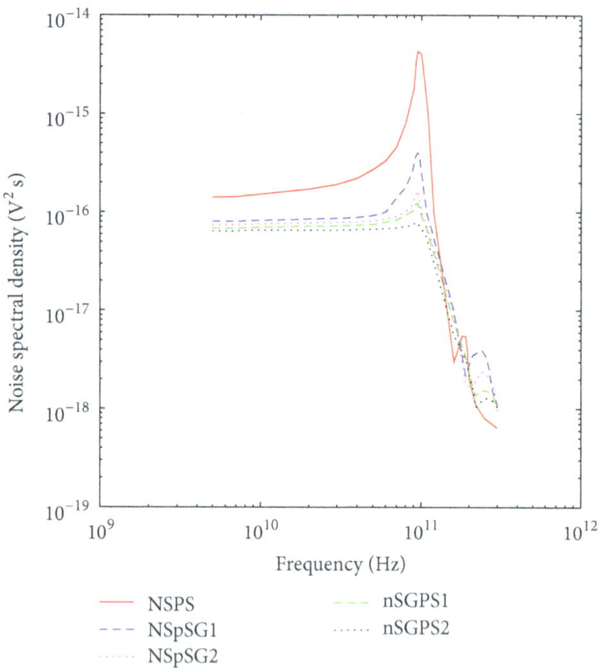


FIGURE 5: Variations of noise spectral densities ($\langle v_n^2 \rangle / df$ $V^2$ sec) of DDR MITATT devices with frequency.

frequency of Si ~ $Si_{1-x}Ge_x$ heterojunction and Si homojunction DDR MITATT devices. It is observed that both NSD and NM are minimum ($0.82 \times 10^{-16}$ $V^2$ sec and 33.09 dB) in $n$-$Si_{0.7}Ge_{0.3}$ ~ $P$-Si heterojunction DDR MITATT device at 94 GHz. The minimum noise level in that particular structure is due to suppression of noisy impact ionization phenomena in the narrowest avalanche zone.

The simulation results presented in this paper are cross checked with the previously reported experimentally obtained results to validate the simulation scheme adopted by the authors. Luy et al. [42] experimentally obtained CW power output of 600 mW at 94 GHz with 6.7% efficiency in Si homojunction flat DDR in 1987. Latter, Dalle et al. [43] measured CW power output of about 500 mW at 94 GHz with 8.0% conversion efficiency in 1990. These experimental results are in close agreement with the simulation results presented in this paper for Si homojunction flat DDR (Table 3). The slight discrepancy in the simulated and experimentally reported values RF power output and DC-to-RF conversion efficiency may be due to the slight difference in the design parameters and DC bias current density. The first experimental results on Si/$Si_{1-x}Ge_x$ heterostructure mixed tunneling avalanche transit time (MITATT) diodes were reported by Luy et al. [44] in 1988. They obtained 25 mW of RF power output with 1.3% of conversion efficiency at 103 GHz. But they used $Si_{0.4}Ge_{0.6}$ alloy to fabricate Si/$Si_{1-x}Ge_x$ heterostructure. Due to this fact and also due to lack of optimization of their design, they obtained such lower power output and lower efficiency, but the simulation results presented in this paper with optimized design of the device predicting the fact that the Si/$Si_{1-x}Ge_x$ heterojunction DDRs especially the $nP$ $Si_{0.7}Ge_{0.3}$/Si heterojunction DDRs are capable of delivering

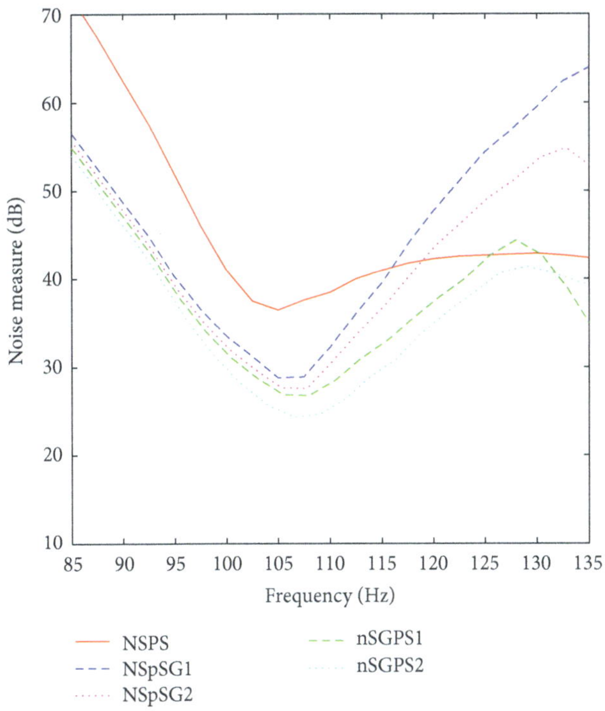


FIGURE 6: Noise measures (NM) versus frequency curves of DDR MITATT devices.

much larger power with much larger conversion efficiency as compared to the experimentally obtained values. Thus the suitable choice of Ge mole fraction ($x$) of $Si_{1-x}Ge_x$, device structure, and proper optimization in design of the device are essential for getting expected RF power output.

## 5. Conclusions

In this paper, the authors have made an attempt to study the millimeter-wave and noise properties of different structures of Si~$Si_{1-x}Ge_x$ anisotype heterojunction DDR MITATT devices. This simulation study clearly indicates that the $n$-$Si_{0.7}Ge_{0.3}$ ~ $P$-Si heterojunction MITATT is the most suitable structure for generation of high RF power with high conversion efficiency and low noise measure. The results are extremely encouraging for the experimentalists to fabricate the $n$-$Si_{0.7}Ge_{0.3}$ ~ $P$-Si heterojunction MITATTs for millimeter-wave applications.

## References

[1] M. E. Hines, "Noise theory for the Read type avalanche diode," *IEEE Transactions on Electron Devices*, vol. 13, no. 1, pp. 158–163, 1966.

[2] H. K. Gummel and J. L. Blue, "A small-signal theory of avalanche noise in IMPATT diodes," *IEEE Transactions on Electron Devices*, vol. 14, pp. 569–580, 1967.

[3] H. A. Haus, H. Statz, and R. A. Pucel, "Optimum noise measure of IMPATT diodes," *IEEE Transactions on Microwave Theory and Techniques*, vol. 19, no. 10, pp. 801–813, 1971.

[4] R. L. Kuvas, "Noise in IMPATT diodes. Intrinsic properties," *IEEE Transactions on Electron Devices*, vol. 19, no. 2, pp. 220–233, 1972.

[5] J. Banerjee, K. Roy, and M. Mitra, "Effect of negative resistance in the noise behavior of Ka-Band IMPATT diodes," *International Journal of Engineering Science and Technology*, vol. 4, no. 7, pp. 3584–3691, 2012.

[6] Z. Pei, C. S. Liang, L. S. Lai et al., "A high-performance SiGe-Si multiple-quantum-well heterojunction phototransistor," *IEEE Electron Device Letters*, vol. 24, no. 10, pp. 643–645, 2003.

[7] A. Acharyya and J. P. Banerjee, "A comparative study on the effect of optical illumination on $Si_{1-x}Ge_x$ and Si based DDR IMPATT diodes at W-band," *Iranian Journal of Electronics and Electrical Engineering*, vol. 7, no. 3, pp. 179–189, 2011.

[8] A. Acharyya and J. P. Banerjee, "A comparative study on the effects of tunneling on W-band Si and $Si_{1-x}Ge_x$ based double-drift IMPATT devices," in *IEEE International Conference on Electronics Computer Technology*, Kanyakumari, India, April 2012.

[9] A. Acharyya and J. P. Banerjee, "Studies on anisotype Si/$Si_{1-x}Ge_x$ heterojunction DDR IMPATTs: efficient millimeter-wave sources at 94 GHz window," *IETE Journal of Research*, vol. 59, no. 3, pp. 1–9, 2013.

[10] "Electronic Archive: New Semiconductor Materials, Characteristics and Properties," 2012, http://www.ioffe.ru/SVA/NSM/Semicond.

[11] W. Shockley, "Negative resistance arising from transit time in semiconductor diodes," *Bell System Technical Journal*, vol. 33, pp. 799–826, 1954.

[12] S. P. Kwok and G. I. Hadded, "Effects of tunnelling on an IMPATT oscillator," *Journal of Applied Physics*, vol. 43, pp. 3824–3860, 1972.

[13] J. Nishizawa, K. Motoya, and Y. Okuno, "GaAs TUNNET diodes," *IEEE Transactions on Microwave Theory and Techniques*, vol. 26, no. 12, pp. 1029–1035, 1978.

[14] E. M. Elta and G. I. Haddad, "High frequency limitations of IMPATT, MITATT and TUNNETT mode devices," *IEEE Transactions on Microwave Theory and Techniques*, vol. 27, no. 5, pp. 442–449, 1979.

[15] E. M. Elta and G. I. Hadded, "Mixed tunnelling and avalanche mechanism in p-n junctions and their effects on microwave transit-time devices," *IEEE Transactions on Electron Devices*, vol. 25, no. 6, pp. 694–702, 1978.

[16] J. F. Luy and R. Kuehnf, "Tunneling-assisted IMPATT operation," *IEEE Transactions on Electron Devices*, vol. 36, no. 3, pp. 589–595, 1989.

[17] S. K. Roy, M. Sridharan, R. Ghosh, and B. B. Pal, "Computer method for the dc field and carrier current profiles in the IMPATT device starting from the field extremum in the depletion layer," in *Proceedings of the 1st Conference on Numerical Analysis of Semiconductor Devices (NASECODE I '79)*, J. H. Miller, Ed., pp. 266–274, Dublin, Ireland, 1979.

[18] S. K. Roy, J. P. Banerjee, and S. P. Pati, "A Computer analysis of the distribution of high frequency negative resistance in the depletion layer of IMPATT Diodes," in *Proceedings of the 4th Conference on Numerical Analysis of Semiconductor Devices (NASECODE IV '85)*, pp. 494–500, Dublin, Ireland, 1985.

[19] G. N. Dash and S. P. Pati, "A generalized simulation method for MITATT-mode operation and studies on the influence of tunnel current on IMPATT properties," *Semiconductor Science and Technology*, vol. 7, no. 2, pp. 222–230, 1992.

[20] M. E. Elta, *The effect of mixed tunneling and avalanche breakdown on microwave transit-time diodes [Ph.D.dissertation]*, Electron Physics Laboratory, University of Michigan, Ann Arbor, Mich, USA, 1978.

[21] E. O. Kane, "Theory of tunneling," *Journal of Applied Physics*, vol. 32, pp. 83–91, 1961.

[22] H. Eisele and G. I. Haddad, "GaAs TUNNETT diodes on diamond heat sink for 100 GHz and above," *IEEE Transactions on Microwave Theory and Techniques*, vol. 43, no. 1, pp. 210–213, 1995.

[23] G. N. Dash, "A new design approach for MITATT and TUNNETT mode devices," *Solid-State Electronics*, vol. 38, no. 7, pp. 1381–1385, 1995.

[24] A. Acharyya, M. Mukherjee, and J. P. Banerjee, "Effects of tunnelling current on mm-wave IMPATT devices," *International Journal of Electronics*. In press.

[25] A. Acharyya, M. Mukherjee, and J. P. Banerjee, "Influence of tunnel current on DC and dynamic properties of silicon based terahertz IMPATT source," *Terahertz Science and Technology*, vol. 4, no. 1, pp. 26–41, 2011.

[26] A. Acharyya, M. Mukherjee, and J. P. Banerjee, "Studies on the millimeter-wave performance of MITATTs from avalanche transit time phase delay," in *Proceedings of the IEEE Applied Electromagnetics Conference*, pp. 1–4, Kolkata, India, December 2011.

[27] S. R. Pattanaik, J. K. Mishra, and G. N. Dash, "A new mm-wave GaAs~$Ga_{0.52}In_{0.48}P$ Heterojunction IMPATT diode," *IETE Journal of Research*, vol. 57, no. 4, pp. 351–356, 2011.

[28] A. Acharyya and J. P. Banerjee, "Prospects of IMPATT devices based on wide bandgap semiconductors as potential terahertz sources," *Applied Nanoscience*, 2012.

[29] A. Acharyya and J. P. Banerjee, "Potentiality of IMPATT devices as terahertz source: an avalanche response time based approach to determine the upper cut-off frequency limits," *IETE Journal of Research*, vol. 59, no. 2, pp. 1–10, 2013.

[30] A. Acharyya, S. Banerjee, and J. P. Banerjee, "Optical control of millimeter-wave lateral double-drift region silicon IMPATT device," *Radioengineering*, vol. 21, no. 4, pp. 1208–1217, 2012.

[31] A. Acharyya and J. P. Banerjee, "Analysis of photo-irradiated double-drift region silicon impact avalanche transit time devices in the millimeter-wave and terahertz regime," *Terahertz Science and Technology*, vol. 5, no. 2, pp. 97–113, 2012.

[32] G. N. Dash, J. K. Mishra, and A. K. Panda, "Noise in mixed tunneling avalanche transit time (MITATT) diodes," *Solid-State Electronics*, vol. 39, no. 10, pp. 1473–1479, 1996.

[33] A. Acharyya, M. Mukherjee, and J. P. Banerjee, "Noise in millimeter-wave mixed tunneling avalanche transit time diodes," *Archives of Applied Science Research*, vol. 3, no. 1, pp. 250–266, 2011.

[34] A. Acharyya, M. Mukherjee, and J. P. Banerjee, "Noise performance of millimeter-wave silicon based mixed tunneling avalanche transit time (MITATT) diode," *International Journal of Electrical and Electronics Engineering*, vol. 4, no. 8, pp. 577–584, 2010.

[35] W. N. Grant, "Electron and hole ionization rates in epitaxial Silicon," *Solid-State Electronics*, vol. 16, no. 10, pp. 1189–1203, 1973.

[36] C. Canali, G. Ottaviani, and A. Alberigi Quaranta, "Drift velocity of electrons and holes and associated anisotropic effects in silicon," *Journal of Physics and Chemistry of Solids*, vol. 32, no. 8, pp. 1707–1720, 1971.

[37] J. Lee, A. L. G. Aitken, S. H. Lee, and P. K. Bhattacharya, "Responsivity and Impact ionisation coefficients of $Si_{1-x}Ge_x$ photodiodes," *IEEE Electron Devices*, vol. 43, no. 6, pp. 977–981, 1996.

[38] K. Yeom, J. M. Hincley, and J. Singh, "Calculation of electron and hole impact ionisation coefficients in SiGe alloys," *Journal of Applied Physics*, vol. 80, no. 12, pp. 6773–6782, 1996.

[39] M. Ershov and V. Ryzhii, "High field electron transport in SiGe alloy," *Japanese Journal of Applied Physics*, vol. 33, no. 3, pp. 1365–1371, 1994.

[40] T. Yamada and D. K. Ferry, "Montecarlo simulation of hole transport in strained $Si_{1-x}Ge_x$," *Solid-State Electronics*, vol. 38, no. 4, pp. 881–890, 1995.

[41] D. L. Scharfetter and H. K. Gummel, "Large-signal analysis of a silicon read diode oscillator," *IEEE Transactions on Electron Devices*, vol. 6, no. 1, pp. 64–77, 1969.

[42] J. F. Luy, A. Casel, W. Behr, and E. Kasper, "A 90-GHz double-drift IMPATT diode made with Si MBE," *IEEE Transactions on Electron Devices*, vol. 34, no. 5, pp. 1084–1089, 1987.

[43] C. Dalle, P. A. Rolland, and G. Lieti, "Flat doping profile double-drift silicon IMPATT for reliable CW high-power high-efficiency generation in the 94-GHz window," *IEEE Transactions on Electron Devices*, vol. 37, no. 1, pp. 227–236, 1990.

[44] J. F. Luy, H. Jorke, H. Kibbel, A. Casel, and E. Kasper, "Si/SiGe heterostructure mitatt diode," *Electronics Letters*, vol. 24, no. 22, pp. 1386–1387, 1988.

# 24

# Analysis of Kink Reduction in SOI MOSFET Using Selective Back Oxide Structure

**M. Narayanan,[1] H. Al-Nashash,[1] Baquer Mazhari,[2] Dipankar Pal,[3] and Mahesh Chandra[4]**

[1] *Department of Electrical Engineering, American University of Sharjah, P.O. Box 26666, Sharjah, UAE*
[2] *Department of Electrical Engineering, Indian Institute of Technology, Kanpur 208016, India*
[3] *NERIST, Nirjuli, 791109, Itanagar, India*
[4] *Birla Institute of Technology, Mesra, 835 215 Ranchi, India*

Correspondence should be addressed to M. Narayanan, mnarayanan@aus.edu

Academic Editor: Daisaburo Takashima

This paper presents a complete analysis of the kink effect in SOI MOSFET and proposes a method for eliminating kink effect observed in the current-voltage output characteristics of a partially depleted SOI MOSFET device. In this method, back oxide for the device is introduced at selected regions below the source and drain and not continuously as in an SOI device giving rise to what is termed a "SELBOX" structure. Selective back oxide structure with different gap lengths and thicknesses was studied. Results obtained through numerical simulations indicate that the proposed structure can significantly reduce the kink while still preserving major advantages offered by conventional SOI structure. Although the new structure is capable of eliminating kink, for narrow gaps the device may still exhibit some kink effect. A device model that explains the kink behavior of the structure for varying gap lengths is also developed.

## 1. Introduction

Silicon on insulator (SOI) MOSFET is fabricated on wafers consisting of a film of crystalline silicon separated by a layer of $SiO_2$ from the bulk substrate as shown in Figure 1(a). SOI MOSFETs offer several advantages as compared to bulk MOSFETs such as excellent lateral and vertical isolation, improved latch up free operation, reduced leakage current, reduced short channel effects, and improved switching speeds due to reduction in the drain-body capacitance [1]. SOI MOSFETs are divided into partially and fully depleted-SOI devices. The depletion region of the partially depleted (PD) SOI devices does not reach through the entire silicon channel/body region. PD SOI devices have silicon thickness greater than $\sim 0.15\,\mu$m. The depletion region extends into the body of the device under the gate at the source body and drain-body junctions and does not deplete all of the charges in the body, resulting in the name partially depleted SOI. Partially depleted devices are often labeled as thick film. In case of fully depleted (FD) SOI devices, the silicon thickness is thinner than the silicon thickness of the partially depleted devices. The background doping of the channel region is lower than the partially depleted devices. Usually the thickness is in the range of few tens of nanometers. These devices are called as thin film [2].

In PD SOI devices, the thickness of top Si layer is around 300 nm. The fabrication process for such devices is totally compatible with that of the bulk silicon technology. As a result, the design for bulk silicon devices can be easily transferred to PD SOI technology [3]. Fabrication of FD SOI devices involves development of ultrathin body and, therefore, needs more sophisticated technology. The threshold voltage in PD SOI devices is relatively less sensitive to the uniformity in the silicon film thickness. In FD SOI device, the threshold voltage depends on the Si film thickness [4]. It is difficult to control the thickness of the ultra thin film body in FD SOI MOSFETS, and as a result, the film thickness becomes nonuniform across the wafer. For these reasons, there has been a growing interest in PD SOI MOSFETs.

In spite of these benefits, PD-SOI MOSFET devices are accompanied with certain undesirable effects such as kink [3] in the output current-voltage characteristics as given in Figure 1(b). The kink effect is observed in partially depleted

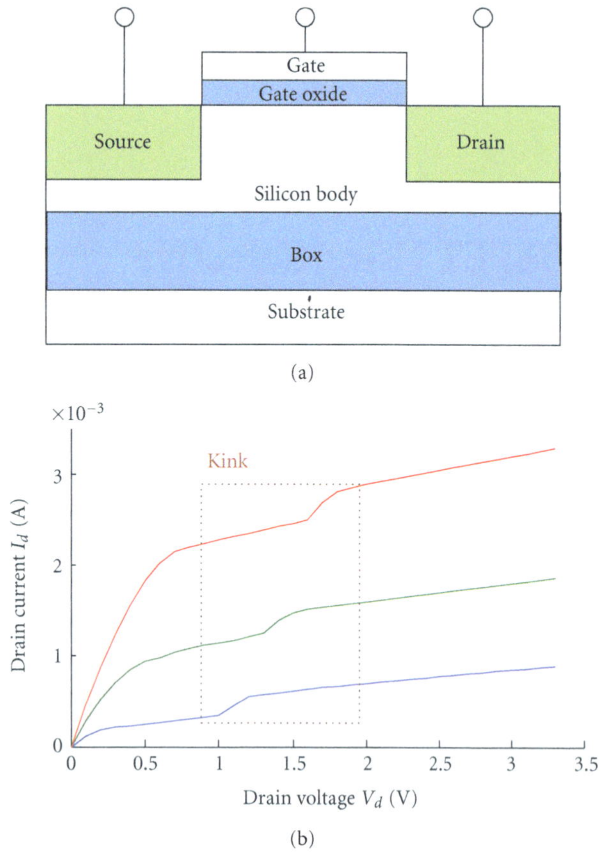


FIGURE 1: (a) SOI MOSFET structure and (b) Kink effect in Output characteristics.

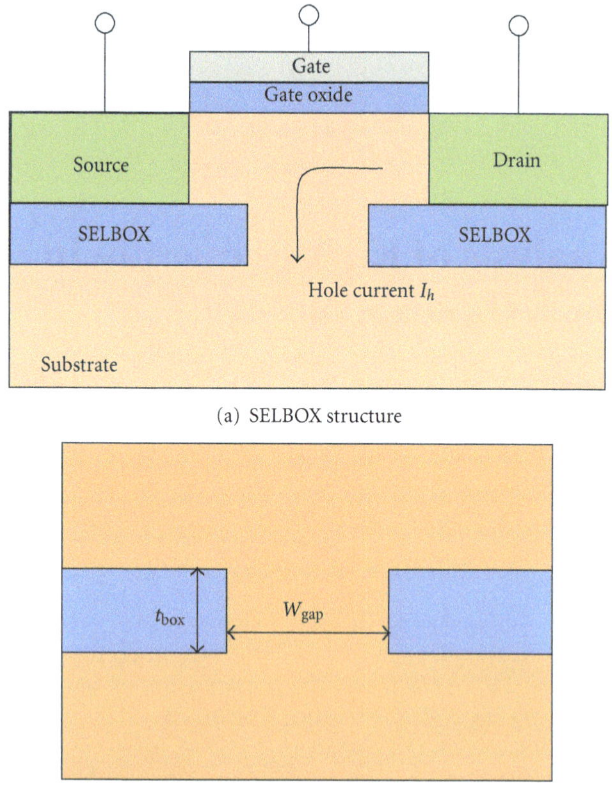


(a) SELBOX structure

(b) Back oxide gap parameters, ($t_{box}$): SELBOX Thickness, ($w_{gap}$): Gap length

FIGURE 2: SELBOX MOSFET structure.

SOI MOSFET devices. In these devices, the electric field near the drain region will be high due to relatively higher drain voltages. Consequently, the channel electrons in the high electric field acquire high energy and create electron-hole pairs by impact ionization mechanism. The generated electrons are collected by the drain and the holes accumulate in the floating body. The accumulation of the holes in the floating body leads to an increase in the body potential, and the associated drop in the threshold voltage leads to a sharp increase in the drain current [2]. The increase in drain current further leads to an increase in the number of holes generated. This cumulative process will continue until the body source junction is forward biased, allowing the holes to exit the device. Consequently, the device will not be suitable for linear applications due to the nonlinearity created by the kink effect.

The second observation is that SOI devices are thermally insulated from the substrate by the buried insulator with very low thermal conductivity. As a result, removal of excess heat generated within the device is less efficient than in bulk devices [3]. This leads to a substantial elevation of device temperature and a reduction in the channel mobility.

A classical remedy for the problems due to floating body is to provide a body contact or a body tie in these devices [2]. Even though body tie can eliminate the kink effect, it will be at the cost of die area [2]. Further, effectiveness of body tie is much less pronounced as these contacts will introduce body charge and discharge time constants. Another method to reduce the floating body effect involves controlling the carrier lifetime in the floating body region of an SOI MOSFET by high-dose silicon implantation. The carrier lifetime in polysilicon is much shorter than that in the single crystal silicon. Ion implantation leads to amorphization of the silicon and creates recombination centers below the channel region close to the interface between the Si layer and the BOX layer. These recombination centers reduce the lifetime of excess carriers generated due to impact ionization during device operation and minimize the hole accumulation in NMOSFET device thus suppressing the kink effect. But, in this method, the drive current is reduced, due to reductions in the channel mobility resulting from the lattice damage [5].

An alternative solution for eliminating the floating body effect is the use of selective back oxide (SELBOX) structure [6] as shown in Figure 2(a). In this structure, the buried oxide covers the channel partially. It offers the prospect of combining the advantages of SOI and bulk devices. With a proper selection of oxide gap length, the kink effect and self-heating effect can be greatly reduced. The structure discussed here can provide enhanced performance by using SELBOX gap ($w_{gap}$) and back oxide thickness as an additional degree of freedom for device optimization. Figure 2(b) indicates the back oxide gap parameters. Oxide thickness ($t_{box}$) is the thickness measured in the $y$-direction. Distance between the oxide block edges in the horizontal direction is taken as the gap length ($w_{gap}$).

The feasibility of the fabrication of MOS devices with SELBOX structure and simulation has been already reported [6–8]. Simulation studies demonstrate the improvements in the threshold voltage roll-off with the SELBOX structure [8]. The present work describes the effectiveness of the SELBOX structure for reducing the kink in the output characteristics of a SOI MOSFET device. In addition, a device circuit model that explains the kink behavior of the SELBOX structure is also proposed.

The paper is divided into five sections. Section 1 explains some of the limitations of the SOI MOS devices and introduces SELBOX MOSFET as a solution to overcome these limitations. The method of implementation of the structure is discussed in Section 2. Observations and results obtained from the studies of the new structure are included in Section 3. Section 4 explains the details of a proposed model for the SELBOX structure. Finally, conclusion and future work is included in Section 5.

## 2. Methodology

A schematic of partially depleted SOI MOSFET structure that was simulated is shown in Figure 3. The dimensions and doping concentration for the device structure considered for simulation are selected on the basis of a typical SOI device with channel length of 2.0 $\mu$m and gate oxide thickness of 0.01 $\mu$m [9]. The p-type substrate is doped with doping concentration of $\sim 2 \times 10^{17}\,\text{cm}^{-3}$ and source and drain regions of $\sim 10^{19}\,\text{cm}^{-3}$. The total device length including source, channel, and drain is 3.5 $\mu$m. Device dimensions and doping concentrations are selected according to the values available in the published work [9] in order to compare and verify the results. Silvaco tools are used for process and device simulation. Silvaco athena provides an easy way to use platform to simulate semiconductor fabrication processes, while Silvaco atlas is used to simulate the electrical characteristics of the semiconductor device under study. Numerical simulation was carried out using athena and atlas tools from Silvaco [10]. Based on literature for SOI devices, field-dependent mobility model, Shockley-Read-Hall recombination, and impact ionization model from Selberherr [11] are used for the simulation. Numeric methods chosen for simulation are Gummel and Newton methods. Figure 4 illustrates a typical output characteristics obtained for partially depleted SOI MOSFET device using atlas simulations. For bulk MOSFET with identical device dimensions and doping concentration, the simulated output characteristics is shown in Figure 4 with dotted lines.

Presence of kink in the output characteristics for SOI structure is clearly visible. The onset of kink takes place at a kink voltage ($V_{kink}$) of 1.5 V for gate-source voltage of 2.2 V.

The simulated output characteristics for a SELBOX device, the structure of which is explained with Figure 2, is given in Figure 5. This structure is identical to the bulk MOSFET device with the only difference being the presence of back oxide partially covering the channel. The SELBOX

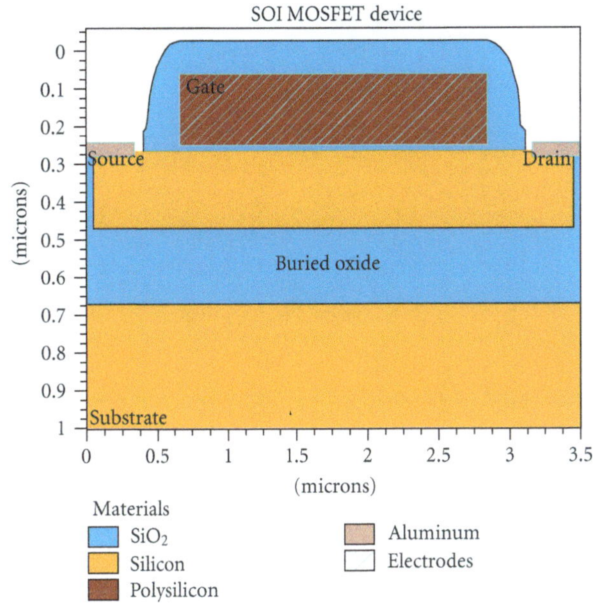


Figure 3: SOI MOSFET structure.

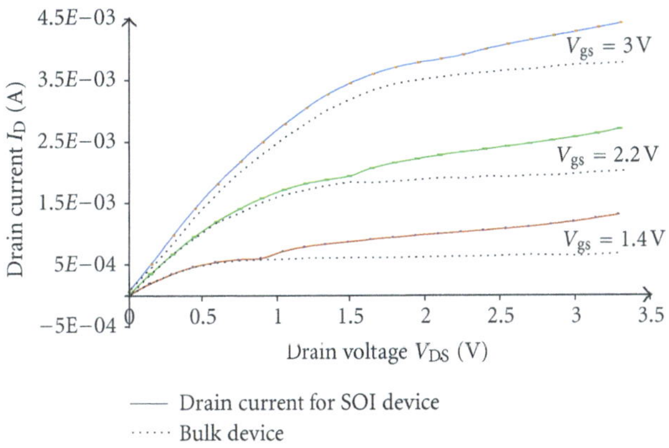


Figure 4: Output characteristics of MOSFET and SOI MOSFET.

thickness is 0.4 $\mu$m. The kink can be effectively reduced even with a small gap in the back oxide. To study the effect of changes in the gap length on the kink voltage, the gap lengths are chosen as 0.004 $\mu$m, 0.01 $\mu$m, and 0.1 $\mu$m. The kink voltage is found to be increasing with increase in the gap length. It is noticed that the plot for a gap length of 0.3 $\mu$m and that of bulk structure coincide and the kink completely disappears.

## 3. Results

*3.1. Kink Effect in SELBOX MOSFETs.* The basic mechanism that leads to the kink in the output characteristics of a SELBOX structure with a gap length of 0.02 $\mu$m and back oxide thickness of 0.4 $\mu$m is illustrated below in Figure 6. The simulation is carried out for $V_{gs}$ = 1.4 V by atlas device simulator, and the results are given in Figures 6(a), 6(b), and 6(c). Figure 6(a) shows the hole current through the gap to the substrate, Figure 6(b) shows the body voltage

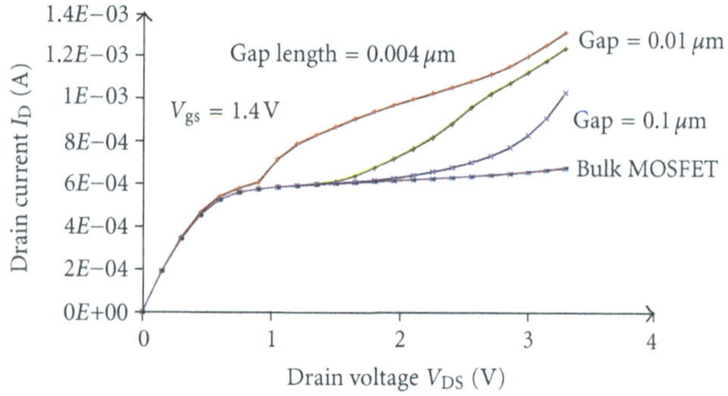


FIGURE 5: SELBOX output characteristics for $V_{gs}$ = 1.4 V with varying gap lengths.

across the gap resulting from the hole current through the gap resistance, and Figure 6(c) shows the kink in the output characteristics, which results due to the increase in the body voltage. At higher drain voltages, the channel electrons acquire energy in the high electric field zone near the drain. These high energy electrons generate electron hole pairs through impact ionization in the region close to the drain [12]. The electrons generated will be collected by the drain, and the holes move towards the substrate through the back oxide gap. Figure 6(a) shows that the hole current starts increasing for a drain voltage close to 1.5 V indicating that the impact ionization starts for $V_{DS}$ close to 1.5 V. The hole current is measured from the hole current density in the SELBOX gap region and by subsequent integration of the current density. For narrow gap lengths, the resistance of the gap for the hole current will be much higher than the lateral channel resistance from drain to the gap. For the hole transport from the region close to drain to the substrate through the gap, the resistance in the device body region is negligibly small compared to the resistance of the narrow gap. The flow of holes through the gap resistance leads to the development of a potential across the gap, which leads to an increase in the body potential as indicated in Figure 6(b). The body potential is measured across the BOX gap. For extremely narrow gap lengths as the gap resistance is large, the hole current through the gap will be associated with a rise in the body potential, which will be sufficient to give rise to a kink in the drain current as indicated in Figure 6(c). This increase in the body potential continues till the body-source junction gets forward biased for the body voltage close to 0.7 V as shown in Figure 6(b).

*3.2. Effect of Changes in Gap Length on Kink Voltage.* The output characteristics of a SELBOX structure for varying gap lengths for a gate to source voltage $V_{gs}$ = 1.4 V. is depicted in Figure 5. The increase in the gap length from 0.004 μm to 0.1 μm results in an increase in the kink voltage. For increased gap lengths, the kink occurs at a higher drain voltage because an increase in the gap length leads to a decrease in the gap resistance. Consequently, the body voltage can rise to the level required for the occurrence of kink for a larger hole current ($I_h$) only, and this is possible at a higher drain voltage. Eventually, for larger values of gap lengths, the kink will completely disappear.

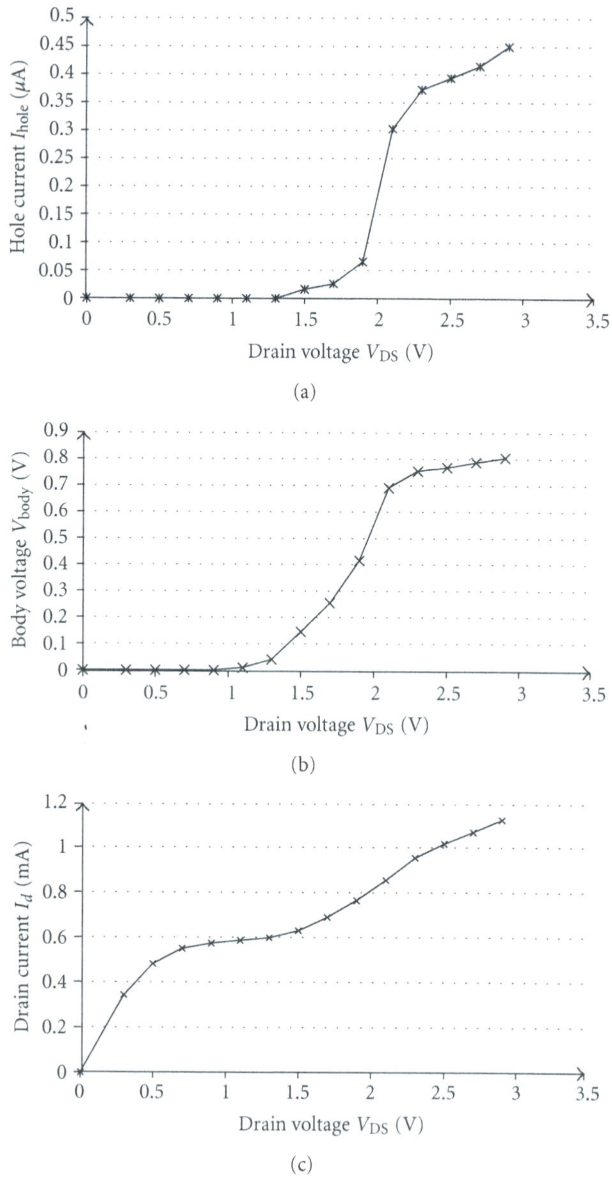


FIGURE 6: Kink effect in SELBOX MOSFET.

The mechanism that leads to the kink in the output characteristics of a SELBOX structure is illustrated with the hole current density (blue shade) as shown in Figure 7(a). The oxide thickness in all cases in Figure 7 is 0.2 μm. For narrow gap lengths, the device still behaves like an SOI device because the narrow gap is inadequate to drain the holes to the substrate. Therefore, the accumulation of holes in the body region will continue till the body source *p-n* junction gets forward biased. With increased gap length, the number of holes getting drained to the substrate increases, which results in an increased kink voltage. This is illustrated in Figure 7(b). For still higher gap lengths as in Figure 7(c), majority of the holes are getting drained to the substrate and hence the kink will be minimum.

For a SELBOX MOSFET, a plot of change in body voltage $\Delta V_b$ versus hole current $I_h$ through the gap for different

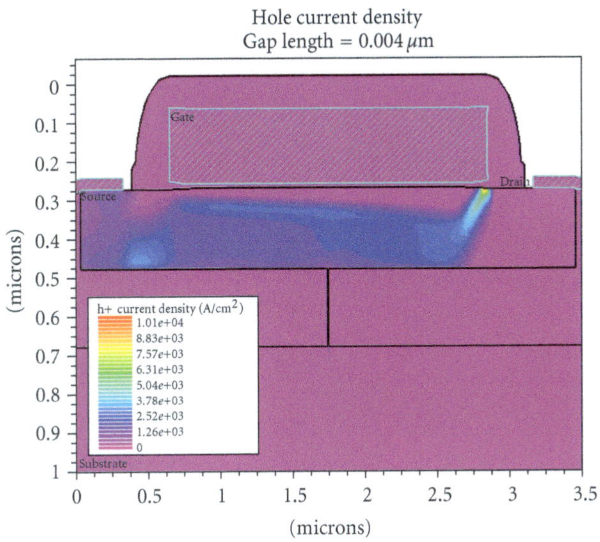


(a) Hole current density for gap width 0.004 $\mu$m

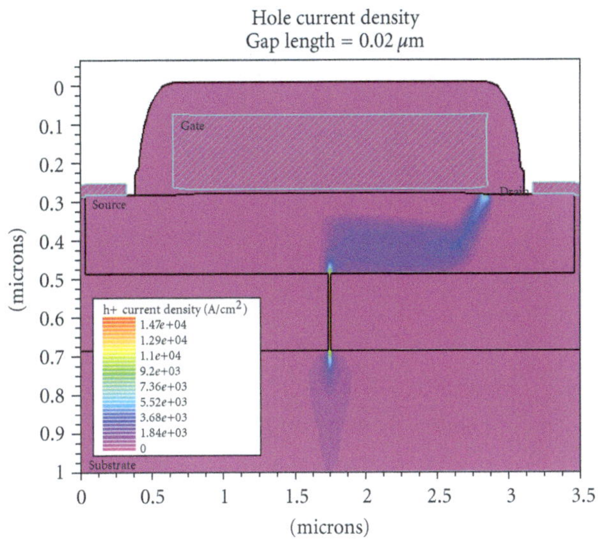


(b) Hole current density for gap width 0.02 $\mu$m

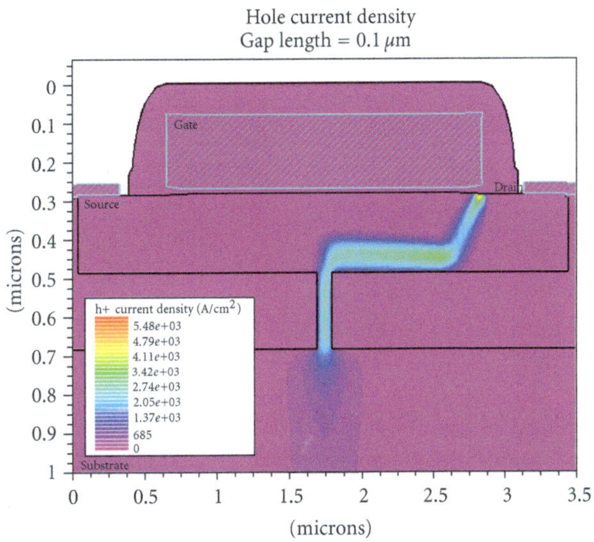


(c) Hole current density for gap width 0.1 $\mu$m

Figure 7: Silvaco simulation results indicating the hole current density in SELBOX MOSFET through the oxide gap for varying gap lengths.

gap lengths is given in Figure 8. The coefficient of $I_h$ in the equations for the lines indicates the gap resistance. The plot illustrates that with increase in the gap length the slope of the line decreases, which is an indication for a proportional reduction in the gap resistance. With decrease in the gap resistance, the rise in the body voltage due to the flow of holes generated during impact ionization is expected to reduce. This is the principle behind the elimination of kink effect with the proposed structure.

Simulation results indicate that the kink voltage increases with increase in the gap length and completely vanishes for higher gap lengths. The dependence of kink voltage on back oxide gap length was investigated keeping all other parameters constant at a gate-source voltage 1.4 V. Figure 9 shows that kink voltage increases with increase in the gap length implying a reduction in the kink effect. The kink voltage becomes larger than 3 V for a gap length of 0.1 $\mu$m. This indicates that with oxide underneath all of source, drain, and more than 95% of the channel, kink effect can be virtually suppressed. For further increase in gap length, the device remains kink-free, tends to look more like a bulk MOSFET and thus less likely to have other advantages associated with SOI.

*3.3. Effect of Changes in Back Oxide Thickness on Kink Voltage.* The kink voltage is found to have a significant dependence on the thickness of the back oxide in a SELBOX structure. Figure 10 shows the dependence of kink voltage on back oxide thickness for a fixed gap length of 0.03 $\mu$m. Keeping the gap length constant, if we vary the back oxide thickness we can expect a change in the gap resistance. For the estimation of gap resistance, back oxide thickness ($t_{box}$) is the length of the resistance and the gap length ($w_{gap}$) is the width of the gap resistance. As we increase the back oxide thickness keeping

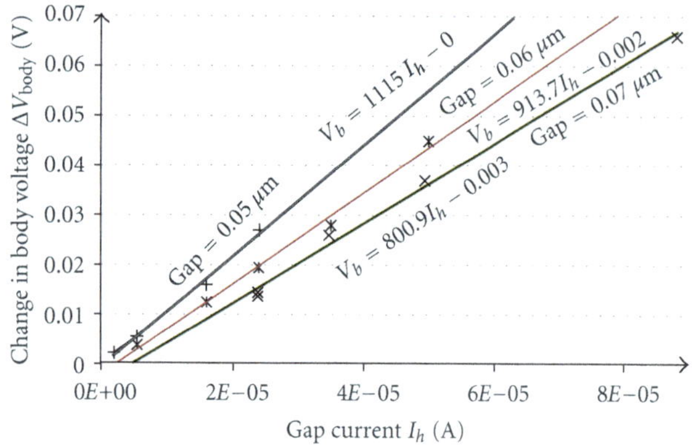


FIGURE 8: Plot of $\Delta V_b$ versus $I_h$.

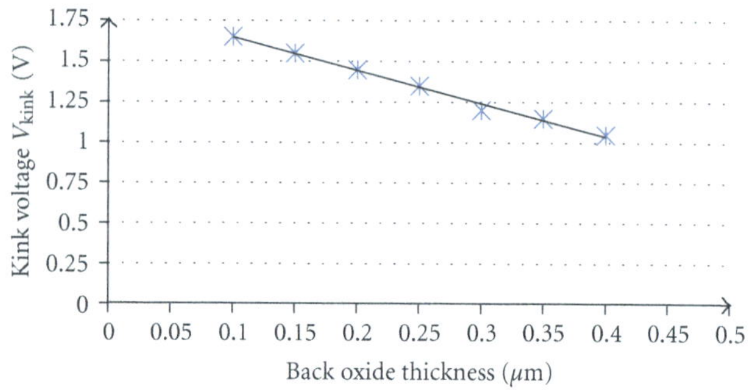


FIGURE 10: Dependence of kink voltage on oxide thickness. (SELBOX Gap is 0.03 μm).

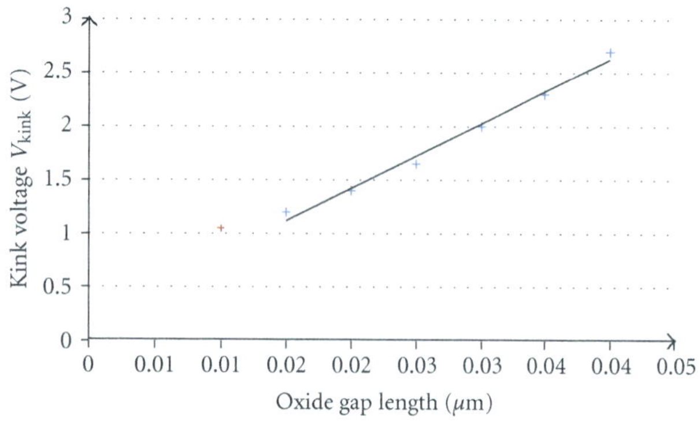


FIGURE 9: Dependence of kink voltage on gap between the oxide regions. (SELBOX thickness is 0.4 μm).

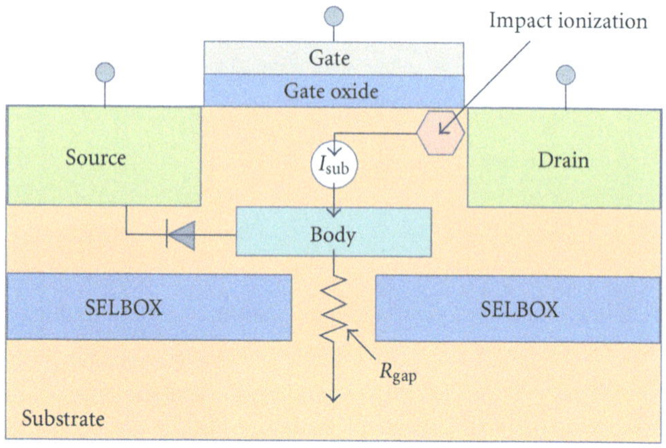


FIGURE 11: Device model units.

the gap length ($w_{gap}$) constant, the effective gap resistance increases. This will lead to an increase in the body potential. Consequently, the body potential at which kink occurs will be reached at a lower drain voltage. With reduced oxide thickness, the back oxide gap resistance also reduces and the rise in the body voltage will be reduced. In this case, the body voltage needed to cause the kink will occur at a higher drain voltage and hence the kink voltage will be higher. Therefore, a device with thinner back oxide is more bulk-like and will be less susceptible to kink phenomenon.

Even though a very small gap length is required to virtually eliminate kink phenomenon from the device characteristics, producing a small gap length can be technologically challenging [9]. Fabrication of devices with large gap length is likely to be easier. But, with increase in the gap length, the device tends to become more bulk-like and thus lose the associated advantages of SOI such as reduced source-body and drain-body capacitances.

## 4. Device Circuit Model

In this section, we introduce device model that explain the mechanism which leads to the rise in the body potential within the SELBOX structure, which leads to the kink in the output characteristics. Figure 7 indicates the possible directions for hole current through the SELBOX structure. When the SELBOX gap is sufficiently wide, the majority of the generated holes will leave the body to lower substrate region through the oxide gap as shown in Figure 7(c). The device under such circumstances behaves like a bulk device as far as the kink is concerned. The events following the impact ionization inside the SELBOX device are shown in Figure 11. The diode between the body region and source represents the body-source *p-n* junction. The current source $I_{sub}$ represents the hole current generated due to the impact ionization, and $R_{gap}$ indicates the back oxide gap resistance. A gradual reduction in the gap length ($w_{gap}$) keeping the gap thickness constant leads to an increase in the gap resistance. Consequently, a proportionate increase in the body voltage is expected for the same $V_{gs}$ and impact ionization current. For larger gap resistance resulting from extremely narrow gap lengths, the rise in the body voltage will be sufficient enough to forward bias the body-source diode, and a portion of the impact ionization holes will pass through the forward biased *p-n* diode to the source terminal.

A plot of gap resistance against ($t_{box}/w_{gap}$) for the SELBOX structure is given in Figure 12. The coefficient $K$ of the ratio ($t_{box}/w_{gap}$) is a characteristic for a given device and represents the gap resistivity. $K$ is the slope of the line given in Figure 12. The value of $K$ for the device is extracted from the graph, and the value depends on the device dimensions and doping concentration. It has unit in Ω.

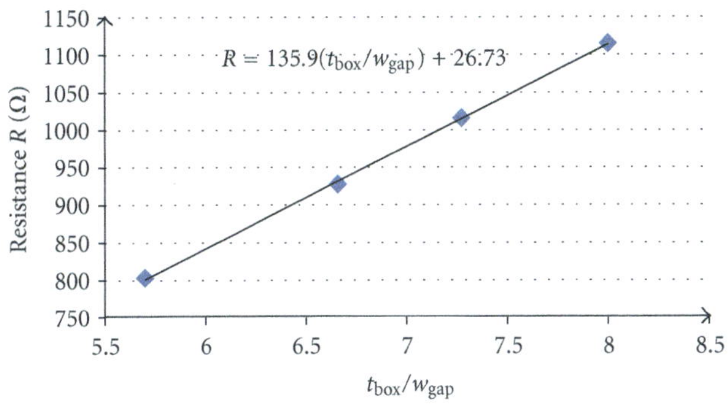


FIGURE 12: Plot of $R$ versus $(t_{box}/w_{gap})$.

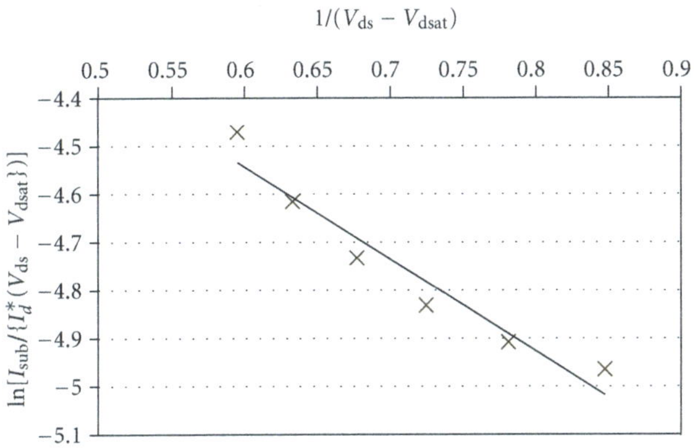


FIGURE 13: Plot of $\ln[I_{sub}/\{I_d^*(V_{ds} - V_{d\,sat})\}]$ versus $[1/(V_{ds} - V_{d\,sat})]$.

TABLE 1: SELBOX parameters.

| | |
|---|---|
| Oxide thickness [$t_{box}$] | 0.4 $\mu$m |
| Oxide gap length [$w_{gap}$] | 0.04 $\mu$m |
| Estimated gap resistance $R_{gap}$ using (2) | 1346 Ω |
| $g_{sb}$ from Figure 14 | $3 \times 10^{-5}$ A/V |

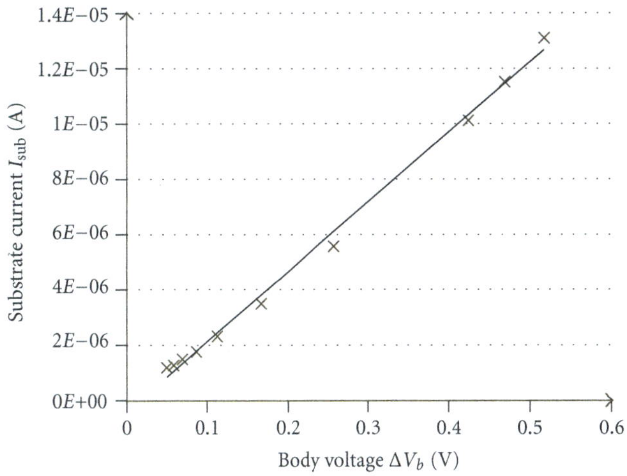


FIGURE 14: Plot for dependence of substrate current on body voltage.

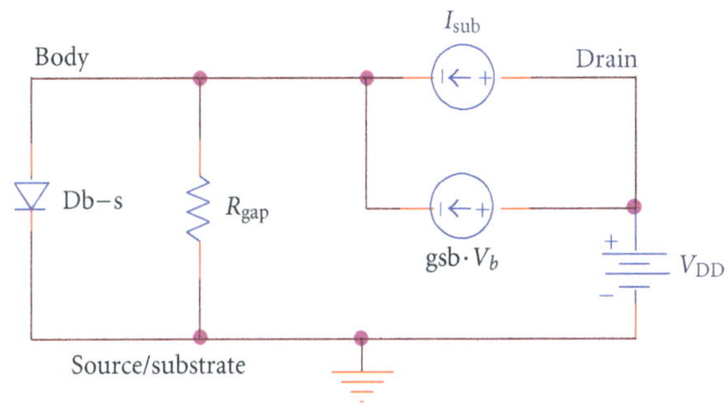


FIGURE 15: Device circuit model.

TABLE 2: Voltage values.

| Applied $V_{ds}$ | Applied $V_{gs}$ | Measured voltage across gap (From simulation results) |
|---|---|---|
| 2.48 V | 1.4 V | 0.015 V |

In the present case, $(t_{box})$ is maintained at 0.4 $\mu$m and the gap length is varied from 0.07 $\mu$m to 0.05 $\mu$m. $K$ for the device is found to be 135.9Ω. The term 26.73 appearing in the expression for $R$ in the graph is due to the nonuniform distribution of the current density at the edges of the gap and also within the gap. It represents an error varying between 2.4–3.3% of the resistance value.

Figure 11 also explains the kink behavior of the device with the variations in the gap length. The hole current through the gap to the substrate results in the development of a voltage across the gap say $\Delta V_b$. This change in the body voltage $\Delta V_b$ across the gap due to the hole current $I_h$ is

$$\Delta V_b = R_{gap} I_h, \tag{1}$$

where $R_{gap}$ is the gap resistance.

Equation (1) helps us to interpret the kink behavior of the SELBOX structure. For large gap lengths $(w_{gap})$ in the back oxide, the gap resistance $R_{gap}$ will be very small and the substrate current will not lead to considerable rise in the body potential. As the body voltage remains fairly constant in this case, the device behaves like a bulk MOSFET. If the gap length $(w_{gap})$ is reduced, the hole transport to the substrate through the gap rises the body voltage. The change in the body voltage $\Delta V_b$ leads to the kink in the output characteristics. Gradual reduction in the gap lengths $(w_{gap})$ while maintaining $(t_{box})$ constant will result in an increase in $R_{gap}$ and the kink will occur at lower drain voltages. This can be seen in Figure 9.

For a certain $V_{gs}$, the resulting hole current is constant at a certain drain voltage. An increase in the gap length $(w_{gap})$ leads to a decrease in the gap resistance $R_{gap}$. The kink will therefore occur at larger drain voltages. In the same way, equation (1) can explain the nature of plot given in Figure 10 for variations in the kink voltage with thickness variations $(t_{box})$. Increasing values of $(t_{box})$ indicate an increase in $R_{gap}$. Therefore, the body voltage required for the occurrence of kink can result from a lower substrate current or in other words the for a lower drain voltage. Hence the kink results at a lower drain voltage.

TABLE 3: Hole current through the gap $I_{sub}$.

| Estimated $I_{sub}$ using (4) | Estimated $\Delta I_{sub}$ due to body voltage from the graph given in Figure 14 | Estimated $I_h$ from gap voltage and gap resistance using (1) | Gap current $I_{sub}$ (from simulation results) |
|---|---|---|---|
| 10.16 $\mu$A | 0.45 $\mu$A | 11.14 $\mu$A | 11.9 $\mu$A |

For a SELBOX structure with a known dimension, we can estimate gap resistance $R_{gap}$ with the following expression:

$$R_{gap} = K \frac{t_{box}}{W_{gap}}. \tag{2}$$

The estimated resistance of the back oxide gap can be further used to find the change in body voltage for a given change in the substrate current.

The body to source diode shown in Figure 11 behaves like a normal $p$-$n$ diode.

The diode current for the body to source diode is given by [13]

$$I_D = I_S e^{(V_{bs}/nV_t)}. \tag{3}$$

$I_s$ for the body to source diode existing in the SELBOX structure used for simulation is found to be $1.3 \times 10^{-15}$; A from simulation and value of $n$ is found to be 1.3. This device will be forward biased, and conduction through this $p$-$n$ junction will only be for extremely narrow gaps.

The substrate current $I_{sub}$ generated due to impact ionization near the drain is empirically related to the drain voltage $V_{ds}$ as given by [14]

$$I_{sub} = C_1 [V_{ds} - V_{d\,sat}] I_d \exp\left[\frac{-C_2}{(V_d - V_{d\,sat})}\right], \tag{4}$$

where $V_{d\,sat}$ is the drain saturation voltage and $C_1$ and $C_2$ are empirical constants [15]. The units of $C_1$ and $C_2$ are $V^{-1}$ and V, respectively. Values of $C_1$ and $C_2$ are extracted from the characteristics of the device. $C_1$ is a function of oxide thickness $t_{ox}$, substrate doping concentration $N_{sub}$, and junction depth $X_j$. $C_2$ is obtained from the slope of the line in Figure 13. $C_1$ is derived from the $y$-intercept of the line in the plot given in Figure 13, which can be also obtained from the equation for the line.

Equation (4) shows the dependence of $I_{sub}$ on the drain voltage. The impact ionization leading to the generation of $I_{sub}$ depends on the electric field near the drain. In addition to the drain voltage, the body voltage of the device will also contribute to the overall electric field near the drain, which determines the impact ionization. The dependence of impact ionization on the body voltage can be estimated in terms of the change in the substrate current due to the body voltage while keeping the other voltages in the device constant. A plot for the change in the substrate current for known changes in the body voltage is given in Figure 14. The slope of the line in the graph $g_{sb}$ can be used to estimate the changes in the substrate current for body voltage variations. For the device under study, $g_{sb}$ is found to be $3 \times 10^{-5}$ A/V.

The above device analysis led to the development of circuit model shown in Figure 15. The current generated by impact ionization is represented by the source $I_{sub}$. The body voltage-dependence of the substrate current is shown by the voltage dependent current source $g_{sb} \cdot v_b$, the diode represents the body-source $p$-$n$ junction, while $R_{gap}$ is the gap resistance.

For oxide gap with small dimensions, the channel resistance will be negligibly small compared to the gap resistance. As the hole current resulting from impact ionization is a function of the drain voltage, for a certain drain voltage the hole current generated will be fixed. For increased values of ($w_{gap}$), there will be a corresponding decrease in $R_{gap}$. Therefore, the required value of body voltage across $R_{gap}$ that can lead to a kink will result at a higher drain voltage. Consequently, the kink voltage increases with gap length ($w_{gap}$) as given in Figure 9. In a similar fashion keeping ($w_{gap}$) constant if ($t_{box}$) is increased, we can expect an increase in $R_{gap}$. With increase in ($t_{box}$) as $R_{gap}$ increases, the required body voltage for the occurrence of kink can result even from smaller hole currents or at lower drain voltages as indicated in Figure 10. Therefore, the drain voltage at which kink occurs reduces with increase in the values of ($t_{box}$).

Results obtained from the verification of the circuit model for the SELBOX structure are summarized in Tables 1, 2, and 3.

The estimated substrate current $I_{sub}$ and the increment in the substrate current due to the changes in the body voltage $\Delta I_{sub}$ is 10.6 $\mu$A. This is very close to the substrate current obtained by using the gap voltage and gap resistance $I_h$ (11.14 $\mu$A) and the $I_{sub}$ (simulation) obtained from the device simulation.

## 5. Conclusion

Partially depleted SOI MOSFET devices exhibit a nonlinearity in the output current voltage characteristics. This abrupt rise in the drain current known as kink results because of the changes in the body voltage of the device due to the accumulation of holes on the floating body. The effectiveness of SELBOX structure in eliminating the kink in the output characteristics of a partially depleted SOI MOSFET device is investigated. Simulation of SELBOX structure using Silvaco Athena and Atlas was performed. It is found that the kink in the output characteristics for a 0.4 $\mu$m thick partially depleted SOI MOSFET can be virtually eliminated with a very narrow gap of ~0.1 $\mu$m in back oxide. This indicates that with oxide underneath all of source, drain, and more than 95% of the channel, kink effect can be virtually suppressed.

The effect of gap length, SELBOX thickness and body voltage on the kink voltage was investigated. A device model is proposed, which explains the basic mechanism that leads to the presence of kink even with SELBOX gap. The model includes the modulation effects of body voltage on the impact ionization and generation of holes. For a given drain voltage, the substrate current generated by impact ionization and the estimated gap current are found to be agreeing well with those found from simulation.

The SELBOX structure has several advantages over the bulk MOS and SOI devices. It is verified in this paper that the kink effect can be completely eliminated while preserving the advantages of SOI. The preliminary results on the frequency response of the SELBOX structure demonstrate that the transition frequency $f_T$ of these devices is larger than that of the bulk MOS devices with similar dimensions. Further, investigations on the thermal characteristics indicate that the peak temperature of SELBOX structures during the device operation is considerably lower than the SOI devices with similar dimensions. However, SELBOX structure has limitations. With increasing gap length, the device will be kink-free but tends to behave more like a bulk MOSFET, and hence, less likely to have other associated advantages of an SOI like reduced drain to substrate and source to substrate capacitances. Introduction of high quality BOX layer below the source and drain regions with narrow gaps under a single MOSFET will require additional process steps. With higher process complexity, there will be an associated increase in the cost as well. Work is in progress on the self-heating effects and frequency characteristics of SELBOX device.

## References

[1] J. B. Kuo and S. C. Lin, *Low-Voltage SOI CMOS VLSI Devices and Circuits*, John Wiley & Sons, 1st edition, 2001.

[2] A. marshall and S. Natarajan, *SOI Design: Analog, Memory and Digital Techniques*, Kluwer Academic Publishers, 2001.

[3] K. Bernstein and j. Norman, *SOI Circuit Design Concepts*, Kluwer Academic Publishers, 2000.

[4] J. Chen, R. Solomon, T. Y. Chan, P. K. Ko, and C. Hu, "Threshold voltage and C-V characteristics of SOI MOSFET's related to Si film thickness variation on SIMOX wafers," *IEEE Transactions on Electron Devices*, vol. 39, no. 10, pp. 2346–2353, 1992.

[5] J. Z. Ren and C. A. T. Salama, "1 V SOI NMOSFET with suppressed floating body effects," *Solid-State Electronics*, vol. 44, no. 11, pp. 1931–1937, 2000.

[6] B. A. Chen, A. Hirsch, S. K. Iyer, N. Rovedo, H. -J. Wann, and Y. Zhang, "Patterned Buried Insulator," US Patent no. 6429091 B1, 2002.

[7] Y. Dong, M. Chen, J. Chen et al., "Patterned buried oxide layers under a single MOSFET to improve the device performance," *Semiconductor Science and Technology*, vol. 19, no. 3, pp. L25–L28, 2004.

[8] C. Pal, B. Mazhari, and S. S. K. Iyer, "Simulation of MOSFET devices and circuits fabricated on selective buried oxide (SELBOX) substrates," in *Proceedings of IEEE Conference on Electron Devices and Solid-State Circuits (EDSSC '05)*, pp. 559–562, Hong Kong, December 2005.

[9] M. Y. Hammad, "Analytical modeling of the partially-depleted SOI MOSFET," *IEEE Transactions on Electron Devices*, vol. 48, no. 2, pp. 252–258, 2001.

[10] *ATHENA User's Manual Device Simulation Software*, Silvaco International, Santa Clara, Calif, USA, 2004.

[11] *ATLAS User's Manual Device Simulation Software*, Silvaco International, Santa Clara, Calif, USA, 2004.

[12] J.-P. Colinge, *Silicon-On-Insulator Technology: Materials to VLSI*, Springer, 3rd edition, 2004.

[13] D. A. Neamen, *Microelectronics*, McGraw-Hill, 3rd edition, 2006.

[14] I. M. Hafez, G. Ghibaudo, and F. Balestra, "Analysis of the kink effect in MOS transistors," *IEEE Transactions on Electron Devices*, vol. 37, no. 3, pp. 818–821, 1990.

[15] T. Y. Chan and P. K. Ko, "A simple method to characterize substrate current in MOSFET's," *IEEE Electron Device Letters*, vol. EDL-5, no. 12, p. 506, 1984.

# A 0.8 V 0.23 nW 1.5 ns Full-Swing Pass-Transistor XOR Gate in 130 nm CMOS

**Nabihah Ahmad and Rezaul Hasan**

*Center for Research in Analog and VLSI microsystem Design (CRAVE), School of Engineering and Advanced Technology (SEAT), Massey University, Auckland 0632, New Zealand*

Correspondence should be addressed to Rezaul Hasan; hasanmic@massey.ac.nz

Academic Editor: Ching Liang Dai

A power efficient circuit topology is proposed to implement a low-voltage CMOS 2-input pass-transistor XOR gate. This design aims to minimize power dissipation and reduce transistor count while at the same time reducing the propagation delay. The XOR gate utilizes six transistors to achieve a compact circuit design and was fabricated using the 130 nm IBM CMOS process. The performance of the XOR circuit was validated against other XOR gate designs through simulations using the same 130 nm CMOS process. The area of the core circuit is only about 56 sq $\cdot\, \mu$m with 1.5659 ns propagation delay and 0.2312 nW power dissipation at 0.8 V supply voltage. The proposed six-transistor implementation thus compares favorably with other existing XOR gate designs.

## 1. Introduction

Low-power circuits have become a major design forethought with the overwhelming growth of portable applications, particularly, for battery operated hand-held devices. In addition, higher power consumption raises the temperature of the chip which in turn affects the reliability of the devices and circuits [1]. Various low-power techniques have been explored to enhance the basic logic gates such as the XOR gate which influence the overall power consumption in many system-on-chip (SOC) implementations. One of the effective ways to reduce the overall power consumption is by reducing the supply voltage. XOR gate optimization in terms of power, speed, and transistor count has significantly improved the performance of larger and complex circuits. Over the years, various 2-input XOR gate designs have been widely reported to enhance the performance of various applications such as full adder, parity generator, encryption processor, and comparator. Traditional eight-transistor static CMOS XOR gate can operate with full output swing but with the drawback of power dissipation and transistor count [2]. On the other hand, XOR circuit based on the transmission gate [3] is used to overcome the signal degradation caused by the PMOS and NMOS devices in pass-transistor logic. However, it has the drawback of the loss of driving capability and requires complementary signals to switch the PMOS and NMOS devices and hence needs more transistors and area. A cross-coupled (CC) XOR gate based on the pass-transistor logic has been reported in [4], which claims to have improved speed and power consumption than the six device pass-transistor XOR gate and works well under a low-voltage regime. A six-transistor XOR gate realized from a modified four-transistor XOR gate by cascading a standard inverter as an output driver can be found in [5], which improves the poor output signal level for certain inputs. Powerless XOR gate (P-XOR) is proposed in [6] using a four-transistor circuit with no power supply connection which consumes less power than other designs but at the expense of a large delay. Another four-transistor XOR gate design was reported in [7], based on the Gate-Diffusion-Input (GDI) cell [8]. A three-transistor XOR gate can be found in [9] using a CMOS inverter and a PMOS pass transistor. It provides low-power-delay product (PDP) but has a voltage degradation with the input combination $A = 1$ and $B = 0$. Elgamel et al. [10] also proposed a similar three-transistor XOR gate, but it consumes high power when $A = 1$ and $B = 0$ and, in addition, produces a poor logic “1” for this input combination. However, it may reach an acceptable logic high voltage level with appropriate transistor sizing. Thus,

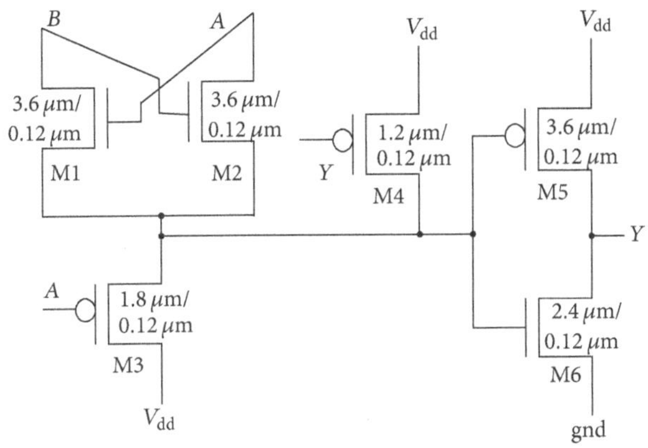


FIGURE 1: Circuit diagram of the proposed six-transistor pass-transistor-based full-swing CMOS XOR gate.

both the circuits [9, 10] may not operate reliably at low-supply voltage. In this brief paper, we present a novel low-power, low-voltage, and full-swing 2-input XOR circuit using 6 devices, implemented in 130 nm IBM CMOS technology.

## 2. Novel XOR Gate Topology for Low-Power CMOS Design

A low-power constrained 2-input XOR gate using six transistors is proposed which provides full output voltage swing for all input combinations and enables low-voltage operation with a small propagation delay. The proposed XOR circuit is based on pass-transistor logic with an inverter as the output driver to achieve perfect output swing. Pass-transistor design enables small transistor count along with smaller input loads (with signal input to source/drain instead of gate) offering very low-power operation with high performance. Since an NMOS device passes a strong "0" but a weak "1," while a PMOS device passes a strong "1" but a weak "0;" the complementary pass transistors are organized to pass a strong output logic level for all input combinations of "1" and "0".

Figure 1 shows the schematic of the proposed XOR circuit along with the device sizes using the 130 nm CMOS process. It performs a perfect full-swing operation for every input pattern. $V_{dd}$ connections to the source terminals of M3 and M4 are used to drive a full-rail output of "1". For $A = B = 0$ condition, transistor M3 is ON, passing a strong "1" from $V_{dd}$ to the inverter input which generates a "0" at the output $Y$ to turn ON M4, when $A = 0$, $B = 1$, and transistors M2 and M3 are ON, which passes signals "0" and "1," respectively. To solve this problem, $W/L$ ratio of M2 is increased making it larger than the $W/L$ ratio of M3 so as to pass only a signal "0" to the inverter input and generate the strong output $Y = 1$. For $A = 1$ and $B = 0$, only the device M1 is ON, and a strong "0" is passed to the inverter generating a full-rail "1" to output $Y$. With $A = B = 1$, transistors M1 and M2 are ON, and a weak signal "1" is passed to the inverter input and, as a result, the output $Y$ will also be degraded. However, the feedback path causes transistor M4 to turn ON when $Y = 0$ thus passing the perfect signal "1" from $V_{dd}$ to the inverter input resulting in the generation of the perfect signal "0" at $Y$. This pass-transistor XOR thus does not suffer from signal level deteriorations like other pass-transistor XOR gates. The transistor sizes are carefully chosen for optimal power-delay performance under various operating conditions.

## 3. XOR Gate Performance Analysis and Simulation Results

Extensive simulations of the proposed XOR gate along with five other existing XOR gates found in the literature have been carried out using the 130 nm IBM CMOS technology in order to analyze the performance comparison. The simulations were carried out on the Cadence Spectre platform and the Synopsys HSPICE platform using the same test environment to measure the propagation delay and the power dissipation in each case. All the simulations were carried out with a 0.6 V to 1.2 V supply voltage range, a load capacitance of 10 fF and a throughput (clocking) rate of 200 MHz. The simulations consist of functional verification, power, timing analysis, design rule checking (DRC), and layout versus schematic (LVS) of the layout verified using Cadence Assura. The performance of all the XOR test circuits has been evaluated in terms of the worst-case propagation delay. Propagation delay is evaluated from the time interval between the 50% input and the 50% output voltage transition points. The figure-of-merit power-delay product (PDP) is calculated from the product of the worst case propagation delay and the average power consumption. Several input patterns that covered all possible cases of input values were applied, and the simulation results verified the correct functionality for every input combination up to the lower end of the supply voltage range. Figure 2 shows the HSPICE transient circuit simulation of the full-swing operation of this pass-transistor XOR gate indicating rise and fall times in the range of few hundred pico seconds (@ supply voltage, $V_{dd} = 0.8$ V). Table 1 summarizes the results of these simulations providing a comparison of the propagation delay, the power dissipation, and the PDP between the proposed circuit and the other recently reported designs. The proposed XOR gate offers lower-propagation delay than the other six-transistor XOR gates as shown in Table 1 and depicted in the line graph in Figure 3. The worst circuit in term of speed is the three-transistor XOR gate. It has the highest propagation delay against voltage scaling. The six-transistor designs by the authors in [4, 5] are close to the proposed XOR gate in term of power dissipation and propagation delay, but the proposed XOR gate provides better overall improvement compared to these other previous designs. The proposed XOR circuit performed satisfactorily for low-supply voltages, with the 0.2312 nW power dissipation compared to the 0.2319 nW dissipation by the design in [5] (@ supply voltage, $V_{dd} = 0.8$ V). The 4T XOR gates in [6, 7] and the 3T XOR gates in [1, 9, 10] have almost identical average power dissipation for all the supply voltages. The line graph in Figure 4 demonstrates the effect of voltage scaling on the average power dissipation. All other circuits have higher PDP than the proposed XOR circuit as evidenced in

TABLE 1: Comparison of the simulation results for different CMOS XOR gates with the proposed XOR gate.

| | $V_{dd}$ (v) | Proposed (6T) | 6T [5] | 6T [4] | 4T [6] | 4T[7] | 3T [1, 9, 10] |
|---|---|---|---|---|---|---|---|
| Delay (ns) | 0.6 | 2.1107 | 2.3113 | 7.0638 | 2.5818 | 4.1910 | 15.48 |
| | 0.8 | 1.5659 | 1.7672 | 3.6837 | 4.5128 | 3.9308 | 8.7314 |
| | 1 | 1.4691 | 1.6475 | 1.6469 | 4.2216 | 8.9244 | 13.038 |
| | 1.2 | 1.4200 | 1.6018 | 1.5500 | 4.1087 | 7.2075 | 12.901 |
| Average power (nW) | 0.6 | 0.1351 | 0.1352 | 0.1385 | 0.2672 | 0.2672 | 0.2672 |
| | 0.8 | 0.2312 | 0.2319 | 0.2401 | 0.4586 | 0.4586 | 0.4586 |
| | 1 | 0.3674 | 0.3689 | 0.3842 | 0.7269 | 0.7269 | 0.7268 |
| | 1.2 | 0.5523 | 0.5574 | 0.5835 | 1.0931 | 1.0931 | 1.0928 |
| PDP (aJ) | 0.6 | 0.2852 | 0.3125 | 0.7526 | 0.6899 | 1.1198 | 4.1363 |
| | 0.8 | 0.3620 | 0.4098 | 0.8845 | 2.0696 | 1.8027 | 4.0042 |
| | 1 | 0.5397 | 0.6078 | 0.6327 | 3.0687 | 6.4871 | 9.4760 |
| | 1.2 | 0.7843 | 0.8928 | 0.9044 | 4.4912 | 7.8785 | 14.098 |

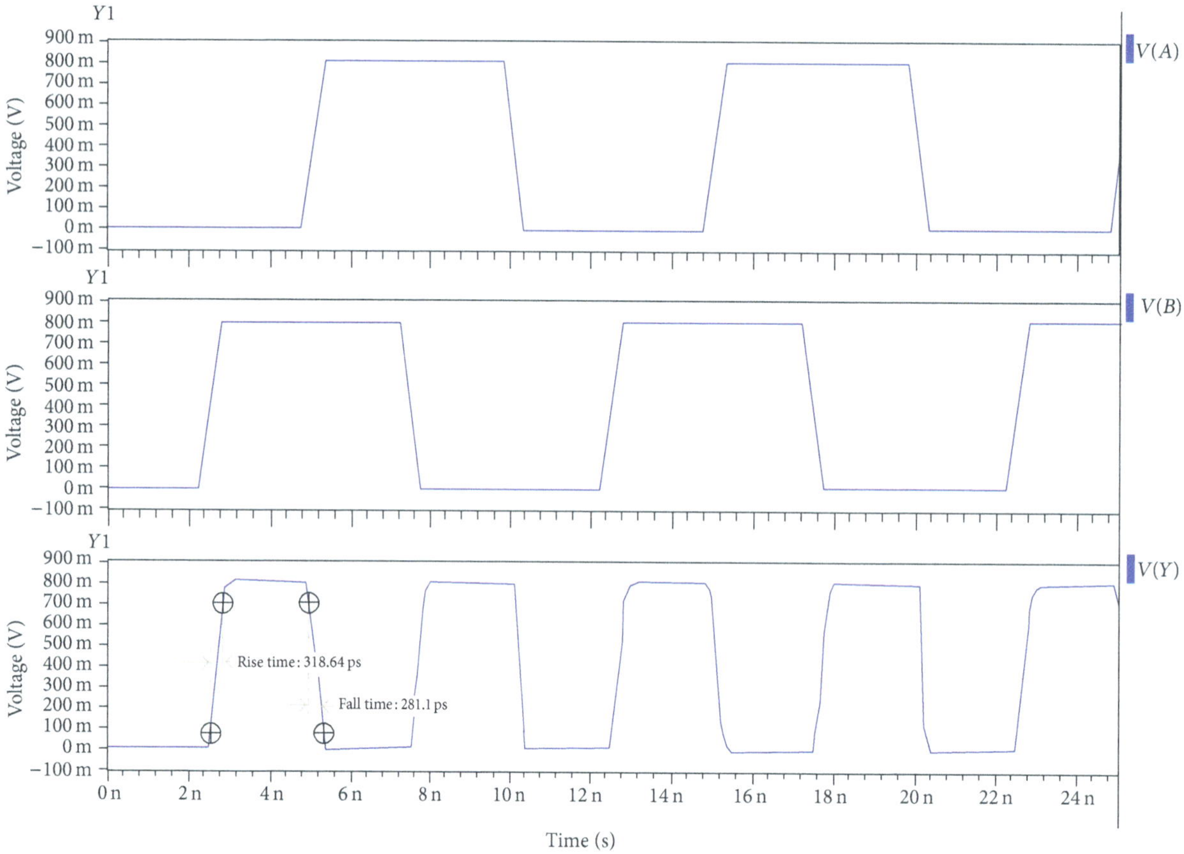


FIGURE 2: HSPICE transient simulation of full-swing output voltage of the novel pass-transistor XOR gate.

Table 1 and illustrated in Figure 5. At 0.8 V supply voltage, the proposed XOR circuit has at least 13.2% improvement in PDP over the design by the authors in [5] and 143.6% over the circuit by the authors in [4]. As output load is one of the parameters that affect the performance of the circuits, we have varied the output load from 10 fF to 50 fF at 0.8 V supply voltage for all circuits to study its effect on the propagation delay. The proposed XOR gate is found to be the best circuit in terms of the propagation delay for all values of output loads as shown in the line graph in Figure 6. Thus, from the simulation results, it is clear that the proposed new XOR gate has the lowest propagation delay as well as the lowest power consumption along with high output driving capability. The improvements attained by the proposed circuit are thus clearly evident when compared to these other XOR gate circuits.

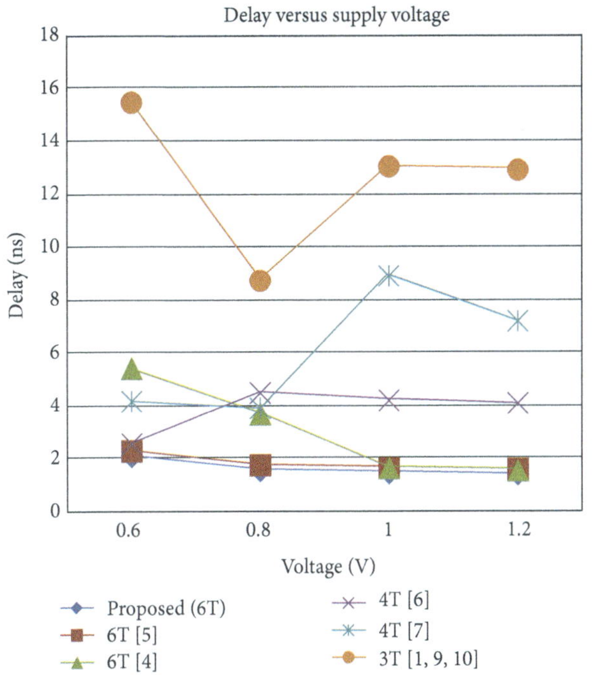


FIGURE 3: Line graph showing the comparison of the propagation delay versus supply voltage scaling for different CMOS XOR gates with the proposed novel CMOS pass-transistor-based full-swing XOR gate.

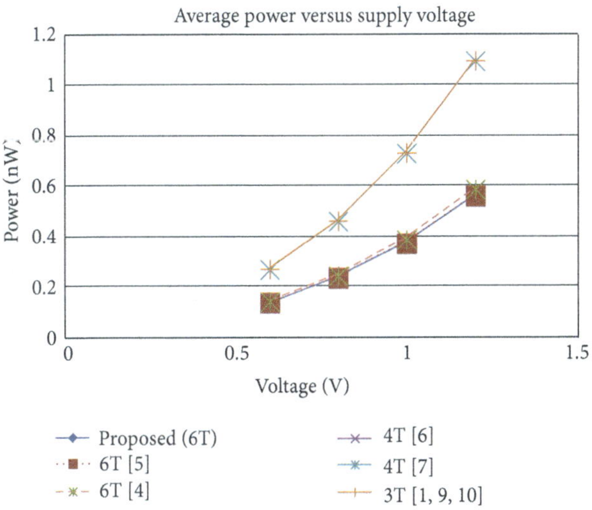


FIGURE 4: Line graph showing the comparison of the power dissipation versus supply voltage scaling for different CMOS XOR gates with the proposed novel CMOS pass-transistor-based full-swing XOR gate.

In order to verify the noise immunity of the proposed XOR gate, the noise margins ($NM_H$ and $NM_L$) of the proposed XOR gate and the other gates were determined based on DC input-output voltage transfer analysis. The noise margins for a 0.8 V supply voltage are shown in Table 2. The proposed XOR gate indicates acceptable values of noise margin when compared with the other XOR gates.

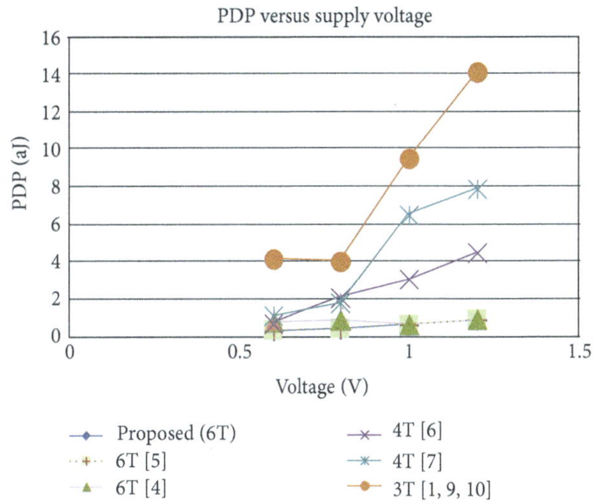


FIGURE 5: Line graph showing the comparison of the power-delay product (PDP) versus the supply voltage scaling for various XOR gates with the proposed novel CMOS pass-transistor-based full-swing XOR gate.

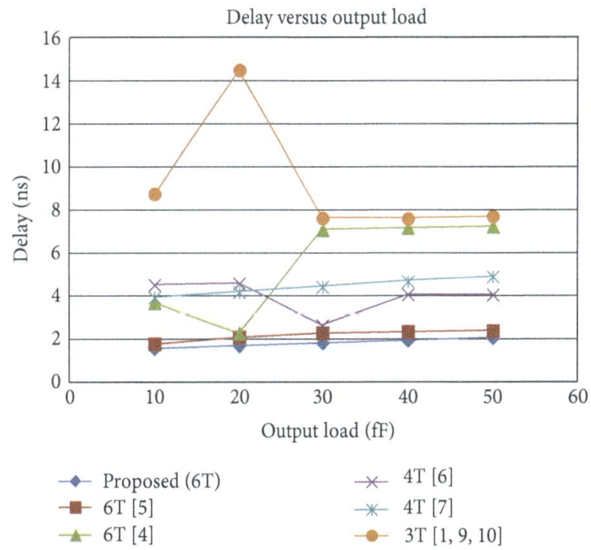


FIGURE 6: Line graph showing the comparison of the propagation delay versus output load for various CMOS XOR gates with the proposed novel CMOS pass-transistor-based full-swing XOR gate.

## 4. XOR Gate Fabrication and Experimental Results

The mask layout of the XOR gate illustrated in Figure 7 was fabricated in the 130 nm IBM CMOS process. Minimum channel length is used for all the devices, and optimum channel width is carefully chosen for each device in order to achieve verified functionality with low-power dissipation and smallest possible propagation delay. The photomicrograph of the fabricated XOR gate along with bonding pads is shown in Figure 8. The silicon area of the XOR gate was 8.02 $\mu$m × 7.03 $\mu$m (≈56 sq·$\mu$m) excluding the bonding pads. Figure 9

TABLE 2: Comparison of noise margins of different CMOS XOR gates with the proposed XOR gate.

| Type of XOR | $V_{dd}$ (V) | $V_{oh}$ (V) | $V_{ih}$ (V) | $V_{il}$ (V) | $V_{ol}$ (V) | $NM_H$ (V) | $NM_L$ (V) |
|---|---|---|---|---|---|---|---|
| Proposed (6T) | 0.800 | 0.7186 | 0.3889 | 0.2867 | 0.0796 | 0.3297 | 0.2071 |
| 6T [5] | 0.800 | 0.7204 | 0.3967 | 0.3022 | 0.0687 | 0.3237 | 0.2335 |
| 6T[4] | 0.800 | 0.7205 | 0.3967 | 0.3022 | 0.0687 | 0.3238 | 0.2335 |
| 4T [6] | 0.800 | 0.9460 | 0.6000 | 0.4489 | 0.1172 | 0.346 | 0.3317 |
| 4T [7] | 0.800 | 1.0809 | 0.6444 | 0.4233 | 0.1423 | 0.4365 | 0.2810 |
| 3T [1, 9, 10] | 0.800 | 1.0797 | 0.5978 | 0.4100 | 0.1332 | 0.4819 | 0.2768 |

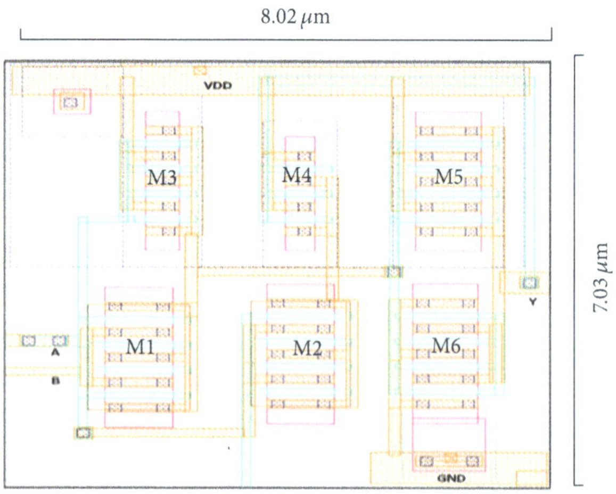


FIGURE 7: Layout of the proposed novel CMOS pass-transistor-based full-swing 2-input XOR gate.

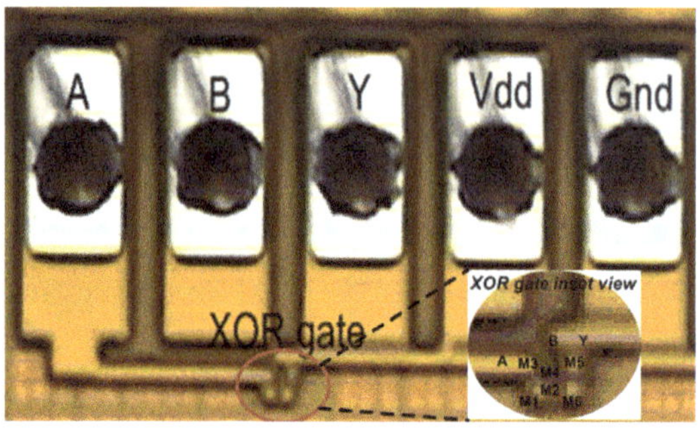


FIGURE 8: Microphotograph of the fabricated novel CMOS pass-transistor-based full-swing XOR gate along with bonding pads.

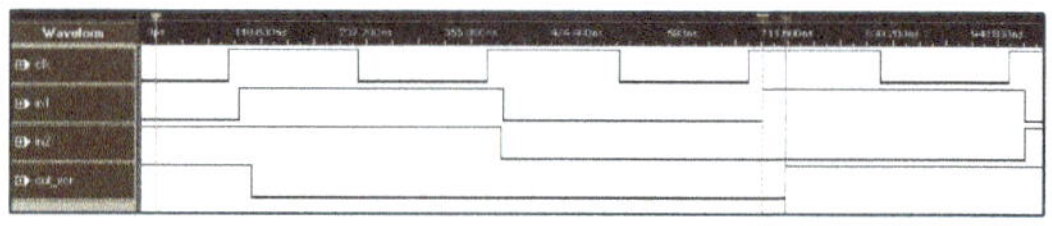

FIGURE 9: Logic analyzer waveform of the input and the output for the fabricated novel CMOS pass-transistor-based full-swing XOR gate.

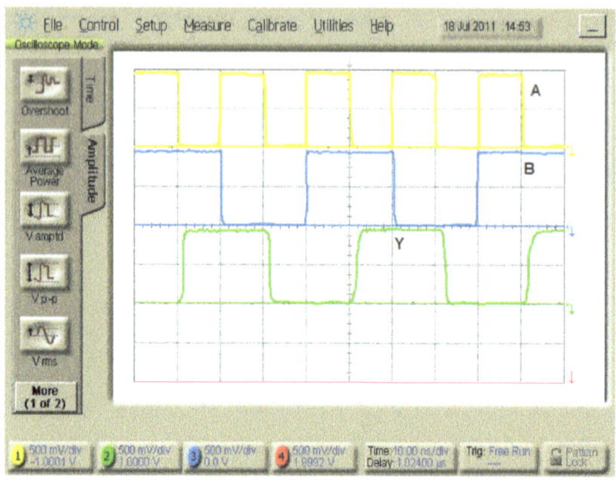


FIGURE 10: Oscilloscope waveforms for the fabricated novel CMOS pass-transistor-based full-swing XOR gate (yellow = input A, blue = input B, and green = output Y).

displays the functional verification of the XOR gate with a snapshot of the XOR input and output waveforms for several input combinations using the Tektronix TLA5202 Logic Analyzer. Level shifter was used to generate low voltage (0.8 V) from high voltage (5 V) of the pattern generator to the XOR gate input. Finally, Figure 10 provides the Agilent DCA-J (86100C Infiniium) oscilloscope electrical waveforms for the fabricated XOR gate (yellow = input A, blue = input B, and green = output Y) indicating a correct operation for a supply voltage of 0.8 V.

## 5. Conclusion

This paper demonstrates a new pass-transistor-based full-swing 2-input XOR gate topology implemented in 130 nm CMOS process suitable for reducing the power and propagation delay of an overall system on chip. The proposed XOR gate was compared to other peer XOR designs, and the results indicate satisfactory performance and improvements in term of power consumption, propagation delay, and power-delay product compared to the other designs. The proposed XOR gate has a lower-gate delay and a lower-power-delay product in comparison to its peer designs. It is thus suitable for small area and low-power applications such as RFID tags.

## References

[1] M. Hosseinghadiry, H. Mohammadi, and M. Nadisenejani, "Two new low power high performance full adders with minimum gates," *International Journal of Electrical and Computer Engineering*, vol. 4, no. 10, pp. 671–678, 2009.

[2] N. Weste and K. Eshraghian, *Principles of CMOS VLSI Design: A Systems Perspective*, Addison-Wesley Longman, Reading, Mass, USA, 2nd edition, 1993.

[3] S. M. Kang and Y. Leblebici, *CMOS Digital Integrated Circuits: Analysis and Design*, McGraw Hill, New York, NY, USA, 3rd edition, 2002.

[4] K. J. Lin and C. W. Wu, "Low-cost realization of multiple-input exclusive-OR gates," in *Proceedings of the 8th Annual IEEE International ASIC Conference and Exhibit*, pp. 307–310, September 1995.

[5] J. M. Wang, S. C. Fang, and W. S. Feng, "New efficient designs for XOR and XNOR functions on the transistor level," *IEEE Journal of Solid-State Circuits*, vol. 29, no. 7, pp. 780–786, 1994.

[6] H. T. Bui, A. K. Al-Sheraidah, and Y. Wang, "New 4-transistor XOR and XNOR designs," in *Proceedings of the 2nd IEEE Asia Pacific Conference on ASICs (AP-ASIC '00)*, pp. 25–28, Cheju, Korea, 2000.

[7] A. K. Nishad and R. Chandel, "Analysis of low power high performance XOR gate using GDI technique," in *Proceedings of the International Conference on Computational Intelligence and Communication Networks*, pp. 187–191, 2011.

[8] A. Morgenshtein, A. Fish, and I. A. Wagner, "Gate-diffusion input (GDI)—a novel power efficient method for digital circuits: a design methodology," in *Proceedings of the 14th Annual IEEE International ASIC/SOC Conference*, pp. 39–43, September 2001.

[9] S. R. Chowdhury, A. Banerjee, A. Roy, and H. Saha, "A high speed 8 transistor full adder design using novel 3 transistor XOR gates," *International Journal of Electrical and Computer Engineering*, vol. 3, no. 12, pp. 784–790, 2008.

[10] M. Elgamel, S. Goel, and M. Bayoumi, "Noise tolerant low voltage XOR-XNOR for fast arithmetic," in *Proceedings of the ACM Great Lakes Symposium on VLSI (GLSVLSI '03)*, pp. 285–288, April 2003.

# A Unified Channel Charges Expression for Analytic MOSFET Modeling

**Hugues Murray and Patrick Martin**

*Normandie Université, Ensicaen, UMR 6508, LaMIPS, Laboratoire Commun CNRS-NXP-PRESTO-ENSICAEN UCBN, 6 boulevard Maréchal Juin, 14050 CAEN Cedex 4, France*

Correspondence should be addressed to Hugues Murray, hugues_murray@presto-eng.com

Academic Editor: Gerard Ghibaudo

Based on a 1D Poissons equation resolution, we present an analytic model of inversion charges allowing calculation of the drain current and transconductance in the Metal Oxide Semiconductor Field Effect Transistor. The drain current and transconductance are described by analytical functions including mobility corrections and short channel effects (CLM, DIBL). The comparison with the Pao-Sah integral shows excellent accuracy of the model in all inversion modes from strong to weak inversion in submicronics MOSFET. All calculations are encoded with a simple C program and give instantaneous results that provide an efficient tool for microelectronics users.

## 1. Introduction

Although MOSFET modeling is now well covered and addressed in BSIM, EKV, and PSP compact models [1], it is always interesting to present a semianalytic resolution of 1D Poissons equation which can be implemented in popular computers with usual software giving most physical results (potential and charges distribution) instantaneously. New approaches of MOSFET surface potential modeling were performed from analytic treatment and have brought a renewal in analytic resolution of surface potential [2–5]. We previously used a similar method in the analytic description of surface potential by Taylor expansion [6].

Oguey and Cserveny [7] proposed as early as 1982 a complete analytic model based on the gate and drain source voltages. An important step was reached in modeling by Enz et al. in 1995 [8], Iniguez et al. [9] in 1996, and Cheng [10] in 1998 who gave analytic expression of the inversion charge. We certainly do not pretend to provide an alternative method to the compact models implemented on the simulators for CAD, but are simply trying to provide analytical support to the understanding of strategic components of microelectronics.

From the analytical expression of inversion charge as a function of gate and drain bias, we attempted to provide a single analytical expression that achieves explicit functions of the drain current $I_D$ $(Vg, V_D)$ and the transconductance $g(Vg, V_D)$. The originality is based on a model in which the threshold voltage does not appears explicitly, but is replaced in the analytical expression by a parameter $b(Vg)$ dependent on the surface potential at zero drain bias.

It became obvious to us that the influence of other parameters could be included in these equations by more complex developments based on quasi two-dimensional analysis that exceeded this paper. Thus, we have not considered the specific effects: ballistic transport, tunneling through the oxide gate, which alone account for modeling of complex developments and led to numerical 2D treatments.

The presentation is made under the Gradual Channel Approximation (GCA) [11] which assumes that the electric field in the direction perpendicular to the channel is much greater than in the direction parallel to the channel and allows a 1D model of Poisson-Boltzmann equation. The different explicit equations (gate voltage and channel potential versus surface potential) are inverted using Taylor expansion, and we solve all equations until the point analytic calculations can be done then calculate the single integrals by Simpson algorithm encoded in simple C programs.

## 2. Basic Assumptions in MOSFET

*2.1. The Surface Potential Equation.* Under the gradual channel approximation [11], with the introduction of the reduced channel voltage $\xi(y) = V(y)/U_T$ as quasi-Fermi potential [12] and the correct charges densities are [4]

$$\begin{aligned} n(x,y) &= n_i e^{u(x,y)-\xi(y)}, \\ p(x,y) &= p_i e^{-u(x,y)}, \\ N_A &= p_i e^{-u_b}, \\ N_D &= n_i e^{u_b-\xi(y)}. \end{aligned} \tag{1}$$

The Poisson-Boltzmann equation can be analytically solved using the 1Dmodel of Nicollian and Brews [13] from the charge density:

$$\frac{dF(x)}{dx} = \frac{q}{\varepsilon_S}[p(x,y) - n(x,y) + N_D - N_A]. \tag{2}$$

The surface electric field, $F_S(y) = F_x(0,y)$, along $y$ (Figure 1) solution of (2) in $x = 0$, is

$$F_S(y) = \sqrt{\frac{2KTn_i}{\varepsilon_S}}\sqrt{e^{-\xi(y)}H[u_S(y)] + G[u_S(y)]}, \tag{3}$$

by setting:

$$\begin{aligned} H(u) &= e^u + e^{u_b}[u_b - u - 1], \\ G(u) &= e^{-u} - e^{-u_b}[u_b - u + 1]. \end{aligned} \tag{4}$$

The gate voltage $Vg$ relative to flat band is

$$\begin{aligned} Vg - V_{fb} = &-\frac{Q_S(y)}{C_0} \\ &+ \Psi_S(y) = \gamma_0\sqrt{e^{-\xi(y)}H[u_S(y)] + G[u_S(y)]} \\ &+ U_T[u_S(y) - u(b)]. \end{aligned} \tag{5}$$

*2.2. The Surface Potential Dependence to Gate and Drain Bias.* The gate voltage is an explicit function of $u_S(y)$ and $\xi(y)$. Several solutions of $Vg = f[u_S(y), \xi(y)]$ were reported to express the band bending $\Psi_S = U_T[u_S(y) - u_b]$ as an analytic function of the gate and channel voltages [3, 14]. Gildenblat et al. have given in [5, 14] a noniterative expression of the surface potential which serves as a reference for surface potential-based models. In the following, we generated $u_S(y)$ by first-order Taylor expansion as previously done in [6].

$u_S(y)$ versus $Vg$ at a constant drain bias $V(y)$ is generated by

$$u_S(l+1) = u_S(l) + \delta\left(\frac{du_s}{dVg}\right)_{u_s[l]}, \tag{6}$$

$Vg(l) = l \cdot \delta$, ($\delta$ sample step and $l$ integer).

And $u_S(y) = u_{S,m}$ versus$V(y)$ at a constant $Vg$ is generated by:

$$u_{S,m+1} = u_{S,m} + h\left[\frac{du_S}{d\xi}\right]_{u_S=u_{S,m}}. \tag{7}$$

$\xi(y) = mh$ ($h$ sample step and $m$ integer) is expressed as a function of $u_S(y)$ at a constant $Vg$ by an analytic model previously developed by Baccarani et al. [15]:

$$\xi(y) = \ln\frac{H[u_S(y)]}{E[u_S(y)] - G[u_S(y)]} \tag{8}$$

with the introduction of the dimensionless quantity:

$$E(u) = \left[\frac{Vg - U_T(u - u_b)}{\gamma_0}\right]^2. \tag{9}$$

## 3. Analytic Model of Inversion Charges

*3.1. The Inversion Charges Dependence to Gate and Drain Bias.* In an $n$-MOSFET, the inversion charges are defined by the integral of electrons density over the "physical" thickness $d_{\text{inv}}$:

$$Q_{\text{inv}} = q\int_0^{d_i} n(x,y)dx. \tag{10}$$

The "physical" inversion starts at the silicon surface with the surface potential $u(0,y) = u_S(y)$ and ends at the abscissa $x = d_{\text{inv}}$ corresponding to $n(x,y) = p(x,y)$ and $u(d_{\text{inv}}, y) = \xi(y)/2$. The inversion charge dependence with channel potential $\xi(y)$ at a constant $Vg$ noted $Q_{\text{inv}}(\xi(y))|_{V(g)}$ can be written in terms of potential as follows:

$$Q_{\text{inv}}(\xi(y))\,|_{V(g)} = \lambda q n_i \int_{\xi(y)/2}^{u(s)} \frac{e^{u-\xi(y)}du}{\sqrt{e^{-\xi(y)}H(u) + G(u)}}. \tag{11}$$

From (8), $Q_{\text{inv}}(\xi(y))|_{V(g)}$ becomes a single integral of $u$ with the limits only dependent of $m$. Figure 2 shows $Q_{\text{inv}}(V(y))|_{V(g)}$ versus $V(y) = \xi(y) \cdot U_T$ with $Vg$ as a parameter in linear (strong inversion) and log scale (weak inversion).

A threshold voltage of inversion charges $V_{DT}|_{V(g)}$ can be defined by the interpolation of the linear part of $Q_{\text{inv}}(V_D)|_{V(g)}$ with the $V_D$ axis. $V_{DT}|_{V(g)} = f(Vg)$ plots (Figure 3) give at $V_D = 0$ a threshold voltage $V_T$ which differs from $V_{T0}$ by a factor $\approx 1.1$. $V_T$ is used in (12) instead of $V_{T0}$.

*3.2. Analytic Expression of the Inversion Charges Dependence to Gate and Drain Bias.* The simplest analytic approximate expression of $Q_{\text{inv}}(Vg, V_D)$ in the whole range of gate and drain bias is well represented by

$$Q_{\text{inv}}(Vg, V_D) = \eta C_0 U_T \ln\left[1 + \exp\left(\frac{Vg - V_T}{\eta_0 U_T} - \frac{V_D}{U_T}\right)\right]. \tag{12}$$

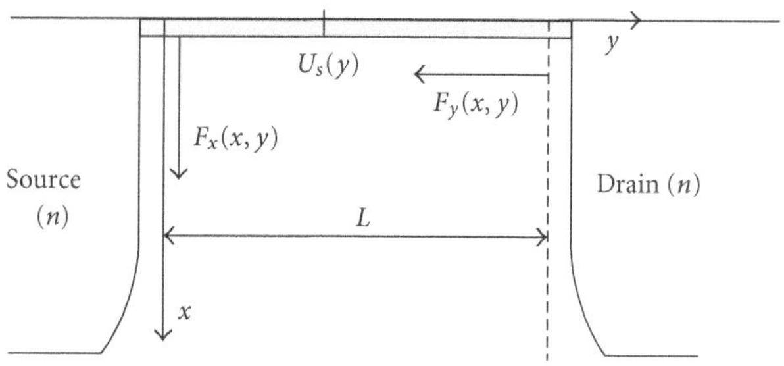


FIGURE 1: The $n$-MOSFET.

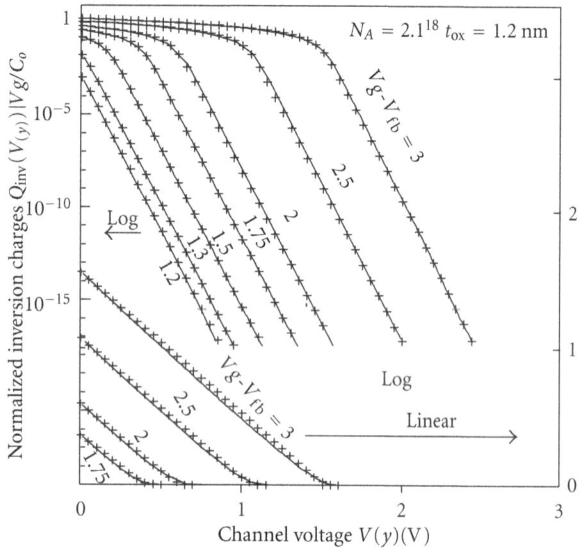


FIGURE 2: Normalized $[Q_{inv}(V(y))|_{Vg}, V(y)]$ plots in logarithm and linear scales. (–) (11), (+) (17).

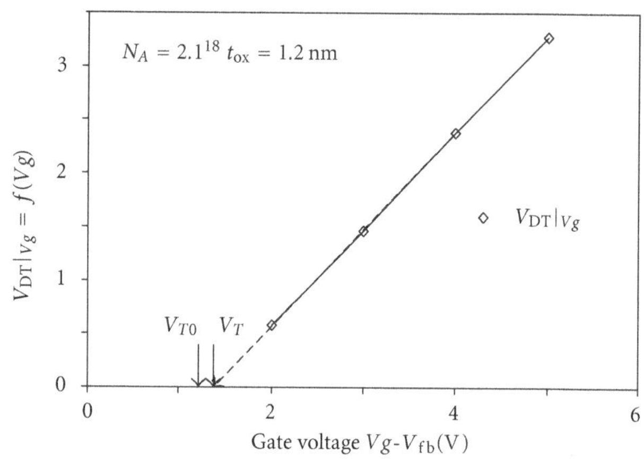


FIGURE 3: The threshold voltages $V_{DT}|_{V(g)} = f(Vg)$.

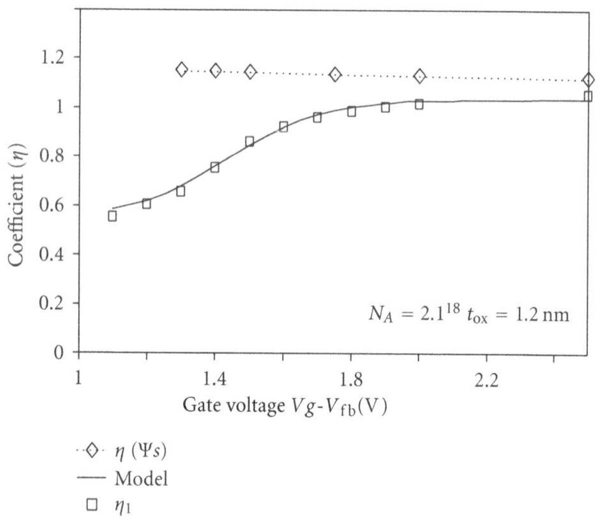


FIGURE 4: $\eta(Vg)$ modeling by exponential functions. ($\eta_{2f} = 1.08$; $N = 5.5$.)

$\eta_0(\approx 1.2)$ is the slope factor defined by the exponential law $Q_{inv}(Vg, 0)$ if $Vg < V_{T0}$.

However, this formulation should contain adjustment coefficients to reduce the error between (11) and (12). This was done in 1996 by Iniguez et al. [9] with the introduction of adjustment coefficients based on the threshold voltage in an expression of inversion charges similar with (12).

We propose an alternative method by introducing a "charge linearization factor," $\eta = \eta(Vg)$ which fits the slope $dQ_{inv}(V_D)|_{V(g)}/dV$ in strong inversion, and a preexponential parameter $a(Vg)$ which fits (12) with (11) in $y = 0$:

$$\widetilde{Q}_{inv\,0}(Vg, V_D) = \eta(Vg) C_0 U_T \ln\left[1 + a(Vg)\exp\left(\frac{Vg - V_T}{\eta_0 U_T} - \frac{V_D}{U_T}\right)\right]. \tag{13}$$

Another definition of the "charge linearization factor" $\eta(\Psi_S)$ was introduced by Sallese et al. [16] in strong inversion and gives results different from $\eta(Vg)$ as shown in Figure 4. $\eta(\Psi_S)(>1)$ increases when $Vg$ decreases. The difference between $\eta(\Psi_S)$ and $\eta(Vg)$ results from $d\Psi_S/dV_D = du_S/d\xi$ which can be calculated from (5). $\eta(Vg)$ and $a(Vg)$ are interdependent and will be estimated in order to minimize the error between $Q_{inv}(V_D)|_{V(g)}$ and $\widetilde{Q}_{inv\,0}(Vg, V_D)$.

Figure 4 shows $[\eta(Vg), Vg]$ plots calculated from $Q_{inv}(v(y))|_{V(g)}$ and $\eta(Vg)$ are well represented by a smoothing function $\eta_{21}$ as follows:

$$\eta(Vg) \approx \eta_{21} = \frac{\eta_{2f}}{2}\frac{1 + \exp((Vg - V_{T0})/NU_T)}{1 + 0.5\exp((Vg - V_{T0})/NU_T)}. \tag{14}$$

$\eta_{2f}$ is the asymptotic value of $\eta(Vg)$ at high gate voltages. $N$ is a slope factor which minimizes the error between $\eta(Vg)$ and $\eta_{21}$ in a large range $[N_A, t_{ox}]$. Under this condition $a(Vg)$ becomes

$$a(Vg)\,|_{V_D=0} = \frac{\exp\left(Q_{inv}(0)|_{Vg}/\eta(Vg)C_0U_T\right) - 1}{\exp((Vg - V_T)/\eta_0 U_T)} \tag{15}$$

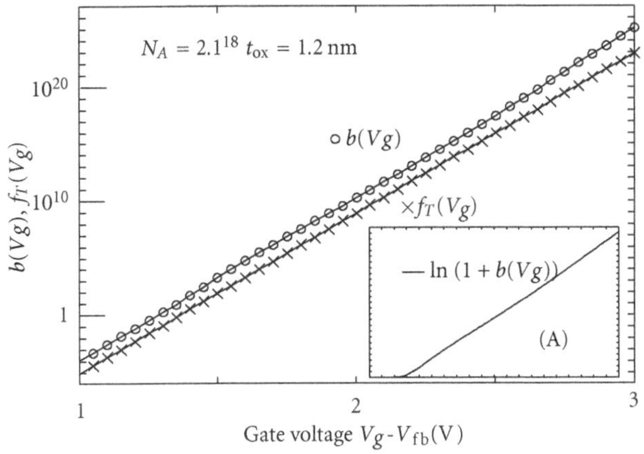


Figure 5: $b(Vg)$ from (18) (○○) and $f_T(Vg) = \exp(Vg - V_T/\eta_0 U_T)$(××). (A): $[\ln(1 + b(Vg)), Vg]$ plots.

with

$$Q_{\text{inv}}(0)|_{Vg} = Q_{\text{inv}}(V_D = 0)|_{Vg} = \lambda q n_i \int_0^{u_S(0)} \frac{e^u}{\sqrt{H(u) + G(u)}} du, \tag{16}$$

and $\tilde{Q}_{\text{inv}}(Vg, V_D)$ is written as follows:

$$\tilde{Q}_{\text{inv}\,0}(Vg, V_D) = \eta(Vg) C_0 U_T \ln\left[1 + b(Vg)\exp\left(-\frac{V_D}{U_T}\right)\right] \tag{17}$$

with

$$b(Vg) = a(Vg)|_{V_D=0} \exp\left(\frac{Vg - V_T}{\eta_0 U_T}\right) = \exp\left(\frac{Q_{\text{inv}}(0)|_{Vg}}{\eta(Vg) C_0 U_T}\right) - 1. \tag{18}$$

The coefficient $b(Vg)$ is dependent on the gate voltage $Vg$ by $u_S(0)$ solution of (5) in $y = 0$. Equation (17) gives, respectively, in strong and weak inversion the simplified expressions:

$$\tilde{Q}_{\text{inv}\,0}(Vg, V(y)) = Q_{\text{inv}}(0)|_{Vg} - \eta(Vg) C_0 V(y),$$
$$\tilde{Q}_{\text{inv}\,0}(Vg, V(y)) = Q_{\text{inv}}(0)|_{Vg} \exp -\frac{V(y)}{U_T}. \tag{19}$$

$b(Vg)$ is a monotonic function in all inversion modes (Figure 5). The originality of the correction by $b(Vg)$ is to give an expression of the inversion charges $\tilde{Q}_{\text{inv}\,0}(Vg, V_D)$ in which the threshold voltage is not explicit but included in $u_S(0)$ and appears in $[\ln(1 + b(Vg)), f(Vg)]$ plots.

The parameter $b(Vg)$ varies from $10^{25}$ to $10^{-5}$ and as shown in Figure 5 is different from $f_T(Vg) = \exp(Vg - V_T/\eta_0 U_T)$. Nevertheless, in usual applications (Section 7), the derivative $db(Vg)/dVg$ can be approximated by

$$\frac{db(Vg)}{dVg} = \frac{b(Vg)}{\eta_0 U_T}. \tag{20}$$

*3.3. Equivalent Expression of Inversion Charge.* By using the mathematical properties of the function:

$$f_2(x) = \frac{2\ln[1 + \exp(x/2)]}{1 + 2\exp(-x/2)} = \frac{2\exp(x/2)}{2 + \exp(x/2)} \ln\left[1 + \exp\left(\frac{x}{2}\right)\right] \tag{21}$$

which has some similarities with $f_1(x) = \ln[1 + \exp(x)]$ in the range $]-\infty, +\infty[$ then (17) can be rewritten by setting

$$X(Vg, V_D) = \sqrt{b(Vg)}\exp\left(-\frac{V_D}{2U_T}\right),$$
$$\Gamma(Vg) = \frac{1 + \sqrt{b(Vg)}}{1 + 0.5\sqrt{b(Vg)}}, \tag{22}$$
$$S(Vg) = \frac{1}{3}\exp\left(\frac{Vg - V_T}{\sigma}\right)^2,$$

$$\tilde{Q}_{\text{inv}\,1}(Vg, V_D) = \eta(Vg) C_0 U_T \frac{\Gamma(Vg)}{1 - S(Vg)} \times \left\{\frac{X(Vg, V_D)}{1 + X(Vg, V_D)} \ln[1 + X(Vg, V_D)]\right\}. \tag{23}$$

(i) $\Gamma(Vg)$ is an adaptive factor which varies between 0.5 ($b(Vg) \gg 1$) and 1 ($b(Vg) \ll 1$);

(ii) $1 - S(Vg)$ and $\sigma^2 = 145$ are fitting factors which minimize the error between $\tilde{Q}_{\text{inv}\,1}(Vg, V_D)$ and $\tilde{Q}_{\text{inv}\,0}(Vg, V_D)$ at $b(Vg) = 1$.

These parameters are available in a large range of $[N_A, t_{ox}]$. Equation (23) gives an expression similar to the Unified MOSFET Channel Charge Model given by (7a) and (7b) in [10] and used in BSIM model [17]. Moreover, (17) and (23) are the synthesis between the expression of inversion charges given in [9, 10] in agreement with the theoretical model (11). Figure 6 shows the normalized expressions of the inversion charges $\tilde{Q}_{\text{inv}\,(0,1)}(Vg, V_D)/\eta_0 C_0 U_T$ at $V(y) = 0$ as a function of the gate voltage.

The term in braces in (23) can be integrated versus $V_D$ and gives an analytic expression of the drain current similar to Oguey and Cserveny model [7].

## 4. Analytic Model of the Drain Current

The general expression for the drain current $I_D(V_D)$ (including drift and diffusion) with a constant mobility $\mu_n$ follows:

$$I_D(V_D) = \mu_n \frac{W}{L} U_T \int_0^{\xi(L)} Q_{\text{inv}}(V(y))|_{Vg} d\xi. \tag{24}$$

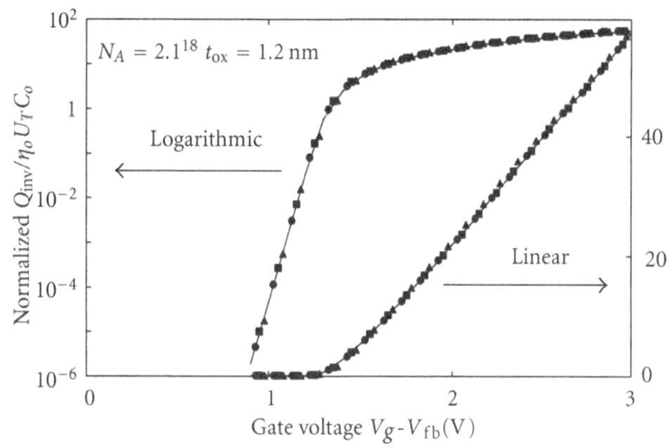


Figure 6: Normalized inversion charges $Q_{inv}(Vg, V_D = 0)\eta_0 C_0 U_T$ versus $Vg$ in logarithm and linear scales. (–) (11), (■) (17), (•) (23), and (▲) (7a) and (7b) in [10].

*4.1. The Pao-Sah Double Integral.* Using the inversion charges dependence to drain bias $Q_{inv}(V(y))|_{Vg}$ (developed in Section 3.1), the Pao-Sah double integral then reads

$$I_{DPS}(V_D) = qn_i\mu_n \frac{W}{L} U_T^2 \int_0^{\xi(L)} \left\{ \int_{\xi(y)/2}^{u_S(y)} \frac{e^{u-\xi(y)}}{F(u,\xi(y))} du \right\} d\xi(y). \tag{25}$$

By substituting $d\xi(y)$ by $d\xi(y) = [d\xi(y)/du_S]du_S$, and grouping $e^{-\xi(y)}$ with $d\xi(y)/du_S$, the Pao-Sah double integral has no singular point and, (25) can be solved into iterated integrals from surface potential $u_S(y)$:

$$\begin{aligned} I_{DPS}(V_D) &= qn_i\mu_n \frac{W}{L} U_T^2 \int_{u_S(0)}^{u_S(L)} \left\{ \int_{\xi(y)/2}^{u_S} \frac{e^u}{F(u,\xi(y))} du \right\} \\ &\quad \times \left[ \frac{d}{du_S} \frac{G(u_S) - E(u_S)}{H(u_S)} \right] du_S. \end{aligned} \tag{26}$$

Equation (26) was previously calculated in a large range of drain and gate voltages and presented in [6].

*4.2. Simplified Expression of the Drain Current.* Equation (17) gives a simplified drain current expression in a single integral:

$$\begin{aligned} I_{\tilde{Q}_{inv0}}(V_D) &= \mu_n \eta(Vg) C_0 U_T \frac{W}{L} \\ &\quad \times \int_0^{V_D} \ln\left[1 + b(Vg)\exp\left(-\frac{V(y)}{U_T}\right)\right] dV(y). \end{aligned} \tag{27}$$

This expression describes the current-voltage characteristics in all inversion modes, insuring a continuous transition between weak and strong inversion. Unfortunately, there is no primitive function for the one defined by (27) which must be numerically calculated by classical integration methods.

*4.3. Explicit Equation of the Drain Current.* Following previous results we propose an analytic expression of the drain current in a square-logarithmic function of $V_D$ based on the adaptive coefficient $b(Vg)$ by integration of (23).

$$\begin{aligned} I_{ap}\left(Vg, \frac{V_{DS}}{V_S}\right) &= I_{DA} \frac{\Gamma(Vg)}{1 - S(Vg)} \\ &\quad \times \left\{ \ln^2[1 + X(Vg, V_S)] - \ln^2[1 + X(Vg, V_D)] \right\}, \end{aligned} \tag{28}$$

$$\begin{aligned} X\left(Vg, \frac{V_D}{V_S}\right) &= \sqrt{b(Vg)} \exp\left(-\frac{V_D/V_S}{2U_T}\right), \\ I_{DA} &= \mu_n \eta(Vg) C_0 U_T^2 \frac{W}{L} \end{aligned} \tag{29}$$

is a dimensional factor.

The drain current, represented by a square-logarithmic function of gate and drain voltage, was proposed as early as 1982 by Oguey and Cserveny [7] in an analytic model based on a control voltage $V_C$ derived from the gate voltage $Vg$ and from drain source functions $f_w(V_D/V_S)$, $f_h(V_D/V_S)$:

$$\begin{aligned} I_{OC}(Vg, V_{DS}) &= \mu_n C_0 U_T^2 \frac{W}{L} [y(V_C, V_S) - y(V_C, V_D)], \\ y\left(V_C, \frac{V_D}{V_S}\right) &= \ln^2\left[1 + \exp f_w\left(\frac{V_D}{V_S}\right) + \exp f_h\left(\frac{V_D}{V_S}\right)\right]. \end{aligned} \tag{30}$$

The inversion charge of this model is given by

$$Q_{inv}|_{OcVg} = \frac{1}{\mu_n (W/L)} \frac{dI_{OC}(Vg, V_{DS})}{dV_D}. \tag{31}$$

Thereafter, the Oguey and Cserveny model has been simplified by Enz et al. [8]. The main difference in this paper is the use of the coefficient $b(Vg)$ instead of $f_T(Vg) = \exp(Vg - V_T/\eta_0 U_T)$.

Equations (27) and (28) (models 2 and 4) coincide with the double integral of Pao-Sah (model 1). The analytic models (Figure 7) are summarized in Table 1.

## 5. Mobility Model

In order to insure carrier drift velocity to be less than the saturation velocity $v_{sat}$ at high electric field, a correction over constant mobility can be implemented in the drain current [18]. In the following, we use the mobility model developed by Roldan et al. [19]:

$$\mu_{neff} = \frac{\mu_1\left(\tilde{F}_x\right)}{\left[1 + \left(\tilde{F}_y/F_{sat}\right)^{\beta}\right]^{1/\beta}}. \tag{32}$$

*5.1. Correction by the Transverse Electric Field $F_x$.*

$$\mu_1\left(\tilde{F}_x\right) = \frac{\mu_0}{\left[1 + \left(\tilde{F}_x/F_0\right)\right]}. \tag{33}$$

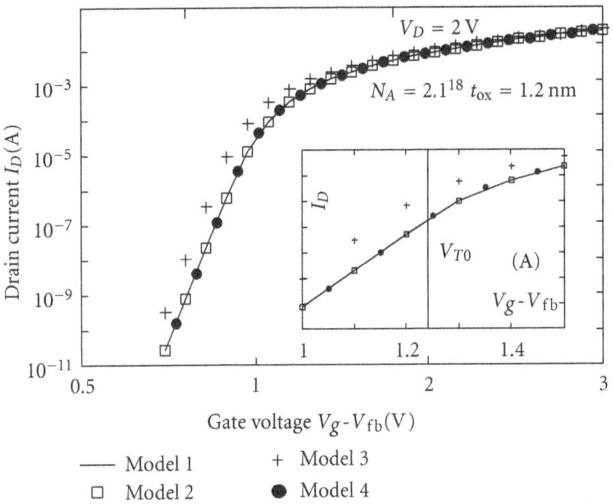


FIGURE 7: The different drain currents $I_D$ versus $Vg$ from models (1 – 4). (A): in the vicinity of the threshold voltage $V_{TO}$.

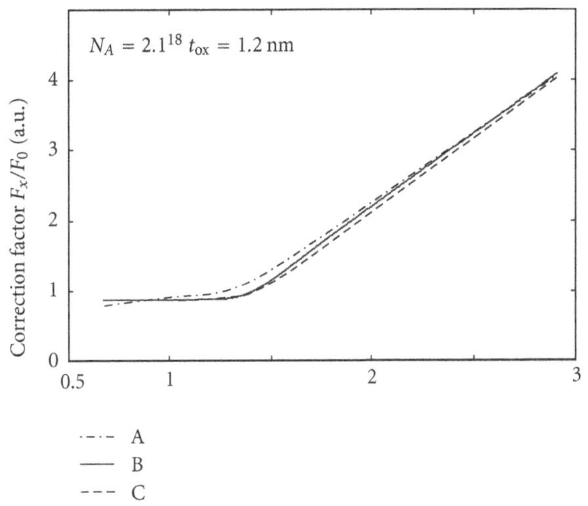


FIGURE 8: The correction factor $F_x/F_0$. (A) $(F_S(u_S(y)) + F_x(0))/2F_0$, (B) $Ua(Vg_{\text{steff}}+2V_{\text{th}})/t_{\text{ox}}$, and (C) $Ua(Vg_{\text{eff}}+2V\text{th})/t_{\text{ox}}$. ($F_0 = 0.67 \cdot 10^6 V \cdot \text{cm}^{-1}$.)

TABLE 1: Analytic models.

| | | Drain current | Inversion charges |
|---|---|---|---|
| Model 1 | Pao-Sah integral | (25) | (11) |
| Model 2 | Single integral | (27) | (13) |
| Model 3 | [7] | (30) | (31) |
| Model 4 | Analytic model | (28) | (23) |

Several expressions are introduced to evaluate the mean electric field $\widetilde{F}_x$ in relation with the channel inversion charges. BSIM models introduce the voltage $Vg_{\text{steff}}$ defined by (7a) and (7b) in [10]. An excellent approximation of $Vg_{\text{steff}}$ can be obtained from the equivalent gate voltage $Vg_{\text{eff}}$ defined from (17) as follows:

$$Vg_{\text{eff}} = \frac{\widetilde{Q}_{\text{inv}}(Vg, 0)}{C_0} = \eta(Vg)U_T \ln[1 + b(Vg)]. \quad (34)$$

The expression of the electric field calculated in (3) allows calculating $\widetilde{F}_x$ as the mean electric field in the inversion region with a dimensionless adaptive coefficient $ua$.

$$\widetilde{F}_x = \frac{ua}{2}\sqrt{\frac{2KTn_i}{\varepsilon_S}}\left(\sqrt{H\left[u_{S(y=0)}\right] + G\left[u_{S(y=0)}\right]} + \sqrt{H(0) + G(0)}\right). \quad (35)$$

Figure 8 shows the correction factors $\widetilde{F}_x$, compared with simplified BSIM 4.6.4 [17].

*5.2. Correction by the Lateral Electric Field $F_y$.* According to $n$-MOSFET models in [20, 21], we use $\beta = 2$:

$$\mu_{\text{neff}} = \frac{\mu_1\left(\widetilde{F}_x\right)}{\sqrt{1 + \left(\widetilde{F}_y/F_{\text{sat}}\right)^2}} \quad (36)$$

with

$$F_{\text{sat}} = \frac{\mu_1\left(\widetilde{F}_x\right)}{v_{\text{sat}}}, \quad (37)$$

and $\widetilde{F}_y$ the average of the lateral electric field:

$$\widetilde{F}_y = \frac{V_D}{L}. \quad (38)$$

The correction over constant mobility is introduced in the general expression of drain current by substituting $\mu_n$ by $\mu_{\text{neff}}$ in (24) as follows:

$$\mu_{\text{neff}} = \frac{\mu_0}{\left[1 + \left(\widetilde{F}_x/F_0\right)\right]\sqrt{1 + \left\{\widetilde{F}_y v_{\text{sat}}\left[1 + \left(\widetilde{F}_x/F_0\right)\right]/\mu_0\right\}^2}}. \quad (39)$$

*5.3. The Saturation Voltage $V_{D\text{sat}}$.* With mobility correction, the models of drain current $[I_D, V_D]$ present a maximum (Figure 9) at a saturation voltage $V_{D\text{sat}}$ defined, according to the mobility model by $I_{D,} = I_{D\,\text{sat}} = WQ_{\text{inv}}v_{\text{sat}}$ [17], $dI_D/d\Psi_S$ [20], or $dI_D/dV_D$ [8]. $[I_D, V_D]$ curves are presented with the same model of correction by transverse electric field. The adaptive parameter in $\widetilde{F}_x$ must then be applied to give the same current and to minimize the error between measured and calculated data.

In this paper, $V_{D\,\text{sat}}$ represented on Figure 10 is calculated from the iterative definition of drain current (Section 4.1) with $dI_D/dV_D$ substituted by $I_{D,m} - I_{D,m-1} = 0$. The saturation voltage is a linear function of $Vg$ in strong inversion and becomes constant in weak inversion.

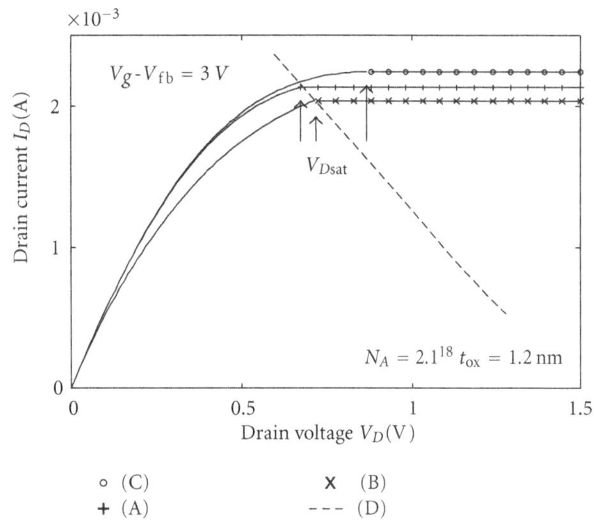


FIGURE 9: $(I_D, V_D)$ plots with velocity saturation. (A) [17] ($\beta = 1$), (B) mobility correction with $\beta = 2$, (C) [20], and (D) saturation current $I_{D\,\mathrm{sat}} = WQ_{\mathrm{inv}}v_{\mathrm{sat}}$.

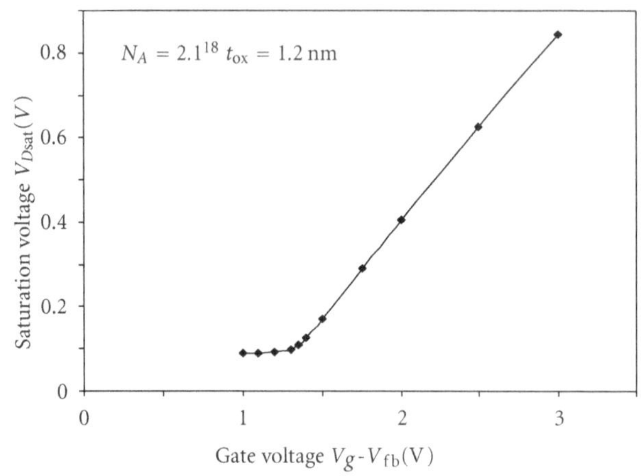


FIGURE 10: The saturation voltage $V_{D\,\mathrm{sat}}$ versus $Vg$. ($L = 90$ nm; $W/L = 10$.)

## 6. Short Channel Drain Current

*6.1. Correction of Saturation Voltage.* The drain current formulation with mobility $\mu_{\mathrm{neff}}$ given from (39) is now written as follows:

$$I_D(V_D) = \mu_{\mathrm{neff}}\frac{W}{L}U_T\int_0^{\xi(L)} \mathrm{Qinv}(V(y))\,|_{Vg}d\xi(y). \quad (40)$$

Equation (40) leads to an unphysical $I_D, V_D$, which must be clamped at $V_{D\,\mathrm{sat}}$. Gildenblat et al. [20] proposed to replace $V_D$ by a smoothing function with a parameter $ax$ : $V_{de} = V_D[1 + (V_D/V_{D\,\mathrm{sat}})^{ax}]^{-1/ax}$. From the analytical and explicit drain current expressions $I_{\tilde{Q}\mathrm{inv}\,0}(V_D)$ and $I_{\mathrm{ap}}(Vg, V_D)$, we can define a new function which includes the effect of velocity saturation by introducing the saturation voltage $V_{D\,\mathrm{sat}}$ in (27):

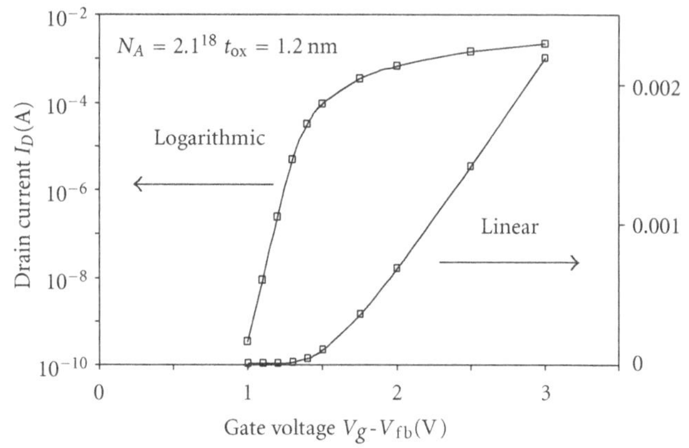


FIGURE 11: Transfer characteristics $I_{D11}(Vg)$ in logarithm and linear scales. $V_S = 0$; $V_D = 2$ V. ($L = 90$ nm; $W/L = 10$.)

$$I_{D11}(V_D) = \frac{W}{L}\Gamma_1\eta_{21}C_0U_T \times \int_0^{V_D} \ln\left[1 + \exp\left(\frac{V_{D\,\mathrm{sat}} - V(y)}{U_T}\right)\right]dV(y). \quad (41)$$

The coefficient $\Gamma_1$ fits $I_{D11}(V_D)$ with $I_{Q\mathrm{inv}}(V_D)$:

$$\Gamma_1 = \mu_{\mathrm{neff}}(V_{D\,\mathrm{sat}}) \times \frac{\int_0^{V_{D\,\mathrm{sat}}} \ln[1 + b(Vg)\exp(-V(y)/U_T)]dV(y)}{\int_0^{V_{D\,\mathrm{sat}}} \ln[1 + \exp((V_{D\,\mathrm{sat}} - V(y))/U_T)]dV(y)} \quad (42)$$

and (28) becomes:

$$I_{D22}(V_{D\,\mathrm{sat}}, V_D) = \frac{W}{L}\Gamma_1\eta_{21}C_0U_T^2\frac{\Gamma(Vg)}{1 - S(Vg)} \times \{Y_{02}^2(V_{D\,\mathrm{sat}}) - Y_{D2}^2(V_{D\,\mathrm{sat}}, V_D)\}, \quad (43)$$

$$Y_{02}(V_{D\,\mathrm{sat}}) = \ln\left[1 + \exp\left(\frac{V_{D\,\mathrm{sat}}}{2U_T}\right)\right], \quad (44)$$

$$Y_{D2}(V_{D\,\mathrm{sat}}, V_D) = \ln\left[1 + \exp\left(\frac{V_{D\,\mathrm{sat}} - V_D}{2U_T}\right)\right]. \quad (45)$$

*6.2. Current-Voltage Characteristics.* Figures 11 and 12 show the simulation results $[I_D, V_D]$ in strong and weak inversion with a mobility model deduced from (39).

The transfer characteristics $I_{D11}(Vg)$ (Figure 11) show linear variations in strong inversion and exponential variations in weak inversion. Figure 12 shows that the smoothing functions (41) and (43) give a unified formulation in the complete range of drain voltage.

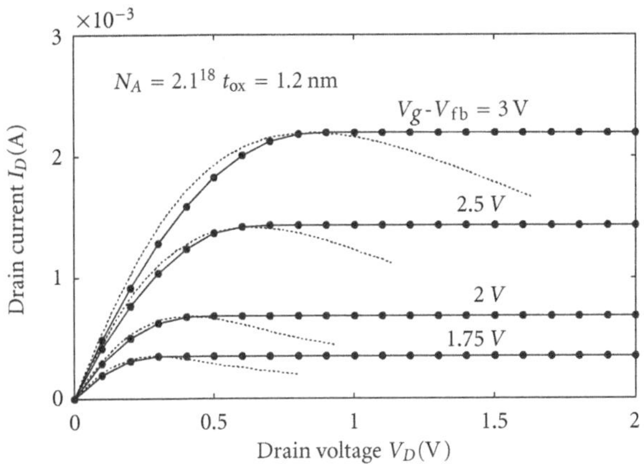


FIGURE 12: $(I_D, V_D)$: (–) (41), (•) (43), (···) (40). ($L$ = 90 nm; $W/L$ = 10.)

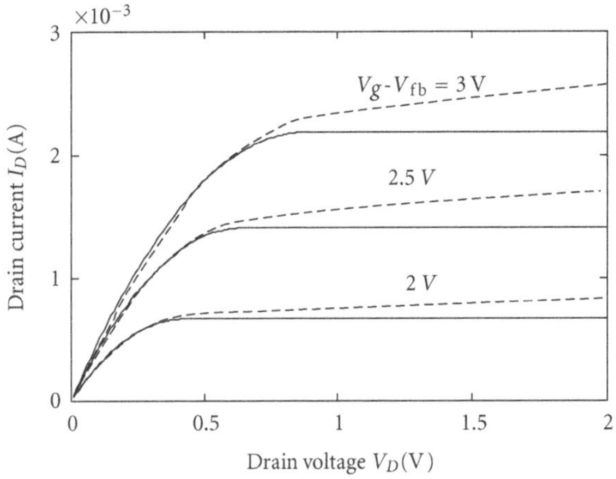


FIGURE 13: $I_D(V_D)$: (–) (43), (- -) (47). ($L$ = 90 nm; $W/L$ = 10.)

*6.3. Channel Length Modulation.* The channel length modulation (CLM) is a shortening of the length of the inverted channel region $L_{\text{eff}} = L - \Delta L$ due to inversion layer in the drain junction. An accurate calculation of $\Delta L$ requires solving the 2D Poisson equation near the drain. An 1D approach may be used for standard expression of the depletion layer in the abrupt junction approximation [22]

$$\Delta L = \sqrt{\frac{2\varepsilon_S}{qN_A}}\left[\sqrt{\frac{\varepsilon_S}{qN_A}\left(\frac{V_D}{L}\right)^2 + V_D} - \sqrt{\frac{\varepsilon_S}{qN_A}\left(\frac{V_D}{L}\right)^2}\right]. \tag{46}$$

Figure 13 shows an illustration of CLM with $I_D(V_D)$ from (41) modified by (46). The drain current formulation is

$$I_{\text{CLM}}(V_D) = \frac{L}{L - \Delta L} I_{D22}(V_D). \tag{47}$$

This approximation is analogous to the early voltage and has the advantage to be described by the single analytic function $I_{\text{CLM}}(V_D)$.

Figure 14 gives a complete summary of the different $I_D(V_D)$ as follows:

(◇) are $I_D(V_D)$ data from Pao-Sah double integral from (40) with correction mobility in the range $0 < V_D < V_{D\text{sat}}$;

(∘) are $I_D(V_D)$ data from the saturation current corrected by the channel length modulation (46) ($V_{D\text{sat}} < V_D$);

(···) are $I_{D11}(V_D)$ data from (41);

The full line shows the single analytic function $I_{\text{CLM}}(V_D)$ from (47).

*6.4. Drain-Induced Barrier Lowering.* The drain-induced barrier lowering (DIBL) was described as soon as 1979 by Troutman et al. [23]. The MOSFET is a three-terminal device in which source-channel drain is a $n - p - n$ (or $p - n - p$) double junction. We described in a previous paper the complete potential distribution in double junction from a 1D resolution of Poissons equation [24]. If this analytic description gives an accurate description of the potential $\varphi(x)$ in an unbiased double junction, the 1D resolution cannot be extrapolated with drain biased, which supposes a 2D device simulation. Most models describe the DIBL by a linear lowering of threshold voltage [21] $V_T = V_{T0} - \sigma V_D$ with the DIBL parameter $\sigma$.

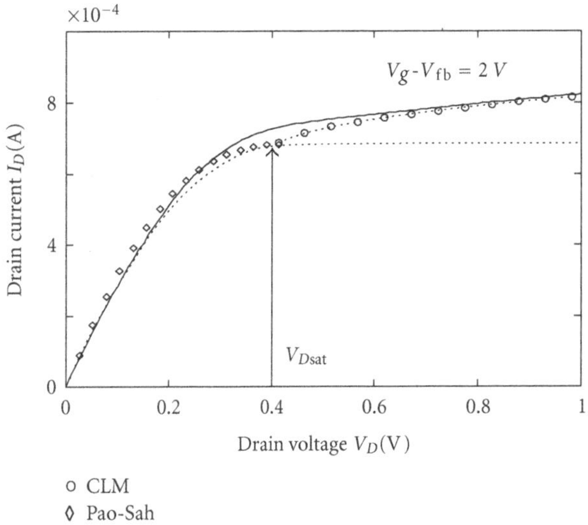


FIGURE 14: The different $(I_D, V_D)$ plots with CLM. ($L$ = 90 nm, $W/L$ = 10.)

In this paper, following the model of DIBL in [25], we propose to insert the increase of inversion charge $\varepsilon_S dF_y(x, y)/dy$ by a quasi 1D calculation. With the same method, Cheng and Hu [26] calculated the threshold shift when $L \gg l = \sqrt{\varepsilon_S t_{\text{ox}} X_{\text{dep}}/\varepsilon_{\text{ox}}\kappa}$

$$\Delta V_{\text{th}} = [2(V_{bi} - 2\varphi_b) + V_D]\left[\exp\left(-\frac{L}{2l}\right) + 2\exp\left(-\frac{L}{l}\right)\right]. \tag{48}$$

In the present paper with $L = 90\,\text{nm}$, $L/l \approx 10$, $\Delta V_{\text{th}}$ is relatively a small correction in $I_D(V_D)$.

In [25], the authors propose, as shown on Figure 15, to add $\Delta V_{\text{th}}$ in the square logarithm with the new expressions:

$$Y_{02DB}(V_{D\,\text{sat}}) = \ln\left[1 + \exp\left(\frac{V_{D\,\text{sat}} + \Delta V_{\text{th}}(0)}{2U_T}\right)\right] \quad (49)$$

$$Y_{D2DB}(V_{D\,\text{sat}}, V_D) = \ln\left[1 + \exp\left(\frac{V_{D\,\text{sat}} - V_D + \Delta V_{\text{th}}(L)}{2U_T}\right)\right], \quad (50)$$

$$I_{\text{CLM+DIBL}}(V_D) = \frac{\Gamma(Vg)}{1 - S(Vg)} \frac{W}{L - \Delta L} \Gamma_1 \eta_{21} C_0 U_T^2 \times \left\{Y_{02DB}^2(V_{D\,\text{sat}}) - Y_{D2DB}^2(V_{D\,\text{sat}}, V_D)\right\}. \quad (51)$$

Due to the simplifying assumptions in the derivative $dF_y(x, y)/dy$, such a model gives a phenomenological description of DIBL, but must include fitting parameter to agree with experimental data. A new study is in progress in order to obtain a more accurate expression of $dF_y(x, y)/dy$ and apply this model to inversion charges in (41) and (43) taking into account the lateral field to provide a complete expression of DIBL.

In the case of $n$-MOSFETs, we have to add the Substrate Current-Induced Body Effect (SCIBE) which is the result of impact ionization by hot electrons coming from the source [23]. The expression of SCIBE is given by

$$I_{\text{sub}} = \frac{A}{B} I_D(V_D - V_{D\,\text{sat}}) \exp\left(-\frac{Bl}{V_D - V_{D\,\text{sat}}}\right). \quad (52)$$

$A$ and $B$ are adaptive parameters resulting from $(I_D, V_D)$ measurements. In the present work, this effect must be added to (51) from $A$ and $B$ and gives the total current. Figure 16 shows an example of SCIBE with $L = 90\,\text{nm}$.

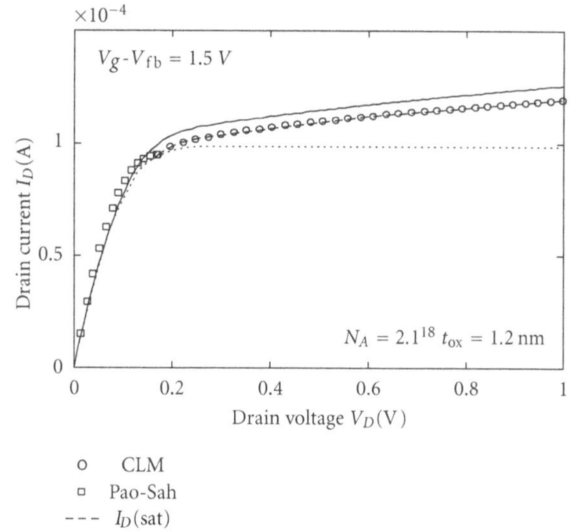


Figure 15: The total current $I_{\text{CLM+DIBL}}(V_D)$ (−) from (51). ($L = 90\,\text{nm}$, $L/l = 9.5$, and $W/L = 10$).

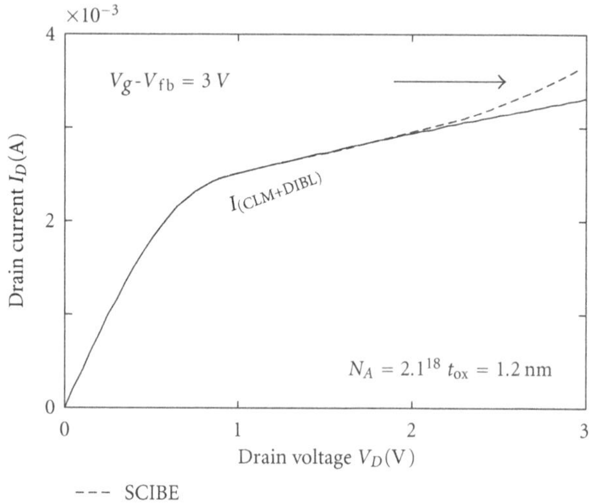


Figure 16: An example of SCIBE ($L = 90\,\text{nm}$; $W/L = 10$).

## 7. Analytic Model of TransConductance

The Pao-Sah double integral gives an expression of the transconductance from the derivative, $g = dI_D/dVg$ in (26) [6]:

$$g_{m\text{PS}} = qn_i\mu_{\text{neff}} U_T^2 \frac{W}{L} \times \int_{u_S(0)}^{u_S(L)} \frac{e^{u_S - \xi}}{F(u_S(y), \xi)} \left(\frac{du_S(y)}{dVg}\right)\left(\frac{d\xi(y)}{du_S(y)}\right) du_S(y). \quad (53)$$

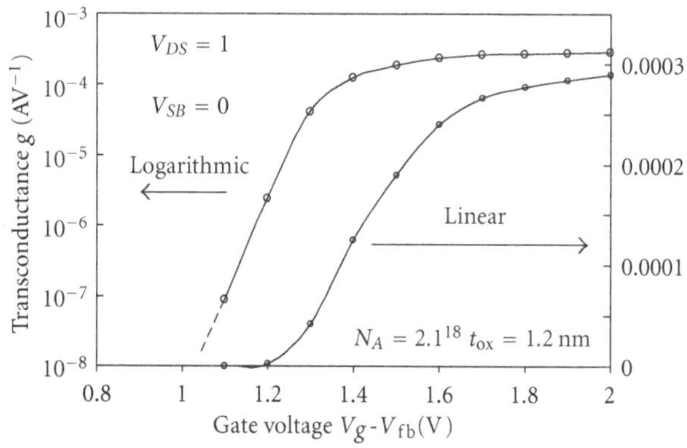


Figure 17: Transconductance versus $Vg$ ($L = 90\,\text{nm}$; $W/L = 4$).

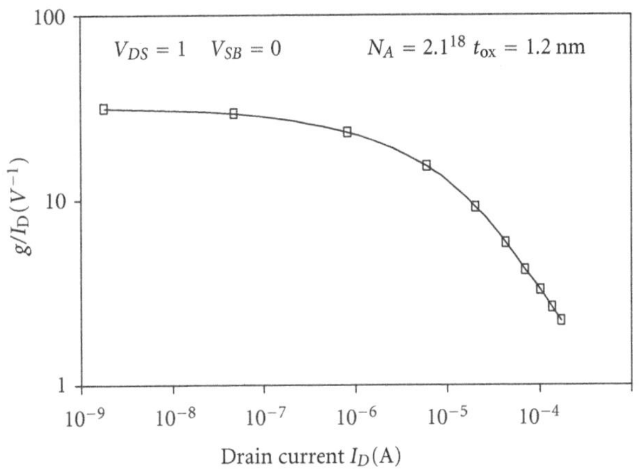


FIGURE 18: Normalized transconductance $[g/I_D, I_D]$. $V_{DS} = 1V$; $Vg - V_{fb} = 1, 1.1, \ldots 2V$ ($L = 90$ nm; $W/L = 4$).

TABLE 2: The simplified analytic drain current models.

| | |
|---|---|
| Physical constants | $\varepsilon_S$, $\varepsilon_{ox}$, $KT/q$ |
| Process parameters | $N_A$, $t_{ox}$, $\mu_n$, $W/L$ |
| Device voltages | $Vg$, $V_D$, $V_S$, $V_{fb}$ |
| Surface potential in $x = 0$, $y = 0$ | $u_S(0)$ |
| Surface potential | $u_S(y) = u_{S,m}$ |
| Mobility | $\mu_{neff}(Vg, V_D)$ |
| Drain current from Pao-sah | $I_{DPS}(V_D)$ |
| Gate voltage factors | $\eta(V_g)$, $b(Vg)$ |
| Analytic expression of drain current | $I_{ap}(Vg, V_D)$ |
| Saturation voltage | $V_{D\,sat}$ |
| Drain current with saturation | $I_{D22}(V_{D\,sat}, V_D)$ |
| Channel length modulation | $I_{CLM}(V_D)$ |
| Drain-induced barrier lowering | $I_{CLM+DIBL}(V_D)$ |

A simplified expression of the transconductance can be obtained from (27) by a derivative under the integral using $db(Vg)/dVg$ from (20) as follows:

$$\begin{aligned} g_{ma} &= \left.\frac{dI_D}{dVg}\right|_{V_D} \\ &= \mu_{neff}\eta_{21} C_0 U_T \frac{W}{L} \\ &\times \int_{V_S}^{V_D} \frac{(b(Vg)/\eta_0 U_T)\exp(-V(y)/U_T)}{1 + b(Vg)\exp(-V(y)/U_T)} dV(y). \end{aligned} \tag{54}$$

In this case, the integral in $g_{ma}$ appears as $f'(v)/f(v)$, and $g_{ma}$ is given by an analytical expression versus $V_D$ and $Vg$:

$$g_{ma} = \mu_{neff} C_0 U_T \frac{\eta_{21}}{\eta} \frac{W}{L} \ln\left\{\frac{1 + b(Vg)\exp(-V_S/U_T)}{1 + b(Vg)\exp(-V_D/U_T)}\right\}. \tag{55}$$

These expressions correspond to a long channel MOSFET with a constant mobility. The mobility model given by (39) introduces a second term in the transconductance due to Vertical Field Mobility Reduction (VFMR):

$$g_{m,\mu} = \frac{d\mu_{neff}}{dVg} \frac{I_D}{\mu_{neff}}. \tag{56}$$

$g_{m,\mu}$ is less than $g_{ma}$ and appears as a corrective term in the transconductance. This contribution, negligible in long channel MOSFET, must be introduced as a corrective factor in the transconductance from

$$\frac{d\mu_{neff}}{dVg} = \frac{d\mu_{neff}}{d\mu_1} \frac{d\mu_1}{d\widetilde{F}_x} \frac{d\widetilde{F}_x}{du_S(0)} \frac{du_S(0)}{dVg}. \tag{57}$$

Each terms of this equation are calculated from (32), (33), (35), and (5). The contribution of CLM and DIBL in transconductance can be, respectively, deduced from (46), (47), (49), and (50).

A simple numerical calculation of the complete transconductance including VFMR, CLM, and DIBL is obtained from (51) by

$$g = \left.\frac{\Delta I_D}{\Delta V_g}\right|_{V_D}. \tag{58}$$

Figure 17 shows transconductance $[g, Vg]$ plots, and Figure 18 shows the "normalized" ratio $[g/I_D]$, versus drain current.

## 8. Conclusion

In this paper, we propose a solution of the Poisson-Boltzmann equation which describes the physical parameters of the MOSFET under gate and drain bias. The Taylor expansion of inverse functions is well suited in the case of implicit functions and gives an accurate solution of the channel potential $\xi(y) = f(u_S(y))$. We introduce an analytic function of the inversion charge giving an expression of the drain current insuring a continuous transition between weak and strong inversion associated with a simple expression of the transconductance. Furthermore, the method gives a good approach of drain current with the velocity saturation. All the equations have been solved with a simple C-encoding program available on all personal computers. This program, associated with a graphic user interface (Figure 19), generates a graph (Figure 20) with different $Vg$ bias. The excellent agreement of the results obtained by an analytic continuous function of the inversion charge compared with those of standard models [1] can be considered as an accurate tool for microelectronics without access to specific CAD software and can provide a comprehensive overview of the complete MOSFET available in all inversion modes (Table 2).

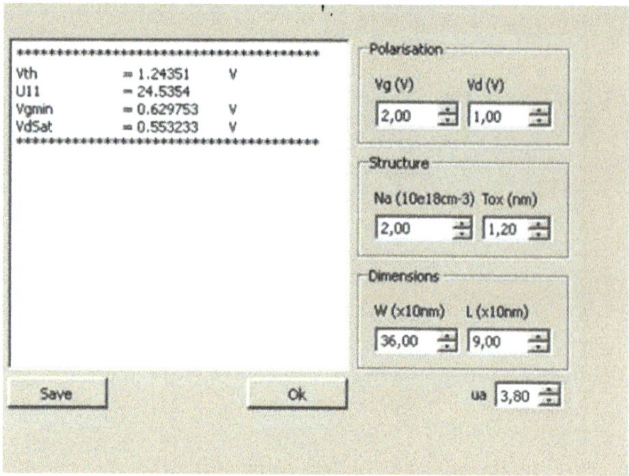


FIGURE 19: The graphic user interface.

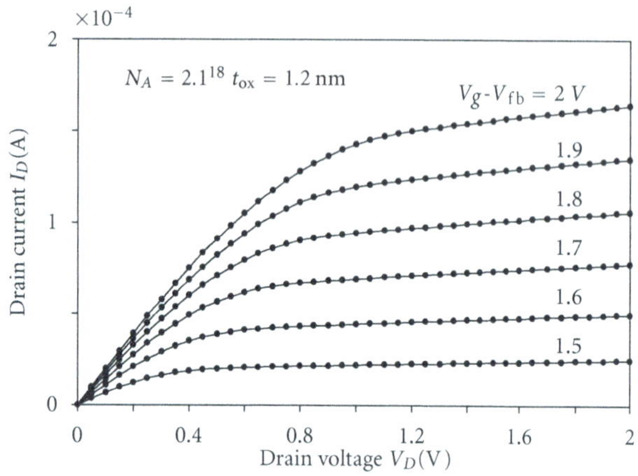


FIGURE 20: $I_D(V_D)$ plots from (51). $Vg - V_{fb} = 1.5, 1.6, \ldots 2$V.

## Nomenclature

$K$ and $T$: Boltzmann constant and temperature (Kelvin)
$\varepsilon_S$ and $\varepsilon_{ox}$: silicon and silicon oxide permittivity
$t_{ox}$: oxide thickness
$C_0 = \varepsilon_{ox}/t_{ox}$: normalized oxide capacitance
$n_i = p_i$: intrinsic carrier concentration in cm$^{-3}$
$N_A$ and $N_D$: dopant concentrations in cm$^{-3}$
$U_T = kT/q$: thermal voltage
$\varphi(b) = -U_T \ln(N_A/n_i)$: bulk potential of p-doped silicon
$u(x) = \varphi(x)/U_T$: reduced potential
$\Psi(x) = \varphi(x) - \varphi(b)$: band bending
$V_T = -2\varphi(b) + \sqrt{2qN_A\varepsilon_S|2\varphi(b)|}/C_0$: charge sheet threshold voltage
$\gamma_0 = \sqrt{2KT\varepsilon_S n_i}/C_0$: intrinsic body factor
$\lambda = \sqrt{KT\varepsilon_S/2q^2 n_i}$: Debye length (cm)
$W, L$ : channel width and channel length.

*Numerical applications use SI units, except for the following:*

$N_A$, $N_D$, $n(x,y)$, and $p(x,y)$, in cm$^{-3}$.
inversion charges $Q_{\text{inv}}(Vg)|_{V(y)}$, $Q_{\text{inv}}(V(y))|_{Vg}$ in C·cm$^{-2}$ and $\varepsilon_S$, $\varepsilon_{ox}$ in farads·cm$^{-1}$.

## References


[1] J. Watts, C. M. Andrew, C. Enz et al., "Advanced compact models for MOSFETs," in *Proceedings of the Workshop on Compact Modeling at Nanotech*, pp. 3–12, Anaheim, Calif, USA, May 2005.

[2] R. Van Langevelde and F. M. Klaassen, "Explicit surface-potential-based MOSFET model for circuit simulation," *Solid-State Electronics*, vol. 44, no. 3, pp. 409–418, 2000.

[3] J. He, M. Chan, X. Zhang, and Y. Wang, "A physics-based analytic solution to the MOSFET surface potential from accumulation to strong-inversion region," *IEEE Transactions on Electron Devices*, vol. 53, no. 9, pp. 2008–2016, 2006.

[4] W. Z. Shangguan, M. Saeys, and X. Zhou, "Surface-potential solutions to the Pao-Sah voltage equation," *Solid-State Electronics*, vol. 50, no. 7-8, pp. 1320–1329, 2006.

[5] G. Gildenblat, Z. Zhu, and C. C. McAndrew, "Surface potential equation for bulk MOSFET," *Solid-State Electronics*, vol. 53, no. 1, pp. 11–13, 2009.

[6] H. Murray, P. Martin, and S. Bardy, "Taylor expansion of surface poten- tial in MOSFET: application to Pao-Sah integral," *Active and Passive Electronics Components*, vol. 2010, Article ID 268431, 11 pages, 2010.

[7] H. Oguey and S. Cserveny, "Modèle du transistor MOS valable dans un grand domaine de courants," *Bulletin SEV/VSE*, vol. 73, pp. 113–116, 1982.

[8] C. C. Enz, F. Krummenacher, and E. A. Vittoz, "An analytical MOS transistor model valid in all regions of operation and dedicated to low-voltage and low-current applications," *Analog Integrated Circuits and Signal Processing*, vol. 8, no. 1, pp. 83–114, 1995.

[9] B. Iniguez, L. F. Ferreira, B. Gentinne, and D. Flandre, "A physically-based C$_\infty$-continuous fully-depleted SOI MOSFET model for analog applications," *IEEE Transactions on Electron Devices*, vol. 43, no. 4, pp. 568–574, 1996.

[10] Y. Cheng, K. Chen, K. Imai, and C. Hu, "A unified mos-fet channel charge model for device modeling in circuit simulation," *IEEE Transactions on Computer-Aided Design of Integrated Circuits and Systems*, vol. 17, no. 8, pp. 641–644, 1998.

[11] C. K. Kim and E. S. Yang, ""On the validity of the gradual-channel approximation for field-effect transistors," *Proceedings of the IEEE*, vol. 58, no. 5, pp. 841–842, 1970.

[12] G. Goudet and C. Meuleau, *Semiconductors: Their Theory and Practice*, Macdonald & Evans, London, UK, 1957, English version by G. King.

[13] E. H. Nicollian and J. R. Brews, *MOS Physics and Technology*, John Wiley & Sons, Hoboken, NJ, USA, 2002.

[14] T. L. Chen and G. Gildenblat, "Analytical approximation for the MOSFET surface potential," *Solid-State Electronics*, vol. 45, no. 2, pp. 335–339, 2001.

[15] G. Baccarani, M. Rudan, and G. Spadini, ""Analytical i.g.f.e.t.model including drift and diffusion currents," *IEE Journal on Solid-State and Electron Devices*, vol. 2, no. 2, pp. 62–68, 1978.

[16] J. M. Sallese, M. Bucher, F. Krummenacher, and P. Fazan, "Inversion charge linearization in MOSFET modeling and rigorous derivation of the EKV compact model," *Solid-State Electronics*, vol. 47, no. 4, pp. 677–683, 2003.

[17] T. H. Morshed et al., "BSIM 4.6.4 MOSFET user manual," University of California, Berkeley, Calif, USA, 2009, http://www-device.eecs.berkeley.edu/bsim/Files/BSIM4/BSIM464/BSIM464_Manual.pdf.

[18] K. Y. Lim and X. Zhou, "Physically-based semi-empirical effective mobility model for MOSFET compact I-V modeling," *Solid-State Electronics*, vol. 45, no. 1, pp. 193–197, 2001.

[19] J. B. Roldan, F. Gamiz, J. A. Lopez-Villanueva, J. E. Carceller, and P. Cartujo, "The dependence of the electron mobility on the longit- udinal electric field in MOSFETs," *Semiconductor Science and Technology*, vol. 12, pp. 321–330, 1997.

[20] G. Gildenblat, X. Li, W. Wu et al., "PSP: an advanced surface-potential-based MOSFET model for circuit simulation," *IEEE Transactions on Electron Devices*, vol. 53, no. 9, pp. 1979–1993, 2006.

[21] N. Arora, *Mosfet Modeling for VLSI Simulation: Theory and Practice*, World Scientific Publishing, Singapore, 2007.

[22] J. J. Liou, A. Ortiz-Conde, and F. Garcia-Sanchez, *Analysis and Design of MOSFETs—Modeling, Simulation, and Parameter Extraction*, Kluwer Academic, Norwell, Mass, USA, 1998.

[23] S. C. Tam, F.-C. Hsu, P.-K. Ko, T.-Y. Chan, and K. W. Terrill, "Hot-electron induced MOSFET degradation—model, monitor, improvement," *IEEE Transactions on Electron Devices*, vol. 32, pp. 375–385, 1985.

[24] H. Murray, "Analytic resolution of Poisson-Boltzmann equation in nanometric semiconductor junctions," *Solid-State Electronics*, vol. 53, no. 1, pp. 107–116, 2009.

[25] C. C. Enz and E. A. Vittoz, *Charge Based MOS Transistor Modeling*, John Wiley & Sons, London, UK, 2006.

[26] Y. Cheng and C. Hu, *MOSFET Modeling & BSIM3 USer's Guide*, Kluwer Academic, New York, NY, USA, 2002.

# GaN-Based High-$k$ Praseodymium Oxide Gate MISFETs with $P_2S_5/(NH_4)_2S_X$ + UV Interface Treatment Technology

**Chao-Wei Lin and Hsien-Chin Chiu**

*Department of Electronics Engineering, Chang Gung University, Taoyuan, Taiwan*

Correspondence should be addressed to Hsien-Chin Chiu, hcchiu@mail.cgu.edu.tw

Academic Editor: Pei-Wen Li

This study examines the praseodymium-oxide- ($Pr_2O_3$-) passivated AlGaN/GaN metal-insulator-semiconductor high electron mobility transistors (MIS-HEMTs) with high dielectric constant in which the AlGaN Schottky layers are treated with $P_2S_5/(NH_4)_2S_X$ + ultraviolet (UV) illumination. An electron-beam evaporated $Pr_2O_3$ insulator is used instead of traditional plasma-assisted chemical vapor deposition (PECVD), in order to prevent plasma-induced damage to the AlGaN. In this work, the HEMTs are pretreated with $P_2S_5/(NH_4)_2S_X$ solution and UV illumination before the gate insulator ($Pr_2O_3$) is deposited. Since stable sulfur that is bound to the Ga species can be obtained easily and surface oxygen atoms are reduced by the $P_2S_5/(NH_4)_2S_X$ pretreatment, the lowest leakage current is observed in MIS-HEMT. Additionally, a low flicker noise and a low surface roughness (0.38 nm) are also obtained using this novel process, which demonstrates its ability to reduce the surface states. Low gate leakage current $Pr_2O_3$ and high-$k$ AlGaN/GaN MIS-HEMTs, with $P_2S_5/(NH_4)_2S_X$ + UV illumination treatment, are suited to low-noise applications, because of the electron-beam-evaporated insulator and the new chemical pretreatment.

## 1. Introduction

Because of their inherent high breakdown voltage ($V_{BR}$), high two-dimensional electron gas (2-DEG) concentration, and high saturation velocity [1, 2], and AlGaN/GaN high electron mobility transistors (HEMTs) are suitable to high-power and low-noise applications. The major factors that limit the performance and reliability of GaN-based HEMTs at radio frequencies (RF) are their high gate leakage current and drain current collapse, which is associated with native oxide-induced surface states [3, 4]. Therefore, AlGaN/GaN metal-insulator-semiconductor HEMTs (MIS-HEMTs), in which $SiO_2$ [5], $Si_3N_4$ [6], $Ga_2O_3$ [7], $Al_2O_3$ [8], and $Sc_2O_3$ [9] are used as the gate dielectrics, are studied, in order to address these problems. Related works focus on the formation of a high-$k$ insulator, which reduces the Schottky gate leakage current at high input signal swings and improves channel modulation. However, the treatment of the interface between AlGaN and the insulator has not been studied systematically. Pretreatment before the deposition of the passivating layer between the source, the drain, and the gate terminals is dominated by the effect of surface traps, which cause flicker noise and current collapse problems. For example, $(NH_4)_2S_X$ sulfide treatment is known to eliminate native $Ga_2O_3$ and $As_2O_3$ dangling bonds on GaAs- and InP-related semiconductors, because of the formation of stable Ga-S and As-S bonds during immersion [10, 11]. In this work, a $P_2S_5/(NH_4)_2S_X$ + UV treatment that suppresses surface traps is studied. Furthermore, to increase the efficiency of the $P_2S_5/(NH_4)_2S_X$ treatment, the treatment is performed in a UV chamber and both the high-$k$ ($\varepsilon \sim 10$) $Pr_2O_3$ gate insulator and the passivating layer are deposited using electron-beam evaporation, which effectively prevents the plasma-induced generation of surface states. A comparison of the flicker noise determined from the pulsed $I$-$V$ of $Pr_2O_3$ AlGaN/GaN MIS-HEMT with that of traditional GaN HEMTs shows that the surface traps are markedly suppressed by $P_2S_5/(NH_4)_2S_X$ + UV treatment. The observed lower surface leakage current also improves the DC-RF dispersion of MIS-HEMT. X-ray photoelectron spectroscopy (XPS) measurement and secondary ion mass spectrometry (SIMS) are used to study the Ga-S energy bonds and the distributions of the depths of oxygen and sulfur atoms, following $P_2S_5/(NH_4)_2S_X$ + UV treatment.

## 2. Device Structure and Fabrication

The AlGaN/GaN HEMT heterostructures used in this study were grown using atmospheric pressure metal organic chemical vapor deposition (AP-MOCVD) on 2 inch sapphire wafers. The 4000 nm-thick undoped GaN was grown first, to form the buffer and channel layers. Then, a 35 nm-thick undoped $Al_{0.25}Ga_{0.75}N$ layer was grown as the Schottky layer. The designed structure had a sheet charge density of 1.65 $\times 10^{13}$ $cm^{-2}$ and a Hall mobility of 1060 $cm^2$/V-s at 300 K. Figure 1 shows the cross-sections of a $Pr_2O_3$/AlGaN/GaN MIS-HEMT with $P_2S_5/(NH_4)_2S_X$ + UV interface pretreatment. During fabrication of the device, the active region was protected by a photoresist and the mesa isolation region was removed using a $BCl_3 + Cl_2$ mixture gas plasma in a reactive ion etching (RIE) chamber. The ohmic contacts of the Ti/Al/Ni/Au (25 nm/125 nm/50 nm/100 nm) metal layers were deposited using electron-beam evaporation and patterned by conventional optical lithography and lift-off method, followed by 850°C RTA annealing for 30 s in an $N_2$ environment. Before the deposition of the high-*k* $Pr_2O_3$ insulator and the passivating layers, the samples were immersed in a standard treatment (dilute HCl), $(NH_4)_2S_X$ and $P_2S_5/(NH_4)_2S_X$ for 15 min. Figure 2 shows the pH value of the $P_2S_5/(NH_4)_2S_X$ solution versus $P_2S_5$ weight. The pH of the saturated $(NH_4)_2S_X$ solution is 11.2 and the pH of the mixed $P_2S_5/(NH_4)_2S_X$ solution was adjusted to a value of 7 by adding 10 g of $P_2S_5$ into 30 mL $(NH_4)_2S_X$ saturated solution. Although the $(NH_4)_2S_X$ treatment effectively removes the native AlGaN surface oxide layer, it cannot prevent the increase in surface roughness that is caused by the alkaline $(NH_4)_2S_X$ solution. After 15 min of pretreatment immersion, the AlGaN surface roughness was 0.24 nm for the standard treatment and 0.51 nm and 0.44 nm for the $(NH_4)_2S_X$ and $P_2S_5/(NH_4)_2S_X$ solution. Therefore, the $P_2S_5/(NH_4)_2S_X$ solution was used as the interface pretreatment solution for the MIS-HEMT. In an earlier study by the authors, the reduction of the number of native oxide-induced dangling bonds on the AlGaN surface, which are difficult to remove at room temperature, was achieved by increasing the temperature of the solution temperature to 60°C, but this increase makes the procedure more complicated and time-consuming. In this study, UV illumination is used during $P_2S_5/(NH_4)_2S_X$ pretreatment immersion, in order to rapidly eliminate the dangling bonds and form Ga-S bonds with high-binding energy. After interface treatment, a 10 nm-thick layer of praseodymium was firstly evaporated, using an optimal oxygen flow rate of 15 sccm. During this stage, the chamber pressure was increased to around $10^{-4}$ Torr. When the chamber pressure was reduced to $3 \times 10^{-6}$ Torr, the conventional Ti/Au (30 nm/150 nm) gate metals were deposited. For comparison, a traditional Ni/Au Schottky gate GaN HEMT was also fabricated. Finally, the Ti/Au (30 nm/300 nm) metals were deposited to form the interconnection and probe pads, and a 200 nm-thick $SiO_2$ layer was deposited to passivate the device. The complete process was also used to fabricate a traditional Ni/Au Schottky gate GaN HEMT, for the purpose of comparison.

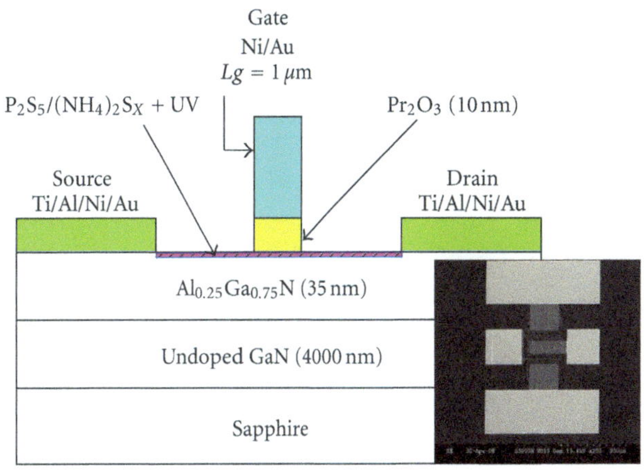


FIGURE 1: The cross-sectional structure of a $P_2S_5/(NH_4)_2S_X$+UV pretreatment $Pr_2O_3$ MISFET.

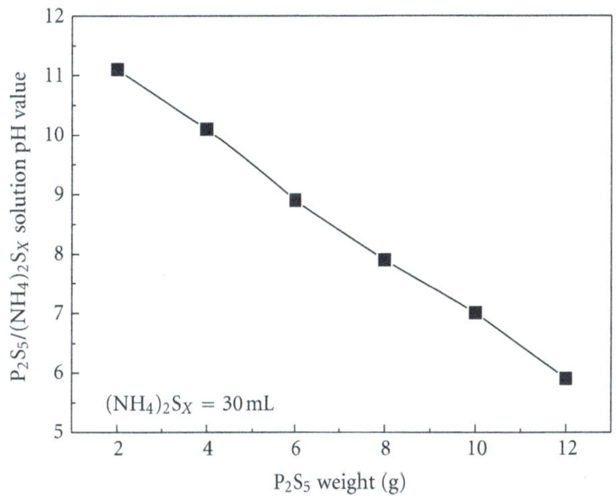


FIGURE 2: pH value of $P_2S_5/(NH_4)_2S_X$ solution versus $P_2S_5$ weight.

The important optimization of the flow rate of the praseodymium oxide high-*k* layer was also considered. The electron-beam evaporation of the high-*k* $Pr_2O_3$ thin film is optimized by adjusting the oxygen flow rate in the chamber. Figure 3 shows the EDX measurements for $Pr_2O_3$ grown in the electron-beam evaporator using various oxygen flow rates. An analysis of the rare-earth metal atomic concentration in the oxide layer demonstrates that the optimal flow rate for the deposition of the praseodymium oxide layer is 15 standard cubic centimeters per minute (sccm), indicating that the dielectric constant is maximized, because a strong dipole is formed at the highest possible rare-earth metal concentration in the high-*k* oxide layer. The thermal stability of the insulator also plays an important role in high-power GaN MIS-HEMTs because the dc power is primarily dissipated near the gate contact, which causes local Joule self-heating [10]. The channel temperature can reach 100°C during high-output-power operation, which

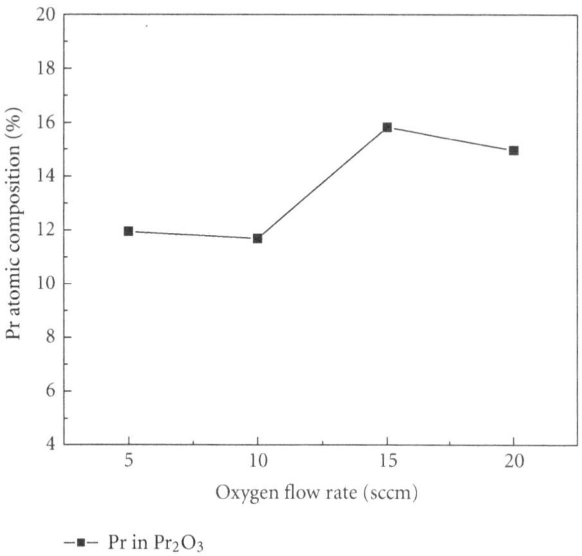


FIGURE 3: EDX analysis of $Pr_2O_3$ films with various oxygen flow rates.

results in increased degradation and failure rates and a reduction in output power. In order to evaluate the thermal stability of $Pr_2O_3$ layer, the X-ray photoelectron spectroscopy (XPS) was used to measure the binding energy of the $Pr_2O_3$ thin films after 400°C, 600°C, and 800°C postannealing. Figure 4(a) shows the XPS $3d$ core-level spectra for $Pr_2O_3$ at various temperatures. The energy of the Ar-gun is 3 KeV, the operation current is 1 mA, and the analysis area is $2 \times 2\ mm^2$. It is seen that the binding energies of the $Pr_2O_3$ after 400°C, 600°C, and 800°C postannealing are close to the standard value of 934 eV in the $3d$ core level, as recorded in the XPS handbook. Figure 4(a) also shows the high signal intensities of $Pr_2O_3$. Therefore, it is concluded that a high-quality and a highly thermally stable high-$k$ insulator is obtained by using electron-beam-evaporated Pr with a high oxygen flow rate. The results of high-resolution cross-sectional transmission electron microscopy (TEM) prove that the $Pr_2O_3$ grows on the GaN in a planar fashion, as shown in Figure 4(b). The $Pr_2O_3$ equivalent oxide thickness, measured by the TEM, is 20 nm. A $(Ga_2O_3)Pr_2O_3$ compound film occurs between the interface of the GaN and $Pr_2O_3$ layers, which is generated by the native GaN surface oxide layer and praseodymium materials.

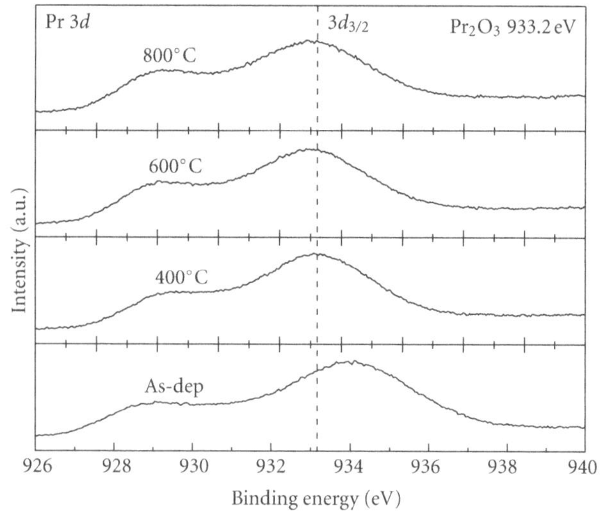


(a)

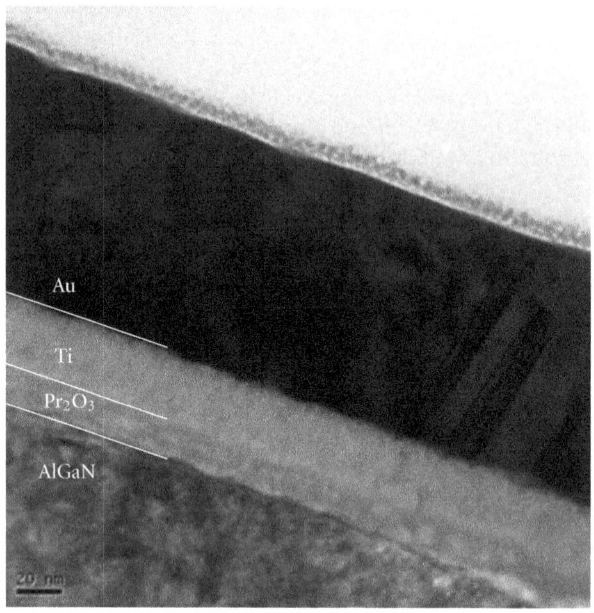


(b)

FIGURE 4: (a) $3d$ core-level XPS spectra of $Pr_2O_3$ versus temperatures. (b) TEM cross-sectional photograph of $Pr_2O_3$ MIS-HEMT.

## 3. Experimental Results for the Device

Atomic force microscopy (AFM) or scanning force microscopy (SFM) has a very high resolution, with a demonstrated resolution of the order of fractions of a nanometer, more than 1000 times better than the optical diffraction limit. It is also one of the most widely used tools for imaging, measuring, and manipulating matter at the nanoscale. The information is gathered by "feeling" the surface with a mechanical probe. Piezoelectric elements that facilitate tiny but accurate and precise movements on (electronic) command enable very precise scanning [11]. Figure 5(a) and Figure 5(b) show the 2D and 3D images of the surface roughness on the AlGaN/GaN surface with different treatments, as measured using a Park Systems XE-70. The $P_2S_5/(NH_4)_2S_X$ + UV-treated surface exhibits a better roughness than the other sulfurization treatments and forms a superior interface between the AlGaN Schottky layer and the $Pr_2O_3$ high-$k$ gate insulator layer.

Table 1 shows the mobility, sheet charge density, and surface roughness for variously treated devices, characterized by Hall measurement at 300 K. The $P_2S_5/(NH_4)_2S_X$ + UV-treated devices have a sheet charge density of $1.403 \times 10^{13}\ cm^{-2}$ and a Hall mobility of $1150\ cm^2/V\text{-}s$ at 300 K; these

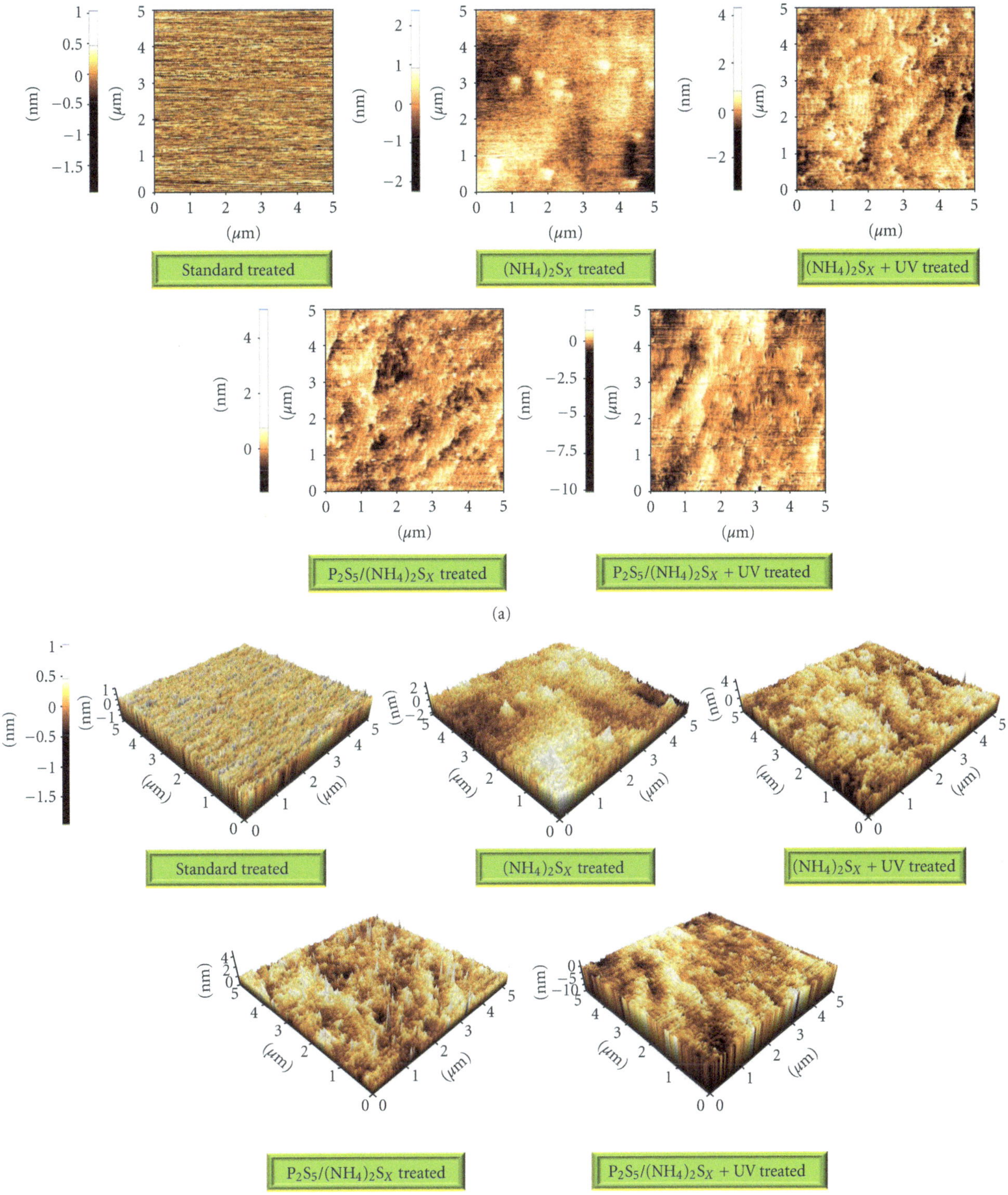


Figure 5: (a) 2D images of the surface roughness on AlGaN/GaN surface evaluated with different treatment. (b) 3D images of the surface roughness on AlGaN/GaN surface evaluated with different treatment.

values are, respectively, $1.648 \times 10^{13}$ cm$^{-2}$, 1060 cm$^2$/V-s and $1.512 \times 10^{13}$, 1087 cm$^2$/V-s for the standard treatment and $(NH_4)_2S_X$ + UV treatment for the devices. These results clearly demonstrate that $P_2S_5/(NH_4)_2S_X$ + UV treatment improves the channel mobility by reducing the number of surface traps.

The surface of composite semiconductor substrates was subject to heat treatment, using an ultraviolet (UV) light,

TABLE 1: The Hall measurement and surface roughness results for various treatment devices.

| Pretreatment | Mobility ($cm^2$/V-sec) | Sheet charge density ($cm^{-2}$) | Surface roughness (nm) |
|---|---|---|---|
| Standard | 1060 | $1.648 \times 10^{13}$ | 0.238 |
| $(NH_4)_2S_X$ | 985 | $1.784 \times 10^{13}$ | 0.514 |
| $(NH_4)_2S_X$ + UV | 1087 | $1.512 \times 10^{13}$ | 0.471 |
| $P_2S_5/(NH_4)_2S_X$ | 1100 | $1.485 \times 10^{13}$ | 0.443 |
| $P_2S_5/(NH_4)_2S_X$ + UV | 1150 | $1.403 \times 10^{13}$ | 0.382 |

in order to enhance the surface reaction [12]. The photoluminescence (PL) measurements in Figure 6 apply to the complete structure of AlGaN/GaN after sulfurization and demonstrate that $P_2S_5/(NH_4)_2S_X$ + UV treatment yields a greater PL intensity for the AlGaN Schottky layer than either $P_2S_5/(NH_4)_2S_X$ treatment or standard treatment. The increase in PL intensity is due to the elimination of surface states, which generates nonradiative recombination centers on the AlGaN surface [13]. No obvious $Al_{0.25}Ga_{0.75}N$ signal is evident, because the laser used is He-Cd laser and the PL signal intensity of $Al_{0.25}Ga_{0.75}N$ is much smaller than that for GaN. Because only 35 nm AlGaN was grown on GaN for the PL evaluation structure after sulfurization, the $Al_{0.25}Ga_{0.75}N$ signals are not obvious in the PL measurements. Although the $P_2S_5/(NH_4)_2S_X$ treatment can suppress the surface state density, as demonstrated by the PL results, a more concentrated $P_2S_5/(NH_4)_2S_X$ solution with UV illumination treatment produces a stable phosphorus oxide layer and more Ga-S bonds, because the phosphorus and sulfur concentrations are higher [14].

Figure 7 shows the $S_{2p}$ and $Ga_{2p}$ XPS spectra of the variously treated devices. The sulfur $2p$ core level and the binding energy of pure sulfur are all 163.8 eV. The binding energy of gallium is 160 eV at the signal peak in the spectrum of the standard treated GaN sample while the signal peak of the $P_2S_5/(NH_4)_2S_X$-treated device is shifted to 160.3 eV, because the AlGaN surface contains Ga-S bonds (Ga-S = 163.2 eV) after this process. However, $P_2S_5/(NH_4)_2S_X$ + UV treatment shifts the signal peak in the spectrum to 160.4 eV, with a linear distribution of intensity from 160 eV to 161 eV. The $Ga_{2p}$ peaks are shifted to a higher binding energy after $P_2S_5/(NH_4)_2S_X$ + UV treatment. This phenomenon is related to the Ga sulfurized states [15]. Therefore, more Ga-S bonds are generated on the AlGaN surface by $P_2S_5/(NH_4)_2S_X$ + UV treatment than by $P_2S_5/(NH_4)_2S_X$ treatment and these high-energy bonds are more stable than the Ga-O bonds that are formed in a moist environment.

In order to investigate the material atomic composition of the AlGaN surface after various sulfide treatments, the samples were subject to secondary ion mass spectroscopy (SIMS). Figure 8 shows the sulfur and oxygen atom concentration profiles of the samples treated using the three aforementioned methods. Based on the measurement results shown in Figure 8(a), the concentration of S atoms in the $P_2S_5/(NH_4)_2S_X$ and $P_2S_5/(NH_4)_2S_X$ + UV-treated samples

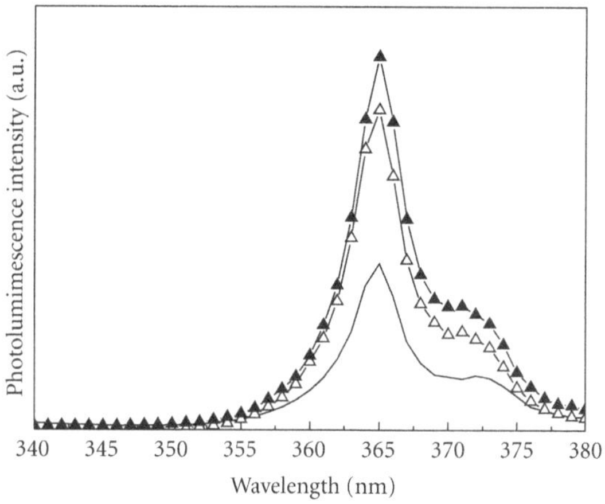


FIGURE 6: PL measurement results for AlGaN/GaN full structure standard, $P_2S_5/(NH_4)_2S_X$ and $P_2S_5/(NH_4)_2S_X$ + UV treatments.

is relatively high near the AlGaN surface, dropping by two to three orders of magnitude in the channel. However, few S atoms are observed after the standard treatment. The S atoms produced from the sulfide solution are incorporated into the sample surface, which is similar to the effect of the immersion of $CF_4$ in plasma, used to enable a GaN HEMT to operate in enhancement mode [16]. In order to further investigate the mechanism for the removal of Ga-O bonds by sulfur treatments, an experiment was conducted to determine the oxygen concentration distribution; Figure 8(b) shows the results. The lowest curve in the figure shows that the $P_2S_5/(NH_4)_2S_X$ + UV treatment removes a huge amount of AlGaN/GaN native oxide and reduces the number of oxygen atoms more effectively than the other treatments, because UV illumination provides enough energy to the sulfur atoms to replace the Ga-O bonds and form stable Ga-S bonds. Therefore, $P_2S_5/(NH_4)_2S_X$ + UV treatment not only produces a smoother Schottky interface, reducing the gate leakage current, but also reduces the number of oxygen atoms more effectively. Both effects reduce the density of the surface states.

Figure 9 plots the gate-to-drain $I$-$V$ curves for a standard GaN HEMT, a $Pr_2O_3$ MIS-HEMT, and a $P_2S_5/(NH_4)_2S_X$ + UV-treated $Pr_2O_3$ MIS-HEMT. A reduction in the number surface states and leakage current in the sulfide-treated sample means that $P_2S_5/(NH_4)_2S_X$ + UV-treated MIS-HEMTs have a higher $V_{ON}$ value than the others. The improved $V_{ON}$ value of 1.71 V for the $P_2S_5/(NH_4)_2S_X$ + UV-treated $Pr_2O_3$-gate device results in a significant reduction in the gate leakage current at a high pumped gate voltage, which improves the linearity and results in a reduction in the signal dispersion of the device. The reversed gate-to-drain breakdown voltages ($V_{BR}$), defined as the voltage at which the gate leakage current is −1 mA/mm, are −131.3 V for the $P_2S_5/(NH_4)_2S_X$ + UV-treated samples, which allows operation at high drain voltage, and only −128.2 V and

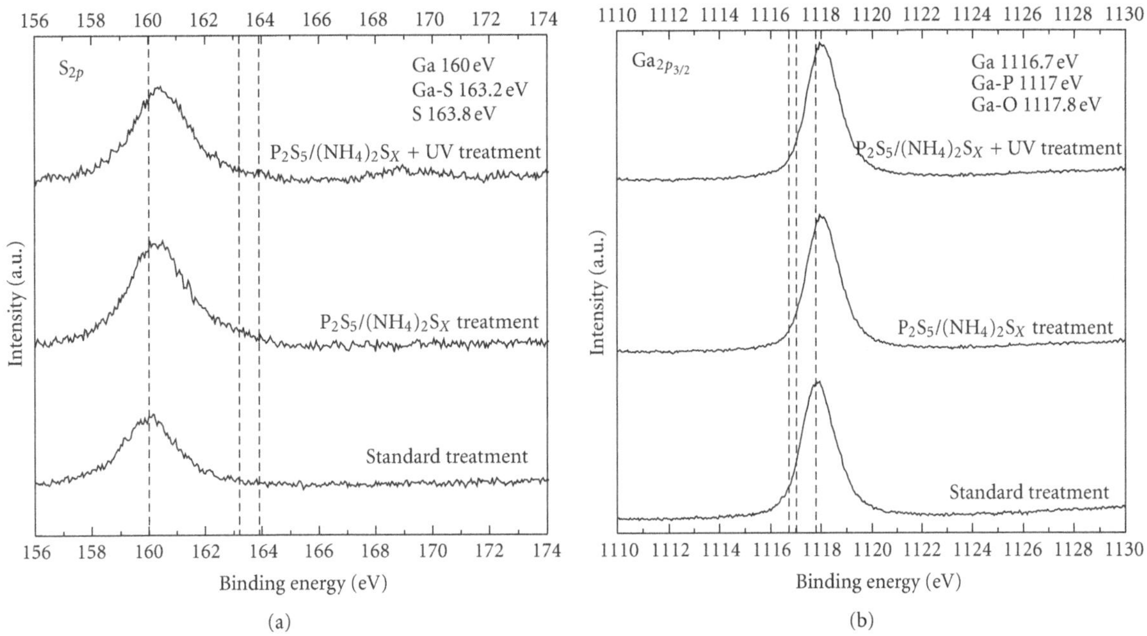


FIGURE 7: The XPS spectra of $S_{2p}$ (a) and $Ga_{2p}$ (b) and core level of various treatment samples.

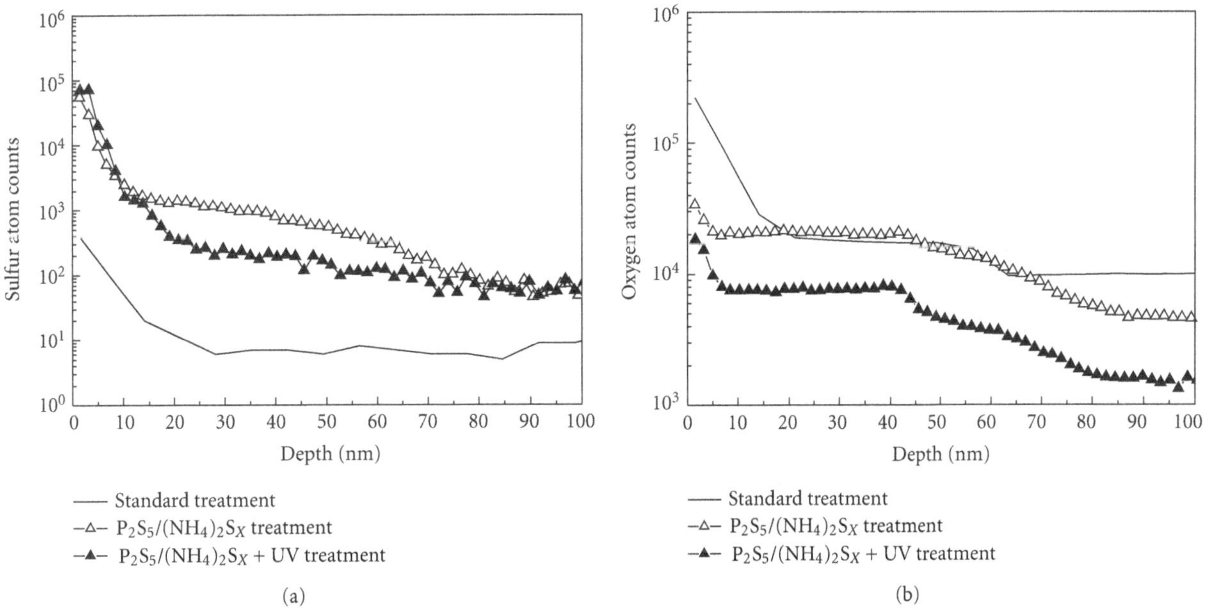


FIGURE 8: The SIMS measurements of the S atoms (a) and O atoms (b) distribution versus AlGaN/GaN depth for various treatment samples.

−107.5 V for the $Pr_2O_3$ MIS-HEMT and standard-treated samples, respectively.

In order to study the dc characteristics, the drain-to-source current ($I_{ds}$) versus drain-to-source voltage ($V_{ds}$) curves for three devices are shown in Figure 10. The drain current does not vary significantly between the treated devices. The turn-on resistance ($R_{ON}$) is 3.958 Ω for the HEMTs subject to standard treatment, at $V_{gs}$ of 0 V, and 4.675 Ω and 4.210 Ω for the $Pr_2O_3$ MIS-HEMTs and the $P_2S_5/(NH_4)_2S_X$ + UV-treated $Pr_2O_3$ MIS-HEMTs, respectively. This is primarily because the sulfide solution influences only the surface states, rather than the intrinsic parameters. However, since the traps within the interface between the gate and channel are suppressed by $P_2S_5/(NH_4)_2S_X$ + UV treatment, the sample with this treatment has the highest output resistance, which results in an increase in device linearity and power gain cut-off frequency ($f_{max}$). The output conductance ($g_0$) of the $P_2S_5/(NH_4)_2S_X$ + UV-treated

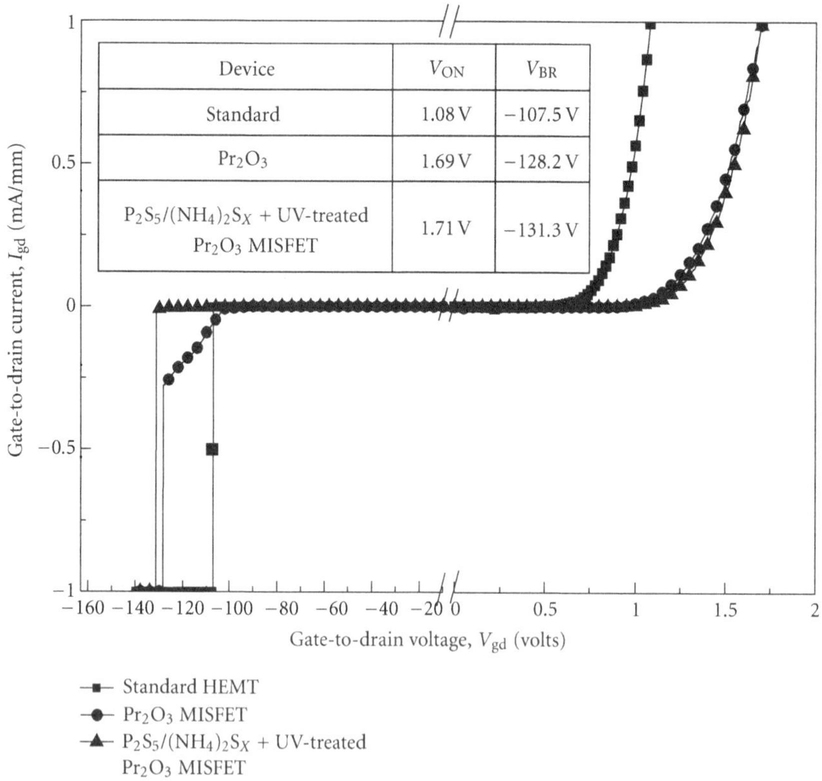


FIGURE 9: The device gate-to-drain $I$-$V$ characteristics of standard HEMT, $Pr_2O_3$ MISFET, and $P_2S_5/(NH_4)_2S_X$ + UV-treated $Pr_2O_3$ MISFET.

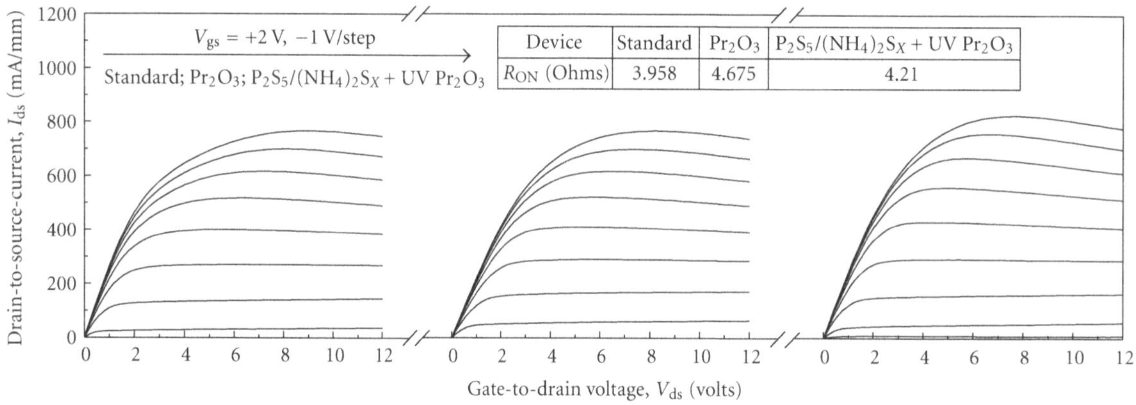


FIGURE 10: The $I_{ds}$-$V_{ds}$ characteristics of various devices.

$Pr_2O_3$ MIS-HEMTs is 0.46 mS/mm; the corresponding values for the $Pr_2O_3$ MIS-HEMTs and the standard device are 0.6 mS/mm and 1 mS/mm, respectively.

Figure 11 plots the transistor transconductance ($g_m$) versus $V_{gs}$ curves for the three devices of interest at a $V_{ds}$ of 8 V. The maximum drain-to-source currents ($I_{d\,max}$) at $V_{gs} = 4$ V are 923 mA/mm, 864 mA/mm, and 920 mA/mm, for the standard HEMTs, the $Pr_2O_3$ MIS-HEMTs, and the $P_2S_5/(NH_4)_2S_X$ + UV-treated $Pr_2O_3$ MIS-HEMTs, respectively. The maximum transconductance values ($g_m$), biased at $V_{ds} = 8$ V, are 144 mS/mm, 121 mS/mm, and 132 mS/mm, respectively. All of these values are reasonably favorable. The standard HEMTs exhibit a higher peak $g_m$, because a high-$k$ insulator is inserted in the MIS-HEMTs structure. The gate-to-channel improves channel modulation and effectively modulates any increase in the depletion region between the metal gate and the channel. Obviously, high-$k$ GaN MIS-HEMTs demonstrate large swing voltage and low gate leakage current.

The measurement of on-wafer microwave S-parameters for $1 \times 200\,\mu m^2$ devices was performed in a common-source configuration using an Agilent E8364C PNA network analyzer, from 0.1 GHz to 20.1 GHz. S-parameter measurements demonstrate a maximum current gain cut-off frequency ($f_T$) of 9.2 GHz and a maximum oscillation frequency ($f_{max}$) of 16.8 GHz for standard HEMTs. These values are 7.9 GHz

TABLE 2: The dc and RF characteristics comparisons of a standard HEMT, a $Pr_2O_3$ MISFET, and a $P_2S_5/(NH_4)_2S_X$ + UV-treated $Pr_2O_3$ MISFET.

| Measurement results | Standard | $Pr_2O_3$ | $P_2S_5/(NH_4)_2S_X$ + UV $Pr_2O_3$ |
|---|---|---|---|
| Schottky turn-on voltage ($V_{ON}$) | 1.08 V | 1.69 V | 1.71 V |
| Breakdown voltage ($V_{BR}$) | −107.5 V | −128.2 V | −131.3 V |
| Pinch-off voltage ($V_p$) | −5.8 V | −6.4 V | −6.8 V |
| Maximum drain current ($I_{d\max}$) | 923.4 mA/mm | 864.4 mA/mm | 920.04 mA/mm |
| Peak transconductance ($g_m$) | 143.58 mS/mm | 121 mS/mm | 132.36 mS/mm |
| Current gain cut-off frequency ($f_T$) | 9.2 GHz | 7.9 GHz | 8.5 GHz |
| Maximum oscillation frequency ($f_{max}$) | 16.8 GHz | 19.2 GHz | 16.2 GHz |

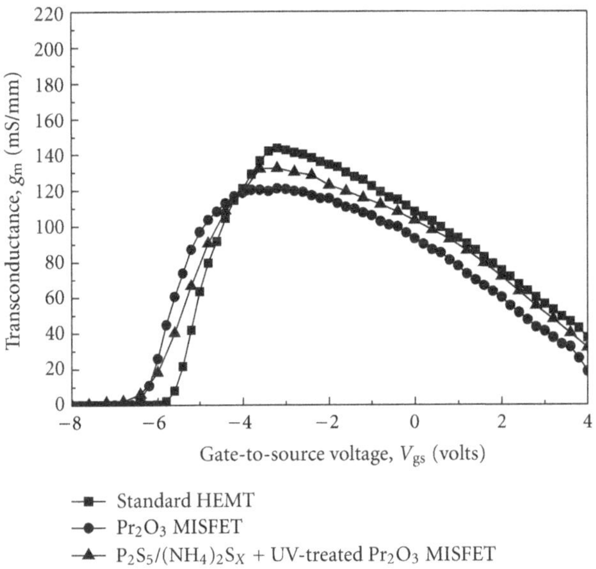


FIGURE 11: The $g_m$-$V_{gs}$ characteristics of various devices.

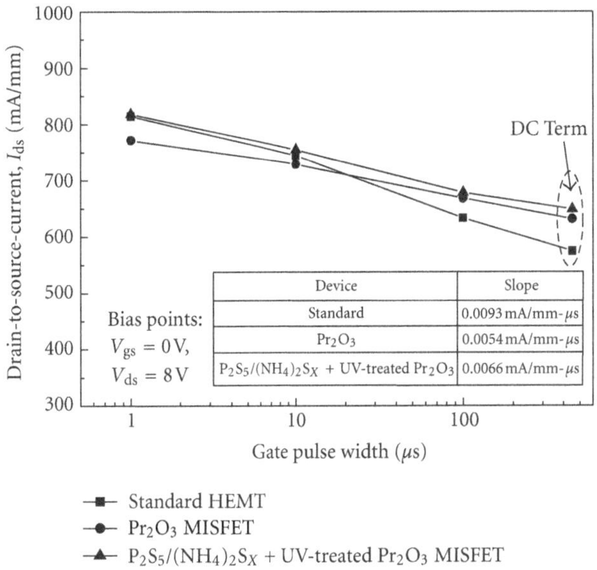


FIGURE 12: Pulse measurement characteristics of various devices.

and 10.7 GHz for $Pr_2O_3$ MIS-HEMTs and 8.5 GHz and 16.2 GHz for $P_2S_5/(NH_4)_2S_X$ + UV-treated $Pr_2O_3$ MIS-HEMTs, at $V_{ds}$ = 8 V and $V_{gs}$ = −3.2 V, respectively. The superior RF characteristics for $P_2S_5/(NH_4)_2S_X$ + UV-treated $Pr_2O_3$ MIS-HEMTs prove their greater power and linearity. Table 2 presents dc and radio RF characteristics for various devices.

Pulse measurements were made to characterize the carrier trapping phenomenon and heating effect in the device. With respect to the trapping carriers, the response time dominates the pulse measurement. The response time for these trapping carriers is typically of the order of $\mu$s longer than the ns of carrier transportation, especially for high-speed and high-power devices. In order to more easily observe the output signals from the drains of each device, a 50 Ω resistor was added between the drain electrode and the power supply, in order to determine the total drain-to-source current and to reduce the damping effect in the output square waveform. Pulse $I$-$V$ measurements for the three devices were also made, in order to confirm their surface trapping effects. Figure 12 plots the dc to 1 $\mu$s pulse $I$-$V$ measurement for 1 × 100 $\mu m^2$ devices at a $V_{gs}$ of 0 V and a $V_{ds}$ of 8 V, for the three devices. The gate width of the measured devices is 100 $\mu$m, so the heating effect could therefore be neglected. Since the surface states determine the dispersion effect, the current density at high current decays as the pulse width decreases. As shown in the figure, the standard GaN HEMTs demonstrate a larger slope with respect to the pulse period than do the $Pr_2O_3$ MIS-HEMTs and $P_2S_5/(NH_4)_2S_X$ + UV-treated $Pr_2O_3$ MIS-HEMTs. It is evident that $P_2S_5/(NH_4)_2S_X$ + UV treatment produces reliable and stable surface performance and less load-line hysteresis and better linearity for the device in high-power applications are which both expected.

To investigate the relationship between the flicker noise and the quantity of the variously treated devices, a low-frequency noise measurement was performed, because, at low frequency, this is sensitive to the semiconductor surface [17]. The bias point for low-frequency noise measurement was selected as $V_{ds}$ = 8 V, which gives an $I_{ds}$ of 100 mA/mm for all devices. Since the measurement is dominated by the series resistance of each device, identical $I_{ds}$ bias point confirms the flicker noise characteristics. As shown in Figure 13, the $P_2S_5/(NH_4)_2S_X$ + UV-treated MIS-HEMT demonstrates a lower 1/f spectral noise than both the $Pr_2O_3$ MIS-HEMT and the GaN HEMT, which proves

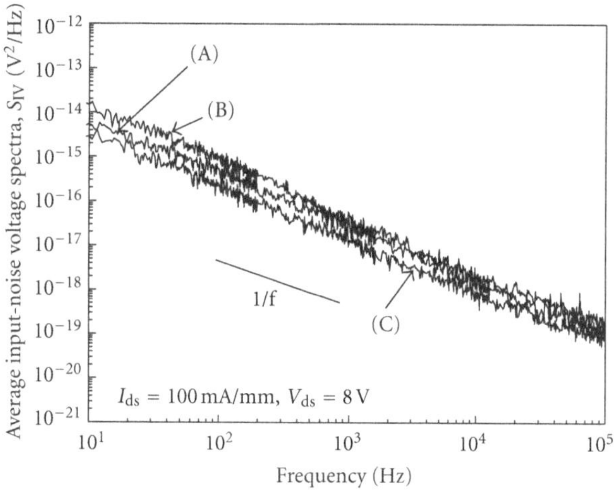


FIGURE 13: The 1/f noise comparisons for various devices.

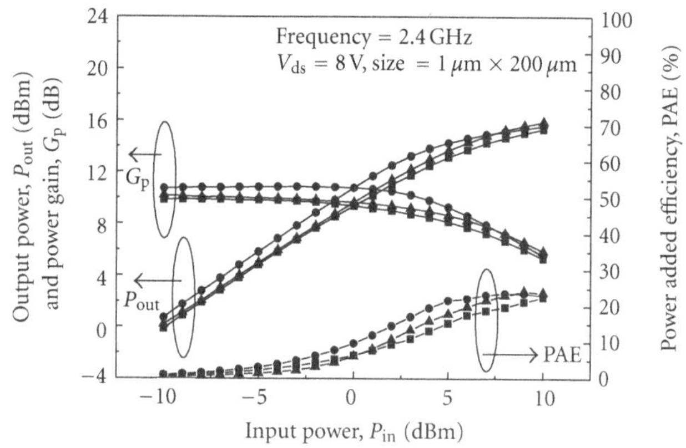


FIGURE 14: Power performances of three devices under 2.4 GHz and $V_{ds}$ of 8 V operation.

the reduction in the number of surface dangling bonds due to $P_2S_5/(NH_4)_2S_X$ + UV treatment.

The fabricated 1 μm long gate GaN HEMT, $Pr_2O_3$ MIS-HEMT, and $P_2S_5/(NH_4)_2S_X$ + UV-treated $Pr_2O_3$MIS-HEMT were tested on-wafer and the microwave power characteristics were evaluated using a load-pull system with automatic tuners, which simultaneously provides conjugate-matched input and load impedances for the maximum output power. The microwave load-pull power performance was conducted at 2.4 GHz, with a drain bias of 8 V, using various devices. The bias points for class AB operation of the standard HEMT, the $Pr_2O_3$ MIS-HEMT, and the $P_2S_5/(NH_4)_2S_X$ + UV-treated $Pr_2O_3$ MIS-HEMT must be biased at −3.1 V, −3.7 V, and −3.4 V (at 1/4 $I_{d\,max}$), respectively. Figure 14 shows the output power ($P_{out}$), power gain ($G_p$), and PAE as a function of the input power ($P_{in}$), for various devices with gate dimensions of 1 × 200 $\mu m^2$. The $P_2S_5/(NH_4)_2S_X$ + UV-treated $Pr_2O_3$ MIS-HEMT has better dc current and lower gate leakage current than that of the standard HEMT or the $Pr_2O_3$ MIS-HEMT, at high-input-power swing. The PAE values are 22.4%, 23.6%, and 24.2%, for the standard HEMT, the $Pr_2O_3$ MIS-HEMT and the $P_2S_5/(NH_4)_2S_X$ + UV-treated $Pr_2O_3$ MIS-HEMT, respectively. As a result, the microwave-power performance is improved by the MIS-gate structure, and the power-gain degradation is also improved in the high-input-power regime. The reduction in gate leakage current in the $P_2S_5/(NH_4)_2S_X$ + UV-treated $Pr_2O_3$ MIS-HEMT allows a significant improvement in device linearity [18].

## 4. Conclusion

In summary, $Pr_2O_3$ MIS-HEMTs with low gate leakage current and low flicker noise, as a result of $P_2S_5/(NH_4)_2S_X$ + UV treatment, were developed and characterized. An electron-beam-evaporated high-*k* insulator and a passivating layer prevent the formation of surface states by plasma. $P_2S_5/(NH_4)_2S_X$ + UV treatment represents a simple and efficient means of reducing the number of surface dangling bonds. Based on the results of Hall, XPS, and SIMS measurements, this $P_2S_5/(NH_4)_2S_X$ + UV treatment prevents the formation of strong Ga-S bonds on the AlGaN surface. This reduction in surface states improves carrier mobility and simultaneously suppresses unstable native Ga-O bonds. This novel pretreatment is therefore proven to be eminently suitable to low-noise GaN MIS-HEMT applications.

## Acknowledgments

The authors would like to thank the Nano Device Labs (NDL) for providing the low frequency noise measurements, and the National Central University (NCU) for providing the thermal image measurements. This work was financially supported by the National Science Council, ROC [NSC-97-2221-E-182-048-MY3 and NSC-100-2221-E-182-009], and High Speed Intelligent Communication (HSIC) Research Center of Chang Gung University, Taoyuan, Taiwan.

## References

[1] T. P. Chow and R. Tyagi, "Wide bandgap compound semiconductors for superior high-voltage unipolar power devices," *IEEE Transactions on Electron Devices*, vol. 41, no. 8, pp. 1481–1483, 1994.

[2] S. J. Pearton, J. C. Zolper, R. J. Shul, and F. Ren, "GaN: processing, defects, and devices," *Journal of Applied Physics*, vol. 86, no. 1, pp. 1–78, 1999.

[3] M. F. Romero, A. Jiménez, J. Miguel-Sánchez et al., "Effects of $N_2$ plasma pretreatment on the SiN passivation of AlGaN/GaN HEMT," *IEEE Electron Device Letters*, vol. 29, no. 3, pp. 209–211, 2008.

[4] C. Rongming, S. Likun, N. Fichtenbaum et al., "Correlation between DC-RF dispersion and gate leakage in deeply recessed GaN/AlGaN/GaN HEMTs," *IEEE Electron Device Letters*, vol. 29, no. 4, pp. 303–305, 2008.

[5] M. A. Khan, X. Hu, A. Tarakji, G. Simin, and J. Yang, "AlGaN/GaN metal-oxide-semiconductor heterostructure field-effect transistors on SiC substrates," *Applied Physics Letters*, vol. 77, no. 9, pp. 1339–1341, 2000.
[6] X. Hu, A. Koudymov, G. Simin et al., "$Si_3N_4$/AlGaN/GaN-metal-insulator-semiconductor heterostructure field-effect transistors," *Applied Physics Letters*, vol. 79, no. 17, pp. 2832–2834, 2001.
[7] C. T. Lee, H. W. Chen, and H. Y. Lee, "Metal-oxide-semiconductor devices using $Ga_2O_3$ dielectrics on *n*-type GaN," *Applied Physics Letters*, vol. 82, no. 24, pp. 4304–4306, 2003.
[8] P. D. Ye, B. Yang, K. K. Ng et al., "GaN metal-oxide-semiconductor high-electron-mobility-transistor with atomic layer deposited $Al_2O_3$ as gate dielectric," *Applied Physics Letters*, vol. 86, no. 6, Article ID 063501, pp. 1–3, 2005.
[9] R. Mehandru, B. Luo, J. Kim et al., "AlGaN/GaN metal-oxide-semiconductor high electron mobility transistors using $Sc_2O_3$ as the gate oxide and surface passivation," *Applied Physics Letters*, vol. 82, no. 15, pp. 2530–2532, 2003.
[10] H. C. Chiu, C. S. Cheng, and Y. J. Shih, "High uniformity $(Al_{0.3}Ga_{0.7})_{0.5}In_{0.5}P$/InGaAs enhancement-mode pseudomorphic HEMTs by selective succinic acid gate recess," *Electrochemical and Solid-State Letters*, vol. 9, no. 2, pp. G59–G61, 2006.
[11] L. B. Chang, C. H. Chang, M. J. Jeng, H. C. Chiu, and H. F. Kuo, "Barrier height enhancement of $Al_xGa_{1-x}N$/GaN schottky diodes prepared by $P_2S_5(NH_4)_2S$ treatments," *Electrochemical and Solid-State Letters*, vol. 10, no. 3, pp. H79–H81, 2007.
[12] M. J. Jeng, H. T. Wang, L. B. Chang, and R. M. Lin, "Surface passivation using $P_2S_5/(NH_4)_2S_x$ and hydrogen fluoride solutions on Ag/n-InAs and Ag/n-InSb Schottky diodes," *Japanese Journal of Applied Physics*, vol. 40, no. 2, pp. 562–564, 2001.
[13] S. L. Chang and J. F. Kauffman, "Excitation power dependence of photoluminescence enhancement from passivated GaAs," *Applied Physics Letters*, vol. 66, no. 25, pp. 3504–3506, 1995.
[14] H. C. Chiu, Y. C. Huang, C. W. Chen, and L. B. Chang, "Electrical characteristics of passivated Pseudomorphic HEMTs with $P_2S_5/(NH_4)_2S_x$ pretreatment," *IEEE Transactions on Electron Devices*, vol. 55, no. 3, pp. 721–726, 2008.
[15] X. Y. Hou, W. Z. Cai, Z. Q. He et al., "Electrochemical sulfur passivation of GaAs," *Applied Physics Letters*, vol. 60, no. 18, pp. 2252–2254, 1992.
[16] Y. Cai, Y. Zhou, K. J. Chen, and K. M. Lau, "High-performance enhancement-mode AlGaN/GaN HEMTs using fluoride-based plasma treatment," *IEEE Electron Device Letters*, vol. 26, no. 7, pp. 435–437, 2005.
[17] Y. J. Chan and D. Pavlidis, "Trap studies in GaInP/GaAs and AlGaAs/GaAs HEMT's by means of low-frequency noise and transconductance dispersion characterizations," *IEEE Transactions on Electron Devices*, vol. 41, no. 5, pp. 637–642, 1994.
[18] C. W. Lin, H. C. Chiu, C. K. Lin, and J. S. Fu, "High-k praseodymium oxide passivated AlGaN/GaN MOSFETs using $P_2S_5/(NH_4)_2S_x$ + UV interface treatment," *Microelectronics Reliability*, vol. 51, no. 2, pp. 381–385, 2011.

# Permissions

The contributors of this book come from diverse backgrounds, making this book a truly international effort. This book will bring forth new frontiers with its revolutionizing research information and detailed analysis of the nascent developments around the world.

We would like to thank all the contributing authors for lending their expertise to make the book truly unique. They have played a crucial role in the development of this book. Without their invaluable contributions this book wouldn't have been possible. They have made vital efforts to compile up to date information on the varied aspects of this subject to make this book a valuable addition to the collection of many professionals and students.

This book was conceptualized with the vision of imparting up-to-date information and advanced data in this field. To ensure the same, a matchless editorial board was set up. Every individual on the board went through rigorous rounds of assessment to prove their worth. After which they invested a large part of their time researching and compiling the most relevant data for our readers. Conferences and sessions were held from time to time between the editorial board and the contributing authors to present the data in the most comprehensible form. The editorial team has worked tirelessly to provide valuable and valid information to help people across the globe.

Every chapter published in this book has been scrutinized by our experts. Their significance has been extensively debated. The topics covered herein carry significant findings which will fuel the growth of the discipline. They may even be implemented as practical applications or may be referred to as a beginning point for another development. Chapters in this book were first published by Hindawi Publishing Corporation; hereby published with permission under the Creative Commons Attribution License or equivalent.

The editorial board has been involved in producing this book since its inception. They have spent rigorous hours researching and exploring the diverse topics which have resulted in the successful publishing of this book. They have passed on their knowledge of decades through this book. To expedite this challenging task, the publisher supported the team at every step. A small team of assistant editors was also appointed to further simplify the editing procedure and attain best results for the readers.

Our editorial team has been hand-picked from every corner of the world. Their multi-ethnicity adds dynamic inputs to the discussions which result in innovative outcomes. These outcomes are then further discussed with the researchers and contributors who give their valuable feedback and opinion regarding the same. The feedback is then collaborated with the researches and they are edited in a comprehensive manner to aid the understanding of the subject.

Apart from the editorial board, the designing team has also invested a significant amount of their time in understanding the subject and creating the most relevant covers. They scrutinized every image to scout for the most suitable representation of the subject and create an appropriate cover for the book.

The publishing team has been involved in this book since its early stages. They were actively engaged in every process, be it collecting the data, connecting with the contributors or procuring relevant information. The team has been an ardent support to the editorial, designing and production team. Their endless efforts to recruit the best for this project, has resulted in the accomplishment of this book. They are a veteran in the field of academics and their pool of knowledge is as vast as their experience in printing. Their expertise and guidance has proved useful at every step. Their uncompromising quality standards have made this book an exceptional effort. Their encouragement from time to time has been an inspiration for everyone.

The publisher and the editorial board hope that this book will prove to be a valuable piece of knowledge for researchers, students, practitioners and scholars across the globe.

# List of Contributors

**Jiun-Wei Horng, Chun-Li Hou, Ching-Yao Tseng, Ryan Chang and Dun-Yih Yang**
Department of Electronic Engineering, Chung Yuan Christian University, Chung-Li 32023, Taiwan

**Shengqi Zhou, Luowei Zhou, Suncheng Liu, Pengju Sun, Quanming Luo and Junke Wu**
State Key Laboratory of Power Transmission Equipment & System Security and New Technology, University of Chongqing, Chongqing 400044, China

**J. H. Yum**
Microelectronics Research Center, Department of Electrical and Computer Engineering, The University of Texas, Austin, TX 78758, USA
SEMATECH, 2706 Montopolis Drive, Austin, TX 78741, USA

**G. Bersuker**
SEMATECH, 2706 Montopolis Drive, Austin, TX 78741, USA

**J. Oh**
School of Integrated Technology, College of Engineering, Yonsei University, 162-1 Songdo-dong, Incheon 406-840, Republic of Korea

**Todd. W. Hudnall**
Department of Chemistry and Biochemistry, Texas State University, 601 University Drive, San Marcos, TX 78666, USA

**C. W. Bielawski**
Department of Chemistry, UT, Austin, TX 78712, USA

**S. K. Banerjee**
Microelectronics Research Center, Department of Electrical and Computer Engineering, The University of Texas, Austin, TX 78758, USA

**Eric Basham and David Parent**
Electrical Engineering Department, San Jose State University, San Jose, CA 95192-0084, USA

**Anand Subramaniam**
Department of Electrical Engineering, University of Texas at Dallas, Richardson, TX 75080, USA

**Eric M. Vogel**
Department of Materials Science and Engineering, University of Texas at Dallas, Richardson, TX 75080, USA
School of Materials Science and Engineering, Georgia Institute of Technology, Atlanta, GA 30332, USA

**Kurtis D. Cantley**
Department of Materials Science and Engineering, University of Texas at Dallas, Richardson, TX 75080, USA

**Neeta Pandey, Deva Nand and Zubair Khan**
Department of Electronics and Communications, Delhi Technological University, Delhi, India

**K. Shah and J. Singh**
Centre for Technology Infusion, La Trobe University, Bundoora, VIC 3086, Australia

**Z. Sauli**
School of Microelectronic Engineering, Universiti Malaysia Perlis, 02600 Perlis, Malaysia

**N. Izza M. Nor**
Centre for Technology Infusion, La Trobe University, Bundoora, VIC 3086, Australia
School of Microelectronic Engineering, Universiti Malaysia Perlis, 02600 Perlis, Malaysia

**G. Jacquemod**
University Nice Sophia Antipolis, IM2NP UMR CNRS 7334, 06410 Biot, France

**F. Ben Abdeljelil and W. Tatinian**
University Nice Sophia Antipolis, LEAT UMR CNRS 6071, 06540 Valbonne, France

**L. Carpineto**
SDRF, Sophia Antipolis, 06410 Biot, France

**M. Borgarino**
University of Modena and Reggio Emilia, 41100 Modena, Italy

**Ming-Kwei Lee and Chih-Feng Yen**
Department of Electronic Engineering, Chung Yuan Christian University, Chung Li City 32023, Taiwan

**Doron Abraham, Avraham Chelly, David Elbaz, Shimron Schiff, Michah Nabozny and Zeev Zalevsky**
Faculty of Engineering, Bar Ilan University, 52900 Ramat Gan, Israel

**Bradley D. Christiansen and Ronald A. Coutu Jr.**
Department of Electrical and Computer Engineering, Air Force Institute of Technology, Wright-Patterson Air Force Base, OH 45433, USA

**Eric R. Heller**
Materials and Manufacturing Directorate, Air Force Research Laboratory, Wright-Patterson Air Force Base, OH 45433, USA

**Ramakrishna Vetury and Jeffrey B. Shealy**
Defense and Power Business Unit, RF Micro Devices, Inc., Charlotte, NC 28269, USA

**Shimul C. Saha**
Department of Electronics and Telecommunications, Norwegian University of Science and Technology, 7491 Trondheim, Norway
Electronics and Nanoscale Engineering Research Division, University of Glasgow, Glasgow G12 8QQ, UK

**Ulrik Hanke**
Institute for Micro and Nano Systems Technology, Vestfold University College, 3103 Tønsberg, Norway

**Hakon Sagberg**
Department of Microsystems and Nanotechnology, SINTEF ICT, P.O. Box 124, Blindern, 0314 Oslo, Norway

**Tor A. Fjeldly and Trond Sæther**
Department of Electronics and Telecommunications, Norwegian University of Science and Technology, 7491 Trondheim, Norway

**Bretislav Sevcik**
Faculty of Electrical Engineering and Communication, Brno University of Technology, 61600 Brno, Czech Republic

**Jyi-Tsong Lin, Yi-Chuen Eng and Po-Hsieh Lin**
Department of Electrical Engineering, National Sun Yat-Sen University, No. 70 Lien-hai Road, Kaohsiung 80424, Taiwan

**Sudhanshu Maheshwari and Rishabh Verma**
Department of Electronics Engineering, Zakir Hussain College of Engineering and Technology, Aligarh Muslim University, Aligarh 202002, India

**Takaya Handa, Yuhei Yoshimoto and Akio Kitagawa**
School of Electrical and Computer Engineering, College of Science and Engineering, Kanazawa University, Kakuma, Kanazawa 920-1192, Japan

**Kazuya Nakayama**
School of Health Sciences, College of Medical, Pharmaceutical and Health Sciences, Kanazawa University, 5-11-80 Kodatsuno, Kanazawa 920-0942, Japan

**Koichi Narahara**
Graduate School of Science and Engineering, Yamagata University, 4-3-16 Jonan, Yonezawa, Yamagata 992-8510, Japan

**P. Villard, T. Thabuis and M. Belleville**
CEA LETI, MINATEC Campus, 38054 Grenoble, France

**G. Sicard**
TIMA Laboratory, 38031 Grenoble, France

**G. Decaens and M. Zecri**
SOFRADIR, 38113 Veurey-Voroize, France

**Yaksh Rawal, Swaroop Ganguly and Maryam Shojaei Baghini**
Department of Electrical Engineering, IIT-Bombay, National Center for Photovoltaic Research and Education (NCPRE), Mumbai 400076, India

**Amjed S. Al-Fahoum, Heba B. Al-Hmoud and Ausaila A. Al-Fraihat**
Biomedical Systems and Informatics Engineering Department, Hijjawi Faculty for Engineering Technology, Yarmouk University, Irbid 21163, Jordan

**Chung-Yen Chien, Jei-Wei Hsu, Pei-Chin Chiu, Jen-Inn Chyi and Pei-Wen Li**
Department of Electrical Engineering, National Central University, Chungli 32001, Taiwan

**Suranjana Banerjee**
Academy of Technology, West Bengal University of Technology, Adisaptagram, Hooghly, West Bengal 712121, India

**Aritra Acharyya and J. P. Banerjee**
Institute of Radio Physics and Electronics, University of Calcutta, 92 APC Road, Kolkata, West Bengal 700009, India

**M. Narayanan and H. Al-Nashash**
Department of Electrical Engineering, American University of Sharjah, P.O. Box 26666, Sharjah, UAE

**Baquer Mazhari**
Department of Electrical Engineering, Indian Institute of Technology, Kanpur 208016, India

**Dipankar Pal**
NERIST, Nirjuli, 791109, Itanagar, India

**Mahesh Chandra**
Birla Institute of Technology, Mesra, 835 215 Ranchi, India

**Nabihah Ahmad and Rezaul Hasan**
Center for Research in Analog and VLSImicrosystemDesign (CRAVE), School of Engineering and Advanced Technology (SEAT), Massey University, Auckland 0632, New Zealand

**Hugues Murray and Patrick Martin**
Normandie Universite, Ensicaen, UMR 6508, LaMIPS, Laboratoire Commun CNRS-NXP-PRESTO ENSICAEN UCBN, 6 boulevard Marechal Juin, 14050 CAEN Cedex 4, France

**Chao-Wei Lin and Hsien-Chin Chiu**
Department of Electronics Engineering, Chang Gung University, Taoyuan, Taiwan